Advances in Optimization and Nonlinear Analysis

Advances in Optimization and Nonlinear Analysis

Editor

Savin Treanţă

MDPI • Basel • Beijing • Wuhan • Barcelona • Belgrade • Manchester • Tokyo • Cluj • Tianjin

Editor
Savin Treanță
University Politehnica of Bucharest
Romania

Editorial Office
MDPI
St. Alban-Anlage 66
4052 Basel, Switzerland

This is a reprint of articles from the Special Issue published online in the open access journal *Fractal and Fractional* (ISSN 2504-3110) (available at: https://www.mdpi.com/journal/fractalfract/special_issues/optimization_nonlinear).

For citation purposes, cite each article independently as indicated on the article page online and as indicated below:

LastName, A.A.; LastName, B.B.; LastName, C.C. Article Title. *Journal Name* **Year**, *Volume Number*, Page Range.

ISBN 978-3-0365-4749-7 (Hbk)
ISBN 978-3-0365-4750-3 (PDF)

© 2022 by the authors. Articles in this book are Open Access and distributed under the Creative Commons Attribution (CC BY) license, which allows users to download, copy and build upon published articles, as long as the author and publisher are properly credited, which ensures maximum dissemination and a wider impact of our publications.
The book as a whole is distributed by MDPI under the terms and conditions of the Creative Commons license CC BY-NC-ND.

Contents

About the Editor vii

Savin Treanţă
Advances in Optimization and Nonlinear Analysis
Reprinted from: *Fractal Fract.* **2022**, *6*, 364, doi:10.3390/fractalfract6070364 1

Madiah Binti Omar, Kishore Bingi, B Rajanarayan Prusty, and Rosdiazli Ibrahim
Recent Advances and Applications of Spiral Dynamics Optimization Algorithm: A Review
Reprinted from: *Fractal Fract.* **2022**, *6*, 27, doi:10.3390/fractalfract6010027 7

Savin Treanţă
Well Posedness of New Optimization Problems with Variational Inequality Constraints
Reprinted from: *Fractal Fract.* **2021**, *5*, 123, doi:10.3390/fractalfract5030123 39

Ananta Thakur, Javid Ali and Rosana Rodríguez-López
Existence of Solutions to a Class of Nonlinear Arbitrary Order Differential Equations Subject to Integral Boundary Conditions
Reprinted from: *Fractal Fract.* **2021**, *5*, 220, doi:10.3390/fractalfract5040220 55

Mohamed A. El Sayed, Mohamed A. El-Shorbagy, Farahat A. Farahat, Aisha F. Fareed and Mohamed A. Elsisy
Stability of Parametric Intuitionistic Fuzzy Multi-Objective Fractional Transportation Problem
Reprinted from: *Fractal Fract.* **2021**, *5*, 233, doi:10.3390/fractalfract5040233 73

Miguel Vivas-Cortez, Muhammad Shoaib Saleem, Sana Sajid, Muhammad Sajid Zahoor and Artion Kashuri
Hermite–Jensen–Mercer-Type Inequalities via Caputo–Fabrizio Fractional Integral for h-Convex Function
Reprinted from: *Fractal Fract.* **2021**, *5*, 269, doi:10.3390/fractalfract5040269 91

Kin Keung Lai, Mohd Hassan, Sanjeev Kumar Singh, Jitendra Kumar Maurya and Shashi Kant Mishra
Semidefinite Multiobjective Mathematical Programming Problems with Vanishing Constraints Using Convexificators
Reprinted from: *Fractal Fract.* **2021**, *6*, 3, doi:10.3390/fractalfract6010003 109

Muhammad Bilal Khan, Savin Treanţă, Mohamed S. Soliman, Kamsing Nonlaopon and Hatim Ghazi Zaini
Some Hadamard–Fejér Type Inequalities for LR-Convex Interval-Valued Functions
Reprinted from: *Fractal Fract.* **2021**, *6*, 6, doi:10.3390/fractalfract6010006 129

Guozeng Yang, Yonggang Li, Jing Wang and Huafei Sun
Application of the Pick Function in the Lieb Concavity Theorem for Deformed Exponentials
Reprinted from: *Fractal Fract.* **2021**, *6*, 20, doi:10.3390/fractalfract6010020 145

Shirin Sultana, Abu Hashan Md Mashud, Yosef Daryanto, Sujan Miah, Adel Alrasheedi and Ibrahim M. Hezam
The Role of the Discount Policy of Prepayment on Environmentally Friendly Inventory Management
Reprinted from: *Fractal Fract.* **2022**, *6*, 26, doi:10.3390/fractalfract6010026 157

Muhammad Bilal Khan, Savin Treanţă and Hüseyin Budak
Generalized *p*-Convex Fuzzy-Interval-Valued Functions and Inequalities Based upon the Fuzzy-Order Relation
Reprinted from: *Fractal Fract.* **2022**, *6*, 63, doi:10.3390/fractalfract6020063 . . . . **179**

Mojtaba Sajjadmanesh, Hassen Aydi, Eskandar Ameer and Choonkil Park
The Method of Fundamental Solutions for the 3D Laplace Inverse Geometric Problem on an Annular Domain
Reprinted from: *Fractal Fract.* **2022**, *6*, 66, doi:10.3390/fractalfract6020066 . . . . **201**

Yijun Zhu and Huilin Shang
Multistability of the Vibrating System of a Micro Resonator
Reprinted from: *Fractal Fract.* **2022**, *6*, 141, doi:10.3390/fractalfract6030141 . . . . **209**

Muhammad Bilal Khan, Jorge E. Macías-Díaz, Savin Treanţă, Mohammed S. Soliman and Hatim Ghazi Zaini
Hermite-Hadamard Inequalities in Fractional Calculus for Left and Right Harmonically Convex Functions via Interval-Valued Settings
Reprinted from: *Fractal Fract.* **2022**, *6*, 178, doi:10.3390/fractalfract6040178 . . . . **225**

Fatima Daqaq, Salah Kamel, Mohammed Ouassaid, Rachid Ellaia, and Ahmed M. Agwa
Non-Dominated Sorting Manta Ray Foraging Optimization for Multi-Objective Optimal Power Flow with Wind/Solar/Small- Hydro Energy Sources
Reprinted from: *Fractal Fract.* **2022**, *6*, 194, doi:10.3390/fractalfract6040194 . . . . **241**

Ahmed H. A. Elkasem, Mohamed Khamies, Mohamed H. Hassan, Ahmed M. Agwa and Salah Kamel
Optimal Design of TD-TI Controller for LFC Considering Renewables Penetration by an Improved Chaos Game Optimizer
Reprinted from: *Fractal Fract.* **2022**, *6*, 220, doi:10.3390/fractalfract6040220 . . . . **279**

About the Editor

Savin Treanţă

Savin Treanţă is currently based in the Department of Applied Mathematics, Faculty of Applied Sciences, University Politehnica of Bucharest, Romania. His research interests include multi-objective optimization, optimal control, mathematical modelling, geometric PDEs, and information theory. Savin has published more than 100 scientific papers on these topics in the most prestigious journals of Mathematics and Engineering.

Editorial

Advances in Optimization and Nonlinear Analysis

Savin Treanţă [1,2]

1. Department of Applied Mathematics, University Politehnica of Bucharest, 060042 Bucharest, Romania; savin.treanta@upb.ro
2. Academy of Romanian Scientists, 54 Splaiul Independentei, 050094 Bucharest, Romania

1. Introduction

There are many applications of optimization and nonlinear analysis in various fields of basic science, engineering, and natural phenomena. In this regard, we have provided the Special Issue "Advances in Optimization and Nonlinear Analysis" to cover the new advances in these mathematical areas. In this Special Issue, we have focused on publishing research studies on optimization and nonlinear analysis by investigating the well-posedness and optimal solutions in new classes of (multiobjective) variational (control) problems governed by multiple and/or path-independent curvilinear integral cost functionals and mixed and/or isoperimetric constraints involving first- and second-order partial differential equations. Additionally, some applications of fractional calculus or related subjects (variational inequalities, equilibrium problems, fixed point problems, evolutionary problems, and so on) have been considered in this Special Issue. In response to our invitation, we received 41 papers from 22 countries (Egypt, Saudi Arabia, Morocco, Pakistan, Mexico, Romania, China, Iran, Tunisia, South Africa, Yemen, Korea, Turkey, Bangladesh, Australia, Indonesia, Thailand, India, Ecuador, Albania, Spain, Malaysia), of which 15 were published and 26 rejected/withdrawn.

2. Brief Overview of the Contributions

In a review conducted by Omar et al. [1], the spiral dynamics optimization (SDO) algorithm was comprehensively reviewed. It is well-known that SDO algorithm is one of the most straightforward physics-based optimization algorithms and it is successfully applied in various broad fields. This review paper describes the recent advances of the SDO algorithm, including its adaptive, improved, and hybrid approaches. The growth of the SDO algorithm and its application in various areas, theoretical analysis, and comparison with its preceding and other algorithms are also described in detail. A detailed description of different spiral paths, their characteristics, and the application of these spiral approaches in developing and improving other optimization algorithms are comprehensively presented. The review concludes the current works on the SDO algorithm, highlighting its shortcomings and suggesting possible future research perspectives.

In [2], Treanţă studies the well posedness for a new class of optimization problems with variational inequality constraints involving second-order partial derivatives. More precisely, by using the notions of lower semicontinuity, pseudomonotonicity, hemicontinuity and monotonicity for a multiple integral functional, and by introducing the set of approximating solutions for the considered class of constrained optimization problems, he establishes some characterization results on well posedness. Furthermore, to illustrate the theoretical developments included in this paper, some examples are presented.

Thakur et al.'s [3] study in this Special Issue investigates the existence of positive solutions for a class of fractional differential equations of arbitrary order $\delta > 2$, subject to boundary conditions that include an integral operator of the fractional type. The consideration of this type of boundary conditions allows to consider heterogeneity on the dependence specified by the restriction added to the equation as a relevant issue for applications. An existence result is

obtained for the sublinear and superlinear case by using the Guo–Krasnosel'skii fixed point theorem through the definition of adequate conical shells that allow to localize the solution. As additional tools in the considered procedure, Thakur et al. obtain the explicit expression of Green's function associated to an auxiliary linear fractional boundary value problem, and study some of its properties, such as the sign and some useful upper and lower estimates. Finally, an example is given to illustrate the results.

A parametric intuitionistic fuzzy multi-objective fractional transportation problem (PIF-MOFTP) is analyzed in El Sayed et al. [4]. The PIF-MOFTP includes a single-scalar parameter in the objective functions and an intuitionistic fuzzy supply and demand. Based on the (α, β)-cut concept, a parametric (α, β)-MOFTP is proposed. Then, a fuzzy goal programming (FGP) approach is utilized to obtain (α, β)-Pareto optimal solution. Moreover, the authors investigates the stability set associated with the first kind (SSFK) corresponding to the solution by extending the Kuhn-Tucker optimality conditions of multi-objective programming problems. Also, an algorithm to crystalize the progressing SSFK for PIF-MOFTP is presented.

Vivas-Cortez et al. [5] use integral inequalities involving many fractional integral operators in order to solve various fractional differential equations. More precisely, the authors generalize the Hermite–Jensen–Mercer-type inequalities for an h-convex function via a Caputo–Fabrizio fractional integral. They develop some novel Caputo–Fabrizio fractional integral inequalities. Also, they establish Caputo–Fabrizio fractional integral identities for differentiable mapping, and these will be used to give estimates for some fractional Hermite–Jensen–Mercer-type inequalities. Some familiar results are recaptured as special cases of these results.

In Lai et al. [6], the authors establish Fritz John stationary conditions for nonsmooth, nonlinear, semidefinite, multiobjective programs with vanishing constraints in terms of convexificator. Also, they introduce generalized Cottle type and generalized Guignard type constraints qualification to achieve strong S—stationary conditions from Fritz John stationary conditions. Further, the authors establish strong S—stationary necessary and sufficient conditions, independently from Fritz John conditions. Some examples are provided to validate the established results.

The purpose of the next paper Khan et al. [7] published in this Special Issue is to introduce a new class of Hermite–Hadamard inequalities for LR-convex interval-valued functions, by means of a pseudo-order relation. This order relation is defined on interval space. Moreover, the interval Hermite–Hadamard–Fejér inequality is also derived for LR-convex interval-valued functions. These inequalities also generalize some new and known results. Useful examples that verify the applicability of the theory developed in this study are presented.

The Lieb concavity theorem, successfully solved in the Wigner–Yanase–Dyson conjecture, is an important application of matrix concave functions. Recently, the Thompson–Golden theorem, a corollary of the Lieb concavity theorem, was extended to deformed exponentials. Hence, it is worthwhile to study the Lieb concavity theorem for deformed exponentials. In Yang [8], the Pick function is used to obtain a generalization of the Lieb concavity theorem for deformed exponentials, and some corollaries associated with exterior algebra are obtained.

Nowadays, more and more consumers consider environmentally friendly products in their purchasing decisions. Companies need to adapt to these changes while paying attention to standard business systems such as payment terms. The purpose of the study realized by Sultana et al. [9] is to optimize the entire profit function of a retailer and to find the optimal selling price and replenishment cycle when the demand rate depends on the price and carbon emission reduction level. This study investigates an economic order quantity model that has a demand function with a positive impact of carbon emission reduction besides the selling price. In this model, the supplier requests payment in advance on the purchased cost while offering a discount according to the payment in the advanced decision. Three different types of payment-in-advance cases are applied: (1) payment in advance with equal numbers of instalments, (2) payment in advance with a single instalment, and (3) the absence of payment in advance. Numerical examples and sensitivity analysis illustrate the proposed model. Here, the total profit increases for all three cases with higher values of carbon emission reduction level. Further, the study finds that the profit becomes maximum for case 2, whereas the selling price and cycle length become minimum. This study considers the sustainable inventory model with payment-in-advance settings when the demand rate depends on the price and carbon emission reduction level.

Convexity is crucial in obtaining many forms of inequalities. As a result, there is a significant link between convexity and integral inequality. Due to the significance of these concepts, the purpose of Khan et al.'s [10] study is to introduce a new class of generalized convex interval-valued functions called (p,s)-convex fuzzy interval-valued functions (for short, (p,s)-convex F-I-V-Fs) in the second sense and to establish Hermite–Hadamard (for short, H–H) type inequalities for (p,s)-convex F-I-V-Fs using fuzzy order relation. In addition, the authors demonstrate that the derived results include a large class of new and known inequalities for (p,s)-convex F-I-V-Fs and their variant forms as special instances. Furthermore, useful examples are given to demonstrate usefulness of the theory produced in this study. These findings and diverse approaches may pave the way for future research in fuzzy optimization, modeling, and interval-valued functions.

In the paper Sajjadmanesh et al. [11], the authors are interested in an inverse geometric problem for the three-dimensional Laplace equation to recover an inner boundary of an annular domain. This work is based on the method of fundamental solutions (MFS) by imposing the boundary Cauchy data in a least-square sense and minimisation of the objective function. This approach can also be considered with noisy boundary Cauchy data. The simplicity and efficiency of this method is illustrated in several numerical examples.

Multiple attractors and their fractal basins of attraction can lead to the loss of global stability and integrity of Micro Electro Mechanical Systems (MEMS). In the paper of Zhu et al. [12], multistability of a class of electrostatic bilateral capacitive micro-resonator is researched in detail. First, the dynamical model is established and made dimensionless. Second, via the perturbating method and the numerical description of basins of attraction, the multiple periodic motions under primary resonance are discussed. It is found that the variation of AC voltage can induce safe jump of the micro resonator. In addition, with the increase of the amplitude of AC voltage, hidden attractors and chaos appear. The results may have some potential value in the design of MEMS devices.

The purpose of the study Khan et al. [13] is to define a new class of harmonically convex functions, which is known as left and right harmonically convex interval-valued functions (for short, LR-H-convex IV-F), and to establish novel inclusions for a newly defined class of interval-valued functions (for short, IV-Fs) linked to Hermite–Hadamard (for short, H-H) and Hermite–Hadamard–Fejér (H-H-Fejér) type inequalities via interval-valued Riemann–Liouville fractional (for short, IV-RL-fractional) integrals. These findings enable the authors to identify a new class of inclusions that may be seen as significant generalizations. Some examples are included in the considered findings that may be used to determine the validity of the results.

The study developed in Daqaq et al. [14] describes a novel manta ray foraging optimization approach based non-dominated sorting strategy, namely (NSMRFO), for solving the multi-objective optimization problems (MOPs). The proposed powerful optimizer can efficiently achieve good convergence and distribution in both the search and objective spaces. In the NSMRFO algorithm, the elitist non-dominated sorting mechanism is followed. Afterwards, a crowding distance with a non-dominated ranking method is integrated for the purpose of archiving the Pareto front and improving the optimal solutions coverage. To judge the NSMRFO performances, a bunch of test functions are carried out including classical unconstrained and constrained functions, a recent benchmark suite known as the completions on evolutionary computation 2020 (CEC2020) that contains twenty-four multimodal optimization problems (MMOPs), some engineering design problems, and also the modified real-world issue known as IEEE 30-bus optimal power flow involving the wind/solar/small-hydro power generations. Comparison findings with multimodal multi-objective evolutionary algorithms (MMMOEAs) and other existing multi-objective approaches with respect to performance indicators reveal the NSMRFO ability to balance between the coverage and convergence towards the true Pareto front (PF) and Pareto optimal sets (PSs). Thus, the competing algorithms fail in providing better solutions while the proposed NSMRFO optimizer is able to attain almost all the Pareto optimal solutions.

The last paper published in the considered Special Issue (see Elkasem et al. [15]) presents an innovative strategy for load frequency control (LFC) using a combination structure of tilt-derivative and tilt-integral gains to form a TD-TI controller. Furthermore, a new improved optimization technique, namely the quantum chaos game optimizer (QCGO) is applied to tune the gains of the proposed combination TD-TI controller in two-area interconnected hybrid power systems, while the effectiveness of the proposed QCGO is validated via a comparison of its performance with the traditional CGO and other optimizers when considering 23 bench functions. Correspondingly, the effectiveness of the proposed controller is validated by comparing its performance with other controllers, such as the proportional-integral-derivative (PID) controller based on different optimizers, the tilt-integral-derivative (TID) controller based on a CGO algorithm, and the TID controller based on a QCGO algorithm, where the effectiveness of the proposed TD-TI controller based on the QCGO algorithm is ensured using different load patterns (i.e., step load perturbation (SLP), series SLP, and random load variation (RLV)). Furthermore, the challenges of renewable energy penetration and communication time delay are considered to test the robustness of the proposed controller in achieving more system stability. In addition, the integration of electric vehicles as dispersed energy storage units in both areas has been considered to test their effectiveness in achieving power grid stability. The simulation results elucidate that the proposed TD-TI controller based on the QCGO controller can achieve more system stability under the different aforementioned challenges.

Funding: This research received no external funding.

Acknowledgments: I am thankful the editors and reviewers of the *Fractal and Fractional* journal for their help and support.

Conflicts of Interest: The author declares no conflict of interest.

References

1. Omar, M.B.; Bingi, K.; Prusty, B.R.; Ibrahim, R. Recent Advances and Applications of Spiral Dynamics Optimization Algorithm: A Review. *Fractal Fract.* **2022**, *6*, 27. [CrossRef]
2. Treanţă, S. Well Posedness of New Optimization Problems with Variational Inequality Constraints. *Fractal Fract.* **2021**, *5*, 123. [CrossRef]
3. Thakur, A.; Ali, J.; Rodríguez-López, R. Existence of Solutions to a Class of Nonlinear Arbitrary Order Differential Equations Subject to Integral Boundary Conditions. *Fractal Fract.* **2021**, *5*, 220. [CrossRef]
4. El Sayed, M.A.; El-Shorbagy, M.A.; Farahat, F.A.; Fareed, A.F.; Elsisy, M.A. Stability of Parametric Intuitionistic Fuzzy Multi-Objective Fractional Transportation Problem. *Fractal Fract.* **2021**, *5*, 233. [CrossRef]
5. Vivas-Cortez, M.; Saleem, M.S.; Sajid, S.; Zahoor, M.S.; Kashuri, A. Hermite–Jensen–Mercer-Type Inequalities via Caputo–Fabrizio Fractional Integral for h-Convex Function. *Fractal Fract.* **2021**, *5*, 269. [CrossRef]
6. Lai, K.K.; Hassan, M.; Singh, S.K.; Maurya, J.K.; Mishra, S.K. Semidefinite Multiobjective Mathematical Programming Problems with Vanishing Constraints Using Convexificators. *Fractal Fract.* **2022**, *6*, 3. [CrossRef]
7. Khan, M.B.; Treanţă, S.; Soliman, M.S.; Nonlaopon, K.; Zaini, H.G. Some Hadamard–Fejér Type Inequalities for LR-Convex Interval-Valued Functions. *Fractal Fract.* **2022**, *6*, 6. [CrossRef]
8. Yang, G.; Li, Y.; Wang, J.; Sun, H. Application of the Pick Function in the Lieb Concavity Theorem for Deformed Exponentials. *Fractal Fract.* **2022**, *6*, 20. [CrossRef]
9. Sultana, S.; Mashud, A.H.M.; Daryanto, Y.; Miah, S.; Alrasheedi, A.; Hezam, I.M. The Role of the Discount Policy of Prepayment on Environmentally Friendly Inventory Management. *Fractal Fract.* **2022**, *6*, 26. [CrossRef]
10. Khan, M.B.; Treanţă, S.; Budak, H. Generalized p-Convex Fuzzy-Interval-Valued Functions and Inequalities Based upon the Fuzzy-Order Relation. *Fractal Fract.* **2022**, *6*, 63. [CrossRef]
11. Sajjadmanesh, M.; Aydi, H.; Ameer, E.; Park, C. The Method of Fundamental Solutions for the 3D Laplace Inverse Geometric Problem on an Annular Domain. *Fractal Fract.* **2022**, *6*, 66. [CrossRef]
12. Zhu, Y.; Shang, H. Multistability of the Vibrating System of a Micro Resonator. *Fractal Fract.* **2022**, *6*, 141. [CrossRef]

13. Khan, M.B.; Macías-Díaz, J.E.; Treanţă, S.; Soliman, M.S.; Zaini, H.G. Hermite-Hadamard Inequalities in Fractional Calculus for Left and Right Harmonically Convex Functions via Interval-Valued Settings. *Fractal Fract.* **2022**, *6*, 178. [CrossRef]
14. Daqaq, F.; Kamel, S.; Ouassaid, M.; Ellaia, R.; Agwa, A.M. Non-Dominated Sorting Manta Ray Foraging Optimization for Multi-Objective Optimal Power Flow with Wind/Solar/Small- Hydro Energy Sources. *Fractal Fract.* **2022**, *6*, 194. [CrossRef]
15. Elkasem, A.H.A.; Khamies, M.; Hassan, M.H.; Agwa, A.M.; Kamel, S. Optimal Design of TD-TI Controller for LFC Considering Renewables Penetration by an Improved Chaos Game Optimizer. *Fractal Fract.* **2022**, *6*, 220. [CrossRef]

Review

Recent Advances and Applications of Spiral Dynamics Optimization Algorithm: A Review

Madiah Binti Omar [1], Kishore Bingi [2,*], B Rajanarayan Prusty [2] and Rosdiazli Ibrahim [3]

[1] Department of Chemical Engineering, Universiti Teknologi PETRONAS, Seri Iskandar 32610, Malaysia; madiah.omar@utp.edu.my
[2] School of Electrical Engineering, Vellore Institute of Technology, Vellore 632014, India; b.r.prusty@ieee.org
[3] Department of Electrical and Electronics Engineering, Universiti Teknologi PETRONAS, Seri Iskandar 32610, Malaysia; rosdiazli@utp.edu.my
* Correspondence: kishore.bingi@vit.ac.in

Abstract: This paper comprehensively reviews the spiral dynamics optimization (SDO) algorithm and investigates its characteristics. SDO algorithm is one of the most straightforward physics-based optimization algorithms and is successfully applied in various broad fields. This paper describes the recent advances of the SDO algorithm, including its adaptive, improved, and hybrid approaches. The growth of the SDO algorithm and its application in various areas, theoretical analysis, and comparison with its preceding and other algorithms are also described in detail. A detailed description of different spiral paths, their characteristics, and the application of these spiral approaches in developing and improving other optimization algorithms are comprehensively presented. The review concludes the current works on the SDO algorithm, highlighting its shortcomings and suggesting possible future research perspectives.

Keywords: advances of SDO; applications of SDO; metaheuristic optimization; nature-inspired algorithms; optimization problems; spiral dynamics optimization; spiral-inspired optimization algorithms; spiral paths

Citation: Omar, M.B.; Bingi, K.; Prusty, B.R.; Ibrahim, R. Recent Advances and Applications of Spiral Dynamics Optimization Algorithm: A Review. *Fractal Fract.* **2022**, *6*, 27. https://doi.org/10.3390/fractalfract6010027

Academic Editor: Savin Treanţă

Received: 16 December 2021
Accepted: 28 December 2021
Published: 2 January 2022

Publisher's Note: MDPI stays neutral with regard to jurisdictional claims in published maps and institutional affiliations.

Copyright: © 2022 by the authors. Licensee MDPI, Basel, Switzerland. This article is an open access article distributed under the terms and conditions of the Creative Commons Attribution (CC BY) license (https:// creativecommons.org/licenses/by/ 4.0/).

1. Introduction

In engineering applications, metaheuristic optimization algorithms are more popular and widely used for computing the optimal solution [1]. This broad application is because:

1. The algorithms are easy to implement and do not require gradient information as they depend on relatively simple concepts;
2. The algorithms can avoid settling at optimal local solutions;
3. The algorithms can be applied to various problems of different fields.

A great variety of nature and population-based metaheuristic optimization algorithms have been published in the literature [2]. As reported in [2], these algorithms are categorized into breeding-based, swarm intelligence-based, physics-based, chemistry-based, social human behavior-based, plant-based, and others. Many developed metaheuristic optimization algorithms published in the literature are swarm intelligence-based algorithms. After swarm intelligence-based algorithms, physics-based algorithms are the most widely proposed and implemented in various applications [3,4]. As the name suggests, in swarm intelligence-based algorithms, some degree of intelligence is present in the algorithm process while finding the optimal solution. However, in physics-based algorithms, the algorithm process is based on specific laws or principles [3,5,6]. The main advantage of physics-based algorithms compared to others is the most straightforwardness. This is because the algorithm's strategy is based on fundamental physical principles. Thus, the algorithms can consistently and accurately represent the dynamics over the entire domain. Further, some physics-based algorithms also take advantage of a nature-inspired ratio, called the golden ratio, which helps to converge quickly and effectively when finding the optimal solution [7].

The most popular physics-based optimization algorithms are harmony search, gravitational search algorithm (GSA), big bang big crunch, electromagnetic field optimization (EFO), galaxy-based search [8], ray optimization, magnetic optimization, spiral dynamics optimization [9], and water cycle optimization [10]. Spiral dynamics optimization (SDO) is one of the most straightforward physics-based algorithms proposed by Tamura and Yasuda in 2011, developed using a logarithmic spiral phenomenon in nature [9]. The algorithm is simple and has few control parameters. Moreover, the algorithm has fast computational speed, local searching capability, diversification in the early phase, and intensification in the later stage.

This review paper provides the origin and concept of the SDO algorithm for an n-dimensional system. The effect of variation of spiral parameters (radius and angle) for two- and three-dimensional systems are analyzed by generating the conventional and hypotrochoid spiral trajectories. Besides, the recent advances in SDO algorithm, including adaptive, improved, and hybrid versions, are highlighted. The current applications of SDO and its variants are also focused. Different types of spirals, coordinates on xy-plane, and trajectories are generated to understand spiral behaviors. Further, various novel optimization algorithms' developments using these spirals are presented comprehensively. Therefore, this review paper helps in guiding multiple researchers who are currently working and willing to work by employing SDO and its variants to solve various engineering problems. Moreover, the review helps in developing or improving existing algorithms using the spiral phenomenon.

The paper's remaining sections are organized as follows: the origin and concept of the SDO algorithm and the effect of the spiral parameter in developing search trajectories are presented in Section 2. Section 3 offers the recent adaptive, improved, and hybrid versions of the SDO algorithm. Section 4 gives the different types of spiral trajectories and a list of novel optimization algorithms created using these trajectories. The applications of SDO and its hybrid versions are presented in Section 5. Finally, the paper is concluded in Section 6.

2. Spiral Dynamics Optimization Algorithm

This section presents the origin and the concept of the SDO algorithm for two-dimensional and three-dimensional systems. A detailed analysis of the effect of varying spiral parameters (radius and angle) is also presented.

2.1. Origin

Tamura and Yoshida developed the SDO algorithm in 2011 to mimic the spiral phenomena in nature [9,11]. Many spirals are available in nature, such as galaxies, aurora, blackbuck horns, hurricanes, tornadoes, seashells, snails, ammonites, cabbage butterflies, Pieris brassicae, chameleon tail, seahorse, and fish vortex [12,13]. The spirals are also seen in ancient art created by humanity during 5000 BC to 1600 AD [12]. Over the years, several researchers have made efforts to understand the spiral sequences and complexities and develop equations and algorithms of the spirals. Moreover, it is worth highlighting that the frequently encountered spiral phenomenon in nature is the logarithmic spiral, which can be seen in galaxies, tropical cyclones, and nautilus shells [14]. The discrete processes of generating a logarithmic spiral have been realized as an effective search behavior in metaheuristics, which inspired the spiral dynamics optimization algorithm to develop.

2.2. Concept

In the SDO algorithm, the multipoint search function for an n-dimensional system is formulated as [15],

$$x_{k+1} = rR^{(n)}(\theta)x_k - (rR^{(n)}(\theta) - I_n)x^*, \qquad (1)$$

where r is the spiral radius, $R^{(n)}(\theta)$ is the rotational matrix of order $n \times n$, θ is the spiral rotation angle, I_n is the identity matrix of order $n \times n$, x^* is the spiral center, x_k and x_{k+1} are the search point positions at iterations k and $k+1$, respectively.

The rotational matrix $R^{(n)}(\theta)$ for an n-dimensional case on an arbitrary $x_i x_j$-plane is given as [9,16,17],

$$R^{(n)}(\theta) = \begin{bmatrix} 1 & 0 & 0 & \cdots & 0 & 0 & 0 \\ 0 & 1 & 0 & \cdots & 0 & 0 & 0 \\ 0 & 0 & \cos(\theta_{i,j}) & \cdots & -\sin(\theta_{i,j}) & 0 & 0 \\ \vdots & \vdots & \vdots & \ddots & \vdots & \vdots & \vdots \\ 0 & 0 & \sin(\theta_{i,j}) & \cdots & \cos(\theta_{i,j}) & 0 & 0 \\ 0 & 0 & 0 & \cdots & 0 & 1 & 0 \\ 0 & 0 & 0 & \cdots & 0 & 0 & 1 \end{bmatrix}, \quad (2)$$

where $\theta_{i,j}$ is the spiral rotation angle around the origin on $x_i x_j$-plane.

From (2), the only one possibility of rotational matrix $R^{(2)}(\theta)$ for a two-dimensional system on $x_1 x_2$-plane is given as follows:

$$R^{(2)}(\theta) = \begin{bmatrix} \cos(\theta) & -\sin(\theta) \\ \sin(\theta) & \cos(\theta) \end{bmatrix}. \quad (3)$$

On the other hand, the three possible combinations of rotational matrix $R^{(3)}(\theta)$ for a three-dimensional system on $x_1 x_2$, $x_2 x_3$, and $x_1 x_3$-planes are respectively given as follows:

$$R^{(3)}_{1,2}(\theta) = \begin{bmatrix} \cos(\theta_{1,2}) & -\sin(\theta_{1,2}) & 0 \\ \sin(\theta_{1,2}) & \cos(\theta_{1,2}) & 0 \\ 0 & 0 & 1 \end{bmatrix}, \quad (4)$$

$$R^{(3)}_{2,3}(\theta) = \begin{bmatrix} 1 & 0 & 0 \\ 0 & \cos(\theta_{2,3}) & -\sin(\theta_{2,3}) \\ 0 & \sin(\theta_{2,3}) & \cos(\theta_{2,3}) \end{bmatrix}, \text{ and} \quad (5)$$

$$R^{(3)}_{1,3}(\theta) = \begin{bmatrix} \cos(\theta_{1,3}) & 0 & -\sin(\theta_{1,3}) \\ 0 & 1 & 0 \\ \sin(\theta_{1,3}) & 0 & \cos(\theta_{1,3}) \end{bmatrix}. \quad (6)$$

From (1), it is to be noted that the model generated the spiral trajectories around the center x^* and these trajectories are classified into two types [18,19]:

- If $r > 1$ and $\theta \in (-\frac{\pi}{2}, \frac{\pi}{2})$, the trajectory is a conventional spiral;
- If $r < 1$ and $\theta \in (-\frac{\pi}{2}, \frac{\pi}{2})$, the trajectory is a hypotrochoid spiral.

From the above classification, the spiral's direction of rotation based on the value of θ is classified as follows:

- If $\theta \in (-\frac{\pi}{2}, 0)$, the rotation of trajectory is clockwise;
- If $\theta \in (0, \frac{\pi}{2})$, the rotation of trajectory is anticlockwise.

The spiral trajectories for a two-dimensional system for various values of $r \in [-1, 1]$ and $\theta = \frac{\pi}{8}$ is shown in Figure 1. Similarly, the trajectories for various values of $\theta \in [-\frac{\pi}{2}, \frac{\pi}{2}]$ and $r = 0.85$ for conventional spiral and $r = -0.85$ for hypotrochoid spiral are shown in Figure 2. Further, the conventional and hypotrochoid spiral trajectories for both positive and negative values of θ are shown in Figure 3. In all these cases, the starting point used in the study is $(25, 25)$.

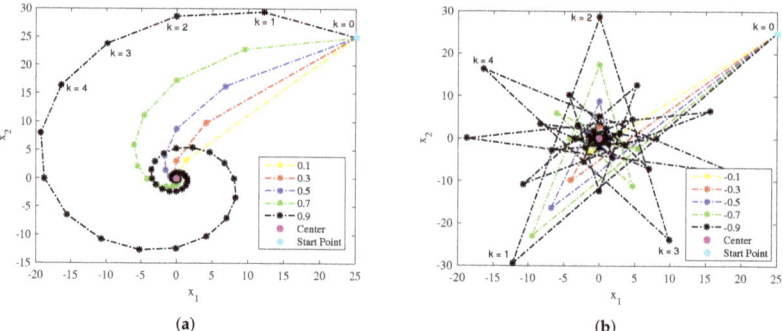

Figure 1. Spiral trajectories for a two-dimensional system for various values of $r \in [-1, 1]$ and $\theta = \frac{\pi}{8}$: (**a**) conventional spiral and (**b**) hypotrochoid spiral.

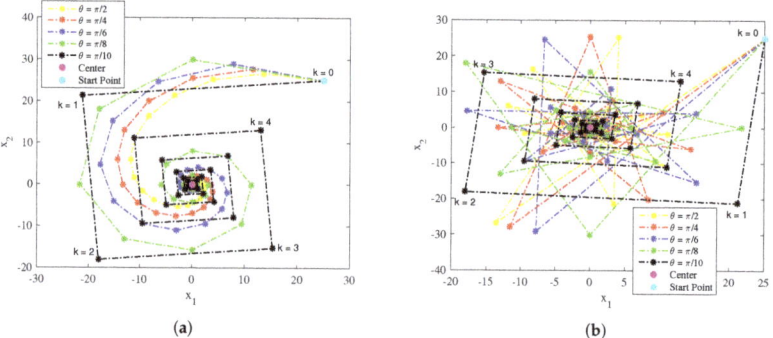

Figure 2. Spiral trajectories for a two-dimensional system for various values of $\theta \in [-\frac{\pi}{2}, \frac{\pi}{2}]$ and $r = 0.85$ for conventional spiral in (**a**) and $r = -0.85$ for hypotrochoid spiral in (**b**).

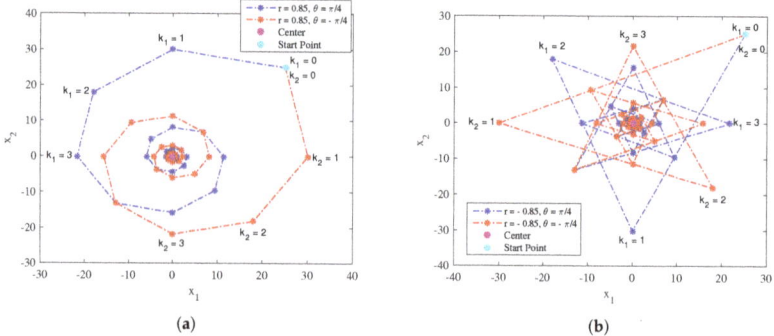

Figure 3. Spiral trajectories for a two-dimensional system for both positive and negative values of θ: (**a**) conventional spiral and (**b**) hypotrochoid spiral.

Observing the notations $k = 0, k = 1, \ldots, k = 4$ on spiral trajectories in Figures 1–3, it can be noted that at each iteration, the spiral point from the starting point moves by an angle θ and then tends towards the center x^*. Thus, the net effect is the spiral movement of the initial point towards the center. The trajectories also depict the angle θ, controlling the spiral curve. A smoother curve is achieved for smaller values of θ, compared to the boxy curved with larger values of θ (refer to Figure 2a). The spiral trajectories in Figure 3 show the clockwise and anticlockwise spiral movement for negative and positive angles, respectively. On the other hand, the spiral radius r controls the spiral movement towards the center x^*. A quick movement of spiral towards the center is achieved for smaller values of r, compared to the slow movement with larger values of r (refer to Figures 1 and 2). The hypotrochoid spirals shown in Figures 1b, 2b, and 3b are internal trajectories which are generated along a circle. The advantage of a hypotrochoid spiral over conventional spirals is it does not exceed the search space and can search most of the area in the search space.

In a similar way, the conventional and hypotrochoid spiral trajectories for a three-dimensional system with $r = 0.95$ and $\theta = \frac{\pi}{4}$ are shown in Figure 4. The trajectory in Figure 4a on the x_1x_2-plane is obtained using the rotational matrix in (4). Similarly, the trajectories in Figure 4b,c on the x_2x_3 and x_1x_3-planes are obtained using the rotational matrices in (5) and (6), respectively. The starting point used is $(25, 25, 25)$ in all of these cases. The trajectories depict the conventional spiral with a positive r value and the hypotrochoid spiral with a negative r value. As the θ value is positive, all the spiral movements are anticlockwise. As mentioned earlier, the advantage of hypotrochoid spirals is they can search most of the area in the search space, as shown in Figure 4. The search space of a conventional spiral is only on the positive plane, while the hypotrochoid spirals search space is both negative and positive. Thus, the trajectories in the figure conclude that the hypotrochoid spirals can search most of the area in the search space.

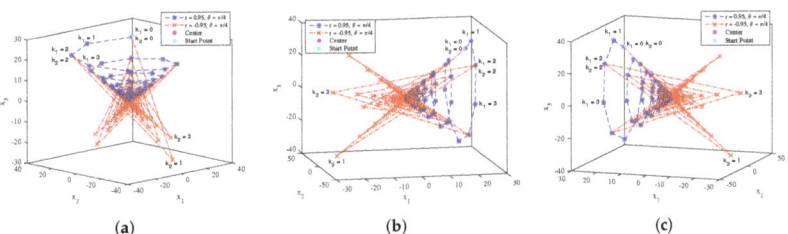

Figure 4. Conventional and hypotrochoid spiral trajectories for a three-dimensional system with $r = 0.95$ and $\theta = \frac{\pi}{4}$: (**a**) on x_1x_2-plane with $R_{1,2}$. (**b**) on x_2x_3-plane with $R_{2,3}$. (**c**) on x_1x_3-plane with $R_{1,3}$.

3. Advances of Spiral Dynamics Optimization Algorithm

This section presents the recent adaptive, improved, and hybrid versions of the SDO algorithm.

3.1. Adaptive Versions of Spiral Dynamics Optimization Algorithm

Researchers have developed the adaptive versions of the SDO algorithm by dynamically varying the spirals' radius and angle based on the fitness value during each iteration. The four types of proposed adaptive approaches in the literature are linear, quadratic, exponential, and fuzzy [16,20,21]. The mathematical functions of spirals' radius and angle using the proposed approaches are given in Figure 5.

In the figure, the notations are defined as follows:

- r_{la} and θ_{la} are the computed radius and angle using linear adaptive approach;
- r_{qa} and θ_{qa} are the obtained radius and angle using quadratic adaptive approach;
- r_{ea} and θ_{ea} are the radius and angle obtained using exponential adaptive approach;
- r_{fa} and θ_{fa} are the calculated radius and angle using fuzzy adaptive approach;
- $r_l \in [0, 1]$ and $r_u \in [0, 1]$ are the minimum and maximum radius of spiral;
- $\theta_l \in [0, 1]$ and $\theta_u \in [0, 1]$ are the minimum and maximum angles of spiral;
- c_1 and c_2 are constants;
- fuzzy$(\cdot)$ is the fuzzy logic mapping;

- Y_{Fit} is the difference between fitness value at a current iteration $f(x_i(k))$ and best fitness $\min(f(x_i(k)))$, is defined as,

$$Y_{\text{Fit}} = f(x_i(k)) - \min(f(x_i(k))). \quad (7)$$

In [17], using the linear adaptive approach in Figure 5, the authors have proposed the adaptive hypotrochoid SDO algorithm. The proposed algorithm performs best on various benchmark functions compared to conventional techniques. On the other hand, in [22], a self-adaptive approach is proposed for the SDO algorithm to update the spiral radius and angle during the optimization. The approach's advantage is that all search points are updated by randomly tuning the parameter values in each iteration. Similarly, the authors of [23] have proposed an adaptive SDO by incorporating three mechanisms, such as (i) bi-considering updation, (ii) self-adaptive radius, and (iii) punish mechanisms. The proposed algorithm boosted the optimization efficiency and avoided trapping at the local optimal minima.

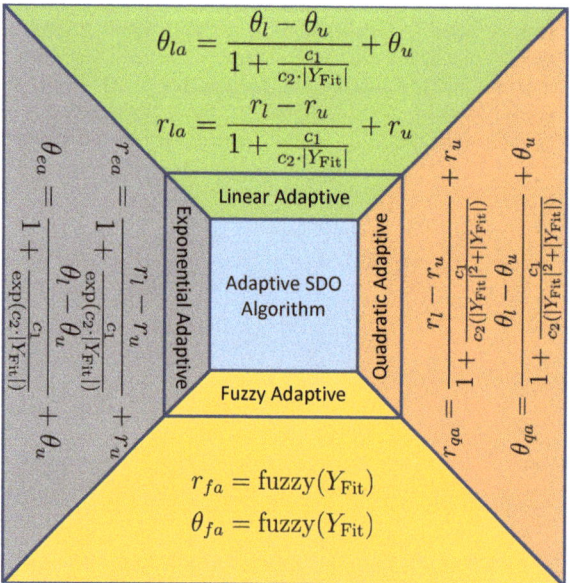

Figure 5. Adaptive versions of the SDO algorithm.

3.2. Improved Versions of Spiral Dynamics Optimization Algorithm

As mentioned earlier in Section 2.2, the algorithm settles into optimal local values at the end of the optimization process due to insufficient exploration of the conventional SDO's search space. Thus, to avoid this problem, Nasir et al. have proposed the improved SDO algorithm using the bacterial foraging algorithms' elimination–dispersal strategy [24,25]. In this enhanced version, the algorithm structure is kept the same. However, two new phases, namely elimination and dispersal, are introduced. Similarly, Hashim et al. have proposed the chaotic SDO algorithm logistic chaotic map patterns in the conventional SDO [26,27]. The chaotic map pattern helps in the initial population distribution rather than randomly in conventional SDO. Moreover, the search strategy of the artificial bee colony optimization algorithm is employed to improve the SDO's exploration capability. The authors have also proposed the greedy SDO algorithm by incorporating the greedy selection stage and chaotic logistic map in the conventional SDO [28]. In this selection stage, the obtained solution is compared to the previous value for updating the spiral positions. The authors of [18,19] have proposed the hypotrochoid SDO algorithm in which the search points follow the hypotrochoid spiral rather than the conventional spiral in SDO.

The proposed hypotrochoid SDO can explore the search space more effectively and explore the whole neighborhood of the optimal center. The experimental validation on optimal triaxial accelerometers placement in the Shanghai Tower in China [19], and sizing and layout of truss structures [18] has shown the better performance of hypotrochoid SDO than its predecessors.

The SDO algorithm in Section 2.2 is developed by utilizing a feature of the logarithmic spiral. This algorithm is also known as a deterministic or direct-solving metaheuristic optimization algorithm. One of the significant drawbacks of this algorithm is the slow convergence. Therefore, the authors of [29–31] have proposed a stochastic SDO algorithm by incorporating some random disturbances at each searching point of the algorithm. Similarly, the authors of [32] have introduced the iterative SDO algorithm for analyzing the information on blurred images. In this algorithm, the model's output is given as an input to the same model iteratively. Thus, the optimization algorithm searches for the sharp image spirally with the blurred vision at the initial stage. On the other hand, the authors of [33] have proposed the distributed SDO algorithm to increase the diversity in the search space. In this conventional SDO algorithm, it is clear that the search points rotate spirally around the optimal center only. Thus, the algorithm falls into the local minimum quickly. However, in the proposed distributed SDO algorithm, the population of search points is split into sub-populations to increase diversity and capture the whole search space. The summary of all these approaches is given in Figure 6.

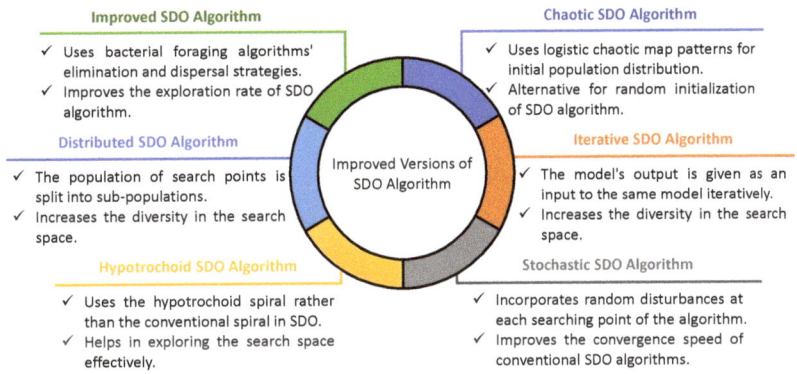

Figure 6. Improved versions of the SDO algorithm.

3.3. Hybrid Versions of Spiral Dynamics Optimization Algorithm

From the literature review, the following points are worth highlighting on the performance of the SDO algorithm. SDO has the advantages of a simple structure, few control parameters, and early diversification and intensification strategies. However, the SDO's performance is poor in searching the whole search space [20,34], and the exploration mechanism of the SDO needs to be improved [35]. The algorithm gets trapped at optimal local minima easily [33].

Thus, to improve the performance of SDO, researchers have proposed the hybridization of SDO with other algorithms. Further, various algorithms' performance has also been enhanced using SDO. The hybrid versions of the SDO algorithm presented in the literature used an artificial bee colony (ABC) [36,37], antlion optimization (ALO) [38], bacterial chemotaxis algorithm (BCA) [20,34,39], bacterial foraging algorithm (BFA) [35,40,41], biogeography-based optimization (BBO) [42], cuckoo search (CS) [43], genetic algorithm (GA) [44], particle swarm optimization (PSO) [45–48], sine-cosine algorithm (SCA) [49], and teaching–learning-based optimization (TLBO) [50], as shown in Figure 7. As shown in the figure, the excellent exploitation strategy of SDO is hybridized with the fast exploration strategy of another algorithm to balance both the exploitation and exploration phases.

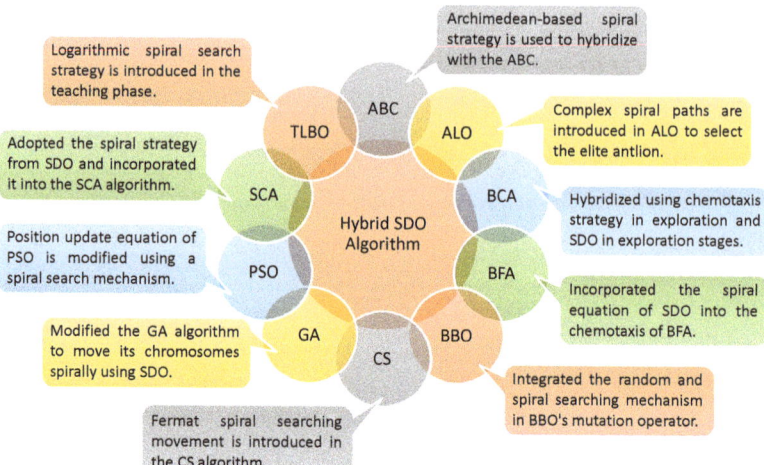

Figure 7. Hybrid versions of the SDO algorithm.

Moreover, there are several other novel optimization algorithms in which spiral behavior or trajectory is used during the development of the algorithm. A detailed description of various spiral paths and a list of novel spiral path-inspired optimization algorithms are discussed in the following section.

4. Spiral Path Inspired Optimization Algorithms

The first part of this section presents the various spiral trajectories used to develop the optimization algorithms. Then, the list of different novel optimization algorithms created using these spirals is shown.

4.1. Spiral Paths

Patterns referred to as visible consistencies found in nature are trees, spirals, waves, etc. Visual patterns in nature are modeled using chaos theory, fractals, spirals, etc. In some natural patterns, the spirals and fractals are related. For instance, a variant of the logarithmic spiral, namely the Fibonacci spiral, is based on the golden ratio and Fibonacci numbers. As it is logarithmic, the curve at every scale appears the same and can be considered a fractal. Romanesco broccoli is an example of such Fractal spirals. The above patterns inspired researchers to develop optimization algorithms. Different types of spiral trajectories used in the research include:

- Archimedes spira;
- Cycloid spiral;
- Epitrochoid spiral;
- Hypotrochoid spiral;
- Logarithmic spiral;
- Rose spiral;
- Inverse spiral; and
- Overshoot spirals.

A detailed description of the five most widely used spirals, including Archimedes, logarithmic, rose, epitrochoid, and hypotrochoid, is provided underneath. This detailed description includes the coordinates on the xy-plane and trajectories showing the effect of each parameter on the xy-plane.

4.1.1. Logarithmic Spiral

The logarithmic spirals often appear in nature. For instance, the nautilus cutaway, Iceland's low-pressure area, galaxies, and tropical cyclones arms usually take a logarithmic spiral shape. The logarithmic spiral is also known as equiangular or growth spiral because the spiral distance increases in geometric progression. The coordinates of a logarithmic spiral on xy-plane are given as follows [13,38]:

$$x(\phi) = a \cdot e^{b\phi} \cdot \cos(\phi), \ y(\phi) = a \cdot e^{b\phi} \cdot \sin(\phi), \tag{8}$$

where ϕ is the angle, a and b are the arbitrary constants.

The logarithmic spiral for $a = 0.18$, ϕ from -4π to 4π, and various b values is shown in Figure 8. The spiral in Figure 8a is obtained for positive values of b, while Figure 8b is obtained for negative values. The trajectories in Figure 8 show that parameter b controls the tightness and the direction of the spiral. The trajectories in Figure 8a also depict the logarithmic spiral proprieties that for positive b values and ϕ tends to $+\infty$, the spiral evolves in an anticlockwise direction. Whereas for the same b values and ϕ tends to $-\infty$, the spiral evolves in a clockwise direction. However, for negative b values, the spiral evolves or twists in the opposite direction.

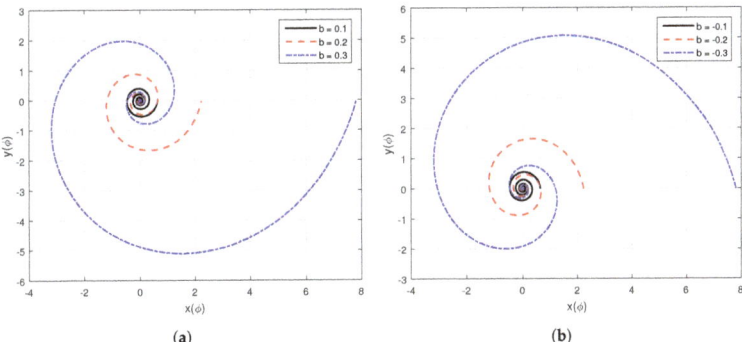

Figure 8. Logarithmic spiral with various values of b: (**a**) logarithmic spiral with positive b values and (**b**) logarithmic spiral with negative b values.

4.1.2. Archimedean Spiral

Archimedean spiral is another famous spiral that has been used in significant applications of engineering, biology, etc. The Archimedean spiral is also known as the arithmetic spiral. This spiral can be seen in nature in ferns, millipedes, and human fingerprints. The spiral trajectory is the locus of a point's position that moves away from the fixed point with a constant speed along a line that rotates with a constant angular velocity. The coordinates of an Archimedean spiral on xy-plane is given as follows [13,38]:

$$x(\psi) = (c + d \cdot \psi) \cdot \cos(\psi), \ y(\psi) = (c + d \cdot \psi) \cdot \sin(\psi), \tag{9}$$

where c and d are constants that define the spirals initial radius and the successive turns difference, respectively.

The Archimedean spiral for $c = 0.5$, ψ from 0 to -7π, and various d values are shown in Figure 9. The trajectory in Figure 9a is obtained for positive values of d, while Figure 9b is obtained for negative values. As the initial radius is $c = 0.5$, all the spirals are starting at this value, as shown in Figure 9. The spiral growth rate d controls the increment per revolution. Thus, the distance between successive turns is constant, which is equal to the value of d. Moreover, the parameter d controls the evolution of the spiral. The spiral in

Figure 9a depicts that for positive d values, and the spiral evolves in an anticlockwise direction. Whereas for negative d values, the spiral evolves clockwise.

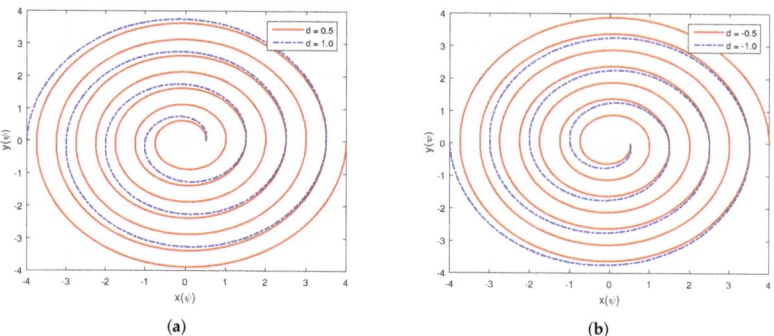

Figure 9. Archimedean spiral with various values of d: (**a**) Archimedean spiral with positive d values and (**b**) Archimedean spiral with negative d values.

Observing the spirals in Figures 8 and 9 shows a difference between the Archimedean and logarithmic spirals worth highlighting. In the Archimedean spiral, the intersection points of a ray from the origin on successive turnings have a constant separation distance. However, in a logarithmic spiral, these distance of intersection points on next turnings from the origin will form a geometric progression.

4.1.3. Rose Spiral

As the name suggests, the rose spiral is often seen in the unfurling of rose petals and holds the properties of symmetric and periodic arc curves. The coordinates of a rose spiral on xy-plane is given as follows [13,38]:

$$x(\xi) = e \cdot \cos(n\xi) \cdot \cos(\xi), \ y(\xi) = e \cdot \cos(n\xi) \cdot \sin(\xi), \tag{10}$$

where e and n are constants that define the pedal length and number, respectively.

The rose spiral with various values of e and n are shown in Figure 10. The spiral in Figure 10a is achieved for $n = 2$ and multiple values of e. Similarly, the spiral in Figure 10b is obtained for $e = 2$ and various values of n. In both cases, ξ ranges from 0 to 2. The spirals in Figure 10a depict that parameter e controls the petal length. It is worth noting that as the value of e increases, the petal length increases. The spirals in Figure 10b also show that n controls petals' number, size, and length. For an even value of n, the number of petals is $2n$. However, for odd values of n, the number of petals is only n.

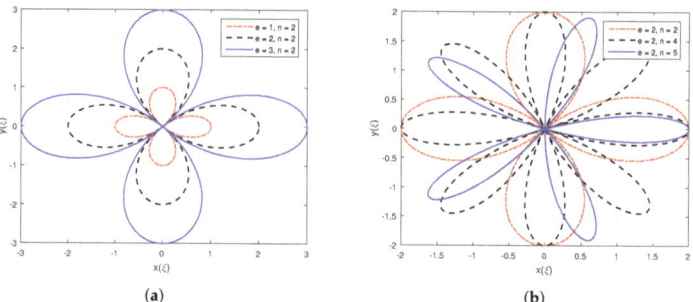

Figure 10. Rose spiral with various values of e and n: (**a**) rose spiral with constant n value and variable e and (**b**) rose spiral with constant e value and variable n.

4.1.4. Epitrochoid and Hypotrochoid Spirals

Epitrochoid and hypotrochoid spirals are a family of curves generated by a point attached to a rolling circle. This rolling circle will roll out around the outside of a fixed circle to form an epitrochoid spiral. On the other hand, to create a hypotrochoid spiral, the rolling one will roll around inside the fixed one. Let ρ_1 and ρ_2 be the radii of rolling and fixed circles, respectively, and f is the distance between the point and rolling circle's center. The coordinates of epitrochoid spiral on xy-plane is given as [13,38],

$$x(\zeta) = (\rho_2 + \rho_1) \cdot \cos(\zeta) - f \cdot \cos\left(\frac{\rho_2 + \rho_1}{\rho_1}\zeta\right), \text{ and}$$
$$y(\zeta) = (\rho_1 + \rho_2) \cdot \sin(\zeta) - f \cdot \sin\left(\frac{\rho_1 + \rho_2}{\rho_1}\zeta\right). \tag{11}$$

Similarly, the coordinates of a hypotrochoid spiral on xy-plane is given as follows:

$$x(\zeta) = (\rho_2 - \rho_1) \cdot \cos(\zeta) + f \cdot \cos\left(\frac{\rho_2 - \rho_1}{\rho_1}\zeta\right), \text{ and}$$
$$y(\zeta) = (\rho_2 - \rho_1) \cdot \sin(\zeta) - f \cdot \sin\left(\frac{\rho_2 - \rho_1}{\rho_1}\zeta\right). \tag{12}$$

The trajectories of epitrochoid and hypotrochoid spirals for $\rho_1 = 0.8, \rho_2 = 3, d = 2.5$, and ζ ranging from 0 to 10π is shown in Figure 11a,b, respectively. In both spirals, it should be noted that ζ significantly affects the spiral's shape. If the considered ζ ranges from 0 to 2π, the rolling circle will revolve only once around the fixed circle. Thus, it is not possible to obtain the whole pattern of the spiral. These spirals can be drawn using Spirograph toys and often appear in nature. For instance, the planets orbit in a geocentric system, and Wankel engines' combustion chambers take these spiral shapes.

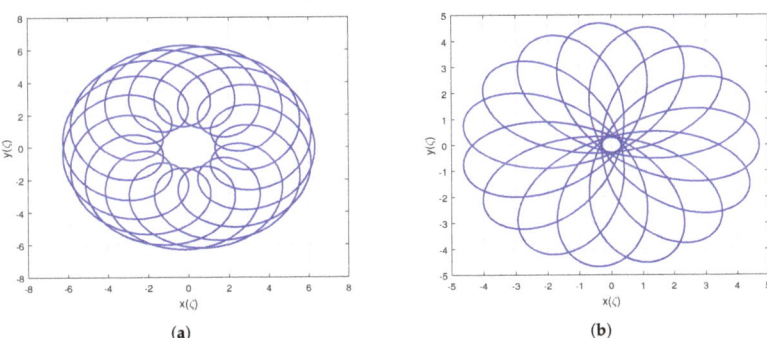

Figure 11. Epitrochoid and hypotrochoid spirals for $\rho_1 = 0.8, \rho_2 = 3$, and $d = 2.5$: (**a**) epitrochoid spiral and (**b**) hypotrochoid spiral.

4.2. Spiral Path-Based Optimization Algorithms

Over the years, researchers have developed various novel optimization algorithms in which the spiral motion has been used while mimicking the system's behavior. Further, an improved version of multiple algorithms is also proposed using spiral trajectories to improve the performance of conventional techniques. Table 1 provides the list of spiral path-inspired optimization techniques, including the inspiration of developing the algorithm, the type of spiral used, and the source code links.

Table 1. List of spiral path-inspired optimization algorithms.

Ref.	Year	Algorithm	Author	Inspiration	Spiral Type	Source Code Link
[40,41]	2010	Spiral Bacterial Foraging Optimization Algorithm	Alireza Kasaiezadeh et al.	*E. coli* bacteria foraging behavior	Spiral	–
[8]	2011	Galaxy-Based Search Algorithm	Hamed Shah-Hosseini	Arms of the spiral galaxy	Spiral	–
[51]	2014	Hurricane-Based Optimization Algorithm	Isamil Rbouh et al.	Behavior of hurricanes, radial wind, and pressure profiles	Logarithmic spiral	–
[52]	2015	Moth–flame Optimization Algorithm	Seyedali Mirjalili	Moths' navigation behavior around the flame	Logarithmic spiral	https://seyedalimirjalili.com/mfo (accessed on 1 December 2021)
[53]	2016	Whale Optimization Algorithm	Seyedali Mirjalili	Whales' hunting bubble net phenomenon	Logarithmic spiral	https://seyedalimirjalili.com/woa (accessed on 1 December 2021)
[54]	2017	Moth Swarm Optimization Algorithm	Al-Attar Ali Mohamed et al.	Moths orientation towards the moonlight	Logarithmic spiral	https://mathworks.com/matlabcentral/fileexchange/57822 (accessed on 1 December 2021)
[50]	2018	Improved Teaching–Learning-Based Optimization Algorithm	Zhuoran Zhang et al.	effect of teacher influence on learners	Logarithmic spiral	–
[55]	2018	Developed Grey Wolf Optimization Algorithm	Mostafa Abdo et al.	Grey wolves leadership and hunting strategies	Logarithmic spiral	–
[56]	2019	Seagull Optimization Algorithm	Gaurav Dhiman et al.	Seagulls' migration and attacking behavior	3D logarithmic spiral	https://mathworks.com/matlabcentral/fileexchange/75180 (accessed on 1 December 2021)
[57]	2019	Emperor Penguins Colony Optimization Algorithm	Sasan Harifi et al.	Emperor penguins behavior	Logarithmic spiral	–
[36,37]	2019	Spiral Artificial Bee Colony Algorithm	Sonal Sharma et al.	Honey bee swarms' intelligent foraging behavior	Logarithmic and Archimedean spirals	–
[58]	2019	Sooty Tern Optimization Algorithm	Gaurav Dhiman et al.	Sooty terns' migration and attacking behaviors	Spiral	https://mathworks.com/matlabcentral/fileexchange/76667 (accessed on 1 December 2021)
[59]	2019	Whirlpool Algorithm	Yuanyang Zou et al.	Physical phenomenon of whirlpool	Spiral	–
[60]	2020	Improved Crow Search Algorithm	Xiaoxia Han et al.	Crows intelligent behavior of searching, hiding and retrieving food	Logarithmic Spiral	–

Table 1. *Cont.*

Ref.	Year	Algorithm	Author	Inspiration	Spiral Type	Source Code Link
[38]	2020	Improved Ant Lion Optimization Algorithm	M. W. Guo et al.	Natural hunting phenomenon of antlions	Archimedes, Cycloid, Epitrochoid, Hypotrochoid, Logarithmic, Rose, Inverse, and Overshoot spirals	–
[61]	2020	Manta Ray Foraging Optimization Algorithm	Weiguo Zhao et al.	Manta rays intelligent behavior	Logarithmic spiral	https://www.mathworks.com/matlabcentral/fileexchange/73130 (accessed on 1 December 2021)
[62]	2020	Bald Eagle Search Optimization Algorithm	H. A. Alsattar et al.	Bald eagles hunting behavior	Spiral	https://mathworks.com/matlabcentral/fileexchange/86862 (accessed on 1 December 2021)
[63]	2020	Improved Firefly Algorithm	Jinran Wu et al.	Fireflies' flashing behavior	Logarithmic spiral	https://github.com/wujrtudou/AdaptiveFireflyAlgorithm (accessed on 1 December 2021)
[64]	2021	Spiral Water Cycle Algorithm	Heba F. Eid et al.	Natural hydrological cycle process	Hyperbolic spiral	–
[65]	2021	Improved Slap Swarm Optimization Algorithm	Diab Mokeddem	Behavior of slap chains	Logarithmic spiral	–
[66]	2021	Spiral Flying Sparrow Search Algorithm	Chengtian Ouyang et al.	Sparrow's behaviors during group wisdom, antipredation, and foraging	Logarithmic spiral	–
[67]	2021	Spiral Grasshopper Optimization Algorithm	Zhangze Xu et al.	Grasshoppers foraging and swarming behavior	Logarithmic spiral	–
[68]	2021	Aquila Optimization Algorithm	Laith Abualigah et al.	Aquilas' behavior during prey catching	Spiral	https://www.mathworks.com/matlabcentral/fileexchange/89381 (accessed on 1 December 2021)
[69]	2021	Spiral Spotted Hyena Optimization Algorithm	Vijay Kumar et al.	Spotted hyenas behavior during hunting	Logarithmic spiral	–
[70]	2021	Spiral Chicken Swarm Optimization Algorithm	Miao Li et al.	Chicken swarms' hierarchical order and its behaviors	Logarithmic spiral	–
[71]	2021	Golden Eagle Optimization Algorithm	Abdolkarim Mohammadi-Balani et al.	Golden eagles' intelligent behavior during hunting	Spiral	https://mathworks.com/matlabcentral/fileexchange/84430 (accessed on 1 December 2021)

For example, a detailed description of four novel optimization algorithms in which spiral trajectory has been used in the development is explained underneath. The chosen novel optimizations algorithms list includes moth–flame, whale, seagull, and Aquila. Further, a detailed description of four improved optimization algorithms using spiral trajectories is also explained in this section. The enhanced optimization algorithms are the water cycle, antlion, slap swarm, and sparrow search. Some of these algorithms have been widely used by various researchers recently, and others have been developed newly, thus selected for the detailed explanation.

4.2.1. Moth–Flame Optimization Algorithm

The moth–flame optimization algorithm was developed in 2015 by Seyedali Mirjalili from the behavior of moths' navigation around the light/flame in a spiral path [52,72,73]. The application of a logarithmic spiral to mimic the moths' transverse orientation property around the flame in this algorithm is explained underneath. In the algorithm, the initial moths' positions will be updated with respect to flames using the logarithmic spiral as follows [52,74]:

$$m_{i,j} = \begin{cases} D_{i,j} \cdot e^{b\tau} \cdot \cos(2\pi\tau) + f_{i,j}, & \text{for } i \leq F_N \\ D_{i,j} \cdot e^{b\tau} \cdot \cos(2\pi\tau) + f_{N,j}, & \text{for } i > F_N \end{cases}, \quad (13)$$

where $m_{i,j}$, $f_{i,j}$, and $D_{i,j}$ are the positions of jth variable of ith moth, flame, and distance between the moth and its corresponding flame, N is the total number of flames. Further, b and τ are the parameters of logarithmic spiral (refer to Section 4.1.1).

The major drawback of this algorithm is the premature convergence at optimal local solutions during the search process. Moreover, they cannot be applied to permutation problems as it is developed for continuous search space [75]. As mentioned in Table 1, the source code of this optimization algorithm created using MATLAB for both single and multiobjective problems is made publicly available by the developer on his website at https://seyedalimirjalili.com/mfo (accessed on 1 December 2021). Further, the links for the source code using other platforms, such as Python, C++, and R studio, are also available on the same website.

4.2.2. Whale Optimization Algorithm

The whale optimization algorithm is a novel metaheuristic algorithm developed in 2016 by Seyedali Mirjalili and Andrew Lewis to mimic whales' hunting bubble net phenomenon in a spiral motion [53,76–79]. The algorithm is a model of capturing whales' behavior during the encircling, attacking, and searching of prey. During the encircling phase, all the whales' positions will be updated to move towards the best whale position, which is near to the target and is given as,

$$\vec{X}(i+1) = \vec{X}^*(i) - \vec{A} \cdot |\vec{C} \cdot \vec{X}^*(i) - \vec{X}(i)|. \quad (14)$$

During the phase of attacking the prey, the whales move spirally using the bubble net movement phenomenon. Thus, position updation of whales during this phenomenon in logarithmic spiral motion is as follows:

$$\vec{X}(i+1) = |\vec{X}^*(i) - \vec{X}(i)| \cdot e^{bl} \cdot \cos(2\pi l)\vec{X}^*(i). \quad (15)$$

Finally, the whales will choose either encircling or attacking during the searching of prey, which can be achieved using the following model:

$$\vec{X}(i+1) = \begin{cases} \vec{X}^*(i) - \vec{A} \cdot |\vec{C} \cdot \vec{X}^*(i) - \vec{X}(i)|, & p < 0.5, \\ |\vec{X}^*(i) - \vec{X}(i)| \cdot e^{bl} \cdot \cos(2\pi l)\vec{X}^*(i), & p \geq 0.5. \end{cases} \quad (16)$$

Therefore, the position updation of all the whales during all three phases is summarized as,

$$\vec{X}(i+1) = \begin{cases} \begin{cases} \vec{X}^*(i) - \vec{A} \cdot |\vec{C} \cdot \vec{X}^*(i) - \vec{X}(i)|, & \vec{A} < 1, \\ \vec{X}_r(i) - \vec{A} \cdot |\vec{C} \cdot \vec{X}_r(i) - \vec{X}(i)|, & \vec{A} \geq 1, \end{cases} & p < 0.5, \\ |\vec{X}^*(i) - \vec{X}(i)| \cdot e^{bl} \cdot \cos(2\pi l)\vec{X}^*(i), & p \geq 0.5, \end{cases} \quad (17)$$

where the vectors $\vec{X}^*(i)$ is the closest whale's position to the prey, $\vec{X}(i)$ and $\vec{X}(i+1)$ are the whales' positions at ith and $i+1^{\text{th}}$ iterations, $\vec{A}$ and $\vec{C}$ are the coefficients, b and l are the parameters of logarithmic spiral (refer to Section 4.1.1). Further, it is to be noted that for $\vec{A} \geq 1$, positions updation has been achieved using $\vec{X}_r(i)$, a random position vector at ith iteration.

This whale optimization algorithm has the drawbacks of lower accuracy, slow convergence, and being trapped into optimal local solutions and cannot solve higher-dimensional problems effectively [80]. As given in Table 1, the source codes of this optimization algorithm for single-objective problems using MATLAB, Python, C++, and R are publicly available at https://seyedalimirjalili.com/woa (accessed on 1 December 2021).

4.2.3. Seagull Optimization Algorithm

Gaurav Dhiman et al. proposed the seagull optimization algorithm in 2019 to mimic the seagulls' migration and hunting behavior [56]. The algorithm is a mathematical model of seagulls' behavior in two stages, namely migration and attack. During the stage of natural attacking, the seagulls maintain spiral behavior in the air. The coordinates of this spiral behavior in x, y, and z planes are modeled as follows:

$$x = u \cdot e^{kv} \cdot \cos(k), \quad y = u \cdot e^{kv} \cdot \sin(k), \quad z = u \cdot e^{kv} \cdot k, \quad (18)$$

where $k \in [0, 2\pi]$ is the spiral angle, u and v are the arbitrary constants.

The seagull optimization algorithm has the significant drawback of weak population diversity during the search process [81]. The link to the MATLAB-based source code of this optimization algorithm is given in Table 1.

4.2.4. Aquila Optimization Algorithm

The Aquila optimization algorithm was proposed in 2021 by Laith Abualigah et al. to mimic Aquila's behavior during prey catching [68]. The algorithm constitutes four stages: (i) expanded exploration, (ii) narrowed exploration, (iii) expanded exploitation, and (iv) narrowed exploitation. During the stage of narrowed exploration, the Aquila rotates over a target prey for a short glide attack. This behavior is modeled as follows:

$$X(t+1) = X_{best}(t) \cdot Levy() + X_r(t) + (y - x) \cdot rand(), \quad (19)$$

where $X_r(t)$ and $X_{best}(t)$ are the random and best solutions at tth iteration, $X(t+1)$ solution at $(t+1)^{\text{th}}$ iteration, $rand() \in (0, 1]$ is the random number, and $Levy()$ is the Lévy distribution. Further, x and y are the Cartesian coordinates of the spiral with radius r and angle l given as follows:

$$x = r\sin(l), \quad y = r\cos(l). \quad (20)$$

From the above, it is to be highlighted that the Levy flight's effect is relatively weak. Thus, the algorithm has insufficient local exploitation ability [82]. The MATLAB and Java-based source code link of this optimization algorithm for single-objective problems is given in Table 1.

4.2.5. Water Cycle Optimization Algorithm

The water cycle optimization algorithm was proposed in 2012 by Eskandar et al. to mimic the natural hydrological cycle process [10,83,84]. The algorithm simulates the stream and river flow, rainfall, and evaporation into the sea. In this algorithm, the position update of (a) streams flow to the rivers, (b) streams flow to the sea, and (c) rivers flow to the sea are respectively given as follows:

$$X_{st}(i+1) = X_{st}(i) + rand() \cdot C \cdot (X_r(i) - X_{st}(i)), \qquad (21)$$

$$X_{st}(i+1) = X_{st}(i) + rand() \cdot C \cdot (X_{se}(i) - X_{st}(i)), \qquad (22)$$

$$X_r(i+1) = X_r(i) + rand() \cdot C \cdot (X_{se}(i) - X_r(i)), \qquad (23)$$

where $X_{st}(i)$, $X_r(i)$ and $X_{se}(i)$ are the positions of stream, river, and sea at ith iteration, $X_{st}(i+1)$, $X_r(i+1)$, and $X_{se}(i+1)$ are the positions of stream, river, and sea at $(i+1)^{th}$ iteration, $C \in [1,2]$ is the constant value and $rand() \in (0,1]$ is the random number.

The MATLAB-based source code of this conventional optimization algorithm for both constrained and unconstrained problems, including several improved versions and multiobjective problems, are made publicly available by the researcher on his website at https://ali-sadollah.com/water-cycle-algorithm-wca/ (accessed on 1 December 2021).

The algorithm has insufficient exploitation ability, and thus, in [64], the authors have integrated the hyperbolic spiral, which helps improve the exploitation ability of the algorithm. Therefore, modified position update equations using the hyperbolic spiral are given as follows:

$$X_{st}(i+1) = X_{st}(i) + |X_r(i) - X_{st}(i)| \cdot \cos(2\pi l)/l, \qquad (24)$$

$$X_{st}(i+1) = X_{st}(i) + |X_{se}(i) - X_{st}(i)| \cdot \cos(2\pi l)/l, \qquad (25)$$

$$X_r(i+1) = X_r(i) + |X_{se}(i) - X_r(i)| \cdot \cos(2\pi l)/l, \qquad (26)$$

where $l \in [-1,1]$ is the parameter of hyperbolic spiral, which is an uniformly distributed random number.

4.2.6. Ant Lion Optimization Algorithm

Seyedali Mirjalili proposed the antlion optimization algorithm in 2015 to mimic the natural hunting phenomenon of antlions [85–88]. The algorithm is a model of capturing the following ants and antlions behaviors: (i) the ants' random walk behavior and gets trapped in antlions pits and (ii) the antlions' hunting behaviors include building traps, sliding ants towards them, catching, rebuilding pits, and elitism. The algorithm retains the best antlion with optimal fitness value, elitism, and the corresponding antlion is called elite antlion. Thus, the elite and selected antlions update their position randomly as follows:

$$Ant_i(t) = \frac{R_e(t) + R_a(t)}{2}, \qquad (27)$$

where $R_e(t)$ and $R_a(t)$ are the elite and selected antlions random walk during tth iteration.

The MATLAB, Python, and R software-based source codes of this conventional optimization algorithm for both single and multiobjective problems are made publicly available by Seyedali Mirjalili on his website at https://seyedalimirjalili.com/alo (accessed on 1 December 2021).

In [38], the authors proposed an improved version of this algorithm. In this enhanced version, the elite and selected antlions update their position using eight spiral complex paths instead of moving in randomly to improve the convergence speed and performance. These spiral trajectories include Archimedes, cycloid, epitrochoid, hypotrochoid, logarithmic, rose, inverse, and overshoot spirals. For an example case, the values of $R_e(t)$ and $R_a(t)$ are computed using logarithmic spiral as,

$$R_e(t) = D_1 \cdot e^{b_1 t} \cos(2\pi t_1), \ R_a(t) = D_1 \cdot e^{b_1 t} \sin(2\pi t_1), \tag{28}$$

where D_1, b_1, and t_1 are the parameters of logarithmic spiral (see Section 4.1.1).

Similarly, using the Archimedes spiral, the values of $R_e(t)$ and $R_a(t)$ are computed as follows:

$$R_e(t) = D_2 + b_2 \cdot t_2 \cdot \cos(2\pi t_2), \ R_a(t) = D_2 + b_2 \cdot t_2 \cdot \sin(2\pi t_2), \tag{29}$$

where D_2, b_2, and t_2 are the parameters of Archimedes spiral (see Section 4.1.2).

4.2.7. Slap Swarm Optimization Algorithm

Slap swarm optimization algorithm was developed in 2017 by Seyedali Mirjalili et al. to mimic the behavior of slap chains, which is searching for target food [89–92]. In the slap chain, the first slap is the leader, and all the other slaps follow the leader. In the algorithm, the update equations for the leader and followers' positions during the searching of target food are as follows:

$$X_i^1 = \begin{cases} F_i + r_1((UB_i - LB_i)r_2 + LB_i), & \text{if } r_3 \geq 0, \\ F_i - r_1((UB_i - LB_i)r_2 + LB_i), & \text{if } r_3 < 0, \end{cases} \tag{30}$$

$$X_i^j = 0.5(X_i^j + X_i^{j-1}), \ j \geq 2, \tag{31}$$

where X_i^1 and X_i^j are the positions of leader and followers, F_i is the target food, LB_i and UB_i are the lower and upper bounds of ith dimension, r_1, r_2, and r_3 are random numbers.

The MATLAB-based source code of this optimization algorithm for both single and multiobjective problems is made publicly available by the developer on his website at https://seyedalimirjalili.com/ssa (accessed on 1 December 2021). Further, the links for the source code using Python and R are also available on the same website.

However, in [65], it is stated that the conventional slap swarm optimization algorithm (SSOA) has a slower convergence and gets trapped at local optima. Thus, the authors have proposed an improved SSOA using a logarithmic spiral. In this improved algorithm, the followers' positions are updated using a logarithmic spiral as follows:

$$X_i^j = 0.5(X_i^j + X_i^{j-1}) \cdot e^{b\theta} \cdot \cos(2\pi\theta), \ j \geq 2, \tag{32}$$

where b and θ are the parameters of logarithmic spiral (refer to Section 4.1.1).

4.2.8. Sparrow Search Optimization Algorithm

Jiankai Xue and Bo Shen proposed a sparrow search optimization algorithm in 2020 to mimic the sparrow's behaviors during group wisdom, antipredation, and foraging [93]. In this algorithm, the sparrows' population is divided into two groups of 20:80 as discovers and followers. The discover have a broad search space to search for the food and guide the followers to move towards the food source. The position update equation for the discover sparrows during the searching of target food is as follows:

$$X_{i,j}(t+1) = \begin{cases} X_{i,j}(t) \cdot \exp(-\frac{h}{\alpha \cdot M}), & \text{if } R_2 < ST, \\ X_{i,j}(t) + Q \cdot L, & \text{if } R_2 \geq ST, \end{cases} \tag{33}$$

where $X_{i,j}(t)$ and $X_{i,j}(t+1)$ are the ith discover sparrows' position of jth dimension tth and $(t+1)^{th}$ iterations, h and M are the current and maximum number of iterations, Q is a uniformly distributed random number, L is a row matrix with all values as one, α and $R_2 \in [0,1]$ are the random numbers, $ST \in [0.5,1]$ is the safety threshold values.

The values of R_2 and ST help indicate the safety of the food source area. Based on these values, the type of environment around the food source area, predators status, and the actions that need to be taken are classified as follows:

$$\text{Condition} = \begin{cases} \text{Safe,} & \text{No predators around, can search for food,} & \text{if } R_2 < ST, \\ \text{Unsafe,} & \text{Predators around, fly to other safe area,} & \text{if } R_2 \geq ST. \end{cases} \quad (34)$$

As some of the followers closely follow the discoverers, they update their positions to move towards the discovered food source area. The position update equation for the follower sparrows towards the food source is as follows:

$$X_{i,j}(t+1) = \begin{cases} Q \cdot \exp\left(\frac{X_{worst}(t) - X_{i,j}(t)}{i^2}\right), & \text{if } i > n/2, \\ X_p(t+1) + |X_{i,j}(t) - X_p(t+1)| \cdot A^T(AA^T)^{-1} \cdot L, & \text{otherwise,} \end{cases} \quad (35)$$

where $X_{worst}(t)$ is the group's worst position at tth iteration, $X_p(t+1)$ is the discovers' optimal position at $(t+1)^{th}$ iteration, A is row matrix of randomly assigned with 1 or -1. Further, $i > n/2$ indicates that the sparrows are in a danger position. Thus, the sparrows make antipredation behavior. The MATLAB-based source code for implementing this algorithm is available for registered users at https://www.mathworks.com/matlabcentral/fileexchange/88788 (accessed on 1 December 2021).

However, in [66,94], the authors proposed a variable spiral search technique for the followers to update their positions better. The position update equation of the followers using this search strategy is as follows:

$$X_{i,j}(t+1) = \begin{cases} e^{zl} \cdot \cos(2\pi l) Q \cdot \exp\left(\frac{X_{worst}(t) - X_{i,j}(t)}{i^2}\right), & \text{if } i > n/2, \\ X_p(t+1) + |X_{i,j}(t) - X_p(t+1)| \cdot A^T(AA^T)^{-1} \cdot L, & \text{otherwise,} \end{cases} \quad (36)$$

where z and l are the parameters of logarithmic spiral (refer to Section 4.1.1). Further, the value of z is varied at every iteration, making the proposed technique a variable spiral search approach.

5. Application of Spiral Dynamics Optimization Algorithm

The conventional and other variants of the SDO algorithm have been applied in various fields for finding the optimal solution, as explained underneath.

5.1. Modeling and Controller Tuning

The application of SDO and its variants in the area of modeling and controller tuning is as follows:

- Controller tuning [95];
- Controlling robotic arm movement [96];
- Flexible manipulator system [14,16,20,26,28,34,35];
- Stair descending in a wheelchair [97,98];
- Inverted pendulum [99];
- Twin rotor systems [25,34];
- Two-wheeled robotic vehicle [39].

Hassan et al. proposed using an SDO algorithm to tune the predictive proportional-integral (PI) controller for wireless networked control systems [95]. Similarly, the authors of [96] have utilized SDO in the tuning of proportional-integral-derivative (PID) in controlling the robotic arm movement. Moreover, for both modeling and control of flexible link manipulator systems, the authors of [14] have used conventional SDO. For the same application, the authors of [20,34,35] proposed the hybridization of the SDO algorithm with BCA and BFA. The improved and adaptive version of SDO is also presented for both modeling and control of a flexible link manipulator system [16,26,28]. In another application, fuzzy control of a stair descending in a wheelchair, an SDO algorithm is used for tuning of controller parameters. In [99], a hybrid algorithm using PSO and SDO is proposed for the tuning of a fuzzy controller designed for the inverted pendulum. Nasir et al. have proposed an improved SDO and hybrid algorithm using SDO and BFA to model twin rotor systems [25,34]. The hybrid SDO and BFA algorithm has also been used for controlling the two-wheeled robotic vehicles [39].

5.2. Electrical Energy Optimization

Similarly, the application of SDO and its variants in the area of optimizing electrical energy systems is as follows:

- Digital filters [100];
- Economic/emission dispatch [14,101,102];
- Hybrid electrical vehicles [23];
- Maximizing power production of a wind farm [103];
- Multigeneration energy system [104];
- Network with power distribution [105].

The economic and emission dispatch problems in power systems have been solved by various researchers using the SDO algorithm [14,101,102]. Similarly, an optimal strategy using the SDO algorithm is proposed for maximum power production in the wind farm [103]. A multiobjective SDO algorithm for a multigeneration energy system is presented for minimizing the total cost while maximizing energy efficiency [104]. In [105], a hybrid algorithm using SDO and BFA is proposed to optimize decentralized generation placement simultaneously. In another application, an optimal sizing strategy using the adaptive version of the SDO algorithm has been presented for hybrid electric air–ground vehicles [23]. The authors of [100] have proposed using SDO for the filter design. The algorithm achieved better performance in achieving the desired magnitude response in the multiobjective optimization task.

5.3. Mechanical Systems Optimization

Over the years, several mechanical systems have been optimized using the SDO algorithm. The list of applications are as follows:

- Micro-channel heat sink [29,30];
- Automation of high-rise buildings [19];
- Planar, spatial truss structures [18];
- Pressure vessel design problems [38,50];
- Welded beam design problems [50].

Cruz et al. proposed the generalized and stochastic SDO algorithms to solve microelectronic thermal management problems [29,30]. The authors of [19] have proposed a hypotrochoid SDO algorithm to optimize the sensor placement in the 632-meter-tall Shanghai Tower and compared the performance with seven optimization algorithms, including its predictors. The authors of [18] also proposed the hypotrochoid SDO algorithm for finding the optimal setting parameters of 10, 37, 52, 72, and 200-bar planar and spatial truss structures. The use of spiral equation in improving the TLBO and antlion optimization algorithms for pressure vessel design problems is presented in [38,50]. The improved TLBO algorithm using logarithmic spiral trajectory is also applied to find the optimal setting parameters for welded beam design problems [50].

5.4. Other Optimization Problems

The application of the SDO algorithm for other types of optimization problems are as follows:

- 2D mesh topologies [106];
- Clustering problems [33];

- Cubic polyhedral cages [107];
- Face image de-blurring [32];
- Neural network training [108,109];
- Sensor pattern sorting [110,111].

The authors of [107] are the first to showcase the problems and scope of spiral dynamics optimization applied to polyhedral cages. Another work before developing the conventional SDO algorithm is reported in [106]. Here, a heuristic spiral mapping algorithm is the first type of SDO applied for 2D mesh network topologies. For clustering problems, distributed SDO is proposed in which the population of search space is split into sub-populations [33]. Hong-Chun Jia et al. have proposed an efficient and intelligent algorithm using SDO for deep neural networks [108]. The network is to find the optimal physical health and fitness level in sports. Recently, James McCaffrey from Microsoft Research has developed the SDO algorithm in Python to train the neural network to find the optimal weights and biases values [112], the real-time implementation of a deterministic SDO algorithm using field-programmable gate arrays for spot patterns sorting in a Shack–Hartmann wavefront sensor [110].

As mentioned earlier, the SDO and its variants have been applied in various applications. The summary of all applications is given in Table 2. The table provides the details of the application system, including the dimension, software tool, cost function, type of optimization problem, and comparison techniques. In the table, SO and MO are optimization problems denoting single objective and multiobjective. The SDO validation and its variants on various benchmark functions are also detailed. It is to highlight that the most widely used error-based cost functions are: mean squared error (MSE), root mean squared error (RMSE), and the sum of squared error (SSE). Similarly, the integral error functions used in the research are integral squared error (ISE) and integral time absolute error (ITAE). The errors are computed as follows:

$$\text{MSE} = \frac{1}{n_s}\sum_{i=1}^{n_s}(Y_a - Y_p)^2, \tag{37}$$

$$\text{RMSE} = \sqrt{\frac{1}{n_s}\sum_{i=1}^{n_s}(Y_a - Y_p)^2}, \tag{38}$$

$$\text{ISE} = \int_{t=0}^{\infty} e(t)\,dt, \tag{39}$$

$$\text{ITAE} = \int_{t=0}^{\infty} t|e(t)|\,dt, \tag{40}$$

where n_s is the total number of samples, Y_a and Y_p are the actual and predicted values, $e(t)$ is the error, the difference between actual and reference values.

Table 2. List of applications of SDO and its variants in various fields over the years.

Ref.	Year	Algorithm	Validation Status	Validation Functions	Application System	Application Dim	Application Tool	Application Cost Function	SO/MO	Comparison Algorithms
[107]	1997	SDO	✗	-	Cubic Polyhedral Cages	-	-	-	SO	-
[106]	2008	Dynamic Spiral Mapping	✗	-	2D Mesh Topologies	3	SMAP	Reconfiguration Time	MO	Partially, Fully DSM
[9]	2010	SDO	✓	Rosenbrock, Minima, Rastrigin	-	-	MATLAB	-	SO	GA, PSO, ALO
[11]	2011	SDO	✓	Schwefel, Minima, Rastrigin, Griewank	-	-	-	-	SO	Differential evolution (DE), PSO
[21]	2012	Adaptive SDO	✓	Sphere, Ackely, Griewank	-	-	-	-	SO	SDO, Linear, Quadratic, and Exponential Adaptive SDO
[20]	2012	Hybrid SDO-BCA	✓	Sphere, Ackley, Rastrigin, Griewank	Flexible Manipulator System	30	MATLAB	ISE	SO	BFA, SDO
[113]	2012	SDO	✓	Sphere, Schwefel, Minima, Rastrigin, Alpine, Levy, Ackely	-	-	-	-	SO	-
[100]	2013	SDO	✗	-	Digital Filters	10	-	Weighted Magnitude, Lp Norm	SO	-
[14]	2014	Adaptive SDO, Hybrid SDO	✓	Rastrigin, Sphere, Griewank, Ackley	Flexible Manipulator System, Economic/Emission Dispatch, Neural Network Training	30, 15, 9	-	RMSE, MSE	MO	SDO
[114]	2014	Cluster-structured SDO	✓	Rosenbrock, Minima	-	-	-	-	SO	SDO
[33]	2014	Distributed SDO	✗	-	Clustering Problems	10	C++	SSE	SO	SDO, Genetic K-Means
[34]	2014	Hybrid SDO-BCA	✗	-	Flexible Manipulator, Twin Rotor Systems	16	MATLAB	RMSE	MO	Recursive Least Squares (RLS), Least Mean squares (LMS), PSO, GA, Hybrid GA RLS

Table 2. Cont.

Ref.	Year	Algorithm	Validation		System	Application			Cost Function	SO/MO	Comparison Algorithms
			Status	Functions		Dim	Tool				
[24]	2014	Improved SDO	✓	Sphere, Rosenbrock, Griewank, Rastrigin	Twin Rotor System	136	MATLAB		Weighted RMSE	SO	BFA, SDO, Improved SDO
[97]	2014	SDO	✗	-	Stair Descending in a Wheelchair	10	MATLAB		Weighted MSE	SO	Trial and Error Method
[101]	2014	SDO	✗	-	Economic/Emission Dispatch	3, 4, 60	-		Min	SO	-
[115]	2014	SDO	✓	Sphere, Rosenbrock, Schwefel, Rastrigin, Ackley, Griewank, Minima, Levy, Six-hump	-	-	MATLAB		-	SO	PSO
[25]	2015	Improved SDO	✓	Sphere, Rosenbrock, Rastrigin, Ackley	Twin Rotor System	50	MATLAB		Weighted RMSE	SO	SDO
[96]	2015	SDO	✗	-	Controlling Robotic Arm Movement Rectangular	3	-		Steady State Error	SO	-
[116]	2015	SDO	✗	-	Microchannel Heat Sink	5	-		Generation Rate	MO	SA, Unified PSO
[26]	2016	Enhanced Chaotic SDO	✓	Sphere, Ackely, Grienwank	Single-Link Flexible Manipulator	50	-		MSE	SO	SDO, ABC
[28]	2016	Greedy SDO	✓	Sphere, Ackely, Grienwank	Single-Link Flexible Manipulator	50	-		MSE	SO	SDO
[16]	2016	Linear Adaptive SDO	✓	Sphere, Rosenbrock, Ackley, Rastrigin, Griewank, Dixon-Price, Goldstien-Price, Six-hump Camel	Flexible Manipulator Rig	16	MATLAB		RMSE	SO	SDO, BFA, Improved BFA
[117]	2016	SDO	✓	Sphere, Schwefel, Ackley, Minima, Bohachevsky, Rosenbrock	-	-	-		-	SO	ABC

Table 2. Cont.

Ref.	Year	Algorithm	Validation Status	Validation Functions	Application System	Application Dim	Application Tool	Cost Function	SO/MO	Comparison Algorithms
[110]	2016	SDO	✗	-	Shack–Hartmann Wavefront Sensor Pattern Sorting	28	MATLAB	RMSE	SO	B-Spline, Zernike
[102]	2016	SDO	✗	-	Economic/Emission Dispatch	40	MATLAB	Optimal Power	SO	BBO, GA, Evolutionary Algorithm (EA) Simulated Annealing (SA), GA, Tabu Search (TS), Hybrid BFA-PSO
[105]	2017	Hybrid BFA-SDO	✗	-	Network With Power Distribution	4	-	Weighted Error	MO	BFA, SDO
[39]	2017	Hybrid SDO-BCA	✗	-	Two-Wheeled Robotic Vehicle	9	-	MSE	MO	BFA, SDO
[95]	2017	SDO	✗	-	Controller Tuning	2	MATLAB	ITAE	SO	-
[15]	2017	SDO	✓	Sphere, Schwefel, Minima, Levy	-	-	MATLAB	-	SO	-
[103]	2017	SDO	✗	-	Maximizing Power Production of Wind Farm	50	-	Maximum Power	SO	PSO, Game Theoretic
[18]	2018	Hypotrochoid SDO	✗	-	10, 37, 52, 72, 200-bar Planar, Spatial Truss Structures	10, 14, 8, 16, 29	MATLAB	Min	SO	SDO, GA, PSO, DE, Hybrid SDO
[50]	2018	Spiral TLBO	✓	Sphere, Schwefel, Rosenbrock, Step, Quartic, Schwefel, Rastrigin, Ackley, Griewank, Penalized	Pressure Vessel, Welded Beam Design Problems	4, 4	MATLAB	Min	SO	TLBO, Whale Optimization Algorithm (WOA), Grey Wolf Optimizer (GWO)
[118]	2018	SDO	✗	-	System of Nonlinear Equations	2, 4, 20	C++	Min	SO	-
[31]	2018	Stochastic SDO	✓	Booth, Chichinadze, Zettl, Dixon–Price, Griewank, Mishra, Wing, Rastrign	-	-	MATLAB	-	MO	Deterministic SDO, EFO, DE, Unified PSO

Table 2. Cont.

Ref.	Year	Algorithm	Validation		Application				SO/MO	Comparison Algorithms
			Status	Functions	System	Dim	Tool	Cost Function		
[37]	2019	Archimedean Spiral-ABC	✓	Griewank, Salomon, Inverted cosine, Neumaier, Beale, Colville, Kowalik, Rosenbrock, spring, Goldstein–Price	-	-	-	Mean Error	SO	Archimedean Spiral-inspired Local Search (ASLS), Modified ABC, Best-So-Far ABC
[42]	2019	Biogeography-based SDO	✓	Sphere, Schwefel, Axis, Quatic, Rosenbrock, Rastrigin, Griewank, Ackley, Step	CEC 2017 Benchmark Problems	-	-	Cluster count	SO	DE, BBO, Slap Swarm Optimization Algorithm (SSOA), GWO, WOA, GSA
[99]	2019	Hybrid PSO-SDO	✗	-	Triple-link Inverted Pendulum on Two-wheels	4	Simwise4D	RMSE	MO	GSA, ABC, GWO, Ant Colony Optimization (ACO), GA
[32]	2019	Iterative SDO	✗	-	Face Image De-blurring, Generative Adversarial Network Model	-	PyTorch	Loss Function	SO	
[36]	2019	Spiral ABC	✓	10 functions of various orders Schwefel, Quartic, Rosenbrock, Sphere, Powell, Brown, Ackley, Griewank	-	-	-	-	SO	ABC, Modified ABC, Best-So-Far ABC
[43]	2019	Spiral CS	✓		Spam Detection	-	Python	-	SO	PSO, DE. GA, CS, Improved CS
[49]	2019	Spiral-based SCA	✓	Sphere, Rosenbrock	-	-	-	-	SO	SDO, SCA
[19]	2020	Hypotrochoid SDO	✗	-	Automated Monitoring of High-rise Buildings	50	-	Modal Assurance Criterion	MO	PSO, ABC, SDO, TLBO

Table 2. Cont.

Ref.	Year	Algorithm	Validation Status	Validation Functions	Application System	Application Dim	Application Tool	Cost Function	SO/MO	Comparison Algorithms
[38]	2020	Improved ALO-Spiral	✓	10 Unimodel, 8 Multimodel, 10 Combinatorial, 6 Multi Objective	Pressure Vessel	4	–	Multicriteria Function	SO, MO	Hypotrochoid, Rose, Logarithmic, Epitrochoid, Archimedes, Cycloid, and Inverse Spiral ALO
[104]	2020	Multiobjective SDO	✗	–	Multigeneration Energy System	–	–	Min Cost	MO	GA, PSO, EA
[17]	2021	Adaptive Hypotrochoid SDO	✓	Shubert, Ackley, Levy, Perm, Sphere, Trid, Booth, Beale, Powell, Shekel	–	–	–	–	SO	SDO, Adaptive SDO, Hypotrochoid SDO
[30]	2021	Reflection-based stochastic SDO	✓	Keane, Hosaki, Branin, Bird, Hansen, Ursem Waves, Damavandi, Giunta, Rana, 2nd Minimum	Microchannel Heat Sink	3	MATLAB	Minimum Entropy Generation	SO	Unified PSO, Deterministic SDO, Cuckoo search
[108]	2021	SDO	✗	–	Physical Fitness Determination Using Deep Neural Networks	20	–	RMSE	SO	GA, PSO
[99]	2021	SDO	✗	–	Fuzzy Control of Inverted Pendulum	20	MATLAB	RMSE	SO	Trial and Error, PSO
[35]	2021	Hybrid SDO-BFA	✓	28 Functions	Fuzzy Control of Flexible Manipulator	103	MATLAB	SO	MAE	SDO, BFA, Hybrid SDO-Bacteria Chemotaxis
[98]	2021	SDO	✗	–	Two-wheeled Wheelchair System	20	MATLAB	RMSE	SO	–
[23]	2022	Adaptive SDO	✗	–	Hybrid Electrical Vehicles	4	MATLAB	Weighted Error	MO	SDO, Enhanced GA, Adaptive PSO

6. Conclusions

6.1. Findings

SDO is a promising and fascinating algorithm that has been greatly appreciated in the literature. The SDO algorithm's advantages over other optimization algorithms lie in its simplicity, ease of implementation, the requirement of few control parameters, and better diversification and intensification strategies. This comprehensive review summarizes the research outcomes published from 1997 until January 2022. The advances and variants of SDO, including adaptive, improved, and hybrid approaches for solving various optimization problems, are critically analyzed. Further, the application of SDO and its variants in multiple fields, including modeling, controller tuning, electrical energy systems, mechanical systems, etc., is comprehensively summarized. Besides, a special interest is devoted to highlighting various nature-inspired optimization algorithms fascinated by the concept of spiral paths. This review is expected to draw the attention of the investigators, experts, and researchers to solve the optimization problems using the SDO algorithm and its variants.

6.2. Future Perspectives

This comprehensive review has helped open up new scopes in the field of spiral-inspired optimization and is highlighted as such underneath.

- Even though the authors have tried to avoid the issue of settling at local optima by the SDO algorithm, the issue is persisting. It requires a careful balance between exploration and exploitation phases.
- The problem of insufficient search space exploration with the conventional SDO, which uses a logarithmic spiral, can be overcome by judiciously selecting spirals. A few such spirals are Fermat, Archimedean, etc., which seem suitable in the present context to solve multiobjective problems. Specifically, the use of Fibonacci and a golden spiral is expected to solve image processing optimization problems effectively as their spiral behavior helps analyze the entire image.
- Dynamically varying control parameters in each iteration of SDO variants is still unresolved, leading to lower accuracy of the optimal solution. The selection of suitable adaptive functions for control parameters is required.
- There is a scope to improve the performance of several existing spiral-inspired optimization algorithms either by utilizing the spiral position update equation of SDO or using other spiral trajectories. Further, the natural behavior of nonspiral-inspired algorithms can be modified using spiral paths for better accuracy in the optimal solution.
- The lack of a mathematical model for complex spiral trajectories, such as the Celtic spiral, limits its use for better search space exploration. Hence, the development of suitable models for such a complex spiral trajectory is expected to enhance the SDO algorithm's exploration performance.

Author Contributions: Conceptualization, K.B.; proofreading, guidance, and regular feedback, B.R.P., M.B.O. and R.I.; writing—original draft preparation, M.B.O. and K.B.; writing—review and editing, B.R.P.; supervision, R.I.; project administration and funding acquisition, M.B.O. and R.I. All authors have read and agreed to the published version of the manuscript.

Funding: This research was funded by Yayasan Universiti Teknologi PETRONAS Fundamental Research Grant (YUTP-FRG) number 015LC0-362.

Institutional Review Board Statement: Not applicable.

Informed Consent Statement: Not applicable.

Data Availability Statement: No new data were created or analyzed in this study. Data sharing is not applicable to this article.

Conflicts of Interest: The authors declare no conflict of interest.

References

1. Abdel-Basset, M.; Abdel-Fatah, L.; Sangaiah, A.K. Metaheuristic algorithms: A comprehensive review. In *Computational Intelligence for Multimedia Big Data on the Cloud with Engineering Applications*; Academic Press: Cambridge, MA, USA, 2018; pp. 185–231.
2. Molina, D.; Poyatos, J.; Del Ser, J.; García, S.; Hussain, A.; Herrera, F. Comprehensive taxonomies of nature-and bio-inspired optimization: Inspiration versus algorithmic behavior, critical analysis recommendations. *Cogn. Comput.* **2020**, *12*, 897–939. [CrossRef]
3. Biswas, A.; Mishra, K.; Tiwari, S.; Misra, A. Physics-inspired optimization algorithms: A survey. *J. Optim.* **2013**, *2013*, 438152. [CrossRef]
4. Siddique, N.; Adeli, H. Physics-based search and optimization: Inspirations from nature. *Expert Syst.* **2016**, *33*, 607–623. [CrossRef]
5. Lindfield, G.; Penny, J. (Eds.) Chapter 8—Physics Inspired Optimization Algorithms. In *Introduction to Nature-Inspired Optimization*; Academic Press: Boston, UK, 2017; pp. 141–170. [CrossRef]
6. Siddique, N.; Adeli, H. *Nature-Inspired Computing: Physics-and Chemistry-Based Algorithms*; Chapman and Hall/CRC: Boca Raton, FL, USA, 2017.
7. Abedinpourshotorban, H.; Shamsuddin, S.M.; Beheshti, Z.; Jawawi, D.N. Electromagnetic field optimization: A physics-inspired metaheuristic optimization algorithm. *Swarm Evol. Comput.* **2016**, *26*, 8–22. [CrossRef]
8. Shah-Hosseini, H. Principal components analysis by the galaxy-based search algorithm: A novel metaheuristic for continuous optimisation. *Int. J. Comput. Sci. Eng.* **2011**, *6*, 132–140.
9. Tamura, K.; Yasuda, K. Primary study of spiral dynamics inspired optimization. *IEEJ Trans. Electr. Electron. Eng.* **2011**, *6*, S98–S100. [CrossRef]
10. Eskandar, H.; Sadollah, A.; Bahreininejad, A.; Hamdi, M. Water cycle algorithm—A novel metaheuristic optimization method for solving constrained engineering optimization problems. *Comput. Struct.* **2012**, *110*, 151–166. [CrossRef]
11. Tamura, K.; Yasuda, K. Spiral multipoint search for global optimization. In Proceedings of the 2011 10th International Conference on Machine Learning and Applications and Workshops, Honolulu, HI, USA, 18–21 December 2011; Volume 1, pp. 470–475.
12. Tsuji, K.; Müller, S.C. *Spirals and Vortices: In Culture, Nature, and Science*; Springer: Berlin/Heidelberg, Germany, 2019.
13. Hammer, Ø. *The Perfect Shape: Spiral Stories*; Springer: Berlin/Heidelberg, Germany, 2016.
14. Siddique, N.; Adeli, H. Spiral dynamics algorithm. *Int. J. Artif. Intell. Tools* **2014**, *23*, 1430001. [CrossRef]
15. Tamura, K.; Yasuda, K. The spiral optimization algorithm: Convergence conditions and settings. *IEEE Trans. Syst. Man Cybern. Syst.* **2017**, *50*, 360–375. [CrossRef]
16. Nasir, A.; Ismail, R.R.; Tokhi, M. Adaptive spiral dynamics metaheuristic algorithm for global optimisation with application to modelling of a flexible system. *Appl. Math. Model.* **2016**, *40*, 5442–5461. [CrossRef]
17. Abishek, R.; Maiti, M.; Sunder, M.; Bingi, K.; Puri, H. Adaptation of Spiral Radius and Angle in Hypotrochoid Spiral Dynamic Algorithm. In Proceedings of the 2021 IEEE Madras Section Conference (MASCON), Chennai, India, 27–28 August 2021; pp. 1–6.
18. Kaveh, A.; Mahjoubi, S. Hypotrochoid spiral optimization approach for sizing and layout optimization of truss structures with multiple frequency constraints. *Eng. Comput.* **2019**, *35*, 1443–1462. [CrossRef]
19. Mahjoubi, S.; Barhemat, R.; Bao, Y. Optimal placement of triaxial accelerometers using hypotrochoid spiral optimization algorithm for automated monitoring of high-rise buildings. *Autom. Constr.* **2020**, *118*, 103273. [CrossRef]
20. Nasir, A.; Tokhi, M.; Abd Ghani, N.; Ahmad, M. A novel hybrid spiral dynamics bacterial chemotaxis algorithm for global optimization with application to controller design. In Proceedings of the 2012 UKACC International Conference on Control, Cardiff, UK, 3–5 September 2012; pp. 753–758.
21. Nasir, A.N.K.; Tokhi, M.O.; Abd Ghani, N.M.; Raja Ismail, R.M.T. Novel adaptive spiral dynamics algorithms for global optimization. In Proceedings of the 2012 IEEE 11th International Conference on Cybernetic Intelligent Systems (CIS), Guildford, UK, 9–11 September 2012; pp. 99–104. [CrossRef]
22. YÜZGEÇ, U.; Tufan, İ. Adaptive Spiral Optimization Algorithm for Benchmark Problems. *Bilecik Şeyh Edebali Üniv. Fen Bilim. Derg.* **2016**, *3*, 8–15.
23. Wang, W.; Chen, Y.; Yang, C.; Li, Y.; Xu, B.; Huang, K.; Xiang, C. An efficient optimal sizing strategy for a hybrid electric air-ground vehicle using adaptive spiral optimization algorithm. *J. Power Sources* **2022**, *517*, 230704. [CrossRef]
24. Nasir, A.; Tokhi, M.; Omar, M.; Ghani, N. An improved spiral dynamic algorithm and its application to fuzzy modelling of a twin rotor system. In Proceedings of the 2014 World Symposium on Computer Applications & Research (WSCAR), Sousse, Tunisia, 18–20 January 2014; pp. 1–6.
25. Nasir, A.; Tokhi, M.O. An improved spiral dynamic optimization algorithm with engineering application. *IEEE Trans. Syst. Man Cybern. Syst.* **2015**, *45*, 943–954. [CrossRef]
26. Hashim, M.; Tokhi, M. Enhanced chaotic spiral dynamic algorithm with application to controller design. In Proceedings of the 2016 IEEE International Conference on Power and Energy (PECon), Melaka, Malaysia, 28–29 November 2016; pp. 752–756.
27. Hashim, M.; Tokhi, M. Chaotic spiral dynamics optimization algorithm. In *Advances in Cooperative Robotics*; World Scientific: Singapore, 2017; pp. 551–558.
28. Hashim, M.; Tokhi, M. Greedy spiral dynamic algorithm with application to controller design. In Proceedings of the 2016 IEEE Conference on Systems, Process and Control (ICSPC), Melaka, Malaysia, 16–18 December 2016; pp. 29–32.

29. Cruz-Duarte, J.M.; Martin-Diaz, I.; Munoz-Minjares, J.; Sanchez-Galindo, L.A.; Avina-Cervantes, J.G.; Garcia-Perez, A.; Correa-Cely, C.R. Primary study on the stochastic spiral optimization algorithm. In Proceedings of the 2017 IEEE International Autumn Meeting on Power, Electronics and Computing (ROPEC), Ixtapa, Mexico, 8–10 November 2017; pp. 1–6.
30. Cruz-Duarte, J.M.; Amaya, I.; Ortíz-Bayliss, J.C.; Correa, R. Solving microelectronic thermal management problems using a generalized spiral optimization algorithm. *Appl. Intell.* **2021**, *51*, 5622–5643. [CrossRef]
31. Matajira-Rueda, D.; Cruz-Duarte, J.M.; Garcia-Perez, A.; Avina-Cervantes, J.G.; Correa-Cely, C.R. A new improvement scheme of spiral algorithm (performance test). In Proceedings of the 2018 IEEE International Autumn Meeting on Power, Electronics and Computing (ROPEC), Ixtapa, Mexico, 14–16 November 2018; pp. 1–6.
32. Ma, Y.; Xu, Y.; Wu, L.; Xu, T.; Zhao, X.; Cai, L. Face Image Deblurring Based on Iterative Spiral Optimazation. In *Chinese Conference on Biometric Recognition*; Springer: Cham, Switzerland, 2019; pp. 163–170.
33. Tsai, C.W.; Huang, B.C.; Chiang, M.C. A novel spiral optimization for clustering. In *Mobile, Ubiquitous, and Intelligent Computing*; Springer: Berlin/Heidelberg, Germany, 2014; pp. 621–628.
34. Nasir, A.; Tokhi, M.O. A novel hybrid bacteria-chemotaxis spiral-dynamic algorithm with application to modelling of flexible systems. *Eng. Appl. Artif. Intell.* **2014**, *33*, 31–46. [CrossRef]
35. Kasruddin Nasir, A.N.; Ahmad, M.A.; Tokhi, M.O. Hybrid spiral-bacterial foraging algorithm for a fuzzy control design of a flexible manipulator. *J. Low Freq. Noise, Vib. Act. Control.* **2021**, 14613484211035646. [CrossRef]
36. Sharma, S.; Kumar, S.; Nayyar, A. Logarithmic spiral based local search in artificial bee colony algorithm. In *International Conference on Industrial Networks and Intelligent Systems*; Springer: Cham, Switzerland, 2019; pp. 15–27.
37. Sharma, S.; Kumar, S.; Sharma, K. Archimedean spiral based artificial bee colony algorithm. *J. Stat. Manag. Syst.* **2019**, *22*, 1301–1313. [CrossRef]
38. Guo, M.; Wang, J.S.; Zhu, L.; Guo, S.S.; Xie, W. Improved ant lion optimizer based on spiral complex path searching patterns. *IEEE Access* **2020**, *8*, 22094–22126. [CrossRef]
39. Goher, K.; Almeshal, A.; Agouri, S.; Nasir, A.; Tokhi, M.; Alenezi, M.; Al Zanki, T.; Fadlallah, S. Hybrid spiral-dynamic bacteria-chemotaxis algorithm with application to control two-wheeled machines. *Robot. Biomim.* **2017**, *4*, 1–15. [CrossRef]
40. Kasaiezadeh, A.; Khajepour, A.; Waslander, S.L. Spiral bacterial foraging optimization method. In Proceedings of the 2010 American Control Conference, Baltimore, MD, USA, 30 June–2 July 2010; pp. 4845–4850.
41. Kasaiezadeh, A.; Khajepour, A.; Waslander, S.L. Spiral bacterial foraging optimization method: Algorithm, evaluation and convergence analysis. *Eng. Optim.* **2014**, *46*, 439–464. [CrossRef]
42. Pal, R.; Saraswat, M. Histopathological image classification using enhanced bag-of-feature with spiral biogeography-based optimization. *Appl. Intell.* **2019**, *49*, 3406–3424. [CrossRef]
43. Pandey, A.C.; Rajpoot, D.S. Spam review detection using spiral cuckoo search clustering method. *Evol. Intell.* **2019**, *12*, 147–164. [CrossRef]
44. Nasir, A.N.; Razak, A.A.; Ismail, R.M.; Ahmad, M.A. A hybrid spiral-genetic algorithm for global optimization. *J. Telecommun. Electron. Comput. Eng. (JTEC)* **2018**, *10*, 93–97.
45. Chen, K.; Zhou, F.Y.; Yuan, X.F. Hybrid particle swarm optimization with spiral-shaped mechanism for feature selection. *Expert Syst. Appl.* **2019**, *128*, 140–156. [CrossRef]
46. Duarte, H.M.M.; de Carvalho, R.L. Hybrid particle swarm optimization with spiral-shaped mechanism for solving high-dimension problems. *Acad. J. Comput. Eng. Appl. Math.* **2020**, *1*, 1–6. [CrossRef]
47. Hassan, S.M.; Ibrahim, R.; Saad, N.; Asirvadam, V.S.; Bingi, K. Hybrid APSO—Spiral dynamic algorithms with application to tuning of filtered PPI controller in a wirelessHART environment. *J. Intell. Fuzzy Syst.* **2019**, *37*, 597–610. [CrossRef]
48. Hassan, S.M.; Ibrahim, R.; Saad, N.; Bingi, K.; Asirvadam, V.S. Hybrid ABFA-APSO Algorithm. In *Hybrid PID Based Predictive Control Strategies for WirelessHART Networked Control Systems*; Springer: Cham, Switzerland, 2020; pp. 121–140.
49. Rizal, N.A.M.; Jusof, M.F.M.; Abd Razak, A.A.; Mohammad, S.; Nasir, A.N.K. Spiral sine-cosine algorithm for global optimization. In Proceedings of the 2019 IEEE 9th Symposium on Computer Applications & Industrial Electronics (ISCAIE), Kota Kinabalu, Malaysia, 27–28 April 2019; pp. 234–238.
50. Zhang, Z.; Huang, H.; Huang, C.; Han, B. An improved TLBO with logarithmic spiral and triangular mutation for global optimization. *Neural Comput. Appl.* **2019**, *31*, 4435–4450. [CrossRef]
51. Rbouh, I.; El Imrani, A.A. Hurricane-based optimization algorithm. *AASRI Procedia* **2014**, *6*, 26–33. [CrossRef]
52. Mirjalili, S. Moth-flame optimization algorithm: A novel nature-inspired heuristic paradigm. *Knowl.-Based Syst.* **2015**, *89*, 228–249. [CrossRef]
53. Mirjalili, S.; Lewis, A. The whale optimization algorithm. *Adv. Eng. Softw.* **2016**, *95*, 51–67. [CrossRef]
54. Mohamed, A.A.A.; Mohamed, Y.S.; El-Gaafary, A.A.; Hemeida, A.M. Optimal power flow using moth swarm algorithm. *Electr. Power Syst. Res.* **2017**, *142*, 190–206. [CrossRef]
55. Abdo, M.; Kamel, S.; Ebeed, M.; Yu, J.; Jurado, F. Solving non-smooth optimal power flow problems using a developed grey wolf optimizer. *Energies* **2018**, *11*, 1692. [CrossRef]
56. Dhiman, G.; Kumar, V. Seagull optimization algorithm: Theory and its applications for large-scale industrial engineering problems. *Knowl.-Based Syst.* **2019**, *165*, 169–196. [CrossRef]
57. Harifi, S.; Khalilian, M.; Mohammadzadeh, J.; Ebrahimnejad, S. Emperor Penguins Colony: A new metaheuristic algorithm for optimization. *Evol. Intell.* **2019**, *12*, 211–226. [CrossRef]

58. Dhiman, G.; Kaur, A. STOA: A bio-inspired based optimization algorithm for industrial engineering problems. *Eng. Appl. Artif. Intell.* **2019**, *82*, 148–174. [CrossRef]
59. Zou, Y. The whirlpool algorithm based on physical phenomenon for solving optimization problems. *Eng. Comput.* **2019**, *36*, 664–690. [CrossRef]
60. Han, X.; Xu, Q.; Yue, L.; Dong, Y.; Xie, G.; Xu, X. An improved crow search algorithm based on spiral search mechanism for solving numerical and engineering optimization problems. *IEEE Access* **2020**, *8*, 92363–92382. [CrossRef]
61. Zhao, W.; Zhang, Z.; Wang, L. Manta ray foraging optimization: An effective bio-inspired optimizer for engineering applications. *Eng. Appl. Artif. Intell.* **2020**, *87*, 103300. [CrossRef]
62. Alsattar, H.; Zaidan, A.; Zaidan, B. Novel meta-heuristic bald eagle search optimisation algorithm. *Artif. Intell. Rev.* **2020**, *53*, 2237–2264. [CrossRef]
63. Wu, J.; Wang, Y.G.; Burrage, K.; Tian, Y.C.; Lawson, B.; Ding, Z. An improved firefly algorithm for global continuous optimization problems. *Expert Syst. Appl.* **2020**, *149*, 113340. [CrossRef]
64. Eid, H.F.; Garcia-Hernandez, L.; Abraham, A. Spiral water cycle algorithm for solving multi-objective optimization and truss optimization problems. *Eng. Comput.* **2021**, 1–11. [CrossRef]
65. Mokeddem, D. A new improved salp swarm algorithm using logarithmic spiral mechanism enhanced with chaos for global optimization. *Evol. Intell.* **2021**, 1–31. [CrossRef]
66. Ouyang, C.; Qiu, Y.; Zhu, D. Adaptive Spiral Flying Sparrow Search Algorithm. *Sci. Program.* **2021**, *2021*, 6505353. [CrossRef]
67. Xu, Z.; Gui, W.; Heidari, A.A.; Liang, G.; Chen, H.; Wu, C.; Turabieh, H.; Mafarja, M. Spiral Motion Mode Embedded Grasshopper Optimization Algorithm: Design and Analysis. *IEEE Access* **2021**, *9*, 71104–71132. [CrossRef]
68. Abualigah, L.; Yousri, D.; Abd Elaziz, M.; Ewees, A.A.; Al-qaness, M.A.; Gandomi, A.H. Aquila Optimizer: A novel meta-heuristic optimization Algorithm. *Comput. Ind. Eng.* **2021**, *157*, 107250. [CrossRef]
69. Kumar, V.; Kaleka, K.K.; Kaur, A. Spiral-inspired spotted hyena optimizer and its application to constraint engineering problems. *Wirel. Pers. Commun.* **2021**, *116*, 865–881. [CrossRef]
70. Li, M.; Li, C.; Huang, Z.; Huang, J.; Wang, G.; Liu, P.X. Spiral-based chaotic chicken swarm optimization algorithm for parameters identification of photovoltaic models. *Soft Comput.* **2021**, *25*, 12875–12898. [CrossRef]
71. Mohammadi-Balani, A.; Nayeri, M.D.; Azar, A.; Taghizadeh-Yazdi, M. Golden eagle optimizer: A nature-inspired metaheuristic algorithm. *Comput. Ind. Eng.* **2021**, *152*, 107050. [CrossRef]
72. Mehne, S.H.H.; Mirjalili, S. Moth-flame optimization algorithm: Theory, literature review, and application in optimal nonlinear feedback control design. In *Nature-Inspired Optimizers*; Springer International Publishing: Cham, Switzerland, 2020; pp. 143–166.
73. Shehab, M.; Abualigah, L.; Al Hamad, H.; Alabool, H.; Alshinwan, M.; Khasawneh, A.M. Moth–flame optimization algorithm: Variants and applications. *Neural Comput. Appl.* **2020**, *32*, 9859–9884. [CrossRef]
74. Bingi, K.; Kulkarni, R.R.; Mantri, R. Development of Hybrid Algorithm Using Moth-Flame and Particle Swarm Optimization. In Proceedings of the 2021 IEEE Madras Section Conference (MASCON), Chennai, India, 27–28 August 2021; pp. 1–6.
75. Helmi, A.; Alenany, A. An enhanced Moth-flame optimization algorithm for permutation-based problems. *Evol. Intell.* **2020**, *13*, 741–764. [CrossRef]
76. Gharehchopogh, F.S.; Gholizadeh, H. A comprehensive survey: Whale Optimization Algorithm and its applications. *Swarm Evol. Comput.* **2019**, *48*, 1–24. [CrossRef]
77. Abd Elaziz, M.; Mirjalili, S. A hyper-heuristic for improving the initial population of whale optimization algorithm. *Knowl.-Based Syst.* **2019**, *172*, 42–63. [CrossRef]
78. Salgotra, R.; Singh, U.; Saha, S. On some improved versions of whale optimization algorithm. *Arab. J. Sci. Eng.* **2019**, *44*, 9653–9691. [CrossRef]
79. Puri, H.; Chaudhary, J.; Bingi, K.; Sivaramakrishnan, U.; Panga, N. Design of Adaptive Weighted Whale Optimization Algorithm. In Proceedings of the 2021 IEEE Madras Section Conference (MASCON), Chennai, India, 27–28 August 2021; pp. 1–6.
80. Fan, Q.; Chen, Z.; Li, Z.; Xia, Z.; Yu, J.; Wang, D. A new improved whale optimization algorithm with joint search mechanisms for high-dimensional global optimization problems. *Eng. Comput.* **2020**, *37*, 1851–1878. [CrossRef]
81. Ma, B.; Lu, P.M.; Liu, Y.G.; Zhou, Q.; Hu, Y.T. Shared seagull optimization algorithm with mutation operators for global optimization. *AIP Adv.* **2021**, *11*, 125217. [CrossRef]
82. Wang, S.; Jia, H.; Abualigah, L.; Liu, Q.; Zheng, R. An improved hybrid aquila optimizer and harris hawks algorithm for solving industrial engineering optimization problems. *Processes* **2021**, *9*, 1551. [CrossRef]
83. Sadollah, A.; Eskandar, H.; Bahreininejad, A.; Kim, J.H. Water cycle algorithm with evaporation rate for solving constrained and unconstrained optimization problems. *Appl. Soft Comput.* **2015**, *30*, 58–71. [CrossRef]
84. Sadollah, A.; Eskandar, H.; Bahreininejad, A.; Kim, J.H. Water cycle, mine blast and improved mine blast algorithms for discrete sizing optimization of truss structures. *Comput. Struct.* **2015**, *149*, 1–16. [CrossRef]
85. Mirjalili, S. The ant lion optimizer. *Adv. Eng. Softw.* **2015**, *83*, 80–98. [CrossRef]
86. Assiri, A.S.; Hussien, A.G.; Amin, M. Ant Lion Optimization: Variants, hybrids, and applications. *IEEE Access* **2020**, *8*, 77746–77764. [CrossRef]
87. Abualigah, L.; Shehab, M.; Alshinwan, M.; Mirjalili, S.; Abd Elaziz, M. Ant lion optimizer: A comprehensive survey of its variants and applications. *Arch. Comput. Methods Eng.* **2021**, *28*, 1397–1416. [CrossRef]

88. Heidari, A.A.; Faris, H.; Mirjalili, S.; Aljarah, I.; Mafarja, M. Ant lion optimizer: Theory, literature review, and application in multi-layer perceptron neural networks. In *Nature-Inspired Optimizers*; Springer: Cham, Switzerland 2020 ; pp. 23–46.
89. Mirjalili, S.; Gandomi, A.H.; Mirjalili, S.Z.; Saremi, S.; Faris, H.; Mirjalili, S.M. Salp Swarm Algorithm: A bio-inspired optimizer for engineering design problems. *Adv. Eng. Softw.* **2017**, *114*, 163–191. [CrossRef]
90. Abualigah, L.; Shehab, M.; Alshinwan, M.; Alabool, H. Salp swarm algorithm: A comprehensive survey. *Neural Comput. Appl.* **2020**, *32*, 11195–11215. [CrossRef]
91. Faris, H.; Mirjalili, S.; Aljarah, I.; Mafarja, M.; Heidari, A.A. Salp swarm algorithm: Theory, literature review, and application in extreme learning machines. In *Nature-Inspired Optimizers*; Springer International Publishing: Cham, Switzerland, 2020 ; pp. 185–199.
92. Kulkarni, R.R.; Sunder, M.; Bingi, K.; Mantri, R.; Selvaraj, K.R. An Inertia Weight Concept-Based Salp Swarm Optimization Algorithm. In Proceedings of the 2021 IEEE Madras Section Conference (MASCON), Chennai, India, 27–28 August 2021; pp. 1–6.
93. Xue, J.; Shen, B. A novel swarm intelligence optimization approach: Sparrow search algorithm. *Syst. Sci. Control Eng.* **2020**, *8*, 22–34. [CrossRef]
94. Ouyang, C.; Zhu, D.; Qiu, Y. Lens Learning Sparrow Search Algorithm. *Math. Probl. Eng.* **2021**, *2021*, 9935090. [CrossRef]
95. Hassan, S.M.; Ibrahim, R.; Saad, N.; Asirvadam, V.S.; Bingi, K. Spiral dynamic algorithm based optimal PPI controller for WirelessHART networked systems. In Proceedings of the 2017 IEEE 3rd International Symposium in Robotics and Manufacturing Automation (ROMA), Kuala Lumpur, Malaysia, 19–21 September 2017; pp. 1–6.
96. Ali, S.K.; Tokhi, M.; Ishak, A.J.; Al Rezage, G. PID and Aaptive Spiral Dynamic Algorithm in Controlling Human Arm Movements. In *Assistive Robotics: Proceedings of the 18th International Conference on CLAWAR 2015*; World Scientific: Singapore, 2015; pp. 87–94.
97. Ghani, N.A.; Nasir, A.K.; Tokhi, M.O. Integrated phases modular fuzzy logic control with spiral dynamic optimization for stair descending in a wheelchair. In Proceedings of the 2014 19th International Conference on Methods and Models in Automation and Robotics (MMAR), Miedzyzdroje, Poland, 2–5 September 2014; pp. 46–51.
98. Razali, N.; Ghani, N.M.A.; Bari, B.S. Lifting and stabilizing of two-wheeled wheelchair system using interval type-2 fuzzy logic control based spiral dynamic algorithm. *Bull. Electr. Eng. Inform.* **2021**, *10*, 3019–3031. [CrossRef]
99. Masrom, M.; A Ghani, N.; Tokhi, M. Particle swarm optimization and spiral dynamic algorithm-based interval type-2 fuzzy logic control of triple-link inverted pendulum system: A comparative assessment. *J. Low Freq. Noise, Vib. Act. Control* **2021**, *40*, 367–382. [CrossRef]
100. Ouadi, A.; Bentarzi, H.; Recioui, A. Optimal multiobjective design of digital filters using spiral optimization technique. *SpringerPlus* **2013**, *2*, 1–13. [CrossRef]
101. Benasla, L.; Belmadani, A.; Rahli, M. Spiral optimization algorithm for solving combined economic and emission dispatch. *Int. J. Electr. Power Energy Syst.* **2014**, *62*, 163–174. [CrossRef]
102. Dosoglu, M.K.; Guvenc, U.; Duman, S.; Sonmez, Y.; Kahraman, H.T. Symbiotic organisms search optimization algorithm for economic/emission dispatch problem in power systems. *Neural Comput. Appl.* **2018**, *29*, 721–737. [CrossRef]
103. Hao, M.R.; Ismail, R.M.T.R.; Ahmad, M.A. Using spiral dynamic algorithm for maximizing power production of wind farm. In Proceedings of the 2017 International Conference on Applied System Innovation (ICASI), Sapporo, Japan, 13–17 May 2017; pp. 1706–1709.
104. Cao, Y.; Rad, H.N.; Jamali, D.H.; Hashemian, N.; Ghasemi, A. A novel multi-objective spiral optimization algorithm for an innovative solar/biomass-based multi-generation energy system: 3E analyses, and optimization algorithms comparison. *Energy Convers. Manag.* **2020**, *219*, 112961. [CrossRef]
105. Kaveh, M.R.; Hooshmand, R.A.; Madani, S.M. Simultaneous optimization of re-phasing, reconfiguration and DG placement in distribution networks using BF-SD algorithm. *Appl. Soft Comput.* **2018**, *62*, 1044–1055. [CrossRef]
106. Mehran, A.; Khademzadeh, A.; Saeidi, S. DSM: A Heuristic Dynamic Spiral Mapping algorithm for network on chip. *IEICE Electron. Express* **2008**, *5*, 464–471. [CrossRef]
107. Brinkmann, G. Problems and scope of spiral algorithms and spiral codes for polyhedral cages. *Chem. Phys. Lett.* **1997**, *272*, 193–198. [CrossRef]
108. Jia, H.C.; Hou, L.H. A spiral optimized deep neural network based adolescence physical fitness determination and training process analysis. *Aggress. Violent Behav.* **2021**, 101561. [CrossRef]
109. Ismail, M.J.; Ibrahim, R.; Ismail, I. Adaptive neural network prediction model for energy consumption. In Proceedings of the 2011 3rd International Conference on Computer Research and Development, Shanghai, China, 11–13 March 2011; Volume 4, pp. 109–113.
110. Mauch, S.; Reger, J. Real-time implementation of the spiral algorithm for Shack-Hartmann wavefront sensor pattern sorting on an FPGA. *Measurement* **2016**, *92*, 63–69. [CrossRef]
111. Altahir, A.A.; Asirvadam, V.S.; Hamid, N.H.B.; Sebastian, P.; Saad, N.B.; Ibrahim, R.B.; Dass, S.C. Optimizing visual surveillance sensor coverage using dynamic programming. *IEEE Sens. J.* **2017**, *17*, 3398–3405. [CrossRef]
112. McCaffrey, J. Spiral Dynamics Optimization with Python. Available online: https://visualstudiomagazine.com/articles/2021/08/02/spiral-dynamics-python.aspx (accessed on 8 February 2021).
113. Tamura, K.; Yasuda, K. Quantitative analysis based tuning law for convergence rate of spiral optimization. In Proceedings of the 2012 IEEE International Conference on Systems, Man, and Cybernetics (SMC), Seoul, Korea, 14–17 October 2012; pp. 767–772.

114. Suzuki, K.; Tamura, K.; Yasuda, K. Study on cluster-structured spiral optimization. In Proceedings of the 2014 IEEE International Conference on Systems, Man, and Cybernetics (SMC), San Diego, CA, USA, 5–8 October 2014; pp. 94–99.
115. Tamura, K.; Yasuda, K. A parameter setting method for spiral optimization from stability analysis of dynamic equilibrium point. *SICE J. Control. Meas. Syst. Integr.* **2014**, *7*, 173–182. [CrossRef]
116. Cruz, J.; Amaya, I.; Correa, R. Optimal rectangular microchannel design, using simulated annealing, unified particle swarm and spiral algorithms, in the presence of spreading resistance. *Appl. Therm. Eng.* **2015**, *84*, 126–137. [CrossRef]
117. Tamura, K.; Yasuda, K. Spiral optimization algorithm using periodic descent directions. *SICE J. Control. Meas. Syst. Integr.* **2016**, *9*, 134–143. [CrossRef]
118. Sidarto, K.A.; Kania, A. Computing Complex Roots of Systems of Nonlinear Equations Using Spiral Optimization Algorithm with Clustering. In Proceedings of the International Conference on Computational Science and Technology, Kuala Lumpur, Malaysia, 29–30 November 2017; Springer: Berlin/Heidelberg, Germany, 2017; pp. 390–398.

fractal and fractional

Article

Well Posedness of New Optimization Problems with Variational Inequality Constraints

Savin Treanţă

Department of Applied Mathematics, University Politehnica of Bucharest, 060042 Bucharest, Romania; savin.treanta@upb.ro

Abstract: In this paper, we studied the well posedness for a new class of optimization problems with variational inequality constraints involving second-order partial derivatives. More precisely, by using the notions of lower semicontinuity, pseudomonotonicity, hemicontinuity and monotonicity for a multiple integral functional, and by introducing the set of approximating solutions for the considered class of constrained optimization problems, we established some characterization results on well posedness. Furthermore, to illustrate the theoretical developments included in this paper, we present some examples.

Keywords: well posedness; constrained variational control problem; monotonicity; pseudomonotonicity; hemicontinuity; multiple integral functional; lower semicontinuity

Citation: Treanţă, S. Well Posedness of New Optimization Problems with Variational Inequality Constraints. *Fractal Fract.* **2021**, *5*, 123. https://doi.org/10.3390/fractalfract5030123

Academic Editor: Hari Mohan Srivastava

Received: 18 August 2021
Accepted: 14 September 2021
Published: 15 September 2021

Publisher's Note: MDPI stays neutral with regard to jurisdictional claims in published maps and institutional affiliations.

Copyright: © 2021 by the authors. Licensee MDPI, Basel, Switzerland. This article is an open access article distributed under the terms and conditions of the Creative Commons Attribution (CC BY) license (https://creativecommons.org/licenses/by/4.0/).

1. Introduction

The notion of well posedness represents a useful mathematical tool by ensuring the convergence of a sequence of approximate solutions to the exact solution of some optimization problems. Starting with the work of Tykhonov [1] for unconstrained optimization problems, various types of well posedness for variational problems have been considered (see, for instance, Levitin-Polyak well posedness [2–5], extended well posedness [6–14]), L-well posedness [15], α-well posedness [16,17]). Moreover, the concept of well posedness can be useful to study some related problems, such as variational inequality and fixed point problems [18–22], hemivariational inequality problems [23], complementary problems [24], equilibrium problems [25,26], Nash equilibrium problems [27] and variational inclusion problems [28]. Recently, the study of well posedness for vector variational inequalities and the associated optimization problems was formulated by Jayswal and Shalini [29]. On the other hand, an important and interesting extension of variational inequality problems is that of multidimensional variational inequality problems and the corresponding multi-time optimization problems (see [30–40]).

Motivated by the aforementioned research works, in this paper we analyze the well posedness of a new class of constrained optimization problems governed by multiple integral functionals involving second-order partial derivatives. To this aim, first we introduce new forms for the concepts of monotonicity, lower semicontinuity, pseudomonotonicity and hemicontinuity associated with a multiple integral functional. Furthermore, we define the set of approximating solutions for the considered optimization problem and establish some characterization theorems on well posedness. The main novelty elements of this paper are represented by the following: the mathematical framework is based on infinite-dimensional function spaces, multiple integral functionals, the presence of second-order partial derivatives, and innovative proofs of the main results. The aforementioned elements are completely new in the area of well-posed variational control problems. Most of the previous works in this field have been studied in classical finite-dimensional spaces, without taking into account the new notions mentioned above.

This paper is organized as follows. Section 2 provides the concepts of monotonicity, pseudomonotonicity, hemicontinuity and the lower semicontinuity of a multiple integral functional, and an auxiliary lemma. Section 3 investigates the well posedness for the considered constrained optimization problem. Concretely, we establish that well posedness and the existence and uniqueness of a solution are equivalent in the aforementioned problem. Furthermore, some examples are formulated throughout the paper to highlight the theoretical elements. Finally, in Section 4, we present the conclusions and provide further developments.

2. Preliminaries

Throughout this work, we consider the following mathematical tools and notations: let Ω be a compact set in $\mathbb{R}^m$ and $\Omega \ni \zeta = (\zeta^\alpha)$, $\alpha = \overline{1,m}$; consider $\mathcal{A}$ as the space of C^4-class *state* functions $a : \Omega \to \mathbb{R}^n$ and $a_\alpha := \frac{\partial a}{\partial \zeta^\alpha}$, $a_{\beta\gamma} := \frac{\partial^2 a}{\partial \zeta^\beta \partial \zeta^\gamma}$ denote the *partial speed* and *partial acceleration*, respectively; also, consider $\mathcal{U}$ ass the space of C^1-class *control* functions $u : \Omega \to \mathbb{R}^k$, and consider $\mathcal{A} \times \mathcal{U}$ as a closed, convex and non-empty subset of $\mathcal{A} \times \mathcal{U}$, $(a,u)|_{\partial \Omega}$ = given, endowed with the scalar product:

$$\langle (a,u), (b,w) \rangle = \int_\Omega \left[a(\zeta) \cdot b(\zeta) + u(\zeta) \cdot w(\zeta) \right] d\zeta$$

$$= \int_\Omega \left[\sum_{i=1}^n a^i(\zeta) b^i(\zeta) + \sum_{j=1}^k u^j(\zeta) w^j(\zeta) \right] d\zeta, \quad \forall (a,u), (b,w) \in \mathcal{A} \times \mathcal{U}$$

and the induced norm, where $d\zeta = d\zeta^1 \cdots d\zeta^m$ denotes the volume element on $\mathbb{R}^m$.

Consider $J^2(\mathbb{R}^m, \mathbb{R}^n)$ as the second-order jet bundle associated with $\mathbb{R}^m$ and $\mathbb{R}^n$. Taking the scalar continuously differentiable function $f : J^2(\mathbb{R}^m, \mathbb{R}^n) \times \mathbb{R}^k \to \mathbb{R}$, we introduce the following multiple integral-type functional:

$$F : \mathcal{A} \times \mathcal{U} \to \mathbb{R}, \quad F(a,u) = \int_\Omega f(\zeta, a(\zeta), a_\alpha(\zeta), a_{\beta\gamma}(\zeta), u(\zeta)) d\zeta.$$

At this moment, we are able to introduce the following *constrained variational control problem* (in short, CVCP), given as follows (we use the notation $(\pi_{a,u}(\zeta)) := (\zeta, a(\zeta), a_\alpha(\zeta), a_{\beta\gamma}(\zeta), u(\zeta))$):

$$\text{(CVCP)} \quad \text{Minimize}_{(a,u)} \int_\Omega f(\pi_{a,u}(\zeta)) d\zeta$$
$$\text{subject to} \quad (a,u) \in \Theta,$$

where Θ is the solution set of the *controlled variational inequality problem* (in short, CVIP): to find a pair $(a,u) \in A \times U$ such that:

$$\text{(CVIP)} \quad \int_\Omega \Big[\frac{\partial f}{\partial a}(\pi_{a,u}(\zeta))(b(\zeta) - a(\zeta)) + \frac{\partial f}{\partial a_\alpha}(\pi_{a,u}(\zeta)) D_\alpha(b(\zeta) - a(\zeta))$$

$$+ \frac{1}{n(\beta,\gamma)} \frac{\partial f}{\partial a_{\beta\gamma}}(\pi_{a,u}(\zeta)) D^2_{\beta\gamma}(b(\zeta) - a(\zeta))$$

$$+ \frac{\partial f}{\partial u}(\pi_{a,u}(\zeta))(w(\zeta) - u(\zeta)) \Big] d\zeta \geq 0, \quad \forall (b,w) \in A \times U,$$

where $D^2_{\beta\gamma} := D_\beta(D_\gamma)$, and $n(\beta,\gamma)$ is the Saunders's multi-index notation (see Saunders [41], Treanţă [40]).

More precisely, the feasible solution set for (CVIP) is given by

$$\Theta = \Big\{(a,u) \in A \times U : \int_\Omega \Big[(b(\zeta) - a(\zeta))\frac{\partial f}{\partial a}(\pi_{a,u}(\zeta))$$
$$+ D_\alpha(b(\zeta) - a(\zeta))\frac{\partial f}{\partial a_\alpha}(\pi_{a,u}(\zeta)) + (w(\zeta) - u(\zeta))\frac{\partial f}{\partial u}(\pi_{a,u}(\zeta))$$
$$+ \frac{1}{n(\beta,\gamma)} D^2_{\beta\gamma}(b(\zeta) - a(\zeta))\frac{\partial f}{\partial a_{\beta\gamma}}(\pi_{a,u}(\zeta))\Big] d\zeta \geq 0,$$
$$\forall (b,w) \in A \times U \Big\}.$$

Next, we define the notions of *monotonicity* and *pseudomonotonicity* for the aforementioned multiple integral functional.

Definition 1. *The multiple integral functional* $F(a,u) = \int_\Omega f(\pi_{a,u}(\zeta)) d\zeta$ *is called monotone on* $A \times U$ *if the following inequality holds:*

$$\int_\Omega \Big[(a(\zeta) - b(\zeta))\Big(\frac{\partial f}{\partial a}(\pi_{a,u}(\zeta)) - \frac{\partial f}{\partial a}(\pi_{b,w}(\zeta))\Big)$$
$$+ (u(\zeta) - w(\zeta))\Big(\frac{\partial f}{\partial u}(\pi_{a,u}(\zeta)) - \frac{\partial f}{\partial u}(\pi_{b,w}(\zeta))\Big)$$
$$+ D_\alpha(a(\zeta) - b(\zeta))\Big(\frac{\partial f}{\partial a_\alpha}(\pi_{a,u}(\zeta)) - \frac{\partial f}{\partial a_\alpha}(\pi_{b,w}(\zeta))\Big)$$
$$+ \frac{1}{n(\beta,\gamma)} D^2_{\beta\gamma}(a(\zeta) - b(\zeta))\Big(\frac{\partial f}{\partial a_{\beta\gamma}}(\pi_{a,u}(\zeta)) - \frac{\partial f}{\partial a_{\beta\gamma}}(\pi_{b,w}(\zeta))\Big)\Big] d\zeta \geq 0,$$
$$\forall (a,u), (b,w) \in A \times U.$$

Definition 2. *The multiple integral functional* $F(a,u) = \int_\Omega f(\pi_{a,u}(\zeta)) d\zeta$ *is called pseudomonotone on* $A \times U$ *if the following implication holds:*

$$\int_\Omega \Big[(a(\zeta) - b(\zeta))\frac{\partial f}{\partial a}(\pi_{b,w}(\zeta)) + (u(\zeta) - w(\zeta))\frac{\partial f}{\partial u}(\pi_{b,w}(\zeta))$$
$$+ D_\alpha(a(\zeta) - b(\zeta))\frac{\partial f}{\partial a_\alpha}(\pi_{b,w}(\zeta))$$
$$+ \frac{1}{n(\beta,\gamma)} D^2_{\beta\gamma}(a(\zeta) - b(\zeta))\frac{\partial f}{\partial a_{\beta\gamma}}(\pi_{b,w}(\zeta))\Big] d\zeta \geq 0$$
$$\Rightarrow \int_\Omega \Big[(a(\zeta) - b(\zeta))\frac{\partial f}{\partial a}(\pi_{a,u}(\zeta)) + (u(\zeta) - w(\zeta))\frac{\partial f}{\partial u}(\pi_{a,u}(\zeta))$$
$$+ D_\alpha(a(\zeta) - b(\zeta))\frac{\partial f}{\partial a_\alpha}(\pi_{a,u}(\zeta))$$
$$+ \frac{1}{n(\beta,\gamma)} D^2_{\beta\gamma}(a(\zeta) - b(\zeta))\frac{\partial f}{\partial a_{\beta\gamma}}(\pi_{a,u}(\zeta))\Big] d\zeta \geq 0, \quad \forall (a,u), (b,w) \in A \times U.$$

Let us give an example of a multiple integral-type functional which is not monotone but is pseudomonotone.

Example 1. *Consider* $n = k = 1$, $m = 2$, *and* $\Omega = [0,3]^2$. *We define:*

$$f(\pi_{a,u}(\zeta)) = 2\sin a(\zeta) + u(\zeta)e^{u(\zeta)}$$

and show, in accordance with Definition 2, that the multiple integral functional $F(a,u) = \int_\Omega f(\pi_{a,u}(\zeta))d\zeta$ is pseudomonotone on $A \times U = C^4(\Omega,[-1,1]) \times C^1(\Omega,[-1,1])$. Indeed, we have:

$$\int_\Omega \Big[(a(\zeta) - b(\zeta))\frac{\partial f}{\partial a}(\pi_{b,w}(\zeta)) + (u(\zeta) - w(\zeta))\frac{\partial f}{\partial u}(\pi_{b,w}(\zeta))$$
$$+ D_\alpha(a(\zeta) - b(\zeta))\frac{\partial f}{\partial a_\alpha}(\pi_{b,w}(\zeta))$$
$$+ \frac{1}{n(\beta,\gamma)}D^2_{\beta\gamma}(a(\zeta) - b(\zeta))\frac{\partial f}{\partial a_{\beta\gamma}}(\pi_{b,w}(\zeta))\Big]d\zeta$$
$$= \int_\Omega \Big[2(a(\zeta) - b(\zeta))\cos b(\zeta) + (u(\zeta) - w(\zeta))(e^{w(\zeta)} + w(\zeta)e^{w(\zeta)})\Big]d\zeta \geq 0,$$
$$\forall (a,u),(b,w) \in A \times U$$

$$\Rightarrow \int_\Omega \Big[(a(\zeta) - b(\zeta))\frac{\partial f}{\partial a}(\pi_{a,u}(\zeta)) + (u(\zeta) - w(\zeta))\frac{\partial f}{\partial u}(\pi_{a,u}(\zeta))$$
$$+ D_\alpha(a(\zeta) - b(\zeta))\frac{\partial f}{\partial a_\alpha}(\pi_{a,u}(\zeta))$$
$$+ \frac{1}{n(\beta,\gamma)}D^2_{\beta\gamma}(a(\zeta) - b(\zeta))\frac{\partial f}{\partial a_{\beta\gamma}}(\pi_{a,u}(\zeta))\Big]d\zeta$$
$$= \int_\Omega \Big[2(a(\zeta) - b(\zeta))\cos a(\zeta) + (u(\zeta) - w(\zeta))(e^{u(\zeta)} + u(\zeta)e^{u(\zeta)})\Big]d\zeta \geq 0,$$
$$\forall (a,u),(b,w) \in A \times U.$$

However, it is not monotone on $A \times U$ in the sense of Definition 1, because:

$$\int_\Omega \Big[(a(\zeta) - b(\zeta))\Big(\frac{\partial f}{\partial a}(\pi_{a,u}(\zeta)) - \frac{\partial f}{\partial a}(\pi_{b,w}(\zeta))\Big)$$
$$+ (u(\zeta) - w(\zeta))\Big(\frac{\partial f}{\partial u}(\pi_{a,u}(\zeta)) - \frac{\partial f}{\partial u}(\pi_{b,w}(\zeta))\Big)$$
$$+ D_\alpha(a(\zeta) - b(\zeta))\Big(\frac{\partial f}{\partial a_\alpha}(\pi_{a,u}(\zeta)) - \frac{\partial f}{\partial a_\alpha}(\pi_{b,w}(\zeta))\Big)$$
$$+ \frac{1}{n(\beta,\gamma)}D^2_{\beta\gamma}(a(\zeta) - b(\zeta))\Big(\frac{\partial f}{\partial a_{\beta\gamma}}(\pi_{a,u}(\zeta)) - \frac{\partial f}{\partial a_{\beta\gamma}}(\pi_{b,w}(\zeta))\Big)\Big]d\zeta$$
$$= \int_\Omega \Big[2(a(\zeta) - b(\zeta))(\cos a(\zeta) - \cos b(\zeta))$$
$$+ (u(\zeta) - w(\zeta))(u(\zeta)e^{u(\zeta)} + e^{u(\zeta)} - w(\zeta)e^{w(\zeta)} - e^{w(\zeta)})\Big]d\zeta \not\geq 0,$$
$$\forall (a,u),(b,w) \in A \times U.$$

Then, in accordance with Usman and Khan [42], we define the concept of *hemicontinuity* for the considered multiple integral-type functional.

Definition 3. *The functional* $F(a,u) = \int_\Omega f(\pi_{a,u}(\zeta))d\zeta$ *is hemicontinuous on* $A \times U$ *if the application:*

$$\lambda \to \Big\langle ((a(\zeta),u(\zeta)) - (b(\zeta),w(\zeta)), \Big(\frac{\delta F}{\delta a_\lambda}(\zeta), \frac{\delta F}{\delta u_\lambda}(\zeta)\Big)\Big\rangle, \quad 0 \leq \lambda \leq 1$$

is continuous at 0^+, for $\forall (a,u), (b,w) \in A \times U$, where:

$$\frac{\delta F}{\delta a_\lambda}(\zeta) := \frac{\partial f}{\partial a}(\pi_{a_\lambda,u_\lambda}(\zeta)) - D_\alpha \frac{\partial f}{\partial a_\alpha}(\pi_{a_\lambda,u_\lambda}(\zeta)) + \frac{1}{n(\beta,\gamma)} D^2_{\beta\gamma} \frac{\partial f}{\partial a_{\beta\gamma}}(\pi_{a_\lambda,u_\lambda}(\zeta)) \in A,$$

$$\frac{\delta F}{\delta u_\lambda}(\zeta) := \frac{\partial f}{\partial u}(\pi_{a_\lambda,u_\lambda}(\zeta)) \in U,$$

$$a_\lambda := \lambda a + (1-\lambda)b, \quad u_\lambda := \lambda u + (1-\lambda)w.$$

The following lemma is an auxiliary result for proving the main results derived in the present paper.

Lemma 1. *Consider $F(a,u) = \int_\Omega f(\pi_{a,u}(\zeta))d\zeta$ is pseudomonotone and hemicontinuous on $A \times U$. The pair $(a,u) \in A \times U$ is a solution for (CVIP) if and only if (a,u) is a solution for the following variational inequality problem:*

$$\int_\Omega \left[(b(\zeta) - a(\zeta)) \frac{\partial f}{\partial a}(\pi_{b,w}(\zeta)) + (w(\zeta) - u(\zeta)) \frac{\partial f}{\partial u}(\pi_{b,w}(\zeta)) \right.$$

$$+ D_\alpha(b(\zeta) - a(\zeta)) \frac{\partial f}{\partial a_\alpha}(\pi_{b,w}(\zeta))$$

$$\left. + \frac{1}{n(\beta,\gamma)} D^2_{\beta\gamma}(b(\zeta) - a(\zeta)) \frac{\partial f}{\partial a_{\beta\gamma}}(\pi_{b,w}(\zeta)) \right] d\zeta \geq 0, \quad \forall (b,w) \in A \times U.$$

Proof. Consider that the pair $(a,u) \in A \times U$ is the solution for (CVIP). As a consequence, it results that:

$$\int_\Omega \left[(b(\zeta) - a(\zeta)) \frac{\partial f}{\partial a}(\pi_{a,u}(\zeta)) + (w(\zeta) - u(\zeta)) \frac{\partial f}{\partial u}(\pi_{a,u}(\zeta)) \right.$$

$$+ D_\alpha(b(\zeta) - a(\zeta)) \frac{\partial f}{\partial a_\alpha}(\pi_{a,u}(\zeta))$$

$$\left. + \frac{1}{n(\beta,\gamma)} D^2_{\beta\gamma}(b(\zeta) - a(\zeta)) \frac{\partial f}{\partial a_{\beta\gamma}}(\pi_{a,u}(\zeta)) \right] d\zeta \geq 0, \quad \forall (b,w) \in A \times U.$$

By using the pseudomonotonicity property of the considered multiple integral functional (see Definition 2), the previous inequality involves:

$$\int_\Omega \left[(b(\zeta) - a(\zeta)) \frac{\partial f}{\partial a}(\pi_{b,w}(\zeta)) + (w(\zeta) - u(\zeta)) \frac{\partial f}{\partial u}(\pi_{b,w}(\zeta)) \right.$$

$$+ D_\alpha(b(\zeta) - a(\zeta)) \frac{\partial f}{\partial a_\alpha}(\pi_{b,w}(\zeta))$$

$$\left. + \frac{1}{n(\beta,\gamma)} D^2_{\beta\gamma}(b(\zeta) - a(\zeta)) \frac{\partial f}{\partial a_{\beta\gamma}}(\pi_{b,w}(\zeta)) \right] d\zeta \geq 0, \quad \forall (b,w) \in A \times U.$$

Conversely, assume that:

$$\int_\Omega \left[(b(\zeta) - a(\zeta)) \frac{\partial f}{\partial a}(\pi_{b,w}(\zeta)) + (w(\zeta) - u(\zeta)) \frac{\partial f}{\partial u}(\pi_{b,w}(\zeta)) \right.$$

$$+ D_\alpha(b(\zeta) - a(\zeta)) \frac{\partial f}{\partial a_\alpha}(\pi_{b,w}(\zeta))$$

$$\left. + \frac{1}{n(\beta,\gamma)} D^2_{\beta\gamma}(b(\zeta) - a(\zeta)) \frac{\partial f}{\partial a_{\beta\gamma}}(\pi_{b,w}(\zeta)) \right] d\zeta \geq 0, \quad \forall (b,w) \in A \times U.$$

For $\lambda \in (0,1]$ and $(b,w) \in A \times U$, we define:

$$(b_\lambda, w_\lambda) = ((1-\lambda)a + \lambda b, (1-\lambda)u + \lambda w) \in A \times U.$$

Thus, the above inequality implies:

$$\int_\Omega \Big[(b_\lambda(\zeta) - a(\zeta))\frac{\partial f}{\partial a}(\pi_{b_\lambda,w_\lambda}(\zeta)) + (w_\lambda(\zeta) - u(\zeta))\frac{\partial f}{\partial u}(\pi_{b_\lambda,w_\lambda}(\zeta))$$

$$+ D_\alpha(b_\lambda(\zeta) - a(\zeta))\frac{\partial f}{\partial a_\alpha}(\pi_{b_\lambda,w_\lambda}(\zeta))$$

$$+ \frac{1}{n(\beta,\gamma)}D^2_{\beta\gamma}(b_\lambda(\zeta) - a(\zeta))\frac{\partial f}{\partial a_{\beta\gamma}}(\pi_{b_\lambda,w_\lambda}(\zeta))\Big]d\zeta \geq 0, \quad (b,w) \in A \times U.$$

By considering $\lambda \to 0$, we obtain:

$$\int_\Omega \Big[(b(\zeta) - a(\zeta))\frac{\partial f}{\partial a}(\pi_{a,u}(\zeta)) + (w(\zeta) - u(\zeta))\frac{\partial f}{\partial u}(\pi_{a,u}(\zeta))$$

$$+ D_\alpha(b(\zeta) - a(\zeta))\frac{\partial f}{\partial a_\alpha}(\pi_{a,u}(\zeta))$$

$$+ \frac{1}{n(\beta,\gamma)}D^2_{\beta\gamma}(b(\zeta) - a(\zeta))\frac{\partial f}{\partial a_{\beta\gamma}}(\pi_{a,u}(\zeta))\Big]d\zeta \geq 0, \quad \forall (b,w) \in A \times U,$$

which proves that (a,u) solves (CVIP). This completes the proof of this lemma. □

Now, we give the definition of *lower semicontinuity* for the multiple integral functional $F(a,u) = \int_\Omega f(\pi_{a,u}(\zeta))d\zeta$.

Definition 4. *The multiple integral functional $F(a,u) = \int_\Omega f(\pi_{a,u}(\zeta))d\zeta$ is called lower semicontinuous at a point $(a_0, u_0) \in A \times U$ if:*

$$\int_\Omega f(\pi_{a_0,u_0}(\zeta))d\zeta \leq \lim_{(a,u) \to (a_0,u_0)} \inf \int_\Omega f(\pi_{a,u}(\zeta))d\zeta.$$

3. Well Posedness Associated with (CVCP)

In this section, by considering the notions introduced in Section 2, we study the well posedness for the considered class of constrained optimization problems (CVCPs). To this aim, we introduce the following definitions and notations.

Denote by $\mathcal{S}$ solution set of (CVCP), namely:

$$\mathcal{S} = \Big\{(a,u) \in A \times U \mid \int_\Omega f(\pi_{a,u}(\zeta))d\zeta \leq \inf_{(b,w) \in \Theta} \int_\Omega f(\pi_{b,w}(\zeta))d\zeta \text{ and}$$

$$\int_\Omega \Big[(b(\zeta) - a(\zeta))\frac{\partial f}{\partial a}(\pi_{a,u}(\zeta)) + (w(\zeta) - u(\zeta))\frac{\partial f}{\partial u}(\pi_{a,u}(\zeta))$$

$$+ D_\alpha(b(\zeta) - a(\zeta))\frac{\partial f}{\partial a_\alpha}(\pi_{a,u}(\zeta))$$

$$+ \frac{1}{n(\beta,\gamma)}D^2_{\beta\gamma}(b(\zeta) - a(\zeta))\frac{\partial f}{\partial a_{\beta\gamma}}(\pi_{a,u}(\zeta))\Big]d\zeta \geq 0, \forall (b,w) \in A \times U\Big\}.$$

Consider the *set of approximating solutions* of (CVCP), for $\sigma, \iota \geq 0$, as follows:

$$S(\sigma,\iota) = \Big\{(a,u) \in A \times U \mid \int_\Omega f(\pi_{a,u}(\zeta))d\zeta \leq \inf_{(b,w)\in\Theta} \int_\Omega f(\pi_{b,w}(\zeta))d\zeta + \sigma \text{ and}$$

$$\int_\Omega \Big[(b(\zeta) - a(\zeta))\frac{\partial f}{\partial a}(\pi_{a,u}(\zeta)) + (w(\zeta) - u(\zeta))\frac{\partial f}{\partial u}(\pi_{a,u}(\zeta))$$

$$+ D_\alpha(b(\zeta) - a(\zeta))\frac{\partial f}{\partial a_\alpha}(\pi_{a,u}(\zeta))$$

$$+ \frac{1}{n(\beta,\gamma)} D^2_{\beta\gamma}(b(\zeta) - a(\zeta))\frac{\partial f}{\partial a_{\beta\gamma}}(\pi_{a,u}(\zeta))\Big]d\zeta + \iota \geq 0,\ \forall (b,w) \in A \times U\Big\}.$$

Remark 1. *For $(\sigma,\iota) = (0,0)$, we obtain $S = S(\sigma,\iota)$, and for $(\sigma,\iota) > (0,0)$, we obtain $S \subseteq S(\sigma,\iota)$.*

Definition 5. *The sequence $\{(a_n,u_n)\}$ is an approximating sequence for (CVCP) if there exists $\iota_n \to 0$ (a sequence of positive real numbers) as $n \to \infty$, such that:*

$$\limsup_{n\to\infty} \int_\Omega f(\pi_{a_n,u_n}(\zeta))d\zeta \leq \inf_{(b,w)\in\Theta} \int_\Omega f(\pi_{b,w}(\zeta))d\zeta$$

and:

$$\int_\Omega \Big[(b(\zeta) - a_n(\zeta))\frac{\partial f}{\partial a}(\pi_{a_n,u_n}(\zeta)) + (w(\zeta) - u_n(\zeta))\frac{\partial f}{\partial u}(\pi_{a_n,u_n}(\zeta))$$

$$+ D_\alpha(b(\zeta) - a_n(\zeta))\frac{\partial f}{\partial a_\alpha}(\pi_{a_n,u_n}(\zeta))$$

$$+ \frac{1}{n(\beta,\gamma)} D^2_{\beta\gamma}(b(\zeta) - a_n(\zeta))\frac{\partial f}{\partial a_{\beta\gamma}}(\pi_{a_n,u_n}(\zeta))\Big]d\zeta + \iota_n \geq 0,\ \forall(b,w) \in A \times U$$

are fulfilled.

Definition 6. *The constrained optimization (CVCP) is well posed if:*

(i) it admits a single solution (a_0, u_0);
and (ii) each approximating sequence of (CVCP) converges to (a_0, u_0).

Furthermore, denote by "diam B" the *diameter* of the set B and it is defined as follows

$$\operatorname{diam} B = \sup_{x,y \in B} \|x - y\|.$$

The next theorem represents a first characterization result on the well posedness for (CVCP).

Theorem 1. *Consider that $F(a,u) = \int_\Omega f(\pi_{a,u}(\zeta))d\zeta$ is lower semicontinuous, monotone and hemicontinuous on $A \times U$. The constrained optimization problem (CVCP) is well posed if and only if:*

$$S(\sigma,\iota) \neq \emptyset, \forall \sigma,\iota > 0 \text{ and } \operatorname{diam} S(\sigma,\iota) \to 0 \text{ as } (\sigma,\iota) \to (0,0).$$

Proof. Consider (CVCP) is well posed. In consequence, it admits a single solution $(\bar{a}, \bar{u}) \in S$. By using the inclusion $S \subseteq S(\sigma,\iota),\ \forall \sigma,\iota > 0$, we obtain $S(\sigma,\iota) \neq \emptyset,\ \forall \sigma,\iota > 0$. Now, contrary to the result, suppose that diam $S(\sigma,\iota) \not\to 0$ as $(\sigma,\iota) \to (0,0)$. Consequently, there exists $r > 0$, a positive integer m, $\sigma_n, \iota_n > 0$ with $\sigma_n, \iota_n \to 0$, and $(a_n, u_n), (a'_n, u'_n) \in S(\sigma_n, \iota_n)$ such that:

$$\|(a_n, u_n) - (a'_n, u'_n)\| > r,\quad \forall n \geq m. \tag{1}$$

Since $(a_n, u_n), (a'_n, u'_n) \in \mathcal{S}(\sigma_n, \iota_n)$, we obtain:

$$\int_\Omega f(\pi_{a_n,u_n}(\zeta))d\zeta \leq \inf_{(b,w)\in\Theta} \int_\Omega f(\pi_{b,w}(\zeta))d\zeta + \sigma_n,$$

$$\int_\Omega \left[(b(\zeta) - a_n(\zeta))\frac{\partial f}{\partial a}(\pi_{a_n,u_n}(\zeta)) + (w(\zeta) - u_n(\zeta))\frac{\partial f}{\partial u}(\pi_{a_n,u_n}(\zeta))\right.$$

$$+ D_\alpha(b(\zeta) - a_n(\zeta))\frac{\partial f}{\partial a_\alpha}(\pi_{a_n,u_n}(\zeta))$$

$$\left.+ \frac{1}{n(\beta,\gamma)} D^2_{\beta\gamma}(b(\zeta) - a_n(\zeta))\frac{\partial f}{\partial a_{\beta\gamma}}(\pi_{a_n,u_n}(\zeta))\right]d\zeta + \iota_n \geq 0, \quad \forall (b,w) \in A \times U$$

and:

$$\int_\Omega f(\pi_{a'_n,u'_n}(\zeta))d\zeta \leq \inf_{(b,w)\in\Theta} \int_\Omega f(\pi_{b,w}(\zeta))d\zeta + \sigma_n,$$

$$\int_\Omega \left[(b(\zeta) - a'_n(\zeta))\frac{\partial f}{\partial a}(\pi_{a'_n,u'_n}(\zeta)) + (w(\zeta) - u'_n(\zeta))\frac{\partial f}{\partial u}(\pi_{a'_n,u'_n}(\zeta))\right.$$

$$+ D_\alpha(b(\zeta) - a'_n(\zeta))\frac{\partial f}{\partial a_\alpha}(\pi_{a'_n,u'_n}(\zeta))$$

$$\left.+ \frac{1}{n(\beta,\gamma)} D^2_{\beta\gamma}(b(\zeta) - a'_n(\zeta))\frac{\partial f}{\partial a_{\beta\gamma}}(\pi_{a'_n,u'_n}(\zeta))\right]d\zeta + \iota_n \geq 0, \quad \forall (b,w) \in A \times U.$$

Clearly, it follows that $\{(a_n, u_n)\}$ and $\{(a'_n, u'_n)\}$ are two approximating sequences for (CVCP) which converge to $(\bar{a}, \bar{u})$ (by hypothesis, the problem (CVCP) is well posed). By computation, we obtain:

$$\|(a_n, u_n) - (a'_n, u'_n)\|$$
$$= \|(a_n, u_n) - (\bar{a}, \bar{u}) + (\bar{a}, \bar{u}) - (a'_n, u'_n)\|$$
$$\leq \|(a_n, u_n) - (\bar{a}, \bar{u})\| + \|(\bar{a}, \bar{u}) - (a'_n, u'_n)\| \leq \iota,$$

which contradicts (1), for some $\iota = r$. It follows that diam $\mathcal{S}(\sigma, \iota) \to 0$ as $(\sigma, \iota) \to (0, 0)$.

Conversely, let $\{(a_n, u_n)\}$ be an approximating sequence for (CVCP). Therefore, there exists a sequence of positive real numbers $\iota_n \to 0$ as $n \to \infty$ such that the inequalities:

$$\limsup_{n\to\infty} \int_\Omega f(\pi_{a_n,u_n}(\zeta))d\zeta \leq \inf_{(b,w)\in\Theta} \int_\Omega f(\pi_{b,w}(\zeta))d\zeta, \qquad (2)$$

$$\int_\Omega \left[(b(\zeta) - a_n(\zeta))\frac{\partial f}{\partial a}(\pi_{a_n,u_n}(\zeta)) + (w(\zeta) - u_n(\zeta))\frac{\partial f}{\partial u}(\pi_{a_n,u_n}(\zeta))\right.$$

$$+ D_\alpha(b(\zeta) - a_n(\zeta))\frac{\partial f}{\partial a_\alpha}(\pi_{a_n,u_n}(\zeta))$$

$$\left.+ \frac{1}{n(\beta,\gamma)} D^2_{\beta\gamma}(b(\zeta) - a_n(\zeta))\frac{\partial f}{\partial a_{\beta\gamma}}(\pi_{a_n,u_n}(\zeta))\right]d\zeta + \iota_n \geq 0, \quad \forall (b,w) \in A \times U \qquad (3)$$

hold, involving that $(a_n, u_n) \in \mathcal{S}(\sigma_n, \iota_n)$ (see $\sigma_n \to 0$ as $n \to \infty$, a sequence of positive real numbers). By considering diam $\mathcal{S}(\sigma_n, \iota_n) \to 0$ as $(\sigma_n, \iota_n) \to (0,0)$, we obtain that $\{(a_n, u_n)\}$ is a Cauchy sequence which converges to some $(\bar{a}, \bar{u}) \in A \times U$ as $A \times U$ is a closed set.

By using the monotonicity property of $\int_\Omega f(\pi_{a,u}(\zeta))d\zeta$ on $A \times U$, for $(\bar{a}, \bar{u}), (b,w) \in A \times U$, we have:

$$\int_\Omega \left[(\bar{a}(\zeta) - b(\zeta))\left(\frac{\partial f}{\partial a}(\pi_{a,\bar{u}}(\zeta)) - \frac{\partial f}{\partial a}(\pi_{b,w}(\zeta))\right)\right.$$

$$\left.+ (\bar{u}(\zeta) - w(\zeta))\left(\frac{\partial f}{\partial u}(\pi_{\bar{a},\bar{u}}(\zeta)) - \frac{\partial f}{\partial a}(\pi_{b,w}(\zeta))\right)\right.$$

$$+D_\alpha(\bar{a}(\zeta)-b(\zeta))\left(\frac{\partial f}{\partial a_\alpha}(\pi_{\bar{a},\bar{u}}(\zeta))-\frac{\partial f}{\partial a_\alpha}(\pi_{b,w}(\zeta))\right)$$

$$+\frac{1}{n(\beta,\gamma)}D^2_{\beta\gamma}(\bar{a}(\zeta)-b(\zeta))\left(\frac{\partial f}{\partial a_{\beta\gamma}}(\pi_{\bar{a},\bar{u}}(\zeta))-\frac{\partial f}{\partial a_{\beta\gamma}}(\pi_{b,w}(\zeta))\right)\bigg]d\zeta\geq 0,$$

or, equivalently:

$$\int_\Omega\bigg[(\bar{a}(\zeta)-b(\zeta))\frac{\partial f}{\partial a}(\pi_{\bar{a},\bar{u}}(\zeta))+(\bar{u}(\zeta)-w(\zeta))\frac{\partial f}{\partial u}(\pi_{\bar{a},\bar{u}}(\zeta))$$

$$+D_\alpha(\bar{a}(\zeta)-b(\zeta))\frac{\partial f}{\partial a_\alpha}(\pi_{\bar{a},\bar{u}}(\zeta))$$

$$+\frac{1}{n(\beta,\gamma)}D^2_{\beta\gamma}(\bar{a}(\zeta)-b(\zeta))\frac{\partial f}{\partial a_{\beta\gamma}}(\pi_{\bar{a},\bar{u}}(\zeta))\bigg]d\zeta$$

$$\geq\int_\Omega\bigg[(\bar{a}(\zeta)-b(\zeta))\frac{\partial f}{\partial a}(\pi_{b,w}(\zeta))+(\bar{u}(\zeta)-w(\zeta))\frac{\partial f}{\partial u}(\pi_{b,w}(\zeta))$$

$$+D_\alpha(\bar{a}(\zeta)-b(\zeta))\frac{\partial f}{\partial a_\alpha}(\pi_{b,w}(\zeta))$$

$$+\frac{1}{n(\beta,\gamma)}D^2_{\beta\gamma}(\bar{a}(\zeta)-b(\zeta))\frac{\partial f}{\partial a_{\beta\gamma}}(\pi_{b,w}(\zeta))\bigg]d\zeta. \tag{4}$$

By considering the limit in inequality (3), we obtain:

$$\int_\Omega\bigg[(\bar{a}(\zeta)-b(\zeta))\frac{\partial f}{\partial a}(\pi_{\bar{a},\bar{u}}(\zeta))+(\bar{u}(\zeta)-w(\zeta))\frac{\partial f}{\partial u}(\pi_{\bar{a},\bar{u}}(\zeta))$$

$$+D_\alpha(\bar{a}(\zeta)-b(\zeta))\frac{\partial f}{\partial a_\alpha}(\pi_{\bar{a},\bar{u}}(\zeta))$$

$$+\frac{1}{n(\beta,\gamma)}D^2_{\beta\gamma}(\bar{a}(\zeta)-b(\zeta))\frac{\partial f}{\partial a_{\beta\gamma}}(\pi_{\bar{a},\bar{u}}(\zeta))\bigg]d\zeta\leq 0. \tag{5}$$

By using (4) and (5), it results that:

$$\int_\Omega\bigg[(b(\zeta)-\bar{a}(\zeta))\frac{\partial f}{\partial a}(\pi_{b,w}(\zeta))+(w(\zeta)-\bar{u}(\zeta))\frac{\partial f}{\partial u}(\pi_{b,w}(\zeta))$$

$$+D_\alpha(b(\zeta)-\bar{a}(\zeta))\frac{\partial f}{\partial a_\alpha}(\pi_{b,w}(\zeta))$$

$$+\frac{1}{n(\beta,\gamma)}D^2_{\beta\gamma}(b(\zeta)-\bar{a}(\zeta))\frac{\partial f}{\partial a_{\beta\gamma}}(\pi_{b,w}(\zeta))\bigg]d\zeta\geq 0.$$

Now, we use Lemma 1 to obtain:

$$\int_\Omega\bigg[(b(\zeta)-\bar{a}(\zeta))\frac{\partial f}{\partial a}(\pi_{\bar{a},\bar{u}}(\zeta))+(w(\zeta)-\bar{u}(\zeta))\frac{\partial f}{\partial u}(\pi_{\bar{a},\bar{u}}(\zeta))$$

$$+D_\alpha(b(\zeta)-\bar{a}(\zeta))\frac{\partial f}{\partial a_\alpha}(\pi_{\bar{a},\bar{u}}(\zeta))$$

$$+\frac{1}{n(\beta,\gamma)}D^2_{\beta\gamma}(b(\zeta)-\bar{a}(\zeta))\frac{\partial f}{\partial a_{\beta\gamma}}(\pi_{\bar{a},\bar{u}}(\zeta))\bigg]d\zeta\geq 0, \tag{6}$$

which implies that $(\bar{a},\bar{u})\in\Theta$.

Since the functional $\int_\Omega f(\pi_{a,u}(\zeta))d\zeta$ is lower semi-continuous, we conclude:

$$\int_\Omega f(\pi_{\bar{a},\bar{u}}(\zeta))d\zeta \leq \liminf_{n\to\infty} \int_\Omega f(\pi_{a_n,u_n}(\zeta))d\zeta \leq \limsup_{n\to\infty} \int_\Omega f(\pi_{a_n,u_n}(\zeta))d\zeta.$$

By (2), the previous inequality can be written as

$$\int_\Omega f(\pi_{\bar{a},\bar{u}}(\zeta))d\zeta \leq \inf_{(b,w)\in\Theta} \int_\Omega f(\pi_{b,w}(\zeta))d\zeta. \tag{7}$$

As a consequence, by (6) and (7), we obtain that $(\bar{a},\bar{u})$ is the solution for (CVCP).

Let us prove that $(\bar{a},\bar{u})$ is the single solution for (CVCP). Suppose that $(a_1,u_1) \neq (a_2,u_2)$ are two solutions for (CVCP). Then:

$$0 < \|(a_1,u_1) - (a_2,u_2)\| \leq \operatorname{diam} S(\sigma,\iota) \to 0 \text{ as } (\sigma,\iota) \to (0,0),$$

which is not possible. The proof is now complete. □

The second main result of this paper is contained in the next theorem.

Theorem 2. *Consider* $F(a,u) = \int_\Omega f(\pi_{a,u}(\zeta))d\zeta$ *is lower semicontinuous, monotone and hemicontinuous on* $A \times U$. *The constrained optimization problem (CVCP) is well posed if and only if it admits a solution.*

Proof. Consider that (CVCP) is well posed. In consequence, it has a solution (a_0,u_0). Conversely, consider that (CVCP) has a solution (a_0,u_0), that is:

$$\int_\Omega f(\pi_{a_0,u_0}(\zeta))d\zeta \leq \inf_{(b,w)\in\Theta} \int_\Omega f(\pi_{b,w}(\zeta))d\zeta,$$

$$\int_\Omega \Big[(b(\zeta) - a_0(\zeta))\frac{\partial f}{\partial a}(\pi_{a_0,u_0}(\zeta)) + (w(\zeta) - u_0(\zeta))\frac{\partial f}{\partial u}(\pi_{a_0,u_0}(\zeta))$$

$$+ D_\alpha(b(\zeta) - a_0(\zeta))\frac{\partial f}{\partial a_\alpha}(\pi_{a_0,u_0}(\zeta))$$

$$+ \frac{1}{n(\beta,\gamma)}D^2_{\beta\gamma}(b(\zeta) - a_0(\zeta))\frac{\partial f}{\partial a_{\beta\gamma}}(\pi_{a_0,u_0}(\zeta))\Big]d\zeta \geq 0, \quad \forall (b,w) \in A \times U, \tag{8}$$

but (CVCP) is not well posed. Therefore, by Definition 6, there exists an approximating sequence $\{(a_n,u_n)\}$ of (CVCP) (which does not converge to (a_0,u_0)), that is the following inequalities hold:

$$\limsup_{n\to\infty} \int_\Omega f(\pi_{a_n,u_n}(\zeta))d\zeta \leq \inf_{(b,w)\in\Theta} \int_\Omega f(\pi_{b,w}(\zeta))d\zeta$$

and:

$$\int_\Omega \Big[(b(\zeta) - a_n(\zeta))\frac{\partial f}{\partial a}(\pi_{a_n,u_n}(\zeta)) + (w(\zeta) - u_n(\zeta))\frac{\partial f}{\partial u}(\pi_{a_n,u_n}(\zeta))$$

$$+ D_\alpha(b(\zeta) - a_n(\zeta))\frac{\partial f}{\partial a_\alpha}(\pi_{a_n,u_n}(\zeta))$$

$$+ \frac{1}{n(\beta,\gamma)}D^2_{\beta\gamma}(b(\zeta) - a_n(\zeta))\frac{\partial f}{\partial a_{\beta\gamma}}(\pi_{a_n,u_n}(\zeta))\Big]d\zeta + \iota_n \geq 0, \quad \forall (b,w) \in A \times U. \tag{9}$$

Furthermore, to prove the boundedness of $\{(a_n,u_n)\}$, we proceed by contradiction. Suppose, in contrast to the result, $\{(a_n,u_n)\}$ is not bounded, that is, $\|(a_n,u_n)\| \to +\infty$ as $n \to +\infty$. Let us consider $\delta_n = \dfrac{1}{\|(a_n,u_n) - (a_0,u_0)\|}$ and $(a_n,u_n) = (a_0,u_0) + \delta_n[(a_n,u_n) - $

$(a_0, u_0)]$. We can see that $\{(a_n, u_n)\}$ is bounded in $A \times U$. If necessary, passing to a subsequence, we may consider that:

$$(a_n, u_n) \to (a, u) \text{ weakly in } A \times U \neq (a_0, u_0).$$

It is easy to check that $(a, u) \neq (a_0, u_0)$ thanks to $\|\delta_n[(a_n, u_n) - (a_0, u_0)]\| = 1$, for all $n \in \mathbb{N}$. Since (a_0, u_0) is a solution of (CVCP), the inequalities in (8) are satisfied. By Lemma 1, we obtain:

$$\int_\Omega f(\pi_{a_0, u_0}(\zeta)) d\zeta \le \inf_{(b,w) \in \Theta} \int_\Omega f(\pi_{b,w}(\zeta)) d\zeta,$$

$$\int_\Omega \left[(b(\zeta) - a_0(\zeta)) \frac{\partial f}{\partial a}(\pi_{b,w}(\zeta)) + (w(\zeta) - u_0(\zeta)) \frac{\partial f}{\partial u}(\pi_{b,w}(\zeta)) \right.$$

$$+ D_\alpha(b(\zeta) - a_0(\zeta)) \frac{\partial f}{\partial a_\alpha}(\pi_{b,w}(\zeta))$$

$$\left. + \frac{1}{n(\beta, \gamma)} D^2_{\beta\gamma}(b(\zeta) - a_0(\zeta)) \frac{\partial f}{\partial a_{\beta\gamma}}(\pi_{b,w}(\zeta)) \right] d\zeta \ge 0, \quad \forall (b, w) \in A \times U. \quad (10)$$

By using the monotonicity property of the multiple integral functional $\int_\Omega f(\pi_{a,u}(\zeta)) d\zeta$ on $A \times U$, for $(a_n, u_n), (b, w) \in A \times U$, we have:

$$\int_\Omega \left[(a_n(\zeta) - b(\zeta)) \left(\frac{\partial f}{\partial a}(\pi_{a_n, u_n}(\zeta)) - \frac{\partial f}{\partial a}(\pi_{b,w}(\zeta)) \right) \right.$$

$$+ (u_n(\zeta) - w(\zeta)) \left(\frac{\partial f}{\partial u}(\pi_{a_n, u_n}(\zeta)) - \frac{\partial f}{\partial a}(\pi_{b,w}(\zeta)) \right)$$

$$+ D_\alpha(a_n(\zeta) - b(\zeta)) \left(\frac{\partial f}{\partial a_\alpha}(\pi_{a_n, u_n}(\zeta)) - \frac{\partial f}{\partial a_\alpha}(\pi_{b,w}(\zeta)) \right)$$

$$\left. + \frac{1}{n(\beta, \gamma)} D^2_{\beta\gamma}(a_n(\zeta) - b(\zeta)) \left(\frac{\partial f}{\partial a_{\beta\gamma}}(\pi_{a_n, u_n}(\zeta)) - \frac{\partial f}{\partial a_{\beta\gamma}}(\pi_{b,w}(\zeta)) \right) \right] d\zeta \ge 0,$$

or, equivalently:

$$\int_\Omega \left[(b(\zeta) - a_n(\zeta)) \frac{\partial f}{\partial a}(\pi_{a_n, u_n}(\zeta)) + (w(\zeta) - u_n(\zeta)) \frac{\partial f}{\partial u}(\pi_{a_n, u_n}(\zeta)) \right.$$

$$+ D_\alpha(b(\zeta) - a_n(\zeta)) \frac{\partial f}{\partial a_\alpha}(\pi_{a_n, u_n}(\zeta))$$

$$\left. + \frac{1}{n(\beta, \gamma)} D^2_{\beta\gamma}(b(\zeta) - a_n(\zeta)) \frac{\partial f}{\partial a_{\beta\gamma}}(\pi_{a_n, u_n}(\zeta)) \right] d\zeta$$

$$\le \int_\Omega \left[(b(\zeta) - a_n(\zeta)) \frac{\partial f}{\partial a}(\pi_{b,w}(\zeta)) + (w(\zeta) - u_n(\zeta)) \frac{\partial f}{\partial u}(\pi_{b,w}(\zeta)) \right.$$

$$+ D_\alpha(b(\zeta) - a_n(\zeta)) \frac{\partial f}{\partial a_\alpha}(\pi_{b,w}(\zeta))$$

$$\left. + \frac{1}{n(\beta, \gamma)} D^2_{\beta\gamma}(b(\zeta) - a_n(\zeta)) \frac{\partial f}{\partial a_{\beta\gamma}}(\pi_{b,w}(\zeta)) \right] d\zeta. \quad (11)$$

Combining with (9) and (11), we have:

$$\int_\Omega \left[(b(\zeta) - a_n(\zeta)) \frac{\partial f}{\partial a}(\pi_{b,w}(\zeta)) + (w(\zeta) - u_n(\zeta)) \frac{\partial f}{\partial u}(\pi_{b,w}(\zeta)) \right.$$

$$+ D_\alpha(b(\zeta) - a_n(\zeta)) \frac{\partial f}{\partial a_\alpha}(\pi_{b,w}(\zeta))$$

$$+\frac{1}{n(\beta,\gamma)}D^2_{\beta\gamma}(b(\zeta)-\mathsf{a}_n(\zeta))\frac{\partial f}{\partial a_{\beta\gamma}}(\pi_{b,w}(\zeta))\bigg]d\zeta \geq -\iota_n, \quad \forall (b,w)\in A\times U.$$

Because of $\delta_n \to 0$ as $n \to \infty$ (by the assumption that $\{(a_n, u_n)\}$ is not bounded), so, we can take $n_0 \in \mathbb{N}$ be large enough such that $\delta_n < 1$, for all $n \geq n_0$. Then, by multiplying the previous inequality and (10) by $\delta_n > 0$ and $1 - \delta_n > 0$, respectively, we obtain:

$$\int_\Omega \bigg[(b(\zeta)-\mathsf{a}_n(\zeta))\frac{\partial f}{\partial a}(\pi_{b,w}(\zeta))+(w(\zeta)-\mathsf{u}_n(\zeta))\frac{\partial f}{\partial u}(\pi_{b,w}(\zeta))$$

$$+D_\alpha(b(\zeta)-\mathsf{a}_n(\zeta))\frac{\partial f}{\partial a_\alpha}(\pi_{b,w}(\zeta))$$

$$+\frac{1}{n(\beta,\gamma)}D^2_{\beta\gamma}(b(\zeta)-\mathsf{a}_n(\zeta))\frac{\partial f}{\partial a_{\beta\gamma}}(\pi_{b,w}(\zeta))\bigg]d\zeta \geq -\iota_n, \quad \forall (b,w)\in A\times U, \forall n\geq n_0.$$

Since $(a_n, u_n) \to (a, u) \neq (a_0, u_0)$ and $(a_n, u_n) = (a_0, u_0) + (\mathbf{a}_n, \mathbf{u}_n)[(a_n, u_n) - (a_0, u_0)]$, we have:

$$\int_\Omega \bigg[(b(\zeta)-\mathsf{a}(\zeta))\frac{\partial f}{\partial a}(\pi_{b,w}(\zeta))+(w(\zeta)-\mathsf{u}(\zeta))\frac{\partial f}{\partial u}(\pi_{b,w}(\zeta))$$

$$+D_\alpha(b(\zeta)-\mathsf{a}(\zeta))\frac{\partial f}{\partial a_\alpha}(\pi_{b,w}(\zeta))$$

$$+\frac{1}{n(\beta,\gamma)}D^2_{\beta\gamma}(b(\zeta)-\mathsf{a}(\zeta))\frac{\partial f}{\partial a_{\beta\gamma}}(\pi_{b,w}(\zeta))\bigg]d\zeta$$

$$= \lim_{n\to\infty} \int_\Omega \bigg[(b(\zeta)-\mathsf{a}_n(\zeta))\frac{\partial f}{\partial a}(\pi_{b,w}(\zeta))+(w(\zeta)-\mathsf{u}_n(\zeta))\frac{\partial f}{\partial u}(\pi_{b,w}(\zeta))$$

$$+D_\alpha(b(\zeta)-\mathsf{a}_n(\zeta))\frac{\partial f}{\partial a_\alpha}(\pi_{b,w}(\zeta))$$

$$+\frac{1}{n(\beta,\gamma)}D^2_{\beta\gamma}(b(\zeta)-\mathsf{a}_n(\zeta))\frac{\partial f}{\partial a_{\beta\gamma}}(\pi_{b,w}(\zeta))\bigg]d\zeta$$

$$\geq -\lim_{n\to\infty} \iota_n = 0, \quad \forall (b,w) \in A\times U.$$

By considering the lower semicontinuity of the considered functional, and taking into account Lemma 1, we have:

$$\int_\Omega f(\pi_{\mathsf{a},\mathsf{u}}(\zeta))d\zeta \leq \inf_{(b,w)\in\Theta} \int_\Omega f(\pi_{b,w}(\zeta))d\zeta,$$

$$\int_\Omega \bigg[(b(\zeta)-\mathsf{a}(\zeta))\frac{\partial f}{\partial a}(\pi_{\mathsf{a},\mathsf{u}}(\zeta))+(w(\zeta)-\mathsf{u}(\zeta))\frac{\partial f}{\partial u}(\pi_{\mathsf{a},\mathsf{u}}(\zeta))$$

$$+D_\alpha(b(\zeta)-\mathsf{a}(\zeta))\frac{\partial f}{\partial a_\alpha}(\pi_{\mathsf{a},\mathsf{u}}(\zeta))$$

$$+\frac{1}{n(\beta,\gamma)}D^2_{\beta\gamma}(b(\zeta)-\mathsf{a}(\zeta))\frac{\partial f}{\partial a_{\beta\gamma}}(\pi_{\mathsf{a},\mathsf{u}}(\zeta))\bigg]d\zeta \geq 0, \quad \forall (b,w)\in A\times U. \qquad (12)$$

We obtain that (a, u) is a solution of (CVCP), which contradicts the uniqueness of (a_0, u_0). In consequence, $\{(a_n, u_n)\}$ is a bounded sequence with a convergent subsequence $\{(a_{n_k}, u_{n_k})\}$ which converges to $(\bar{a}, \bar{u}) \in A \times U$ as $k \to \infty$. Now, by Definition 1, for (a_{n_k}, u_{n_k}), $(b, w) \in A \times U$, we have (see (11)):

$$\int_\Omega \bigg[(b(\zeta)-a_{n_k}(\zeta))\frac{\partial f}{\partial a}(\pi_{a_{n_k},u_{n_k}}(\zeta))+(w(\zeta)-u_{n_k}(\zeta))\frac{\partial f}{\partial u}(\pi_{a_{n_k},u_{n_k}}(\zeta))$$

$$+D_\alpha(b(\zeta)-a_{n_k}(\zeta))\frac{\partial f}{\partial a_\alpha}(\pi_{a_{n_k},u_{n_k}}(\zeta))$$

$$+\frac{1}{n(\beta,\gamma)}D^2_{\beta\gamma}(b(\zeta)-a_{n_k}(\zeta))\frac{\partial f}{\partial a_{\beta\gamma}}(\pi_{a_{n_k},u_{n_k}}(\zeta))\bigg]d\zeta$$

$$\leq \int_\Omega \bigg[(b(\zeta)-a_{n_k}(\zeta))\frac{\partial f}{\partial a}(\pi_{b,w}(\zeta))+(w(\zeta)-u_{n_k}(\zeta))\frac{\partial f}{\partial u}(\pi_{b,w}(\zeta))$$

$$+D_\alpha(b(\zeta)-a_{n_k}(\zeta))\frac{\partial f}{\partial a_\alpha}(\pi_{b,w}(\zeta))$$

$$+\frac{1}{n(\beta,\gamma)}D^2_{\beta\gamma}(b(\zeta)-a_{n_k}(\zeta))\frac{\partial f}{\partial a_{\beta\gamma}}(\pi_{b,w}(\zeta))\bigg]d\zeta. \tag{13}$$

Furthermore, on behalf of (9), we can write that:

$$\lim_{k\to\infty}\int_\Omega \bigg[(b(\zeta)-a_{n_k}(\zeta))\frac{\partial f}{\partial a}(\pi_{a_{n_k},u_{n_k}}(\zeta))+(w(\zeta)-u_{n_k}(\zeta))\frac{\partial f}{\partial u}(\pi_{a_{n_k},u_{n_k}}(\zeta))$$

$$+D_\alpha(b(\zeta)-a_{n_k}(\zeta))\frac{\partial f}{\partial a_\alpha}(\pi_{a_{n_k},u_{n_k}}(\zeta))$$

$$+\frac{1}{n(\beta,\gamma)}D^2_{\beta\gamma}(b(\zeta)-a_{n_k}(\zeta))\frac{\partial f}{\partial a_{\beta\gamma}}(\pi_{a_{n_k},u_{n_k}}(\zeta))\bigg]d\zeta \geq 0. \tag{14}$$

Combining (13) and (14), we have:

$$\lim_{k\to\infty}\int_\Omega \bigg[(b(\zeta)-a_{n_k}(\zeta))\frac{\partial f}{\partial a}(\pi_{b,w}(\zeta))+(w(\zeta)-u_{n_k}(\zeta))\frac{\partial f}{\partial u}(\pi_{b,w}(\zeta))$$

$$+D_\alpha(b(\zeta)-a_{n_k}(\zeta))\frac{\partial f}{\partial a_\alpha}(\pi_{b,w}(\zeta))$$

$$+\frac{1}{n(\beta,\gamma)}D^2_{\beta\gamma}(b(\zeta)-a_{n_k}(\zeta))\frac{\partial f}{\partial a_{\beta\gamma}}(\pi_{b,w}(\zeta))\bigg]d\zeta \geq 0$$

$$\Rightarrow \int_\Omega \bigg[(b(\zeta)-\bar{a}(\zeta))\frac{\partial f}{\partial a}(\pi_{b,w}(\zeta))+(w(\zeta)-\bar{u}(\zeta))\frac{\partial f}{\partial u}(\pi_{b,w}(\zeta))$$

$$+D_\alpha(b(\zeta)-\bar{a}(\zeta))\frac{\partial f}{\partial a_\alpha}(\pi_{b,w}(\zeta))$$

$$+\frac{1}{n(\beta,\gamma)}D^2_{\beta\gamma}(b(\zeta)-\bar{a}(\zeta))\frac{\partial f}{\partial a_{\beta\gamma}}(\pi_{b,w}(\zeta))\bigg]d\zeta \geq 0.$$

By considering the lower semicontinuity of the considered functional, in accordance with Lemma 1, we have:

$$\int_\Omega f(\pi_{\bar{a},\bar{u}}(\zeta))d\zeta \leq \inf_{(b,w)\in\Theta}\int_\Omega f(\pi_{b,w}(\zeta))d\zeta,$$

$$\int_\Omega \bigg[(b(\zeta)-\bar{a}(\zeta))\frac{\partial f}{\partial a}(\pi_{\bar{a},\bar{u}}(\zeta))+(w(\zeta)-\bar{u}(\zeta))\frac{\partial f}{\partial u}(\pi_{\bar{a},\bar{u}}(\zeta))$$

$$+D_\alpha(b(\zeta)-\bar{a}(\zeta))\frac{\partial f}{\partial a_\alpha}(\pi_{\bar{a},\bar{u}}(\zeta))$$

$$+\frac{1}{n(\beta,\gamma)}D^2_{\beta\gamma}(b(\zeta)-\bar{a}(\zeta))\frac{\partial f}{\partial a_{\beta\gamma}}(\pi_{\bar{a},\bar{u}}(\zeta))\bigg]d\zeta \geq 0,$$

implying that $(\bar{a},\bar{u})$ is a solution for (CVCP). Therefore, $(a_{n_k},u_{n_k})\to(\bar{a},\bar{u})$, that is, $(a_{n_k},u_{n_k})\to(a_0,u_0)$, involving $(a_n,u_n)\to(a_0,u_0)$ and the proof is complete. □

In the following illustrative example, we present an application of Theorems 1 and 2, as well.

Example 2. Let us consider $n = k = 1$, $m = 2$, and $\Omega = [0,3]^2$. We define:

$$f(\pi_{a,u}(\zeta)) = 3u^2(\zeta) + e^{a(\zeta)} - a(\zeta)$$

and consider the following constrained variational control problem:

(CVCP-1) $\text{Minimize}_{(a,u)} \int_\Omega f(\pi_{a,u}(\zeta))d\zeta$

subject to

$$\int_\Omega \Big[6(w(\zeta) - u(\zeta))u(\zeta) + (b(\zeta) - a(\zeta))(e^{a(\zeta)} - 1)\Big]d\zeta \geq 0, \ \forall (b,w) \in A \times U,$$

$$(a,u)|_{\partial\Omega} = 0.$$

We have $\mathcal{S} = \{(0,0)\}$. Moreover, we have that the functional $\int_\Omega f(\pi_{a,u}(\zeta))d\zeta$ is monotone, hemicontinuous and lower semicontinuous on $A \times U = C^4(\Omega, [-10,10]) \times C^1(\Omega, [-10,10])$. In consequence, all hypotheses in Theorem 2 hold, therefore the optimization problem (CVCP-1) is well posed. Furthermore, since $\mathcal{S}(\sigma, \iota) = \{(0,0)\}$, we obtain $\mathcal{S}(\sigma, \iota) \neq \emptyset$ and diam $\mathcal{S}(\sigma, \iota) \to 0$ as $(\sigma, \iota) \to (0,0)$. Consequently, by using Theorem 1, we obtain that the constrained optimization problem (CVCP-1) is well posed.

4. Conclusions and Further Developments

In this paper, we investigated the well posedness for a new class of constrained optimization problems governed by second-order partial derivatives. Concretely, by using the monotonicity, lower semicontinuity, pseudomonotonicity and hemicontinuity of multiple integral functional, we proved that the well posedness of the constrained optimization problem under study is characterized in terms of existence and uniqueness of solution. Furthermore, the theoretical developments have been accompanied by some examples.

Some further developments associated with the present paper are the following: to formulate the necessary and sufficient optimality conditions for the considered optimization problems, to establish some duality results, and to study the well posedness for similar classes of control problems by using fractional derivatives.

Funding: This research received no external funding.

Institutional Review Board Statement: Not applicable.

Informed Consent Statement: Not applicable.

Data Availability Statement: Not applicable.

Conflicts of Interest: The author declares no conflict of interest.

References

1. Tykhonov, A.N. On the stability of the functional optimization problems. *USSR Comput. Math. Math. Phys.* **1966**, *6*, 631–634. [CrossRef]
2. Hu, R.; Fang, Y.P. Levitin-Polyak well-posedness by perturbations of inverse variational inequalities. *Optim. Lett.* **2013**, *7*, 343–359. [CrossRef]
3. Jiang, B.; Zhang, J.; Huang, X.X. Levitin-Polyak well-posedness of generalized quasivariational inequalities with functional constraints. *Nonlinear Anal. TMA* **2009**, *70*, 1492–1503. [CrossRef]
4. Lalitha, C.S.; Bhatia, G. Levitin-Polyak well-posedness for parametric quasivariational inequality problem of the Minty type. *Positivity* **2012**, *16*, 527–541. [CrossRef]
5. Levitin, E.S.; Polyak, B.T. Convergence of minimizing sequences in conditional extremum problems. *Sov. Math. Dokl.* **1996**, *7*, 764–767.
6. Čoban, M.M.; Kenderov, P.S.; Revalski, J.P. Generic well-posedness of optimization problems in topological spaces. *Mathematika* **1989**, *36*, 301–324. [CrossRef]
7. Dontchev, A.L.; Zolezzi, T. *Well-Posed Optimization Problems*; Springer: Berlin, Germany, 1993.

8. Furi, M.; Vignoli, A. A characterization of well-posed minimum problems in a complete metric space. *J. Optim. Theory Appl.* **1970**, *5*, 452–461. [CrossRef]
9. Huang, X.X. Extended and strongly extended well-posedness of set-valued optimization problems. *Math. Methods Oper. Res.* **2001**, *53*, 101–116. [CrossRef]
10. Huang, X.X.; Yang, X.Q. Generalized Levitin-Polyak well-posedness in constrained optimization. *SIAM J. Optim.* **2006**, *17*, 243–258. [CrossRef]
11. Lignola, M.B.; Morgan, J. Well-posedness for optimization problems with constraints defined by variational inequalities having a unique solution. *J. Glob. Optim.* **2000**, *16*, 57–67. [CrossRef]
12. Lin, L.J.; Chuang, C.S. Well-posedness in the generalized sense for variational inclusion and disclusion problems and well-posedness for optimization problems with constraint. *Nonlinear Anal.* **2009**, *70*, 3609–3617. [CrossRef]
13. Lucchetti, R. *Convexity and Well-Posed Problems*; Springer: New York, NY, USA, 2006.
14. Zolezzi, T. Extended well-posedness of optimization problems. *J. Optim. Theory Appl.* **1996**, *91*, 257–266. [CrossRef]
15. Lignola, M.B. Well-posedness and L-well-posedness for quasivariational inequalities. *J. Optim. Theory Appl.* **2006**, *128*, 119–138. [CrossRef]
16. Lignola, M.B.; Morgan, J. *Approximate Solutions and α-Well-Posedness for Variational Inequalities and Nash Equilibria, Decision and Control in Management Science*; Zaccour, G., Ed.; Kluwer Academic Publishers: Dordrecht, The Netherlands, 2002; pp. 367–378.
17. Virmani, G.; Srivastava, M. On Levitin-Polyak α-well-posedness of perturbed variational-hemivariational inequality. *Optimization* **2015**, *64*, 1153–1172. [CrossRef]
18. Ceng, L.C.; Yao, J.C. Well-posedness of generalized mixed variational inequalities, inclusion problems and fixed-point problems. *Nonlinear Anal.* **2008**, *69*, 4585–4603. [CrossRef]
19. Ceng, L.C.; Hadjisavvas, N.; Schaible, S.; Yao, J.C. Well-posedness for mixed quasivariational-like inequalities. *J. Optim. Theory Appl.* **2008**, *139*, 109–125. [CrossRef]
20. Fang, Y.P.; Hu, R. Estimates of approximate solutions and well-posedness for variational inequalities. *Math. Meth. Oper. Res.* **2007**, *65*, 281–291. [CrossRef]
21. Hammad, H.A.; Aydi, H.; la Sen, M.D. New contributions for tripled fixed point methodologies via a generalized variational principle with applications. *Alex. Eng. J.* **2021**, in press. [CrossRef]
22. Lalitha, C.S.; Bhatia, G. Well-posedness for parametric quasivariational inequality problems and for optimization problems with quasivariational inequality constraints. *Optimization* **2010**, *59*, 997–1011. [CrossRef]
23. Xiao, Y.B.; Yang, X.M.; Huang, N.J. Some equivalence results for well-posedness of hemivariational inequalities. *J. Glob. Optim.* **2015**, *61*, 789–802. [CrossRef]
24. Heemels, P.M.H.; Camlibel, M.K.C.; Schaft, A.J.V.; Schumacher, J.M. Well-posedness of the complementarity class of hybrid systems. In Proceedings of the 15th IFAC Triennial World Congress, Barcelona, Spain, 21–26 July 2002.
25. Chen, J.W.; Wang, Z.; Cho, Y.J. Levitin-Polyak well-posedness by perturbations for systems of set-valued vector quasi-equilibrium problems. *Math. Meth. Oper. Res.* **2013**, *77*, 33–64. [CrossRef]
26. Fang, Y.P.; Hu, R.; Huang, N.J. Well-posedness for equilibrium problems and for optimization problems with equilibrium constraints. *Comput. Math. Appl.* **2008**, *55*, 89–100. [CrossRef]
27. Lignola, M.B.; Morgan, J. α-Well-posedness for Nash equilibria and for optimization problems with Nash equilibrium constraints. *J. Glob. Optim.* **2006**, *36*, 439–459. [CrossRef]
28. Agarwal, P.; Filali, D.; Akram, M.; Dilshad, M. Convergence Analysis of a Three-Step Iterative Algorithm for Generalized Set-Valued Mixed-Ordered Variational Inclusion Problem. *Symmetry* **2021**, *13*, 444. [CrossRef]
29. Jayswal, A.; Jha, S. Well-posedness for generalized mixed vector variational-like inequality problems in Banach space. *Math. Commun.* **2017**, *22*, 287–302.
30. Treanţă, S. A necessary and sufficient condition of optimality for a class of multidimensional control problems. *Optim. Control Appl. Meth.* **2020**, *41*, 2137–2148. [CrossRef]
31. Treanţă, S.; Arana-Jiménez, M.; Antczak, T. A necessary and sufficient condition on the equivalence between local and global optimal solutions in variational control problems. *Nonlinear Anal.* **2020**, *191*, 111640. [CrossRef]
32. Treanţă, S. On a modified optimal control problem with first-order PDE constraints and the associated saddle-point optimality criterion. *Eur. J. Control* **2020**, *51*, 1–9. [CrossRef]
33. Treanţă, S. Some results on (ρ, b, d)-variational inequalities. *J. Math. Ineq.* **2020**, *14*, 805–818. [CrossRef]
34. Treanţă, S. On weak sharp solutions in (ρ, b, d)-variational inequalities. *J. Ineq. Appl.* **2020**, *2020*, 54. [CrossRef]
35. Treanţă, S.; Singh, S. Weak sharp solutions associated with a multidimensional variational-type inequality. *Positivity* **2021**, *25*, 329–351. [CrossRef]
36. Treanţă, S.; Das, K. On robust saddle-point criterion in optimization problems with curvilinear integral functionals. *Mathematics* **2021**, *9*, 1790. [CrossRef]
37. Treanţă, S. On well-posed isoperimetric-type constrained variational control problems. *J. Differ. Eq.* **2021**, *298*, 480–499. [CrossRef]
38. Treanţă, S. Second-order PDE constrained controlled optimization problems with application in mechanics. *Mathematics* **2021**, *9*, 1472. [CrossRef]
39. Treanţă, S. On a class of second-order PDE&PDI constrained robust modified optimization problems. *Mathematics* **2021**, *9*, 1473.
40. Treanţă, S. On a class of isoperimetric constrained controlled optimization problems. *Axioms* **2021**, *10*, 112. [CrossRef]

41. Saunders, D.J. *The Geometry of Jet Bundles, London Math. Soc. Lecture*; Notes Series, 142; Cambridge University Press: Cambridge, UK, 1989.
42. Usman, F.; Khan, S.A. A generalized mixed vector variational-like inequality problem. *Nonlinear Anal.* **2009**, *71*, 5354–5362. [CrossRef]

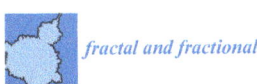 *fractal and fractional*

Article

Existence of Solutions to a Class of Nonlinear Arbitrary Order Differential Equations Subject to Integral Boundary Conditions

Ananta Thakur [1], Javid Ali [1] and Rosana Rodríguez-López [2,*]

[1] Department of Mathematics, Aligarh Muslim University, Aligarh 202002, India; anantathakur557@gmail.com (A.T.); javid.mm@amu.ac.in (J.A.)

[2] Departamento de Estatística, Instituto de Matemáticas, Análise Matemática e Optimización, Universidade de Santiago de Compostela, 15705 Santiago de Compostela, Spain

* Correspondence: rosana.rodriguez.lopez@usc.es; Tel.: +34-881-813368

Abstract: We investigate the existence of positive solutions for a class of fractional differential equations of arbitrary order $\delta > 2$, subject to boundary conditions that include an integral operator of the fractional type. The consideration of this type of boundary conditions allows us to consider heterogeneity on the dependence specified by the restriction added to the equation as a relevant issue for applications. An existence result is obtained for the sublinear and superlinear case by using the Guo–Krasnosel'skii fixed point theorem through the definition of adequate conical shells that allow us to localize the solution. As additional tools in our procedure, we obtain the explicit expression of Green's function associated to an auxiliary linear fractional boundary value problem, and we study some of its properties, such as the sign and some useful upper and lower estimates. Finally, an example is given to illustrate the results.

Keywords: fractional differential equations; fractional derivative of Riemann–Liouville type; integral boundary value problems; Green's functions; Guo–Krasnosel'skii fixed point theorem in cones; sublinearity and superlinearity; Arzelà-Ascoli Theorem

MSC: Primary 26A33; Secondary 34B15

Citation: Thakur, A.; Ali, J.; Rodríguez-López, R. Existence of Solutions to a Class of Nonlinear Arbitrary Order Differential Equations Subject to Integral Boundary Conditions. *Fractal Fract.* **2021**, 5, 220. https://doi.org/10.3390/fractalfract5040220

Academic Editor: Savin Treanţă

Received: 17 October 2021
Accepted: 8 November 2021
Published: 15 November 2021

Publisher's Note: MDPI stays neutral with regard to jurisdictional claims in published maps and institutional affiliations.

Copyright: © 2021 by the authors. Licensee MDPI, Basel, Switzerland. This article is an open access article distributed under the terms and conditions of the Creative Commons Attribution (CC BY) license (https://creativecommons.org/licenses/by/4.0/).

1. Introduction

Differential equations for non-integer order play an important role to describe the physical phenomena more accurately than classical integer order differential equations. The need for fractional order differential equations stems in part from the fact that many phenomena cannot be modeled by differential equations with integer derivatives. Therefore, the existence results for solutions to fractional differential equations have received considerable attention in recent years.

Some relevant monographs on fractional calculus and fractional differential equations are, for instance [1–3]. The work [4] gives some fundamental ideas on initial value problems for fractional differential equations from the point of view of Riemann–Liouville operators, discussing local and global existence, or extremal solutions, and the monograph [5] includes different theoretical results as well as developments related to applications in the field of fractional calculus.

There are several papers dealing with the existence and uniqueness of solution to initial and boundary value problems for fractional order differential equations. For instance, in 2009, some impulsive problems for Caputo-type differential equations with $\delta \in (1,2]$ and boundary conditions given by $x(0) + x'(0) = 0$, $x(1) + x'(1) = 0$, were studied (see [6]). Later, in 2010, initial value problems and periodic boundary value problems for linear fractional differential equations were analyzed in [7] by giving some comparison results. The authors of [8] studied the existence of positive solutions for fractional differential equations of order $\delta \in (1,2)$, whose nonlinearity depended on a fractional derivative of the unknown function, subject to Dirichlet boundary conditions.

They completed their study by calculating the associated Green's function and by applying the compressive version of the Guo–Krasnosel'skii fixed point theorem. Green's function, Banach contraction mapping and fixed point index theory are the main tools used in [9] for the analysis of a nonlocal problem for fractional differential equations. In [10], a result that guarantees the existence of a unique fixed point for a mixed monotone operator was used to provide the existence of a unique positive solution to an initial value problem for fractional differential equations of general order $n - 1 < \delta \leq n$, with $n \geq 2$, whose nonlinearity depends on the classical derivatives of the unknown function up to order $n - 2$.

On the other hand, the development of the monotone iterative technique for periodic boundary value problems associated with impulsive fractional differential equations with Riemann–Liouville sequential derivatives was made in [11], and [12] was devoted to boundary value problems for fractional differential inclusions. We refer also to [13] for a monograph devoted to the positive solutions for differential, difference and integral equations.

Integral boundary value problems for differential equations with integer and non-integer order have been studied by several researchers [1,2,4,12,14,15]. To mention some related references, in [16], first-order problems were considered by using the method of upper and lower solutions, and, in [17], the Guo–Krasnosel'skii fixed point result was applied to study the existence of positive solutions to integral boundary value problems for classical second-order differential equations.

These kind of problems were also considered in [18], where some results were derived as a consequence of the nonlinear alternative of Leray–Schauder type. On the other hand, the monotone iterative technique was developed in [19] for integral boundary value problems relative to first-order integro-differential equations with deviating arguments. See also [20] for a similar study on analogous differential systems. Very recently, the results in [21] were devoted to the study of first-order problems with multipoint and integral boundary conditions by applying Banach or Schaefer's fixed point theorem.

In the fractional case, some sufficient conditions were established in [22] for the existence of solutions to nonlocal boundary value problems associated to Caputo-type fractional differential equations by using Banach and Schaefer's fixed point theorems. A related problem with integral boundary conditions in the context of Banach spaces was analyzed in [23] by using Green's functions and the Kuratowski measure of noncompactness.

The authors of [24] studied fractional differential equations subject to a nonlocal strip condition of integral type that, in the limit, approaches the usual integral boundary condition, and some results were derived by applying fixed point results and the Leray–Schauder degree theory. In [25], the authors considered boundary value problems for a class of fractional differential equations of order $\delta \in (1,2]$ with three-point fractional integral boundary conditions by means of Schaefer's fixed point theorem.

In [26], the contractive mapping principle and the monotone iterative technique were the basic tools and procedures used in the study of a class of Riemann–Liouville fractional differential equations with integral boundary conditions. On the other hand, in [27], Lyapunov-type results were used to study the nonexistence, the uniqueness and the existence and uniqueness of solutions to fractional boundary value problems.

More recently, in [28], a fractional problem subject to Stieltjes and generalized fractional integral boundary conditions was analyzed by applying the Banach contraction mapping principle. An analogous method was applied in [29], where the authors studied a Cauchy problem for Caputo–Fabrizio fractional differential equations in Banach spaces, imposing an initial condition that involves an integral operator, and they deduced the existence and uniqueness of solutions by applying the Banach fixed point theorem.

Some results for Hilfer fractional differential equations subject to boundary conditions involving Riemann–Liouville fractional integral operators were given in [30], and the study was completed by applying a nonlinear alternative of Leray–Schauder type and the Nadler theorem. Classical fixed point theory was also the tool used in [31] for the analysis of sequential ψ-Hilfer fractional boundary value problems. In particular, one of the results applied was the Krasnosel'skii fixed point theorem for the addition of a contractive mapping and a compact mapping.

Several other recent papers include, for instance, [32], where the type of derivative considered was Caputo fractional derivatives with respect to a fixed function, and, under this framework the authors studied an impulsive problem subject to integral boundary conditions based on the Riemann–Stieltjes fractional integral through Leray–Schauder's nonlinear alternative; or [33], where ψ-Caputo operators were considered in the differential equation and in the integral boundary conditions, and the method of upper and lower solutions coupled with the monotone iterative technique were the main tools used.

More specifically, in 2012, Cabada and Wang [15] considered the following boundary value problem for fractional order differential equations with classical integral boundary conditions:

$$\begin{cases} {}^c D^\delta u(t) + f(t, u(t)) = 0, \ 0 < t < 1, \\ u(0) = u''(0) = 0, \ u(1) = \lambda \int_0^1 u(s)ds, \end{cases}$$

where $2 < \delta < 3, 0 < \lambda < 2$, ${}^c D^\delta$ is the Caputo fractional derivative and $f : [0,1] \times [0, \infty) \to [0, \infty)$ is a continuous function.

In 2014, Cabada and Hamdi [14] discussed, by defining a suitable cone on a Banach space and by applying Guo–Krasnosel'skii fixed point theorem, the existence of positive solutions for the following class of nonlinear fractional differential equations with integral boundary conditions:

$$\begin{cases} D^\delta u(t) + f(t, u(t)) = 0, \ 0 < t < 1, \\ u(0) = u'(0) = 0, \ u(1) = \lambda \int_0^1 u(s)ds, \end{cases}$$

where $2 < \delta \leq 3, 0 < \lambda, \lambda \neq \delta$, D^δ is the Riemann–Liouville fractional derivative of order δ and f is a continuous function.

The large collection of research works existing on the topic shows the increasing interest that the study of integral boundary value problems for fractional differential equations has received in the recent times, due to their applicability to the modeling of various processes for which hereditary or memory properties leave a footprint in the performance of the phenomena, and because, in many occasions, the restrictions on the real problem make it adequate to consider boundary conditions that consider the influence that the state on a certain interval has on the evolution of the system.

It is worthwhile to devote efforts to study the existence of positive solutions, since controlling the sign of the solutions is a relevant issue in many fields of application for which negative values are not admissible (populations, amount of substances etc.). In this sense, in comparison with the above mentioned works, we are interested in the consequences, in terms of the properties of the solutions, that the application of the Guo–Krasnosel'skii fixed point theorem may present for a fractional problem with a boundary condition including a fractional operator.

Motivated by the above-mentioned work [14] and its approach, this paper deals with the existence of positive solutions for the following fractional differential equation of general order $\delta > 2$ with fractional integral boundary conditions:

$$\begin{cases} D_{0+}^\delta w(t) + f(t, w(t)) = 0, \ 0 < t < 1, \\ w(0) = w'(0) = w''(0) = w'''(0) = \cdots = w^{(n-2)}(0) = 0, \\ w(1) = \lambda I_{0+}^\gamma w(\zeta), \ 0 < \zeta < 1, \ n-1 < \delta \leq n, \end{cases} \qquad (1)$$

where $n \in \mathbb{N}, n \geq 3, \lambda > 0$ and D_{0+}^{δ} denotes the Riemann–Liouville fractional derivative of order δ, I^{γ} is the Riemann–Liouville fractional integral operator of order $\gamma > 0$ and $f : [0, 1] \times [0, \infty) \to [0, \infty)$ is a continuous function.

As original contributions of the paper, we mention the consideration of a boundary value problem that involves an integral operator of fractional type, which allows us to consider heterogeneity on the dependence specified by the restriction added to the equation and also the subsequent explicit calculation of the Green's function for this general problem, which is not easy to handle due to the high order of the equation and the introduction of fractional operators in the boundary conditions considered.

These novelties in the problem considered add more complexity to the study of the particular properties of the Green's function that are essential to build the mathematical constructs required for the application of the fixed point result, namely, the establishment of estimates, which allow us to define an appropriate cone that is mapped into itself through the integral operator corresponding to the boundary value problem.

To prove the existence of positive solutions to (1), we apply the Guo–Krasnosel'skii fixed point theorem in cones, used in [14] in the context of fractional problems with boundary conditions involving a classical integral term but different from the techniques followed in the discussed works dealing with boundary conditions involving integral operators of a fractional type. The main reason to use this fixed point result is its potential to provide a localization of the solution by handling conical shells whose boundary is defined by the boundaries of two sets, which can be, in this case, more general than open balls [34,35].

Then, it is not only possible to deduce the existence of a positive solution but also we can give an upper bound for its maximum value and establish a certain positive number that is exceeded by the values of the solution at some points. Having, at our disposal, a contractive and an expansive version of the hypotheses, it is possible to deduce the existence of a positive solution under different types of restrictions on the function defining the equation—namely, the sublinear and the superlinear case.

The organization of the paper is as follows. In Section 2, we recall some basic notations and concepts concerning fractional calculus as well as the fixed point result that we apply as a fundamental tool in our procedure. In Section 3, we explicitly obtain the Green's function for a modified linear fractional boundary value problem, and we deduce some estimates for its expression.

The study of the sign of the Green's function is relevant too, as well as the comparison between its value at different points, which is also useful to our reasoning. Then, in Section 4, we present our main result, which allows us to derive the existence of a positive solution for the nonlinear problem (1) in the sublinear and superlinear cases. The proof of the main result provides details regarding the conical shells to which the mentioned solution belongs in each case. In Section 5, an example is included, and, finally, Section 6 shows our conclusions.

2. Materials and Methods

In this section, we recall some notations, definitions and results that are essential to prove our main result.

Definition 1. *The fractional derivative of Riemann–Liouville type and fractional order $\delta > 0$ is defined for a function f as*

$$D_{0+}^{\delta} f(t) = \frac{1}{\Gamma(n-\delta)} \left(\frac{d}{dt} \right)^n \int_0^t (t-s)^{n-\delta-1} f(s) ds,$$

where $n = [\delta] + 1$, and $[\delta]$ is the integer part of δ, provided that the integral on the right-hand side converges pointwise on $(0, \infty)$.

Definition 2. *The fractional integral of Riemann–Liouville type and fractional order $\delta > 0$ is defined for a function f as*

$$I_{0+}^{\delta} f(t) = \frac{1}{\Gamma(\delta)} \int_0^t (t-s)^{\delta-1} f(s) ds,$$

provided that the integral on the right-hand side converges pointwise on $(0, \infty)$.

Lemma 1 ([1]). *Let $\delta > 0$, and then the solutions to $D_{0+}^{\delta} w(t) + y(t) = 0$ are given by*

$$w(t) = -\int_0^t \frac{(t-s)^{\delta-1}}{\Gamma(\delta)} y(s) ds + c_1 t^{\delta-1} + c_2 t^{\delta-2} + \cdots + c_n t^{\delta-n}.$$

Without loss of generality, we assume in this and later results that the fractional derivatives are developed taking 0 as base point. For a discussion on other types of conditions, we refer to Kilbas et al. [1] and Samko et al. [3].

Definition 3. *Let E be a real Banach space. A nonempty closed and convex set $K \subset E$ is called a cone if it satisfies the following two conditions:*
(i) $x \in K, \lambda \geq 0$ implies $\lambda x \in K$;
(ii) $x \in K, -x \in K$ implies $x = 0$, where 0 denotes the zero element of E.

Theorem 1 ([34]). *Let E be a Banach space, and let $K \subset E$ be a cone. Assume that Ω_1, Ω_2 are open and bounded subsets of E with $0 \in \Omega_1 \subset \overline{\Omega_1} \subset \Omega_2$, and let $T : K \cap (\overline{\Omega_2} \setminus \Omega_1) \longrightarrow K$ be a completely continuous mapping such that one of the following conditions holds:*
(i) $\|Tu\| \geq \|u\|$, $u \in K \cap \partial \Omega_1$, and $\|Tu\| \leq \|u\|$, $u \in K \cap \partial \Omega_2$; or
(ii) $\|Tu\| \leq \|u\|$, $u \in K \cap \partial \Omega_1$, and $\|Tu\| \geq \|u\|$, $u \in K \cap \partial \Omega_2$.

Then, the mapping T has at least one fixed point in $K \cap (\overline{\Omega_2} \setminus \Omega_1)$.

We define the mapping $T : C[0,1] \to C[0,1]$ as $[Tu](t) = \int_0^1 G(t,s) f(s, u(s)) ds$, with G a certain Green's function whose expression is given as indicated below (see (3)). This Green's function will be built in such a way that the fixed points of the mapping T coincide with the solutions to problem (1), and, hence, by Theorem 1, we will deduce the existence of positive solutions to problem (1).

3. Some Auxiliary Results

First, we prove the following lemma, relative to the expression of the explicit solution for a linear fractional problem subject to integral boundary conditions of fractional type.

Lemma 2. *Let $\delta > 0$, $n - 1 < \delta \leq n$, $0 < \zeta < 1$, $y \in C[0,1]$, and suppose that $P := 1 - \frac{\lambda \Gamma(\delta)}{\Gamma(\delta + \gamma)} \zeta^{\delta + \gamma - 1} \neq 0$. Then, the problem*

$$\begin{cases} D_{0+}^{\delta} w(t) + y(t) = 0, \ 0 < t < 1, \\ w(0) = w'(0) = w''(0) = w'''(0) = \cdots = w^{(n-2)}(0) = 0, \\ w(1) = \lambda I_{0+}^{\gamma} w(\zeta), \ 0 < \zeta < 1, \ n-1 < \delta \leq n, \end{cases} \quad (2)$$

has a unique solution $w \in C^1[0,1]$, given by $w(t) = \int_0^1 G(t,s) y(s) ds$, where

$$G(t,s) = \begin{cases} \frac{-P\Gamma(\delta+\gamma)(t-s)^{\delta-1} + \Gamma(\delta+\gamma)(1-s)^{\delta-1} t^{\delta-1} - \Gamma(\delta)\lambda(\zeta-s)^{\delta+\gamma-1} t^{\delta-1}}{P\Gamma(\delta)\Gamma(\delta+\gamma)}, & 0 \leq s \leq t \leq 1, s \leq \zeta, \\ \frac{\Gamma(\delta+\gamma)(1-s)^{\delta-1} t^{\delta-1} - \Gamma(\delta)\lambda(\zeta-s)^{\delta+\gamma-1} t^{\delta-1}}{P\Gamma(\delta)\Gamma(\delta+\gamma)}, & 0 \leq t \leq s \leq \zeta \leq 1, \\ \frac{-P\Gamma(\delta+\gamma)(t-s)^{\delta-1} + \Gamma(\delta+\gamma)(1-s)^{\delta-1} t^{\delta-1}}{P\Gamma(\delta)\Gamma(\delta+\gamma)}, & 0 \leq \zeta \leq s \leq t \leq 1, \\ \frac{\Gamma(\delta+\gamma)(1-s)^{\delta-1} t^{\delta-1}}{P\Gamma(\delta)\Gamma(\delta+\gamma)}, & 0 \leq t \leq s \leq 1, s \geq \zeta. \end{cases} \quad (3)$$

Here, $G(t,s)$ is called the Green's function associated to the boundary value problem (1). Note that $G(t,s)$ is a continuous function on $[0,1] \times [0,1]$.

Proof. The first equation in problem (2) is equivalent to the following integral equation:

$$w(t) = -I_{0+}^{\delta} y(t) + c_1 t^{\delta-1} + c_2 t^{\delta-2} + \cdots + c_n t^{\delta-n}.$$

By using

$$w(0) = w'(0) = \cdots = w^{(n-2)}(0) = 0,$$

we obtain that

$$w(t) = -I_{0+}^{\delta} y(t) + c_1 t^{\delta-1}.$$

It follows from

$$w(1) = \lambda I_{0+}^{\gamma} w(\zeta),$$

combined with

$$w(1) = -I_{0+}^{\delta} y(1) + c_1$$

and

$$\lambda I_{0+}^{\gamma} w(\zeta) = -\lambda I_{0+}^{\delta+\gamma} y(\zeta) + \lambda c_1 \frac{\Gamma(\delta)}{\Gamma(\delta+\gamma)} \zeta^{\delta+\gamma-1},$$

that

$$w(t) = -\frac{1}{\Gamma(\delta)} \int_0^t (t-s)^{\delta-1} y(s) ds + \frac{t^{\delta-1}}{P\Gamma(\delta)} \int_0^1 (1-s)^{\delta-1} y(s) ds$$
$$- \frac{\lambda t^{\delta-1}}{P\Gamma(\delta+\gamma)} \int_0^{\zeta} (\zeta-s)^{\delta+\gamma-1} y(s) ds.$$

For $t \leq \zeta$, we have

$$w(t) = \frac{-1}{\Gamma(\delta)} \int_0^t (t-s)^{\delta-1} y(s) ds + \frac{t^{\delta-1}}{P\Gamma(\delta)} \left\{ \int_0^t + \int_t^{\zeta} + \int_{\zeta}^1 \right\} (1-s)^{\delta-1} y(s) ds$$
$$- \frac{\lambda t^{\delta-1}}{P\Gamma(\delta+\gamma)} \left\{ \int_0^t + \int_t^{\zeta} \right\} (\zeta-s)^{\delta+\gamma-1} y(s) ds$$
$$= \int_0^t \frac{-P\Gamma(\delta+\gamma)(t-s)^{\delta-1} + \Gamma(\delta+\gamma)(1-s)^{\delta-1} t^{\delta-1} - \Gamma(\delta)\lambda(\zeta-s)^{\delta+\gamma-1} t^{\delta-1}}{P\Gamma(\delta)\Gamma(\delta+\gamma)} y(s) ds$$
$$+ \int_t^{\zeta} \frac{\Gamma(\delta+\gamma)(1-s)^{\delta-1} t^{\delta-1} - \Gamma(\delta)\lambda(\zeta-s)^{\delta+\gamma-1} t^{\delta-1}}{P\Gamma(\delta)\Gamma(\delta+\gamma)} y(s) ds$$
$$+ \int_{\zeta}^1 \frac{\Gamma(\delta+\gamma)(1-s)^{\delta-1} t^{\delta-1}}{P\Gamma(\delta)\Gamma(\delta+\gamma)} y(s) ds$$
$$= \int_0^1 G(t,s) y(s) ds.$$

For $t \geq \zeta$, we deduce that

$$w(t) = -\frac{1}{\Gamma(\delta)}\left\{\int_0^\zeta + \int_\zeta^t\right\}(t-s)^{\delta-1}y(s)ds + \frac{t^{\delta-1}}{P\Gamma(\delta)}\left\{\int_0^\zeta + \int_\zeta^t + \int_t^1\right\}(1-s)^{\delta-1}y(s)ds$$
$$- \frac{\lambda t^{\delta-1}}{P\Gamma(\delta+\gamma)}\int_0^\zeta (\zeta-s)^{\delta+\gamma-1}y(s)ds$$
$$= \int_0^\zeta \frac{-P\Gamma(\delta+\gamma)(t-s)^{\delta-1} + \Gamma(\delta+\gamma)(1-s)^{\delta-1}t^{\delta-1} - \Gamma(\delta)\lambda(\zeta-s)^{\delta+\gamma-1}t^{\delta-1}}{P\Gamma(\delta)\Gamma(\delta+\gamma)}y(s)ds$$
$$+ \int_\zeta^t \frac{-P\Gamma(\delta+\gamma)(t-s)^{\delta-1} + \Gamma(\delta+\gamma)(1-s)^{\delta-1}t^{\delta-1}}{P\Gamma(\delta)\Gamma(\delta+\gamma)}y(s)ds$$
$$+ \int_t^1 \frac{\Gamma(\delta+\gamma)(1-s)^{\delta-1}t^{\delta-1}}{P\Gamma(\delta)\Gamma(\delta+\gamma)}y(s)ds$$
$$= \int_0^1 G(t,s)y(s)ds.$$

□

A careful analysis of the Green's function G allows us to prove some of its properties that will be useful to our procedure, such as the nonnegativity or the establishment of upper and lower estimates.

Lemma 3. *Let G be the Green's funtion corresponding to the problem* (2), *which is given in Lemma 2. Then, for all $\delta \in (n-1, n]$, and $\lambda > 0$ with $P := 1 - \frac{\lambda \Gamma(\delta)}{\Gamma(\delta+\gamma)}\zeta^{\delta+\gamma-1} > 0$, the following properties hold:*

(I) $G(t,s) \geq \frac{\lambda t^{\delta-1}\zeta^{\delta+\gamma-1}}{P\Gamma(\delta+\gamma)}[(1-s)^{\delta-1} - (1-s)^{\delta+\gamma-1}]$ *for all $t, s \in (0,1)$.*

(II) $G(t,s) \leq \frac{(1-s)^{\delta-1}t^{\delta-1}}{P\Gamma(\delta)}$ *for all $t, s \in (0,1)$.*

(III) $G(t,s) > 0$ *for all $t, s \in (0,1)$.*

(IV) $G(1,s) > 0$ *for all $s \in (0,1)$.*

(V) $G(t,s)$ *is a continuous function for all $t, s \in (0,1)$.*

Proof. We start by proving (I) and (II) simultaneously. First, assume that $0 \leq s \leq t \leq 1$, $s \leq \zeta$. Since $0 < \frac{\lambda \Gamma(\delta)\zeta^{\delta+\gamma-1}}{\Gamma(\delta+\gamma)} < 1$, then we obtain

$$P\Gamma(\delta)\Gamma(\delta+\gamma)G(t,s)$$
$$= -P\Gamma(\delta+\gamma)(t-s)^{\delta-1} + \Gamma(\delta+\gamma)(1-s)^{\delta-1}t^{\delta-1} - \Gamma(\delta)\lambda(\zeta-s)^{\delta+\gamma-1}t^{\delta-1}$$
$$= \lambda\Gamma(\delta)\zeta^{\delta+\gamma-1}(t-s)^{\delta-1} + [-\Gamma(\delta+\gamma)(t-s)^{\delta-1} + \Gamma(\delta+\gamma)(1-s)^{\delta-1}t^{\delta-1}]$$
$$- \Gamma(\delta)\lambda(\zeta-s)^{\delta+\gamma-1}t^{\delta-1}$$
$$\geq \lambda\Gamma(\delta)\zeta^{\delta+\gamma-1}(t-s)^{\delta-1} - \lambda\Gamma(\delta)\zeta^{\delta+\gamma-1}(t-s)^{\delta-1}$$
$$+ \lambda\Gamma(\delta)\zeta^{\delta+\gamma-1}(1-s)^{\delta-1}t^{\delta-1} - \Gamma(\delta)\lambda(\zeta-s)^{\delta+\gamma-1}t^{\delta-1}$$
$$= \lambda\Gamma(\delta)\zeta^{\delta+\gamma-1}(1-s)^{\delta-1}t^{\delta-1} - \Gamma(\delta)\lambda(\zeta-s)^{\delta+\gamma-1}t^{\delta-1}$$
$$\geq \lambda\Gamma(\delta)\zeta^{\delta+\gamma-1}t^{\delta-1}[(1-s)^{\delta-1} - (1-s)^{\delta+\gamma-1}],$$

and

$$P\Gamma(\delta)\Gamma(\delta+\gamma)G(t,s)$$
$$= -P\Gamma(\delta+\gamma)(t-s)^{\delta-1} + \Gamma(\delta+\gamma)(1-s)^{\delta-1}t^{\delta-1} - \Gamma(\delta)\lambda(\zeta-s)^{\delta+\gamma-1}t^{\delta-1}$$
$$= \lambda\Gamma(\delta)\zeta^{\delta+\gamma-1}(t-s)^{\delta-1} - \Gamma(\delta+\gamma)(t-s)^{\delta-1}$$
$$+ \Gamma(\delta+\gamma)(1-s)^{\delta-1}t^{\delta-1} - \Gamma(\delta)\lambda(\zeta-s)^{\delta+\gamma-1}t^{\delta-1}$$
$$\leq \Gamma(\delta+\gamma)(1-s)^{\delta-1}t^{\delta-1} - \Gamma(\delta)\lambda(\zeta-s)^{\delta+\gamma-1}t^{\delta-1}$$
$$\leq \Gamma(\delta+\gamma)(1-s)^{\delta-1}t^{\delta-1}.$$

For $0 \leq t \leq s \leq \zeta \leq 1$, we have

$$P\Gamma(\delta)\Gamma(\delta+\gamma)G(t,s)$$
$$= \Gamma(\delta+\gamma)(1-s)^{\delta-1}t^{\delta-1} - \Gamma(\delta)\lambda(\zeta-s)^{\delta+\gamma-1}t^{\delta-1}$$
$$\geq \lambda\Gamma(\delta)\zeta^{\delta+\gamma-1}(1-s)^{\delta-1}t^{\delta-1} - \Gamma(\delta)\lambda(\zeta-s)^{\delta+\gamma-1}t^{\delta-1}$$
$$\geq \lambda\Gamma(\delta)\zeta^{\delta+\gamma-1}t^{\delta-1}[(1-s)^{\delta-1} - (1-s)^{\delta+\gamma-1}],$$

and

$$P\Gamma(\delta)\Gamma(\delta+\gamma)G(t,s)$$
$$= \Gamma(\delta+\gamma)(1-s)^{\delta-1}t^{\delta-1} - \Gamma(\delta)\lambda(\zeta-s)^{\delta+\gamma-1}t^{\delta-1}$$
$$\leq \Gamma(\delta+\gamma)(1-s)^{\delta-1}t^{\delta-1}.$$

For $0 \leq \zeta \leq s \leq t \leq 1$, we find

$$P\Gamma(\delta)\Gamma(\delta+\gamma)G(t,s)$$
$$= -P\Gamma(\delta+\gamma)(t-s)^{\delta-1} + \Gamma(\delta+\gamma)(1-s)^{\delta-1}t^{\delta-1}$$
$$= \lambda\Gamma(\delta)\zeta^{\delta+\gamma-1}(t-s)^{\delta-1} - \Gamma(\delta+\gamma)(t-s)^{\delta-1} + \Gamma(\delta+\gamma)(1-s)^{\delta-1}t^{\delta-1}$$
$$\geq \lambda\Gamma(\delta)\zeta^{\delta+\gamma-1}(t-s)^{\delta-1} - \lambda\Gamma(\delta)\zeta^{\delta+\gamma-1}(t-s)^{\delta-1} + \lambda\Gamma(\delta)\zeta^{\delta+\gamma-1}(1-s)^{\delta-1}t^{\delta-1}$$
$$\geq \lambda\Gamma(\delta)\zeta^{\delta+\gamma-1}t^{\delta-1}[(1-s)^{\delta-1} - (1-s)^{\delta+\gamma-1}],$$

and

$$P\Gamma(\delta)\Gamma(\delta+\gamma)G(t,s)$$
$$= -P\Gamma(\delta+\gamma)(t-s)^{\delta-1} + \Gamma(\delta+\gamma)(1-s)^{\delta-1}t^{\delta-1}$$
$$= \lambda\Gamma(\delta)\zeta^{\delta+\gamma-1}(t-s)^{\delta-1} - \Gamma(\delta+\gamma)(t-s)^{\delta-1} + \Gamma(\delta+\gamma)(1-s)^{\delta-1}t^{\delta-1}$$
$$\leq \Gamma(\delta+\gamma)(1-s)^{\delta-1}t^{\delta-1}.$$

For $0 \leq t \leq s \leq 1\, s \geq \zeta$, we have

$$P\Gamma(\delta)\Gamma(\delta+\gamma)G(t,s)$$
$$= \Gamma(\delta+\gamma)(1-s)^{\delta-1}t^{\delta-1}$$
$$\geq \lambda\Gamma(\delta)\zeta^{\delta+\gamma-1}t^{\delta-1}[(1-s)^{\delta-1} - (1-s)^{\delta+\gamma-1}].$$

Property (III) is derived from (I). On the other hand, for the validity of (IV), we observe that

$$G(1,s) = \begin{cases} \frac{(1-P)\Gamma(\delta+\gamma)(1-s)^{\delta-1} - \Gamma(\delta)\lambda(\zeta-s)^{\delta+\gamma-1}}{P\Gamma(\delta)\Gamma(\delta+\gamma)} = \frac{\lambda[\zeta^{\delta+\gamma-1}(1-s)^{\delta-1} - (\zeta-s)^{\delta+\gamma-1}]}{P\Gamma(\delta+\gamma)}, & s \leq \zeta, \\ \frac{(1-P)\Gamma(\delta+\gamma)(1-s)^{\delta-1}}{P\Gamma(\delta)\Gamma(\delta+\gamma)} = \frac{\lambda\zeta^{\delta+\gamma-1}(1-s)^{\delta-1}}{P\Gamma(\delta+\gamma)}, & \zeta \leq s, \end{cases}$$

which is obviously positive for $s \in (0,1)$. Finally, (V) is trivially derived. □

The previous result is consistent with those obtained in [14] for the problem with $2 < \delta \leq 3$. In fact, for $\gamma = 1$, we have $P = 1 - \frac{\lambda}{\delta}\zeta^\delta$, and thus the assumption $\lambda \in (0, \delta)$ (as considered in [14]) guarantees that $P > 0$.

Corollary 1. *For all $\delta \in (n-1, n]$, and $\lambda > 0$ with $P := 1 - \frac{\lambda \Gamma(\delta)}{\Gamma(\delta+\gamma)}\zeta^{\delta+\gamma-1} > 0$, the Green's function $G(t,s)$ satisfies*

$$t^{\delta-1} w_1(s) \leq G(t,s) \leq t^{\delta-1} w_2(s), \quad \forall\, t, s \in (0,1), \qquad (4)$$

where

$$w_1(s) = \frac{\lambda \zeta^{\delta+\gamma-1}}{P\Gamma(\delta+\gamma)}[(1-s)^{\delta-1} - (1-s)^{\delta+\gamma-1}],$$

$$w_2(s) = \frac{(1-s)^{\delta-1}}{P\Gamma(\delta)}.$$

Similarly to [14], we derive the following Lemma, which expresses a correspondence between the values $G(t,s)$ and $G(1,s)$. This relation will be essential in the proof of the main result.

Lemma 4. *For all $\delta \in (n-1, n]$, and $\lambda > 0$ with $P := 1 - \frac{\lambda \Gamma(\delta)}{\Gamma(\delta+\gamma)}\zeta^{\delta+\gamma-1} > 0$, the Green's function $G(t,s)$ also satisfies*

$$t^{\delta-1} G(1,s) \leq G(t,s) \leq \frac{1}{1-P} G(1,s) = \frac{\Gamma(\delta+\gamma)}{\lambda \Gamma(\delta)\zeta^{\delta+\gamma-1}} G(1,s), \quad \forall\, t, s \in (0,1). \qquad (5)$$

Proof. By Lemma 3 (IV), the sought inequality is equivalent to prove that

$$t^{\delta-1} \leq \frac{G(t,s)}{G(1,s)} \leq \frac{1}{1-P} = \frac{\Gamma(\delta+\gamma)}{\lambda \Gamma(\delta)\zeta^{\delta+\gamma-1}}, \quad \forall\, t, s \in (0,1). \qquad (6)$$

Note also that, under the hypotheses imposed, $G(t,s) > 0$ for all $t, s \in (0,1)$.
First, we consider the case $0 < s \leq t < 1$, with $s \leq \zeta$, and then

$$\varphi(t,s) := \frac{G(t,s)}{G(1,s)}$$

$$= \frac{-P\Gamma(\delta+\gamma)(t-s)^{\delta-1} + \Gamma(\delta+\gamma)(1-s)^{\delta-1} t^{\delta-1} - \Gamma(\delta)\lambda(\zeta-s)^{\delta+\gamma-1} t^{\delta-1}}{-P\Gamma(\delta+\gamma)(1-s)^{\delta-1} + \Gamma(\delta+\gamma)(1-s)^{\delta-1} - \Gamma(\delta)\lambda(\zeta-s)^{\delta+\gamma-1}}$$

$$= t^{\delta-1} \frac{-P\Gamma(\delta+\gamma)(1-\frac{s}{t})^{\delta-1} + \Gamma(\delta+\gamma)(1-s)^{\delta-1} - \Gamma(\delta)\lambda(\zeta-s)^{\delta+\gamma-1}}{-P\Gamma(\delta+\gamma)(1-s)^{\delta-1} + \Gamma(\delta+\gamma)(1-s)^{\delta-1} - \Gamma(\delta)\lambda(\zeta-s)^{\delta+\gamma-1}}$$

$$= t^{\delta-1} \frac{-P\frac{(1-\frac{s}{t})^{\delta-1}}{(1-s)^{\delta-1}} + 1 - \frac{\Gamma(\delta)\lambda(\zeta-s)^{\delta+\gamma-1}}{\Gamma(\delta+\gamma)(1-s)^{\delta-1}}}{-P + 1 - \frac{\Gamma(\delta)\lambda(\zeta-s)^{\delta+\gamma-1}}{\Gamma(\delta+\gamma)(1-s)^{\delta-1}}}$$

$$\in \left[t^{\delta-1}, t^{\delta-1} \frac{1 - \frac{\Gamma(\delta)\lambda(\zeta-s)^{\delta+\gamma-1}}{\Gamma(\delta+\gamma)(1-s)^{\delta-1}}}{1 - P - \frac{\Gamma(\delta)\lambda(\zeta-s)^{\delta+\gamma-1}}{\Gamma(\delta+\gamma)(1-s)^{\delta-1}}}\right] \subseteq \left[t^{\delta-1}, \frac{1}{1-P}\right] = \left[t^{\delta-1}, \frac{\Gamma(\delta+\gamma)}{\lambda \Gamma(\delta)\zeta^{\delta+\gamma-1}}\right].$$

For $0 < t \leq s \leq \zeta < 1$, we have

$$\varphi(t,s) := \frac{G(t,s)}{G(1,s)}$$

$$= t^{\delta-1} \frac{\Gamma(\delta+\gamma)(1-s)^{\delta-1} - \Gamma(\delta)\lambda(\zeta-s)^{\delta+\gamma-1}}{-P\Gamma(\delta+\gamma)(1-s)^{\delta-1} + \Gamma(\delta+\gamma)(1-s)^{\delta-1} - \Gamma(\delta)\lambda(\zeta-s)^{\delta+\gamma-1}}$$

$$\geq t^{\delta-1}.$$

Next, we prove that $\varphi(t,s) \leq \frac{1}{1-P}$, for $0 < t \leq s \leq \zeta < 1$. We study the behavior of the auxiliary one-variable function

$$\psi(s) := \frac{\Gamma(\delta+\gamma)(1-s)^{\delta-1} - \Gamma(\delta)\lambda(\zeta-s)^{\delta+\gamma-1}}{(1-P)\Gamma(\delta+\gamma)(1-s)^{\delta-1} - \Gamma(\delta)\lambda(\zeta-s)^{\delta+\gamma-1}}$$

in the interval $[t,\zeta]$, with $t \in (0,\zeta]$ fixed. The sign of $\psi'(s)$ coincides with the sign of

$$\phi(s) := \left(-\Gamma(\delta+\gamma)(\delta-1)(1-s)^{\delta-2} + \Gamma(\delta)\lambda(\delta+\gamma-1)(\zeta-s)^{\delta+\gamma-2}\right)$$
$$\times \left((1-P)\Gamma(\delta+\gamma)(1-s)^{\delta-1} - \Gamma(\delta)\lambda(\zeta-s)^{\delta+\gamma-1}\right)$$
$$- \left(\Gamma(\delta+\gamma)(1-s)^{\delta-1} - \Gamma(\delta)\lambda(\zeta-s)^{\delta+\gamma-1}\right)$$
$$\times \left(-(1-P)\Gamma(\delta+\gamma)(\delta-1)(1-s)^{\delta-2} + \Gamma(\delta)\lambda(\delta+\gamma-1)(\zeta-s)^{\delta+\gamma-2}\right)$$
$$= \Gamma(\delta+\gamma)\Gamma(\delta)(1-s)^{\delta-2}\lambda(\zeta-s)^{\delta+\gamma-2}P\{(\delta-1)(\zeta-1)-(1-s)\gamma\},$$

which is, clearly, nonpositive for $s \in [t,\zeta]$. Hence, $\psi(s) \leq \psi(t)$, for $s \in [t,\zeta]$. Since $\varphi(t,s) = t^{\delta-1}\psi(s)$, this proves that, in the case $0 < t \leq s \leq \zeta < 1$, we have

$$\varphi(t,s) \leq t^{\delta-1}\psi(t) = \frac{\Gamma(\delta+\gamma)t^{\delta-1}(1-t)^{\delta-1} - \Gamma(\delta)\lambda t^{\delta-1}(\zeta-t)^{\delta+\gamma-1}}{(1-P)\Gamma(\delta+\gamma)(1-t)^{\delta-1} - \Gamma(\delta)\lambda(\zeta-t)^{\delta+\gamma-1}} =: \mathcal{M}(t).$$

We now check that $\mathcal{M}(t) \leq \frac{1}{1-P}$, for $t \in (0,\zeta]$, which is equivalent to

$$(1-P)\Gamma(\delta+\gamma)(1-t)^{\delta-1}(1-t^{\delta-1}) \geq \Gamma(\delta)\lambda(\zeta-t)^{\delta+\gamma-1}(1-(1-P)t^{\delta-1}), \ t \in (0,\zeta].$$

By substituting the value of P, the previous condition is equivalent to the nonnegativity on the interval $(0,\zeta]$ of the function

$$R(t) := \zeta^{\delta+\gamma-1}(1-t)^{\delta-1}(1-t^{\delta-1}) - (\zeta-t)^{\delta+\gamma-1}\left(1 - \frac{\Gamma(\delta)\lambda\zeta^{\delta+\gamma-1}}{\Gamma(\delta+\gamma)}t^{\delta-1}\right).$$

Indeed, $R(0) = \zeta^{\delta+\gamma-1} - \zeta^{\delta+\gamma-1} = 0$, $R(\zeta) = \zeta^{\delta+\gamma-1}(1-\zeta)^{\delta-1}(1-\zeta^{\delta-1}) > 0$, and

$$R'(t) = \zeta^{\delta+\gamma-1}(\delta-1)(1-t)^{\delta-2}\left(1 - t^{\delta-1} - (1-t)t^{\delta-2}\right)$$
$$+ (\zeta-t)^{\delta+\gamma-1}\left\{(\delta+\gamma-1)\left(1 - \frac{\Gamma(\delta)\lambda\zeta^{\delta+\gamma-1}}{\Gamma(\delta+\gamma)}t^{\delta-1}\right) + \frac{\Gamma(\delta)\lambda\zeta^{\delta+\gamma-1}}{\Gamma(\delta+\gamma)}(\delta-1)t^{\delta-2}\right\},$$

which is clearly positive on $(0,\zeta]$, since

$$\frac{\Gamma(\delta)\lambda\zeta^{\delta+\gamma-1}}{\Gamma(\delta+\gamma)}t^{\delta-1} < \frac{\Gamma(\delta)\lambda\zeta^{\delta+\gamma-1}}{\Gamma(\delta+\gamma)} < 1,$$

and $S(t) := 1 - t^{\delta-1} - (1-t)t^{\delta-2}$ satisfies $S(0) = 1$, $S(1) = 0$, and $S'(t) = t^{\delta-3}(2-\delta) < 0$ for $t \in (0,1)$; thus, $S > 0$ on $(0,\zeta]$. This proves that $R > 0$ on $(0,\zeta]$.

For $0 < \zeta \leq s \leq t < 1$,

$$\varphi(t,s) := \frac{G(t,s)}{G(1,s)}$$

$$= \frac{-P\Gamma(\delta+\gamma)(t-s)^{\delta-1} + \Gamma(\delta+\gamma)(1-s)^{\delta-1}t^{\delta-1}}{-P\Gamma(\delta+\gamma)(1-s)^{\delta-1} + \Gamma(\delta+\gamma)(1-s)^{\delta-1}}$$

$$= t^{\delta-1} \frac{-P\Gamma(\delta+\gamma)(1-\frac{s}{t})^{\delta-1} + \Gamma(\delta+\gamma)(1-s)^{\delta-1}}{-P\Gamma(\delta+\gamma)(1-s)^{\delta-1} + \Gamma(\delta+\gamma)(1-s)^{\delta-1}}$$

$$= t^{\delta-1} \frac{-P \frac{(1-\frac{s}{t})^{\delta-1}}{(1-s)^{\delta-1}} + 1}{-P + 1}$$

$$\in \left[t^{\delta-1}, \frac{1}{1-P} \right] = \left[t^{\delta-1}, \frac{\Gamma(\delta+\gamma)}{\lambda\Gamma(\delta)\zeta^{\delta+\gamma-1}} \right].$$

Finally, for $0 < t \leq s < 1, s \geq \zeta$,

$$\varphi(t,s) := \frac{G(t,s)}{G(1,s)}$$

$$= t^{\delta-1} \frac{\Gamma(\delta+\gamma)(1-s)^{\delta-1}}{-P\Gamma(\delta+\gamma)(1-s)^{\delta-1} + \Gamma(\delta+\gamma)(1-s)^{\delta-1}}$$

$$= t^{\delta-1} \frac{1}{-P+1} \in \left[t^{\delta-1}, \frac{1}{1-P} \right] = \left[t^{\delta-1}, \frac{\Gamma(\delta+\gamma)}{\lambda\Gamma(\delta)\zeta^{\delta+\gamma-1}} \right].$$

□

4. Main Results

This section of the paper is focused on the study of the existence of at least one positive solution to the nonlinear boundary value problem specified in expression (1). The main tool used is the fixed point result by Guo and Krasnosel'skii [34], i.e., Theorem 1.

The base space of interest is $E = C[0,1]$, which is a Banach space if we consider the usual supremum norm $\|\cdot\|$.

Next, similarly to [14], we consider the cone $K \subset E$ defined in the following way:

$$K := \left\{ u \in E : u(t) \geq 0 \text{ for all } t \in [0,1], u(t) \geq t^{\delta-1}(1-P)\|u\|, \text{ for all } t \in \left[\frac{1}{2}, 1\right] \right\}, \quad (7)$$

and develop, in the rest of the section, a procedure similar to that in the mentioned reference [14]. Hence, one of the assumptions that will be used is specified below:

(a) The function $f : [0,1] \times [0,\infty) \to [0,\infty)$ is continuous.

We take the following finite or infinite values:

$$f_0 := \lim_{h \to 0^+} \left\{ \min_{t \in [\frac{1}{2},1]} \frac{f(t,h)}{h} \right\}, \quad f_\infty := \lim_{h \to \infty} \left\{ \min_{t \in [\frac{1}{2},1]} \frac{f(t,h)}{h} \right\},$$

$$f^0 := \lim_{h \to 0^+} \left\{ \max_{t \in [0,1]} \frac{f(t,h)}{h} \right\}, \text{ and } f^\infty := \lim_{h \to \infty} \left\{ \max_{t \in [0,1]} \frac{f(t,h)}{h} \right\}.$$

Then, it is possible to extend Theorem 3.2 [14] to the context of the general-order problem (1). This fact is the main conclusion of this paper.

Theorem 2. *Suppose that the hypothesis (a) is satisfied, and that one of the following assumptions also holds:*

(i) $f_0 = \infty$ and $f^\infty = 0$ (that is, the sublinear case).
(ii) $f^0 = 0$ and $f_\infty = \infty$ (that is, the superlinear case).

Then, for all $\delta \in (n-1, n]$, and $\lambda > 0$ with $P := 1 - \frac{\lambda \Gamma(\delta)}{\Gamma(\delta+\gamma)} \zeta^{\delta+\gamma-1} > 0$, the problem (1) has a positive solution that belongs to the cone K given by (7).

Proof. We consider the mapping T defined by $[Tu](t) := \int_0^1 G(t,s)f(s,u(s))\,ds$, where G is the Green's function given in expression (3). In the first place, we check that the mapping $T : K \to K$ is a self-mapping and that T is also completely continuous. Indeed, using the continuity and the nonnegative character of the functions G and f on $[0,1] \times [0,1]$ and $[0,1] \times [0,\infty)$, respectively, it is clear that, if $u \in K$, then Tu is continuous and nonnegative on $[0,1]$.

To prove that T is self-mapping, let $u \in K$, and then, by Lemma 4, we have

$$[Tu](t) = \int_0^1 G(t,s)f(s,u(s))\,ds$$
$$\geq t^{\delta-1} \int_0^1 G(1,s)f(s,u(s))\,ds$$
$$\geq t^{\delta-1}(1-P) \int_0^1 \max_{t \in [0,1]} G(t,s)f(s,u(s))\,ds$$
$$\geq t^{\delta-1}(1-P) \max_{t \in [0,1]} \left\{ \int_0^1 G(t,s)f(s,u(s))\,ds \right\}$$
$$= t^{\delta-1}(1-P)\|Tu\|.$$

It is clear that the mapping $T : K \to K$ is continuous, since G and f are both continuous.

Next, to check that T is completely continuous, let $\mathcal{B} \subset K$ be a bounded set, i.e., such that there exists a positive constant $N > 0$ with $\|u\| \leq N$ for all $u \in \mathcal{B}$. Consider the compact set $[0,1] \times [0,N]$, and take $L := \max_{(t,u) \in [0,1] \times [0,N]} |f(t,u)| + 1 > 0$.

Now we check that $T(\mathcal{B})$ is a bounded set. Indeed, for an arbitrary $u \in \mathcal{B}$, we have, by Corollary 1, that

$$\|[Tu](t)\| \leq \max_{t \in [0,1]} \int_0^1 G(t,s)|f(s,u(s))|\,ds \leq L \max_{t \in [0,1]} \int_0^1 t^{\delta-1} \frac{(1-s)^{\delta-1}}{P\Gamma(\delta)}\,ds \leq \frac{L}{P\Gamma(\delta)},$$

for every $t \in [0,1]$, so that $T(\mathcal{B})$ is a bounded subset of E.

On the other hand, we seek an estimate for the derivative of the functions in $T(\mathcal{B})$. Given an arbitrary $u \in \mathcal{B}$, we have, from the calculations in Lemma 2, that

$$[Tu](t) = \int_0^1 G(t,s)f(s,u(s))\,ds$$
$$= -\frac{1}{\Gamma(\delta)} \int_0^t (t-s)^{\delta-1} f(s,u(s))\,ds + \frac{t^{\delta-1}}{P\Gamma(\delta)} \int_0^1 (1-s)^{\delta-1} f(s,u(s))\,ds$$
$$- \frac{\lambda t^{\delta-1}}{P\Gamma(\delta+\gamma)} \int_0^\zeta (\zeta-s)^{\delta+\gamma-1} f(s,u(s))\,ds,$$

so that

$$|(Tu)'(t)| = \left| -\frac{1}{\Gamma(\delta-1)} \int_0^t (t-s)^{\delta-2} f(s,u(s)) ds \right.$$
$$\left. + \frac{t^{\delta-2}}{P\Gamma(\delta-1)} \int_0^1 (1-s)^{\delta-1} f(s,u(s)) ds - \frac{(\delta-1)\lambda t^{\delta-2}}{P\Gamma(\delta+\gamma)} \int_0^\zeta (\zeta-s)^{\delta+\gamma-1} f(s,u(s)) ds \right|$$
$$\leq \frac{1}{\Gamma(\delta-1)} \int_0^t (t-s)^{\delta-2} |f(s,u(s))| ds$$
$$+ \frac{t^{\delta-2}}{|P|\Gamma(\delta-1)} \int_0^1 (1-s)^{\delta-1} |f(s,u(s))| ds + \frac{(\delta-1)\lambda t^{\delta-2}}{|P|\Gamma(\delta+\gamma)} \int_0^\zeta (\zeta-s)^{\delta+\gamma-1} |f(s,u(s))| ds$$
$$\leq \frac{Lt^{\delta-1}}{\Gamma(\delta)} + \frac{Lt^{\delta-2}}{|P|\Gamma(\delta-1)\delta} + \frac{(\delta-1)\lambda L t^{\delta-2} \zeta^{\delta+\gamma}}{|P|\Gamma(\delta+\gamma)(\delta+\gamma)}$$
$$\leq \frac{Lt^{\delta-1}}{\Gamma(\delta)} + \frac{t^{\delta-2}L}{|P|\Gamma(\delta)} + \frac{(\delta-1)Lt^{\delta-2}\zeta^{\delta+\gamma}\lambda}{|P|\Gamma(\delta+\gamma+1)} \leq \frac{L}{\Gamma(\delta)} + \frac{L}{|P|\Gamma(\delta)} + \frac{(\delta-1)L\zeta^{\delta+\gamma}\lambda}{|P|\Gamma(\delta+\gamma+1)} =: M.$$

Therefore, for every $t_1, t_2 \in [0,1]$ with $t_1 < t_2$, we obtain

$$|[Tu](t_2) - [Tu](t_1)| \leq M(t_2 - t_1),$$

and we deduce that $T(\mathcal{B})$ is an equicontinuous set in E.

With these ingredients, the application of the Arzelà–Ascoli Theorem proves that $T(\mathcal{B})$ is relatively compact. As a consequence, $T: K \to K$ is completely continuous.

Once we have proven some relevant properties of the mapping T, we distinguish two cases and complete the proof following the ideas in [14]. We include the explanations and adaptations here for completeness.

Case (i): $(f_0 = \infty$ and $f^\infty = 0)$.

We choose $\tilde{\delta} > 0$ to be sufficiently large such that

$$\tilde{\delta}(1-P) \max_{t \in [0,1]} \left\{ \int_{\frac{1}{2}}^1 s^{\delta-1} G(t,s) ds \right\} \geq 1. \tag{8}$$

Since $f_0 = \infty$, we can affirm the existence of a constant $\tilde{\rho} > 0$ such that $f(t,h) \geq \tilde{\delta} h$ for every $t \in [\frac{1}{2}, 1]$ and every $0 < h \leq \tilde{\rho}$.

Then, for an arbitrary $u \in K$ with $\|u\| = \tilde{\rho}$, we have that $u(t) > 0$ for $t \in [\frac{1}{2}, 1]$ and, using the selection for $\tilde{\delta}$, we obtain that

$$\|Tu\| = \max_{t \in [0,1]} \left\{ \int_0^1 G(t,s) f(s,u(s)) ds \right\}$$
$$\geq \tilde{\delta} \max_{t \in [0,1]} \left\{ \int_{\frac{1}{2}}^1 G(t,s) u(s) ds \right\}$$
$$\geq \tilde{\delta} \|u\| (1-P) \max_{t \in [0,1]} \left\{ \int_{\frac{1}{2}}^1 s^{\delta-1} G(t,s) ds \right\}$$
$$\geq \|u\|.$$

By the continuity of $f(t, \cdot)$ on the interval $[0, \infty)$, we can consider the function:

$$\tilde{f}(t,h) = \max_{z \in [0,h]} f(t,z),$$

which is clearly a nondecreasing function on $[0, \infty)$. By the hypothesis $f^\infty = 0$, it is deduced that

$$\lim_{h \to \infty} \left\{ \max_{t \in [0,1]} \frac{\tilde{f}(t,h)}{h} \right\} = 0.$$

Next, we select $\delta^* > 0$ small enough such that $\frac{\delta^*}{P\Gamma(\delta)} \leq 1$.

By virtue of the previous limit, we can prove the existence of a constant $\rho^* > \tilde{\rho} > 0$ such that $\tilde{f}(t,h) \leq \delta^* h$ for every $t \in [0,1]$ and all $h \geq \rho^*$.

If we take $u \in K$ such that $\|u\| = \rho^*$, then, using the nondecreasing character of $\tilde{f}$ and Lemma 3 (II) (or Corollary 1), the next inequalities are satisfied:

$$\|Tu\| = \max_{t \in [0,1]} \left\{ \int_0^1 G(t,s) f(s,u(s)) \, ds \right\} \leq \max_{t \in [0,1]} \left\{ \int_0^1 G(t,s) \tilde{f}(s,\|u\|) \, ds \right\}$$

$$\leq \delta^* \|u\| \max_{t \in [0,1]} \left\{ \int_0^1 G(t,s) \, ds \right\} \leq \frac{\delta^*}{P\Gamma(\delta)} \|u\| \leq \|u\|.$$

Therefore, by part (i) in Theorem 1, we can affirm that problem (1) has at least one positive solution u with $\tilde{\rho} \leq \|u\| \leq \rho^*$.

Case (ii): $f^0 = 0$ and $f_\infty = \infty$.

We take $\delta^* > 0$ with $\frac{\delta^*}{P\Gamma(\delta)} \leq 1$.

Using $f^0 = 0$, it is possible to find a constant $r^* > 0$ such that $f(t,h) \leq \delta^* h$ for every $t \in [0,1]$ and $0 < h \leq r^*$. From $f^0 = 0$, it is clear that $\lim_{h \to 0^+} \frac{f(t,h)}{h} = 0$ for every $t \in [0,1]$; hence, $\lim_{h \to 0^+} f(t,h) = 0$, and thus, by the continuity of f, $f(t,0) = 0$, for every $t \in [0,1]$. This, together with the previous inequality, implies that $f(t,h) \leq \delta^* h$ for every $t \in [0,1]$ and $0 \leq h \leq r^*$.

Then, for every $u \in K$ with $\|u\| = r^*$, we deduce that

$$\|Tu\| = \max_{t \in [0,1]} \left\{ \int_0^1 G(t,s) f(s,u(s)) \, ds \right\} \leq \delta^* \|u\| \max_{t \in [0,1]} \left\{ \int_0^1 G(t,s) \, ds \right\}$$

$$\leq \frac{\delta^*}{P\Gamma(\delta)} \|u\| \leq \|u\|.$$

Finally, we select $\hat{\delta} > 0$ large enough such that

$$\frac{\hat{\delta}}{2^{\delta-1}} (1-P) \max_{t \in [0,1]} \left\{ \int_{\frac{1}{2}}^1 G(t,s) \, ds \right\} \geq 1.$$

Since $f_\infty = \infty$, we can affirm the existence of $\hat{r} > r^* > 0$, which can be taken satisfying the additional condition $\hat{r} 2^{\delta-1} > r^*(1-P)$, such that $f(t,h) \geq \hat{\delta} h$ for all $t \in [\frac{1}{2}, 1]$ and all $h \geq \hat{r}$.

Next, we choose a convenient shell, in particular, we take an arbitrary $u \in K$ with $\|u\| = \frac{\hat{r}}{1-P} 2^{\delta-1}$. The definition of the cone K implies that $u(t) \geq \hat{r}$ for every $t \in [\frac{1}{2}, 1]$.

In summary, in this case, we obtain that

$$\|Tu\| = \max_{t \in [0,1]} \left\{ \int_0^1 G(t,s) f(s,u(s)) \, ds \right\}$$

$$\geq \max_{t \in [0,1]} \left\{ \int_{\frac{1}{2}}^1 G(t,s) f(s,u(s)) \, ds \right\}$$

$$\geq \hat{\delta} \max_{t \in [0,1]} \left\{ \int_{\frac{1}{2}}^1 G(t,s) u(s) \, ds \right\}$$

$$\geq \frac{\hat{\delta}}{2^{\delta-1}} (1-P) \|u\| \max_{t \in [0,1]} \left\{ \int_{\frac{1}{2}}^1 G(t,s) \, ds \right\}$$

$$\geq \|u\|.$$

In consequence, by case (ii) in Theorem 1, we deduce that problem (1) has at least one positive solution such that $r^* \leq \|u\| \leq \frac{\hat{r}}{1-P} 2^{\delta-1}$. □

5. Example

In this section, we discuss an example to show the applicability of our result.

Example 1. *Consider the following fractional integral boundary value problem on the interval $[0,1]$:*

$$\begin{cases} D_{0+}^{\frac{5}{2}} u(t) + f(t, u(t)) = 0 \\ u(0) = u'(0) = 0, \ u(1) = 2 I_{0+}^{\frac{1}{2}} u(\zeta), \end{cases} \quad (9)$$

where $f(t, u(t)) = u^{\frac{1}{3}}(t) + \log(1 + u^2(t)) + \sin^2(e^{u(t)})$, $D_{0+}^{\frac{5}{2}}$ denotes the Riemann–Liouville fractional derivative operator of order $\delta = \frac{5}{2}$, $I_{0+}^{\frac{1}{2}}$ is the Riemann–Liouville fractional integral operator of order $\gamma = \frac{1}{2}$ and $0 < \zeta < 1$. Here, $f : [0,1] \times [0, \infty) \to [0, \infty)$ is a continuous function. It is clear that $f_0 = \infty$, $f^\infty = 0$, and thus the function f is sublinear. Note that, since $\frac{2\Gamma(\frac{5}{2})}{\Gamma(3)} > 1$, $P := 1 - \frac{2\Gamma(\frac{5}{2})}{\Gamma(3)} \zeta^2$ vanishes at a certain $\zeta \in (0,1)$, exactly at $\zeta_* := \sqrt{\frac{\Gamma(3)}{2\Gamma(\frac{5}{2})}}$. Therefore, we must impose that $\zeta \in (0, \zeta_*)$ in order to guarantee $P > 0$. Under this restriction, from case (i) in Theorem 2, the particular problem (9) has, at least, a positive solution.

6. Conclusions

In this paper, we extended the results in [14] to general fractional problems of order greater than 2, dealing with the existence of positive solutions for differential equations of arbitrary order with fractional integral boundary conditions of the type (1). The introduction of a boundary condition that involves an integral operator of fractional type is interesting from the point of view of applications, since it allows for the mathematical expression of heterogeneity that may affect the dependence specified by the restriction added to the equation—a fact that is consistent with many physical problems.

The main tool used in the paper was Guo–Krasnosel'skii fixed point theorem in cones. In particular, in Lemma 2, we obtained, by imposing some adequate restrictions on the parameters, the integral expression of the solution to a modified linear fractional boundary value problem, which provides the Green's function of interest. Then, in Lemma 3, we studied some properties of the Green's function, including its positivity on $(0,1) \times (0,1)$ under some restrictions on the parameters, as well as some upper and lower estimates for its expression.

Another useful result is Lemma 4, which establishes the relation between the value of the Green's function at an arbitrary point and the value at the point with the same ordinate and abscise 1. The explicit calculations for this general problem were developed in detail due to the high order of the equation and the difficulty generated by the introduction of fractional operators in the boundary conditions.

Theorem 2 provides the existence of a positive solution to (1) by assuming that the nonlinearity f is sublinear or superlinear. The proof, based on the Guo–Krasnosel'skii fixed point theorem, makes a selection of the conical shells that allow localization of the solution in each case. Then, we have not only deduced the existence of a positive solution but the details of the proof also provide the procedure to obtain an estimate for its maximum value and to determine positive numbers that are not upper bounds for the solution.

Since the fixed point theorem used has two contexts of application (a contractive and expansive case), it is possible to consider the problem under two types of hypotheses; that is, two types of restrictions on the function defining the equation. The consideration of other types of restrictions on the function f can be one of the possible future lines of research.

Finally, an example was presented.

Author Contributions: Conceptualization, A.T., J.A. and R.R.-L.; methodology, A.T., J.A. and R.R.-L.; formal analysis, A.T., J.A. and R.R.-L.; investigation, A.T., J.A. and R.R.-L.; writing—review and editing, A.T., J.A. and R.R.-L. All authors have read and agreed to the published version of the manuscript.

Funding: The research of R. Rodríguez-López was partially supported by AEI/FEDER, UE, grant numbers PID2020-113275GB-I00 and MTM2016-75140-P, and by GRC Xunta de Galicia grant number ED431C 2019/02.

Institutional Review Board Statement: Not applicable.

Informed Consent Statement: Not applicable.

Data Availability Statement: Not applicable.

Acknowledgments: The authors are grateful to the anonymous Referees for their helpful comments and suggestions towards the improvement of the paper.

Conflicts of Interest: The authors declare no conflict of interest.

References

1. Kilbas, A.A.; Srivastava, H.M.; Trujillo, J.J. *Theory and Applications of Fractional Differential Equations*; Elsevier: Amsterdam, The Netherlands, 2006.
2. Podlubny, I. *Fractional Differential Equations*; Mathematics in Science and Engineering; Academic Press: New York, NY, USA, 1999.
3. Samko, S.G.; Kilbas, A.A.; Marichev, O.I. *Fractional Integrals and Derivatives, Theory and Applications*; Gordon and Breach: Yverdon, Switzerland, 1993.
4. Lakshmikantham, V.; Vatsala, A.S. Basic theory of fractional differential equations. *Nonlinear Anal.* **2008**, *69*, 2677–2682. [CrossRef]
5. Sabatier, J.; Agrawal, O.P.; Machado, J.A.T. *Advances in Fractional Calculus: Theoretical Developments and Applications in Physics and Engineering*; Springer: Dordrecht, The Netherlands, 2007.
6. Ahmad, B.; Sivasundaram, S. Existence of solutions for impulsive integral boundary value problems of fractional order. *Nonlinear Anal. Hybrid Syst.* **2010**, *4*, 134–141. [CrossRef]
7. Nieto, J.J. Maximum principles for fractional differential equations derived from Mittag-Leffler functions. *Appl. Math. Lett.* **2010**, *23*, 1248–1251. [CrossRef]
8. Agarwal, R.P.; O'Regan, D.; Stanek, S. Positive solutions for Dirichlet problems of singular nonlinear fractional differential equations. *J. Math. Anal. Appl.* **2010**, *371*, 57–68. [CrossRef]
9. Bai, Z.B. On positive solutions of a nonlocal fractional boundary value problem. *Nonlinear Anal.* **2010**, *72*, 916–924. [CrossRef]
10. Zhang, S.Q. Positive solutions to singular boundary value problem for nonlinear fractional differential equation. *Comput. Math. Appl.* **2010**, *59*, 1300–1309. [CrossRef]
11. Bai, C. Impulsive periodic boundary value problems for fractional differential equation involving Riemann–Liouville sequential fractional derivative. *J. Math. Anal. Appl.* **2011**, *384*, 211–231. [CrossRef]
12. Agarwal, R.P.; Benchohra, M.; Hamani, S. Boundary value problems for differential inclusions with fractional order. *Adv. Stud. Contemp. Math.* **2008**, *16*, 181–196.
13. Agarwal, R.P.; O'Regan, D.; Wong, P.J.Y. *Positive Solutions of Differential, Difference and Integral Equations*; Kluwer Academic Publishers: Dordrecht, The Netherlands, 1999.
14. Cabada, A.; Hamdi, Z. Nonlinear fractional differential equations with integral boundary value conditions. *Applied Math. Comp.* **2014**, *228*, 251–257. [CrossRef]
15. Cabada, A.; Wang, G. Positive solutions of nonlinear fractional differential equations with integral boundary value conditions. *J. Math. Anal. Appl.* **2012**, *389*, 403–411. [CrossRef]
16. Jankowski, T. Differential equations with integral boundary conditions. *J. Comput. Appl. Math.* **2002**, *147*, 1–8. [CrossRef]
17. Boucherif, A. Second-order boundary value problems with integral boundary conditions. *Nonlinear Anal.* **2009** *70*, 364–371. [CrossRef]
18. Benchohra, M.; Nieto, J.J.; Ouahab, A. Second-order boundary value problem with integral boundary conditions. *Bound. Value Probl.* **2011**, *2011*, 260309. [CrossRef]
19. Wang, G.; Song, G.; Zhang, L. Integral boundary value problems for first order integro-differential equations with deviating arguments. *J. Comput. Appl. Math.* **2009**, *225*, 602–611. [CrossRef]
20. Wang, G. Boundary value problems for systems of nonlinear integro-differential equations with deviating arguments. *J. Comput. Appl. Math.* **2010**, *234*, 1356–1363. [CrossRef]
21. Mardanov, M.J.; Sharifov, Y.A.; Gasimov, Y.S.; Cattani, C. Non-linear first-order differential boundary problems with multipoint and integral conditions. *Fractal Fract.* **2021**, *5*, 15. [CrossRef]
22. Benchohra, M.; Hamani, S.; Ntouyas, S.K. Boundary value problems for differential equations with fractional order and nonlocal conditions. *Nonlinear Anal.* **2009**, *71*, 2391–2396. [CrossRef]
23. Benchohra, M.; Cabada, A.; Seba, D. An existence result for nonlinear fractional differential equations on Banach spaces. *Bound. Value Probl.* **2009**, *2000*, 628916. [CrossRef]

24. Ahmad, B.; Agarwal, R.P. On nonlocal fractional boundary value problems. *Dyn. Contin. Discrete Impuls. Syst. Ser. A Math. Anal.* **2011**, *18*, 535–544.
25. Sudsutad, W.; Tariboon, J. Boundary value problems for fractional differential equations with three-point fractional integral boundary conditions. *Adv. Differ. Equ.* **2012**, *2012*, 93. [CrossRef]
26. Wang, X.; Wang, L.; Zeng, Q. Fractional differential equations with integral boundary conditions. *J. Nonlinear Sci. Appl.* **2015**, *8*, 309–314. [CrossRef]
27. Dhar, S.; Kong, Q.; McCabe, M. Fractional boundary value problems and Lyapunov-type inequalities with fractional integral boundary conditions. *Electron. J. Qual. Theory Differ. Equ.* **2016**, *43*, 1–16. [CrossRef]
28. Ahmad, B.; Alghanmi, M.; Ntouyas, S.K.; Alsaedi, A. Fractional differential equations involving generalized derivative with Stieltjes and fractional integral boundary conditions. *Appl. Math. Lett.* **2018**, *84*, 111–117. [CrossRef]
29. Keten, A.; Yavuz, M.; Baleanu, D. Nonlocal Cauchy problem via a fractional operator involving power kernel in Banach spaces. *Fractal Fract.* **2019**, *3*, 27. [CrossRef]
30. Wongcharoen, A.; Ntouyas, S.K.; Tariboon, J. Boundary value problems for Hilfer fractional differential inclusions with nonlocal integral boundary conditions. *Mathematics* **2020**, *8*, 1905. [CrossRef]
31. Sitho, S.; Ntouyas, S.K.; Samadi, A.; Tariboon, J. Boundary value problems for ψ-Hilfer type sequential fractional differential equations and inclusions with integral multi-point boundary conditions. *Mathematics* **2021**, *9*, 1001. [CrossRef]
32. Asawasamrit, S.; Thadang, Y.; Ntouyas, S.K.; Tariboon, J. Non-instantaneous impulsive boundary value problems containing Caputo fractional derivative of a function with respect to another function and Riemann-Stieltjes fractional integral boundary conditions. *Axioms* **2021**, *10*, 130. [CrossRef]
33. Boutiara, A.; Benbachir, M.; Alzabut, J.; Samei, M.E. Monotone Iterative and upper-lower solution techniques for solving nonlinear ψ-Caputo fractional boundary value problem. *Fractal Fract.* **2021**, *5*, 194. [CrossRef]
34. Guo, D.; Lakshmikantham, V. *Nonlinear Problems in Abstract Cones*; Academic Press: New York, NY, USA, 1988.
35. Krasnosel'skii, M.A. Fixed points of cone-compressing or cone-extending operators. *Soviet Math. Dokl.* **1960**, *1*, 1285–1288.

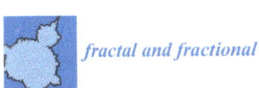

Article

Stability of Parametric Intuitionistic Fuzzy Multi-Objective Fractional Transportation Problem

Mohamed A. El Sayed [1,2,*], Mohamed A. El-Shorbagy [3,4], Farahat A. Farahat [5], Aisha F. Fareed [2] and Mohamed A. Elsisy [2,6]

1. Basic Sciences Department, Faculty of Engineering, Badr University in Cairo BUC, Cairo 11829, Egypt
2. Department of Basic Engineering Sciences, Faculty of Engineering, Benha University, Benha 13511, Egypt; Aisha.farid@yahoo.com (A.F.F.); elsisy.mohamed@yahoo.com (M.A.E.)
3. Department of Mathematics, College of Science and Humanities in Al-Kharj, Prince Sattam bin Abdulaziz University, Al-Kharj 11942, Saudi Arabia; mohammed_shorbagy@yahoo.com or ma.hassan@psau.edu.sa
4. Department of Basic Engineering Science, Faculty of Engineering, Menoufia University, Shebin El-Kom 32511, Egypt
5. Higher Technological Institute, Tenth of Ramadan City 44637, Egypt; Farahat_math@yahoo.com
6. Department of Basic Engineering Sciences, Menoufia Higher Institute of Engineering and Technologey, El-Bagour 32821, Egypt
* Correspondence: Mohamed.ali@bhit.bu.edu.eg or mohamedali20008@yahoo.com

Citation: El Sayed, M.A.; El-Shorbagy, M.A.; Farahat, F.A.; Fareed, A.F.; Elsisy, M.A. Stability of Parametric Intuitionistic Fuzzy Multi-Objective Fractional Transportation Problem. *Fractal Fract.* 2021, 5, 233. https://doi.org/10.3390/fractalfract5040233

Academic Editor: Savin Treanţă

Received: 14 October 2021
Accepted: 8 November 2021
Published: 19 November 2021

Publisher's Note: MDPI stays neutral with regard to jurisdictional claims in published maps and institutional affiliations.

Copyright: © 2021 by the authors. Licensee MDPI, Basel, Switzerland. This article is an open access article distributed under the terms and conditions of the Creative Commons Attribution (CC BY) license (https://creativecommons.org/licenses/by/4.0/).

Abstract: In this study, a parametric intuitionistic fuzzy multi-objective fractional transportation problem (PIF-MOFTP) is proposed. The current PIF-MOFTP has a single-scalar parameter in the objective functions and an intuitionistic fuzzy supply and demand. Based on the (α, β)-cut concept a parametric (α, β)-MOFTP is established. Then, a fuzzy goal programming (FGP) approach is utilized to obtain (α, β)-Pareto optimal solution. We investigated the stability set of the first kind (SSFK) corresponding to the solution by extending the Kuhn-Tucker optimality conditions of multi-objective programming problems. An algorithm to crystalize the progressing SSFK for PIF-MOFTP as well as an illustrative numerical example is presented.

Keywords: multi-objective programming; fractional transportation problem; intuitionistic fuzzy set; parametric programming

1. Introduction

Transportation issues (TP) have been studied in various writings [1–7]. These issues and their solution processes postulate a worthy task in logistics and supply chain organization for reducing expenses, further developing service quality, etc. [3,8]. Nonetheless, TP is described by multiple, incommensurable, and clashing objective functions, being known as the multi-objective transportation problem (MO-TP). Accordingly, in MO-TP, the idea of an ideal solution offers spot to the idea of the best compromise solution or the non-dominated solutions. Optimization of the ratio of two functions is called fractional programming (ratio optimization) [7,9]. To be sure, in such circumstances, it is often a question of optimizing a ratio of benefit/cost, stock/deals, specialist/patient, and so on, subject to some constraints [7,9].

One of the significant issues looked at by specialists is that involving the exact values of parameters [7]. In this way, this might involve thinking about vagueness, or specifying the fundamental parameters of the model, which are the coefficients of the objective function and the constrains [4,8]. Accordingly, it might be naturalistic to take the distinct adjectival information on specialists and leaders about the parameters which can be exemplified as fuzzy data [7,10]. Uncertainty may happen because of the accompanying unrestrained factors. In this study the main hypotheses are that the transportation charge has a parametric nature, and the supply and the demand parameters are intuitionistic fuzzy numbers (IFNs). The main hypotheses have not been presented in the literature, and the basic question is how we can get the SSFK for such PIF-MOFTP.

2. Literature Review

The research on MO-TP is improved by fusing the diverse numerical models and procedures. James et al. [11] examined transportation administration quality dependent on data combination. A lot of examination that deals with transportation wellbeing was created by Ergun et al. [12], Sheu and Chen [13]. Recently, MO-TP under different circumstances has been discussed by Roy et al. [14,15], Roy and Mahapatra [16], Roy [17], Maity and Roy [18,19].

Although fuzzy set theory (FST) is novel tool in handling uncertainties, it cannot tackle special kinds of uncertainties, as it is difficult to depict the membership degree using one specific value. To overcome the lack of knowledge of non-membership degrees, intuitionistic fuzzy set (IFS) was presented in 1986 by Atanassov [20] as an extension of FST. In IFS, each element in a set is attached with two grades: membership and non-membership, where the sum of these two grades is restricted to less or equal to one. Moreover, many creators have been utilized IFS for addressing various sorts of TPs [21,22]. The study of MO-TP with vague numbers has been presented by Ammar and Youness [1]. The fuzzy programming strategy was acquainted with tackle MO-TP with various non-linear membership functions [23]. IFS has additionally been utilized by several scientists to tackle different types of TPs [10,24]. One more strategy for thoroughly considering linear MO-TPs with vague nature has been suggested by Gupta and Kumar [25]. Recently, MO-TP under various types of uncertainty has been discussed by Roy and Mahapatra [16], Maity and Roy [26], and Ebrahimnejad and Verdegay [10]. Mahajan and Gupta [27] proposed a fully IF MO-TP utilizing various membership functions. Achievement stability set for parametric linear FGP problems has been introduced by El Sayed and Farahat [28]. The neutrosophic goal programming approach for solving the multi-objective fractional transportation problem was introduced by Veeramani et al., [29]. Pramanik and Banerjee [30] proposed a chance-constrained capacitated MO-TP with two fuzzy goals, and a consensus solution was found. Edalatpanah [31] developed a nonlinear framework for neutrosophic linear programming. Furthermore, Rizk-Allah et al. [32] developed a compromise solution framework for the MO-TP based on the neutrosophic environment. A fuzzy approach using generalized dinkelbach's algorithm for linear multi-objective fractional transportation problem (MOFTP) has been presented by Cetin and Tiryaki [3]. A fuzzy mathematical programming approach for solving fuzzy linear fractional programming problem has been demonstrated by Veeramani and Sumathi [33]. El Sayed and Abo-Sinna [7] introduced the intuitionistic fuzzy multi-objective fractional transportation problem (IF-MOFTP).

Parametric programming examines the impact of preordained continuous varieties in the objective function coefficients and the right-hand side of the constraints on the ideal solution [34–36]. In parametric analysis the objective function and the right-hand side vectors are replaced with the parameterized function $c(\vartheta)$ and $b(\alpha, \beta)$, where ϑ and α, β are the parameter of variation. The general idea of parametric analysis is to start with the α-Pareto optimal solution at $\vartheta = \vartheta^*$, $\alpha = \alpha^*$, $\beta = \beta^*$. Then by applying KKT optimality the SSFK is determined [35,37]. The concept of the stability set of the first kind (SSFK) has been introduced by Osman [35], and extended by Saad [38], Saad and Hughes [39], Osman et al. [36], Saad et al. [40].

In prior examinations, the MO-TP was created with the presumption that the supply, demand, and cost boundaries were known. Nonetheless, applications, every one of the parameters of the TP are not for the most part characterized definitively. It might have IF values. Comparable contemplations might be taken for supply and demand parameters in TP of this paper. Keeping this perspective, the primary commitments are concerned with two unique viewpoints: one is to find a (α, β)-Pareto optimal solution for the PIF-MOFTP, and another is to investigate the SSFK for PIF-MOFTP. First, based on the (α, β)-cut methodology a parametric (α, β)-MOFTP is established. Then, A FGP approach is used to get (α, β)-Pareto optimal solution. Finally, the KKT optimality conditions applied to get the SSFK. An algorithm to clarify the developed SSFK for PIF-MOFTP as well as an illustrative numerical example are given.

The rest of this study is organized as follows: after the introduction and literature review, Section 3 introduces some basic concepts. Modelling of the PIF-MOFTP is presented in Section 4. Section 5 demonstrates the FGP methodology for tackling the PIF-MOFTP. In the next section the SSFK is investigated. An algorithm for obtaining the SSFK for PIF-MOFTP is introduced in Section 6. An illustrative example, discussion and limitations is given in Section 7. This paper ends with some concluding remarks.

3. Preliminaries

This part presents the concept of IFS [20,21,41,42].

Definition 1. *An IFS $\widetilde{A}^I$ in X is a set of ordered triples $\widetilde{A}^I = \{(x, \mu_{\widetilde{A}^I}(x), v_{\widetilde{A}^I}(x)) | x \in X\}$, where $\mu_{\widetilde{A}^I}(x), v_{\widetilde{A}^I}(x) : X \to [0,1]$ are functions such that $0 \leq \mu_{\widetilde{A}^I}(x) + v_{\widetilde{A}^I}(x) \leq 1, \forall x \in X$. The value of $\mu_{\widetilde{A}^I}(x)$ acts as the grade of membership and $v_{\widetilde{A}^I}(x)$ acts as the grade of non-membership of the element $x \in X$ being in $\widetilde{A}^I$. $h(x) = 1 - \mu_{\widetilde{A}^I}(x) - v_{\widetilde{A}^I}(x)$ represents the grade of hesitation for the element x in $\widetilde{A}^I$ [20,41].*

Definition 2. *An IFN of the form $\widetilde{A}^I = \left(a_1, a_2, a_3; \overline{a}_1, a_2, \overline{a}_3\right)$ is said to be triangular IFN (TIFN) with membership and non-membership functions defined as [41,43]:*

$$\mu_{\widetilde{A}^I}(x) = \begin{cases} \frac{x-a_1}{a_2-a_1}, & a_1 \leq x \leq a_2, \\ \frac{a_3-x}{a_3-a_2}, & a_2 \leq x \leq a_3, \\ 0, & otherwise \end{cases} \tag{1}$$

$$v_{\widetilde{A}^I}(x) = \begin{cases} \frac{a_2-x}{a_2-\overline{a}_1}, & \overline{a}_1 \leq x \leq a_2 \\ \frac{x-a_2}{\overline{a}_3-a_2}, & a_2 \leq x \leq \overline{a}_3, \\ 1 & otherwise \end{cases} \tag{2}$$

where $\frac{x-a_1}{a_2-a_1}$, and $\frac{x-a_2}{\overline{a}_3-a_2}$ are continuous monotone increasing functions, $\frac{a_3-x}{a_3-a_2}$ and $\frac{a_2-x}{a_2-\overline{a}_1}$ are continuous monotone decreasing functions. $\frac{x-a_1}{a_2-a_1}$, $\frac{a_3-x}{a_3-a_2}$, $\frac{a_2-x}{a_2-\overline{a}_1}$ and $\frac{x-a_2}{\overline{a}_3-a_2}$ are the left and the right basis functions of the membership function and the non-membership function (see Figure 1), respectively. $\overline{a}_1 \leq a_1 \leq a_2 \leq a_3 \leq \overline{a}_3$ and $0 \leq \mu_{\widetilde{A}^I}(x) + v_{\widetilde{A}^I}(x) \leq 1, \forall x \in X$.

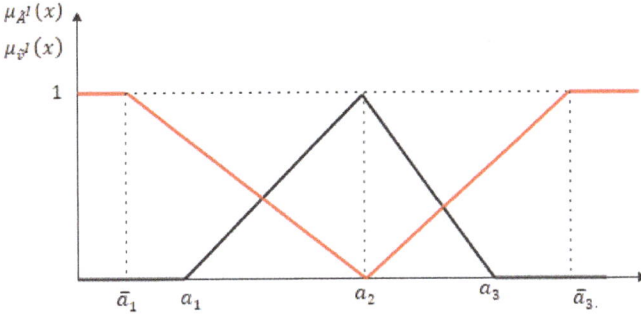

Figure 1. Triangular Intuitionistic Fuzzy number.

Definition 3. *A TIFNs $\widetilde{A}^I = \left(a_1, a_2, a_3; \overline{a}_1, a_2, \overline{a}_3\right)$ is assumed to be a non-negative TIFN iff, $\overline{a}_1 \geq 0$ [41,43].*

Definition 4. Two TIFNs $\widetilde{A}^I = \left(a_1, a_2, a_3; \bar{a}_1, a_2, \bar{a}_3\right)$ and $\widetilde{A}^I = \left(b_1, b_2, b_3; \bar{b}_1, b_2, \bar{b}_3\right)$ are equivalent to one another, $\widetilde{A}^I = \widetilde{B}^I$ iff, $a_i = b_i$ and $\bar{a}_i = \bar{b}_i \ \forall \ i = 1, 2, 3$ [7,41,43].

Definition 5. (α, β)-cut of an IFS $\widetilde{A}^I$ is defined by: $\widetilde{A}^I_{(\alpha,\beta)} = \{x : \mu_{\widetilde{A}^I}(x) \geq \alpha, \nu_{\widetilde{A}^I}(x) \leq \beta, \alpha + \beta \leq 1, x \in X\}$; where $\alpha, \beta \in (0, 1]$.

Definition 6. (α, β)-cut of a TIFN $\widetilde{A}^I = \left(a_1, a_2, a_3; \bar{a}_1, a_2, \bar{a}_3\right)$ is the set of all x whose degree of membership is greater than or equal to α and degree of non-membership is less than or equal to β, i.e., $\widetilde{A}^I_{(\alpha,\beta)} = \{x : \mu_{\widetilde{A}^I}(x) \geq \alpha, \nu_{\widetilde{A}^I}(x) \leq \beta, \alpha + \beta \leq 1, x \in X\}$.

The (α, β)-cut of a TIFN is shown in Figure 2, is defined as the crisp set of elements x which belong to $\widetilde{A}^I$ at least to the degree α and which does belong to $\widetilde{A}^I$ at most to the degree β.

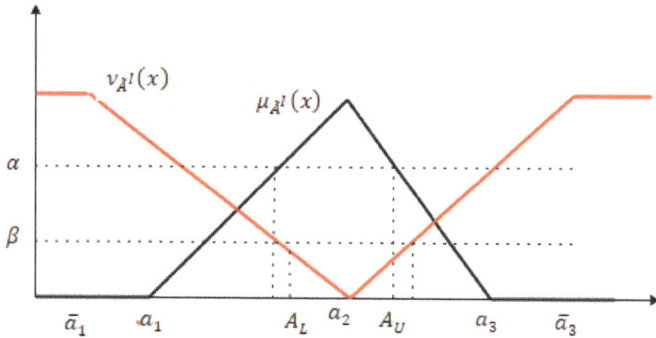

Figure 2. The (α, β)-cut of a TIFN.

Now, $\mu_{\widetilde{A}^I}(x) \geq \alpha \Rightarrow \frac{x-a_1}{a_2-a_1} \geq \alpha$ and $\frac{a_3-x}{a_3-a_2} \geq \alpha$, or $x \geq a_1 + \alpha(a_2 - a_1)$ and $x \leq a_3 - \alpha(a_3 - a_2)$ again, $\nu_{\widetilde{A}^I}(x) \leq \beta \Rightarrow \frac{a_2-x}{a_2-\bar{a}_1} \leq \beta$ and $\frac{x-a_2}{\bar{a}_3-a_2} \leq \beta$, or $x \geq a_2 - \beta(a_2 - \bar{a}_1)$ and $x \leq a_2 + \beta(\bar{a}_3 - a_2)$ [43]. Thus, referring to Figure 2 $\widetilde{A}^I_{(\alpha,\beta)} = [A_L, A_U]$, where $A_L = \max\{a_1 + \alpha(a_2 - a_1), a_2 - \beta(a_2 - \bar{a}_1)\}$ and $A_U = \min\{a_3 - \alpha(a_3 - a_2), a_2 + \beta(\bar{a}_3 - a_2)\}$.

4. Mathematical Formulation

In genuine case TP, during the modeling process, the transportation parameters are not precise on account of insufficient information the variance of the market situation. To deal quantitatively with such unclear information, we deemed parametric IF-MOFTP in which single-scalar parameter $\vartheta \in \mathbb{R}$ in the objective functions and an intuitionistic fuzzy supply and demand. Suppose that there are m sources and n destinations. Thus, modelling of the parametric IF-MOFTP can be obtained as [3,7,9]:

$$\text{Max } Z_q(x, \vartheta) = \frac{\sum_{i=1}^{m} \sum_{j=1}^{n} \left(c_{ij} + \vartheta \omega_{ij}\right)^{(q)} x_{ij}^{(q)} + \delta^{(q)}}{\sum_{i=1}^{m} \sum_{j=1}^{n} d_{ij}^{(q)} x_{ij}^{(q)} + \rho^{(q)}}, \quad q = 1, 2, \ldots, Q, \quad (3)$$

Subject to:

$$\sum_{j=1}^{n} x_{ij} \leq \widetilde{a}_i^I, \quad i = 1, 2, \ldots, m, \quad (4)$$

$$\sum_{i=1}^{m} x_{ij} \geq \widetilde{b}_j^I, \quad j = 1, 2, \ldots, n, \quad (5)$$

$$x_{ij} \geq 0, \ i = 1, 2, \ldots, m, \ j = 1, 2, \ldots, n. \tag{6}$$

where $c_{ij}^{(q)} = (c_{ij} + \vartheta\omega_{ij})^{(q)}$ denotes the parametric profit gained from shipment of ith source to j^{th} destination. Also, $d_{ij}^{(q)}$ denotes the expense per unit of shipment from i^{th} source to j^{th} destination. $\delta^{(q)}, \rho^{(q)}$ are some constant profit and cost, respectively. $x_{ij}^{(q)}$ is the quantity shipped from i^{th} source to j^{th} destination. $\tilde{a}_i^I = \left(a_i^1, a_i^2, a_i^3; \overline{a}_i^1, a_i^2, \overline{a}_i^3\right)$ stands for the available intuitionistic fuzzy supply at i^{th} source and $\tilde{b}_j^I = \left(b_j^1, b_j^2, b_j^3; \overline{b}_j^1, b_j^2, \overline{b}_j^3\right)$ alludes to the accessible intuitionistic fuzzy demand at j^{th} destination. Further, we postulate that $\sum_{i=1}^m \sum_{j=1}^n d_{ij}^{(q)} x_{ij}^{(q)} + \rho^{(q)} > 0, \ q = 1, 2, \ldots, Q; \ \tilde{a}_i^I > 0^I, \ \tilde{b}_j^I > 0^I, \ \forall j; (c_{ij} + \vartheta\omega_{ij})^{(q)} > 0^I, \ \delta^{(q)}, \rho^{(q)} > 0$ for all i, j, and the gross supply is greater than or equal the gross demand [3,7].

$$\sum_{i=1}^m \left(\tilde{a}_i^I\right)_{(\alpha,\beta)} \geq \sum_{j=1}^n \left(\tilde{b}_j^I\right)_{(\alpha,\beta)}. \tag{7}$$

The disparity (7) is considered as a necessary and sufficient condition for the existence of a feasible solution to PIF-MOFTP.

For a certain degree of (α, β)-cut the PIF-MOFTP could be transformed into parametric (α, β)-MOFTP as:

$$\text{Max } Z_q(x, \vartheta) = \frac{\sum_{i=1}^m \sum_{j=1}^n (c_{ij} + \vartheta\omega_{ij})^{(q)} x_{ij}^{(q)} + \delta^{(q)}}{\sum_{i=1}^m \sum_{j=1}^n d_{ij}^{(q)} x_{ij}^{(q)} + \rho^{(q)}}, \ q = 1, 2, \ldots, Q, \tag{8}$$

Subject to:

$$\sum_{j=1}^n x_{ij} \leq (a_i)_{(\alpha,\beta)} \ \ i = 1, 2, \ldots, m, \tag{9}$$

$$\sum_{i=1}^m x_{ij} \geq (b_j)_{(\alpha,\beta)} \ \ j = 1, 2, \ldots, n, \tag{10}$$

$$x_{ij} \geq 0, \ i = 1, 2, \ldots, m, \ j = 1, 2, \ldots, n, \tag{11}$$

$$a_i^L \leq (a_i)_{(\alpha,\beta)} \leq a_i^U, \ \ i = 1, 2, \ldots, m, \tag{12}$$

$$b_j^L \leq (b_j)_{(\alpha,\beta)} \leq b_j^U, \ \ j = 1, 2, \ldots, n. \tag{13}$$

Based on the concept of a convex linear combination method proposed in [40] parametric (α, β)-MOFTP can be rewritten as:

$$\text{Max } Z_q(x, \vartheta) = \frac{\sum_{i=1}^m \sum_{j=1}^n (c_{ij} + \vartheta\omega_{ij})^{(q)} x_{ij}^{(q)} + \delta^{(q)}}{\sum_{i=1}^m \sum_{j=1}^n d_{ij}^{(q)} x_{ij}^{(q)} + \rho^{(q)}}, \ q = 1, 2, \ldots, Q, \tag{14}$$

Subject to:

$$\sum_{j=1}^n x_{ij} \leq \lambda a_i^L + (1-\lambda) a_i^U \ \ i = 1, 2, \ldots, m, \tag{15}$$

$$\sum_{i=1}^m x_{ij} \geq \lambda b_j^L + (1-\lambda) b_j^U \ \ j = 1, 2, \ldots, n, \tag{16}$$

$$x_{ij} \geq 0, \ \lambda \in [0,1], \ i = 1, 2, \ldots, m, \ j = 1, 2, \ldots, n, \tag{17}$$

Let $M_{(\alpha,\beta)}$ denote the set of constraints in Equations (15)–(17), the parametric (α, β)-MOFTP has an (α, β)-Pareto optimal solution x_{ij}^* at ϑ^*.

Definition 7. (α, β)-Pareto optimal solution. $x_{ij}^* \in M_{(\alpha,\beta)}$ is said to be an (α, β)-Pareto optimal solution to (α, β)-MOFTP if and only if there does not exist another $\mathring{x}_{ij} \in M_{(\alpha,\beta)}$, $a_i \in (a_i)_{(\alpha,\beta)}$, $b_j \in (b_j)_{(\alpha,\beta)}$, such that $Z_q\left(x_{ij}^\circ, \vartheta^*\right) \geq Z_q\left(x_{ij}^*, \vartheta^*\right)$ with at least one strict inequality hold for q $(q = 1, 2, \ldots, Q)$.

5. FGP Methodology for PIF-MOFTP

In this section the FGP approach is applied to obtain the compromise solution of the parametric (α, β)-MOFTP. The objective functions are modeled as fuzzy goals characterized by its' membership function $\mu_{(z_q(x,\vartheta^*))}$ [36,44–46]. The model formulation and solution process are carried out at $\vartheta = \vartheta^*$. The membership functions of the qth fuzzy goals [36,44], is defined as:

$$\mu_{(z_q(x,\vartheta^*))} = \begin{cases} 1, & if\ Z_q(x, \vartheta^*) \geq u_q^*, \\ \frac{Z_q(x,\vartheta^*) - g_q^*}{u_q^* - g_q^*}, & if\ g_q^* \leq Z_q(x, \vartheta^*) \leq u_q^*, \\ 0, & if\ Z_q(x, \vartheta^*) \leq g_q^*, \end{cases} \quad q = 1,2,\ldots,Q \quad (18)$$

where $u_q^* = max\ Z_q(x, \vartheta^*)$, $g_q^* = min\ Z_q(x, \vartheta^*)$, and denotes the upper and lower tolerance limit for the membership function of qth objective, respectively. In the FGP approach, the most extensive level of membership is unity. So, the membership goals having the aspired level unity follows as [44]:

$$\mu_q\left(Z_q(x, \vartheta^*)\right) + d_q^- - d_q^+ = 1, \quad q = 1,2,\ldots,Q, \quad (19)$$

where $d_q^-, d_q^+ \geq 0$, with $d_q^- \times d_q^+ = 0$, denote the under- and over-deviations, respectively, from the aspired levels [36,44]. The final FGP model of the parametric (α, β)-MOFTP can be obtained as:

$$\text{Min } AF = \sum_{q=1}^{Q} w_q^- d_q^-, \quad (20)$$

Subject to:

$$\frac{Z_q(x, \vartheta^*) - g_q^*}{u_q^* - g_q^*} + d_q^- - d_q^+ = 1, \quad q = 1,2,\ldots,Q, \quad (21)$$

$$\sum_{j=1}^{n} x_{ij} \leq \lambda\ a_i^L + (1-\lambda) a_i^U \quad i = 1,2,\ldots,m, \quad (22)$$

$$\sum_{i=1}^{m} x_{ij} \geq \lambda\ b_j^L + (1-\lambda) b_j^U \quad j = 1,2,\ldots,n, \quad (23)$$

$$x_{ij} \geq 0,\ \lambda \in [0,1],\ i = 1,2,\ldots,m,\ j = 1,2,\ldots,n, \quad (24)$$

$$d_q^- \times d_q^+ = 0,\ and\ d_q^-, d_q^+ \geq 0,\ q = 1,2,\ldots,Q, \quad (25)$$

where w_q^- represents the relative importance of achieving the aspired levels of the respective fuzzy goals which given by [44,47]:

$$w_q^- = \frac{1}{u_q^* - g_q^*}, \quad q = 1,2,\ldots,Q \quad (26)$$

Extension of Pal's Method to Linearize the Membership Goals

It can be easily realized that the parametric membership goals in Equation (19) are non-linear fractional in nature. To avoid such problem, Pal et al. [45] method is extended here to linearize the qth membership goals with single-scalar parameter $\vartheta = \vartheta^*$ as:

$$\mu_q(Z_q(x, \vartheta^*)) + d_q^- - d_q^+ = 1, \quad q = 1, 2, \ldots, Q, \tag{27}$$

$$L_q(Z_q(x, \vartheta^*)) - L_q g_q^* + d_q^- - d_q^+ = 1; \quad L_q = \frac{1}{u_q^* - g_{ij}^*}, \tag{28}$$

$$Z_q(x, \vartheta^*) = \frac{\sum_{i=1}^m \sum_{j=1}^n (c_{ij} + \vartheta^* \omega_{ij})^{(q)} x_{ij}^{(q)} + \delta^{(q)}}{\sum_{i=1}^m \sum_{j=1}^n d_{ij}^{(q)} x_{ij}^{(q)} + \rho^{(q)}}, \quad q = 1, 2, \ldots, Q, \tag{29}$$

Substituting from Equation (29) in Equation (28), we obtain:

$$L_q \frac{\sum_{i=1}^m \sum_{j=1}^n (c_{ij} + \vartheta^* \omega_{ij})^{(q)} x_{ij}^{(q)} + \delta^{(q)}}{\sum_{i=1}^m \sum_{j=1}^n d_{ij}^{(q)} x_{ij}^{(q)} + \rho^{(q)}} - L_q g_q^* + d_q^- - d_q^+ = 1, \tag{30}$$

$$L_q \left[\sum_{i=1}^m \sum_{j=1}^n (c_{ij} + \vartheta^* \omega_{ij})^{(q)} x_{ij}^{(q)} + \delta^{(q)} \right] - L_q g_q^* \left[\sum_{i=1}^m \sum_{j=1}^n d_{ij}^{(q)} x_{ij}^{(q)} + \rho^{(q)} \right] + d_q^- \left[\sum_{i=1}^m \sum_{j=1}^n d_{ij}^{(q)} x_{ij}^{(q)} + \rho^{(q)} \right] \\ - d_q^+ \left[\sum_{i=1}^m \sum_{j=1}^n d_{ij}^{(q)} x_{ij}^{(q)} + \rho^{(q)} \right] = \left[\sum_{i=1}^m \sum_{j=1}^n d_{ij}^{(q)} x_{ij}^{(q)} + \rho^{(q)} \right], \tag{31}$$

$$\left(\begin{array}{c} L_q \left[\sum_{i=1}^m \sum_{j=1}^n (c_{ij} + \vartheta^* \omega_{ij})^{(q)} x_{ij}^{(q)} + \delta^{(q)} \right] \\ + d_q^- \left[\sum_{i=1}^m \sum_{j=1}^n d_{ij}^{(q)} x_{ij}^{(q)} + \rho^{(q)} \right] \\ - d_q^+ \left[\sum_{i=1}^m \sum_{j=1}^n d_{ij}^{(q)} x_{ij}^{(q)} + \rho^{(q)} \right] \end{array} \right) = (1 + L_q g_q^*) \left[\sum_{i=1}^m \sum_{j=1}^n d_{ij}^{(q)} x_{ij}^{(q)} + \rho^{(q)} \right], \tag{32}$$

$$\left(\begin{array}{c} L_q \left[\sum_{i=1}^m \sum_{j=1}^n (c_{ij} + \vartheta^* \omega_{ij})^{(q)} x_{ij}^{(q)} + \delta^{(q)} \right] \\ + d_q^- \left[\sum_{i=1}^m \sum_{j=1}^n d_{ij}^{(q)} x_{ij}^{(q)} + \rho^{(q)} \right] \\ - d_q^+ \left[\sum_{i=1}^m \sum_{j=1}^n d_{ij}^{(q)} x_{ij}^{(q)} + \rho^{(q)} \right] \end{array} \right) = L_q^\circ \left[\sum_{i=1}^m \sum_{j=1}^n d_{ij}^{(q)} x_{ij}^{(q)} + \rho^{(q)} \right]; \quad L_q^0 = (1 + L_q g_q^*) \tag{33}$$

$$\left[L_q \sum_{i=1}^m \sum_{j=1}^n (c_{ij} + \vartheta^* \omega_{ij})^{(q)} - L_q^0 \sum_{i=1}^m \sum_{j=1}^n d_{ij}^{(q)} \right] x_{ij}^{(q)} + d_q^- \left[\sum_{i=1}^m \sum_{j=1}^n d_{ij}^{(q)} x_{ij}^{(q)} + \rho^{(q)} \right] - d_q^+ \left[\sum_{i=1}^m \sum_{j=1}^n d_{ij}^{(q)} x_{ij}^{(q)} + \rho^{(q)} \right] \\ = \left[L_q^0 \rho^{(q)} - L_q \delta^{(q)} \right], \tag{34}$$

$$C_{ij}^{(q)} x_{ij}^{(q)} + d_q^- \left[\sum_{i=1}^m \sum_{j=1}^n d_{ij}^{(q)} x_{ij}^{(q)} + \rho^{(q)} \right] - d_q^+ \left[\sum_{i=1}^m \sum_{j=1}^n d_{ij}^{(q)} x_{ij}^{(q)} + \rho^{(q)} \right] = G_q; \tag{35}$$

where

$$C_{ij}^{(q)} = \left[L_q \sum_{i=1}^m \sum_{j=1}^n (c_{ij} + \vartheta^* \omega_{ij})^{(q)} - L_q^0 \sum_{i=1}^m \sum_{j=1}^n d_{ij}^{(q)} \right], \tag{36}$$

$$G_q = \left[L_q^0 \rho^{(q)} - L_q \delta^{(q)} \right], \tag{37}$$

Considering Pal et al. [45], the goal expression in Equation (35) can be linearized as follows. Letting $D_q^- = d_q^- \left[\sum_{i=1}^m \sum_{j=1}^n d_{ij}^{(q)} x_{ij}^{(q)} + \rho^{(q)}\right]$ and $D_q^+ = d_q^+ \left[\sum_{i=1}^m \sum_{j=1}^n d_{ij}^{(q)} x_{ij}^{(q)} + \rho^{(q)}\right]$, then the linear form of expression in Equation (32) is obtained as:

$$C_{ij}^{(q)} x_{ij}^{(q)} + D_q^- - D_q^+ = G_q, \tag{38}$$

with D_q^-, $D_q^+ \geq 0$; and $D_q^- \times D_q^+ = 0$, since d_q^-, $d_q^+ \geq 0$, and $\sum_{i=1}^m \sum_{j=1}^n d_{ij}^{(q)} x_{ij}^{(q)} + \rho^{(q)} > 0$. So, minimization of d_q^- means minimization of $D_q^- = d_q^- \left[\sum_{i=1}^m \sum_{j=1}^n d_{ij}^{(q)} x_{ij}^{(q)} + \rho^{(q)}\right]$ which is also non-linear. So, involvement of $d_q^- \leq 1$, in the solution leads to impose the following constraint in the model:

$$\frac{D_q^-}{\left[\sum_{i=1}^m \sum_{j=1}^n d_{ij}^{(q)} x_{ij}^{(q)} + \rho^{(q)}\right]} \leq 1. \tag{39}$$

Now, the final FGP model of the parametric (α, β)-MOFTP in model (20)–(25) becomes:

$$\text{Min } AF = \sum_{q=1}^Q w_q^- d_q^-, \tag{40}$$

Subject to:

$$\left[L_q \sum_{i=1}^m \sum_{j=1}^n (c_{ij} + \vartheta^* \omega_{ij})^{(q)} - L_q^0 \sum_{i=1}^m \sum_{j=1}^n d_{ij}^{(q)}\right] x_{ij}^{(q)} + D_q^- - D_q^+ = \left[L_q^0 \rho^{(q)} - L_q \delta^{(q)}\right], \tag{41}$$

$$\sum_{i=1}^m \sum_{j=1}^n -d_{ij}^{(q)} x_{ij}^{(q)} + D_q^- \leq \rho^{(q)}, \ q = 1, 2, \ldots, Q, \ \forall i, j, \tag{42}$$

$$\sum_{j=1}^n x_{ij} \leq \lambda a_i^L + (1-\lambda) a_i^U, \ i = 1, 2, \ldots, m, \tag{43}$$

$$\sum_{i=1}^m x_{ij} \geq \lambda b_j^L + (1-\lambda) b_j^U \ j = 1, 2, \ldots, n, \tag{44}$$

$$x_{ij} \geq 0, \ \lambda \in [0, 1], \ i = 1, 2, \ldots, m, \ j = 1, 2, \ldots, n, \tag{45}$$

$$D_q^- \times D_q^+ = 0, \text{ and } D_q^-, D_q^+ \geq 0, \ q = 1, 2, \ldots, Q. \tag{46}$$

Thus, the above FGP model provides the satisfactory solution x_{ij}^* for the parametric (α, β)-MOFTP.

6. The SSFK for Parametric (α, β) -MOFTP

The main area of inquiry is as follows: having solved the parametric (α, β)-MOFTP, to what extent can its data with respect to α, β and ϑ be changed without invalidating the efficiency of its (α, β)-Pareto optimal solution? The set of feasible parameters, the solvability set, and the SSFK for parametric (α, β)-MOFTP are defined as:

Definition 8. *The set of feasible parameters for the parametric (α, β)-MOFTP is defined by:*

$$\mathcal{F} = \left\{ \begin{array}{l} a \in R^m, \\ b \in R^n \end{array} \middle| \begin{array}{l} a_i \in L_{\alpha, \beta}(\tilde{a}_i^I), \ i = 1, 2, \ldots m; \ b_j \in L_{\alpha, \beta}(\tilde{b}_j^I), \ j = 1, 2, \ldots, n; \\ \alpha, \beta \in [0, 1]; \text{ and } M_{(\alpha, \beta)}(x_{ij}, a, b) \neq \varnothing \end{array} \right\}$$

Definition 9. *The solvability set $\mathcal{M}$ of the parametric (α, β)-MOFTP is defined by:*

$$\mathcal{M} = \left\{ (\vartheta, a, b) \in R \times R^m \times R^n \,\middle|\, \begin{array}{l} parametric(\alpha, \beta) - \text{MOFTP has} \\ an\ (\alpha, \beta) - \text{Pareto optimal solution.} \end{array} \right\}.$$

Definition 10. *Suppose that x_{ij}^* be an (α, β)-Pareto optimal solution of the parametric (α, β)-MOFTP, then the SSFK $S_1\left(x_{ij}^*, \alpha, \beta\right)$ corresponding to x_{ij}^* is defined by:*

$$S_1\left(x_{ij}^*, \alpha, \beta\right) = \left\{ (\vartheta, a, b) \in R \times R^m \times R^n \,\middle|\, \begin{array}{l} x_{ij}^* \text{ is an } (\alpha, \beta) - \text{Pareto optimal solution of} \\ parametric\ (\alpha, \beta) - \text{MOFTP} \end{array} \right\}.$$

The SSFK of the parametric (α, β)-MOFTP is the set of all parameters corresponding to one (α, β)-Pareto optimal solution [35,36]. It is easy to see that the stability of the parametric (α, β)-MOFTP model (14)–(17) implies the stability of the parametric FGP model which is defined as follows:

$$\text{Min } AF = \sum_{q=1}^{Q} w_q^- d_q^- \tag{47}$$

Subject to:

$$\left[L_q \sum_{i=1}^{m} \sum_{j=1}^{n} (c_{ij} + \vartheta \omega_{ij})^{(q)} - L_q^0 \sum_{i=1}^{m} \sum_{j=1}^{n} d_{ij}^{(q)} \right] x_{ij}^{(q)} + D_q^- - D_q^+ = \left[L_q^0 \rho^{(q)} - L_q \delta^{(q)} \right], \tag{48}$$

$$\sum_{i=1}^{m} \sum_{j=1}^{n} -d_{ij}^{(q)} x_{ij}^{(q)} + D_q^- \leq \rho^{(q)}, \ q = 1, 2, \ldots, Q, \ \forall i, j \tag{49}$$

$$\sum_{j=1}^{n} x_{ij} \leq \lambda\, a_i^L + (1-\lambda) a_i^U, \ i = 1, 2, \ldots, m, \tag{50}$$

$$\sum_{i=1}^{m} x_{ij} \geq \lambda\, b_j^L + (1-\lambda) b_j^U, \ j = 1, 2, \ldots, n, \tag{51}$$

$$x_{ij} \geq 0, \lambda \in [0,1], \vartheta \in R, \ i = 1, 2, \ldots, m, \ j = 1, 2, \ldots, n, \tag{52}$$

$$D_q^- \times D_q^+ = 0, \text{ and } D_q^-, D_q^+ \geq 0, \ q = 1, 2, \ldots, Q. \tag{53}$$

6.1. KKT Optimality Conditions for Parametric FGP Model

The Lagrangian function of parametric FGP model (47)–(53) follows as [36,37]:

$$\begin{aligned} L = &\left[\sum_{q=1}^{Q} w_q^- D_q^- \right] + \xi_q \left[\left[L_q \sum_{i=1}^{m} \sum_{j=1}^{n} (c_{ij} + \vartheta \omega_{ij})^{(q)} - L_q^0 \sum_{i=1}^{m} \sum_{j=1}^{n} d_{ij}^{(q)} \right] x_{ij}^{(q)} + D_q^- - D_q^+ - \left[L_q^0 \rho^{(q)} - L_q \delta^{(q)} \right] \right] \\ &+ v_q \left[\sum_{i=1}^{m} \sum_{j=1}^{n} -d_{ij}^{(q)} x_{ij}^{(q)} + D_q^- - \rho^{(q)} \right] + \tau_i \left[\sum_{j=1}^{n} x_{ij} - \left(\lambda\, a_i^L + (1-\lambda) a_i^U \right) \right] \\ &+ \eta_j \left[-\sum_{i=1}^{m} x_{ij} + \left(\lambda\, b_j^L + (1-\lambda) b_j^U \right) \right] + \varphi_{ij}[-x_{ij}] + \psi_i[-a_i^L] + \phi_j[-b_j^L] + \omega_i[-a_i^U] + \epsilon_j[-b_j^U] \\ &+ \zeta_q[-D_q^-] + \pi_q[-D_q^+], \end{aligned} \tag{54}$$

where $\xi, v, \tau, \eta, \varphi, \psi, \phi, \omega, \epsilon, \zeta$ and π are the Lagrange multipliers. Thus, KKT optimality conditions [28,36,37,39] have the following form:

$$\frac{\partial L}{\partial x_{ij}} = \xi_q \left[L_q \sum_{i=1}^{m} \sum_{j=1}^{n} (c_{ij} + \vartheta \omega_{ij})^{(q)} - L_q^0 \sum_{i=1}^{m} \sum_{j=1}^{n} d_{ij}^{(q)} \right] + v_q \left[\sum_{i=1}^{m} \sum_{j=1}^{n} -d_{ij}^{(q)} \right] + \tau_i - \eta_j - \varphi_{ij} = 0, i = 1, 2, \ldots, m, \ j = 1, 2, \ldots, n, \tag{55}$$

$$\frac{\partial L}{\partial a_i^L} = -\lambda \tau_i - \psi_i = 0, \ i = 1, 2, \ldots m, \tag{56}$$

$$\frac{\partial L}{\partial a_i^U} = -(1-\lambda)\tau_i - \varpi_i = 0, \ i=1,2,\ldots m, \tag{57}$$

$$\frac{\partial L}{\partial b_j^L} = \lambda \eta_j - \phi_j = 0, \ i=1,2,\ldots m, \tag{58}$$

$$\frac{\partial L}{\partial b_j^U} = (1-\lambda)\eta_j - \epsilon_j = 0, \ i=1,2,\ldots m, \tag{59}$$

$$\frac{\partial L}{\partial D_q^-} = \sum_{q=1}^{Q} w_q^- + \xi_q + v_q - \zeta_q = 0, \ q=1,2,\ldots,Q, \tag{60}$$

$$\frac{\partial L}{\partial D_q^+} = -\xi_q - \pi_q = 0, \ q=1,2,\ldots,Q, \tag{61}$$

$$\left[L_q \sum_{i=1}^{m}\sum_{j=1}^{n}(c_{ij}+\vartheta\omega_{ij})^{(q)} - L_q^0 \sum_{i=1}^{m}\sum_{j=1}^{n} d_{ij}^{(q)} \right] x_{ij}^{(q)} + D_q^- - D_q^+ - \left[L_q^0 \rho^{(q)} - L_q \delta^{(q)} \right] = 0, \tag{62}$$

$$\sum_{i=1}^{m}\sum_{j=1}^{n} -d_{ij}^{(q)} x_{ij}^{(q)} + D_q^- - \rho^{(q)} \leq 0, \ q=1,2,\ldots,Q, \ \forall i,j \tag{63}$$

$$\sum_{j=1}^{n} x_{ij} - \left[\lambda a_i^L + (1-\lambda)a_i^U \right] \leq 0, \ i=1,2,\ldots,m, \tag{64}$$

$$\left[\lambda b_j^L + (1-\lambda)b_j^U \right] - \sum_{i=1}^{m} x_{ij} \leq 0, \ j=1,2,\ldots,n, \tag{65}$$

$$x_{ij} \geq 0, \ i=1,2,\ldots,m, \ j=1,2,\ldots,n, \tag{66}$$

$$D_{ij}^-, D_{ij}^+ \geq 0, \ q=1,2,\ldots,Q, \tag{67}$$

$$v_q \left[\sum_{i=1}^{m}\sum_{j=1}^{n} -d_{ij}^{(q)} x_{ij}^{(q)} + D_q^- - \rho^{(q)} \right] = 0, \ q=1,2,\ldots,Q, \ \forall i,j \tag{68}$$

$$\tau_i \left[\sum_{j=1}^{n} x_{ij} - \left(\lambda a_i^L + (1-\lambda)a_i^U\right) \right] = 0, \ i=1,2,\ldots,m, \tag{69}$$

$$\eta_j \left[-\sum_{i=1}^{m} x_{ij} + \left(\lambda b_j^L + (1-\lambda)b_j^U\right) \right] = 0, \ j=1,2,\ldots,n, \tag{70}$$

$$\varphi_{ij}[x_{ij}] = 0, \tag{71}$$

$$\psi_i\left[a_i^L\right] = 0, \tag{72}$$

$$\phi_j\left[b_j^L\right] = 0, \tag{73}$$

$$\varpi_i\left[a_i^U\right] = 0, \tag{74}$$

$$\epsilon_j\left[b_j^U\right] = 0, \tag{75}$$

$$\zeta_q\left[D_q^-\right] = 0, \tag{76}$$

$$\pi_q\left[D_q^+\right] = 0, \tag{77}$$

$$v, \tau, \eta, \varphi, \psi, \phi, \varpi, \epsilon, \zeta, \pi \geq 0, \text{ and } \vartheta, \xi \in R; \tag{78}$$

where the KKT conditions (55)–(78) are evaluated at x_{ij}^*. Solving the system of Equations (55)–(78), the SSFK $S_1\left(x_{ij}^*, \alpha, \beta\right)$ for parametric IF-MOFTP is obtained.

6.2. Algorithm for Determination of the SSFK S_1 (x_{ij}^*, a, b)

Following the above discussion, the algorithm for obtaining the SSFK $S_1\left(x_{ij}^*, \alpha, \beta\right)$ for parametric (α, β)-MOFTP van be described as follows (Algorithms 1 and 2):

Algorithm 1 Phase I: Obtain an (α, β)-Pareto Optimal Solution of the Problem

---

1: Set the value of α, and β.
2: Presume that $\vartheta = \vartheta^*$.
3: Calculate the sole maximum and minimum values of $Z_q(x, \vartheta^*)$, $q = 1, 2, \ldots, Q$.
4: Set the goals and the upper tolerance limits for $Z_q(x, \vartheta^*)$, $q = 1, 2, \ldots, Q$.
5: Formulate $\mu_{(Z_q(x,\vartheta^*))}$, $q = 1, 2, \ldots, Q$ as in Equation (18).
6: Evaluate the weights w_{ij}^- as defined in Equation (26).
7: Do the linearization procedures at $\vartheta = \vartheta^*$ for each parametric membership goal according to Equations (35)–(38).
8: Formulate and solve the FGP model (Equations (40)–(46)) to get (α, β)-Pareto optimal solution x_{ij}^*.

Algorithms 2 Phase II: Determination of the SSFK $S_1(x_{ij}^*, \alpha, \beta)$

1: Formulate the parametric FGP model (Equations (47)–(53)).
2: Obtain the Lagrangian function, for the final FGP model, as in Equation (54).
3: Apply the KKT optimality conditions to find the SSFK (Equations (55)–(78)).
4: Reduce and solve the system of Equations (55)–(78), to obtain $S_1(x_{ij}^*, \alpha, \beta)$ and stop.

7. Numerical Example

To demonstrate the proposed algorithm for finding the SSFK, consider the following parametric IF-MOFTP:

$$\text{Max} \begin{pmatrix} Z_1(x, \vartheta) = \frac{\vartheta x_{11} + (2+\vartheta)x_{12} + (3+2\vartheta)x_{21} + 6x_{22} + 4}{x_{11} + 3x_{12} + x_{21} + 2x_{22} + 2}, \\ Z_2(x, \vartheta) = \frac{2x_{11} + (3+\vartheta)x_{12} + (4+2\vartheta)x_{21} + (5+\vartheta)x_{22} + 6}{x_{11} + 2x_{12} + 3x_{21} + x_{22} + 4} \end{pmatrix},$$

Subject to:
Supply constraints:
$$x_{11} + x_{12} \leq \tilde{a}_1^I, x_{21} + x_{22} \leq \tilde{a}_2^I,$$

Demand constraints:
$$x_{11} + x_{21} \geq \tilde{b}_1^I, x_{12} + x_{22} \geq \tilde{b}_2^I,$$

where the membership functions $\mu_{\tilde{a}_1^I}(x), \mu_{\tilde{a}_2^I}(x), \mu_{\tilde{b}_1^I}(x), \mu_{\tilde{b}_2^I}(x)$ and the non-membership functions $\gamma_{\tilde{a}_1^I}(x), \gamma_{\tilde{a}_2^I}(x), \gamma_{\tilde{b}_1^I}(x), \gamma_{\tilde{b}_2^I}(x)$ of the supplies and demands are described as follows:

$$\mu_{\tilde{a}_1^I}(x) = \begin{cases} \frac{x-140}{20} & if\ 140 \leq x \leq 160, \\ \frac{180-x}{20} & if\ 160 \leq x \leq 180, \\ 0 & otherwise, \end{cases} \gamma_{\tilde{a}_1^I}(x) = \begin{cases} \frac{160-x}{30} & if\ 130 \leq x \leq 160, \\ \frac{x-160}{40} & if\ 160 \leq x \leq 200, \\ 1 & otherwise, \end{cases}$$

$$\mu_{\tilde{a}_2^l}(x) = \begin{cases} \frac{x-220}{20} & if\ 220 \leq x \leq 240, \\ \frac{250-x}{10} & if\ 240 \leq x \leq 250, \\ 0 & otherwise, \end{cases} \gamma_{\tilde{a}_2^l}(x) = \begin{cases} \frac{240-x}{20} & if\ 210 \leq x \leq 240, \\ \frac{x-240}{30} & if\ 240 \leq x \leq 270, \\ 1 & otherwise, \end{cases}$$

$$\mu_{\tilde{b}_1^l}(x) = \begin{cases} \frac{x-40}{10} & if\ 40 \leq x \leq 50, \\ \frac{60-x}{10} & if\ 50 \leq x \leq 60, \\ 0 & otherwise, \end{cases} \gamma_{\tilde{b}_1^l}(x) = \begin{cases} \frac{50-x}{20} & if\ 30 \leq x \leq 50, \\ \frac{x-50}{30} & if\ 50 \leq x \leq 80, \\ 1 & otherwise, \end{cases}$$

$$\mu_{\tilde{b}_2^l}(x) = \begin{cases} \frac{x-310}{10} & if\ 310 \leq x \leq 320, \\ \frac{350-x}{30} & if\ 320 \leq x \leq 350, \\ 0 & otherwise, \end{cases} \gamma_{\tilde{b}_2^l}(x) = \begin{cases} \frac{320-x}{20} & if\ 300 \leq x \leq 320, \\ \frac{x-320}{60} & if\ 320 \leq x \leq 380, \\ 1 & otherwise, \end{cases}$$

Phase I: Finding an (α, β)-Pareto optimal solution of the parametric IF-MOFTP.

For a desired values of $\alpha = 0.6$, and $\beta = 0.2$, then applying the concept of (α, β)-cut of the IFN we formulate the (α, β)-MOFTP at $\vartheta = \vartheta^* = 3$.

$$\text{Max} \begin{pmatrix} Z_1(x) = \frac{3x_{11}+ 5x_{12}+ 9x_{21}+ 6x_{22}+ 8}{x_{11}+ 3x_{12}+ x_{21}+ 2x_{22}+ 2}, \\ Z_2(x) = \frac{2x_{11}+ 6x_{12}+ 10x_{21}+ 8x_{22}+ 6}{x_{11}+ 2x_{12}+ 3x_{21}+ x_{22}+ 4} \end{pmatrix},$$

Subject to:
Supply constraints:

$$x_{11} + x_{12} \leq [154,\ 168], x_{21} + x_{22} \leq [234,\ 244].$$

Demand constraints:

$$x_{11} + x_{21} \geq [46,\ 54], x_{12} + x_{22} \geq [316,\ 332].$$

Based on the concept of convex linear combination on the constraints, then we obtain the MOFTP:

$$\text{Max} \begin{pmatrix} Z_1(x) = \frac{3x_{11}+ 5x_{12}+ 9x_{21}+ 6x_{22}+ 8}{x_{11}+ 3x_{12}+ x_{21}+2x_{22}+ 2}, \\ Z_2(x) = \frac{2x_{11}+ 6x_{12}+ 10x_{21}+ 8x_{22}+ 6}{x_{11}+ 2x_{12}+ 3x_{21}+ x_{22}+ 4} \end{pmatrix},$$

Subject to:

$$x_{11} + x_{12} \leq 165.2,\ x_{21} + x_{22} \leq 240,\ x_{11} + x_{21} \geq 51.6,\ x_{12} + x_{22} \geq 328.8.$$

A FGP approach is utilized to solve the MOFTP according to the model of Equations (40)–(46). Firstly, the coefficients of the linearized membership goals are obtained in Table 1.

Table 1. The coefficient of the linearized membership goals $\left(c^{ij}\right)^T$ and G_{ij}.

	$Z_1(x)$	$Z_2(x)$
$\left(c_{ij}^q\right)^T$	$\begin{pmatrix} 0.682 \\ -10.22 \\ 19.081 \\ 1.364 \end{pmatrix}^T$	$\begin{pmatrix} -2.8628 \\ -4.048 \\ -5.234 \\ 2.1688 \end{pmatrix}^T$
G_{ij}	-7.497	13.128

$$\text{Min}\ AF = 3.0665 D_1^- + 0.8386 D_2^-,$$

Subject to:

$$0.682 x_{11} - 10.22 x_{12} + 19.081 x_{21} + 1.364 x_{22} + D_1^- - D_1^+ = -7.497,$$

$$-2.8628 x_{11} - 4.048 x_{12} - 5.234 x_{21} + 2.169 x_{22} + D_2^- - D_2^+ = 13.128,$$

$$-x_{11} - 3x_{12} - x_{21} - 2x_{22} + D_1^- \leq 2,$$
$$-x_{11} - 2x_{12} - 3x_{21} - x_{22} + D_2^- \leq 4,$$
$$x_{11} + x_{12} \leq 165.2,$$
$$x_{21} + x_{22} \leq 240,$$
$$x_{11} + x_{21} \geq 51.6,$$
$$x_{12} + x_{22} \geq 328.8,$$
$$x_{11}, x_{12}, x_{21}, x_{22}, D_1^-, D_1^+, D_2^-, D_2^+ \geq 0.$$

Using Lingo programming, the (α, β)-Pareto optimal solution of the parametric IF-MOFTP is obtained at $(x_{11}^*, x_{12}^*, x_{21}^*, x_{22}^*, D_1^-, D_1^+, D_2^-, D_2^+) = (0, 165.88, 76.39, 163.61, 0, 0, 726.78, 0)$.

Phase II: determination of the SSFK $S_1(x^*, \alpha, \beta)$.

To determine the SSFK $S_1(x^*, a, b)$ of the parametric IF-MOFTP, the coefficients of the linearized membership goals in the parametric form are recalculated in Table 2.

Table 2. The coefficients of the linearized membership goals $\left[\left(c_{ij} + \vartheta\omega_{ij}\right)^{(q)}\right]^T$ and G_{ij}.

	$Z_1(x, \vartheta)$	$Z_2(x, \vartheta)$
$\left[\left(c_{ij} + \vartheta\omega_{ij}\right)^{(q)}\right]^T$	$\begin{pmatrix} -8.518 + 3.067\vartheta \\ -19.42 + 3.067\vartheta \\ 0.682 + 6.133\vartheta \\ 1.364 \end{pmatrix}^T$	$\begin{pmatrix} -2.863 \\ -6.564 + 0.839\vartheta \\ -10.266 + 1.677\vartheta \\ -0.347 + 0.839\vartheta \end{pmatrix}^T$
G_{ij}	-7.497	13.128

Therefore, the stability of parametric IF-MOFTP implies the stability of the parametric FGP model which is defined as:

$$\text{Min } AF = 3.067 D_1^- + 0.839 D_2^-,$$

Subject to:

$$(-8.518 + 3.067\vartheta)x_{11} + (-19.42 + 3.067\vartheta)x_{12} + (0.682 + 6.133\vartheta)x_{21} + 1.364 x_{22}$$
$$+ D_1^- - D_1^+ = -7.497,$$
$$-2.8628 x_{11} + (-6.564 + 0.839\vartheta)x_{12} + (-10.266 + 1.677\vartheta)x_{21}$$
$$+ (-0.347 + 0.839\vartheta)x_{22} + D_2^- - D_2^+ = 13.128,$$
$$-x_{11} - 3x_{12} - x_{21} - 2x_{22} + D_1^- \leq 2,$$
$$-x_{11} - 2x_{12} - 3x_{21} - x_{22} + D_2^- \leq 4,$$
$$x_{11} + x_{12} \leq 0.2 a_1^L + 0.8 a_1^U,$$
$$x_{21} + x_{22} \leq 0.4 a_2^L + 0.6 a_2^U,$$
$$x_{11} + x_{21} \geq 0.3 b_1^L + 0.7 b_1^U,$$
$$x_{12} + x_{22} \geq 0.2 b_2^L + 0.8 b_2^U,$$
$$x_{11}, x_{12}, x_{21}, x_{22}, a_1^L, a_1^U, a_2^L, a_2^U, b_1^L, b_1^U, b_2^L, b_2^U \geq 0,$$
$$D_1^-, D_1^+, D_2^-, D_2^+ \geq 0; \vartheta \in R$$

The Lagrangean function of the above parametric FGP model follows as:

$$\begin{aligned}
L &= 3.067D_1^- + 0.839D_2^- + \xi_1 \begin{bmatrix} (-8.518 + 3.067\vartheta)x_{11} + (-19.42 + 3.067\vartheta)x_{12} \\ +(0.682 + 6.133\vartheta)x_{21} + 1.364x_{22} + D_1^- - D_1^+ + 7.497 \end{bmatrix} \\
&+ \xi_2 \begin{bmatrix} -2.8628x_{11} + (-6.564 + 0.839\vartheta)x_{12} + (-10.266 + 1.677\vartheta)x_{21} \\ +(-0.347 + 0.839\vartheta)x_{22} + D_2^- - D_2^+ - 13.128 \end{bmatrix} \\
&+ \vartheta_1 \left[-x_{11} - 3x_{12} - x_{21} - 2x_{22} + D_1^- - 2 \right] + \vartheta_2 \left[-x_{11} - 2x_{12} - 3x_{21} - x_{22} + D_2^- - 4 \right] \\
&+ \tau_1 \left[x_{11} + x_{12} - 0.2a_1^L - 0.8a_1^U \right] + \tau_2 \left[x_{21} + x_{22} - 0.4a_2^L - 0.6a_2^U \right] \\
&+ \eta_1 \left[-x_{11} - x_{21} + 0.3b_1^L + 0.7b_1^U \right] + \eta_2 \left[-x_{12} - x_{22} + 0.2b_2^L + 0.8b_2^U \right] + \varphi_1[-x_{11}] \\
&+ \varphi_2[-x_{12}] + \varphi_3[-x_{21}] + \varphi_4[-x_{22}] + \psi_1[-a_1^L] + \psi_2[-a_2^L] + \phi_1[-b_1^L] + \phi_2[-b_2^L] \\
&+ \omega_1[-a_1^U] + \omega_2[-b_2^U] + \epsilon_1[-b_1^U] + \epsilon_2[-b_2^U] + \zeta_1[-D_1^-] + \zeta_2[-D_2^-] + \pi_1[-D_1^+] \\
&+ \pi_2[-D_2^+]
\end{aligned}$$

where $\vartheta, \xi_1, \xi_2 \in R$, and $v_1, v_2, \tau_1, \tau_2, \eta_1, \eta_2, \varphi_1, \varphi_2, \varphi_3, \varphi_4, \psi_1, \psi_2, \phi_1, \phi_2, \omega_1, \omega_2, \epsilon_1, \epsilon_2 \geq 0$, and $\zeta_1, \zeta_2, \pi_1, \pi_2 \geq 0$, are the Lagrange multipliers. Therefore, KKT optimality conditions follows as:

$$\frac{\partial L}{\partial x_{11}} = (-8.518 + 3.067\vartheta)\xi_1 - 2.863\xi_2 - v_1 - v_2 + \tau_1 - \eta_1 - \varphi_1 = 0$$

$$\frac{\partial L}{\partial x_{12}} = (-19.42 + 3.067\vartheta)\xi_1 + (-6.564 + 0.839\vartheta)\xi_2 - 3v_1 - 2v_2 + \tau_1 - \eta_2 - \varphi_2 = 0,$$

$$\frac{\partial L}{\partial x_{21}} = (0.682 + 6.133\vartheta)\xi_1 + (-10.266 + 1.677\vartheta)\xi_2 - v_1 - 3v_2 + \tau_2 - \eta_1 - \varphi_3 = 0,$$

$$\frac{\partial L}{\partial x_{22}} = 1.364\xi_1 + (-0.347 + 0.839\vartheta)\xi_2 - 2v_1 - v_2 + \tau_2 - \eta_2 - \varphi_4 = 0,$$

$$\frac{\partial L}{\partial a_1^L} = -0.2\tau_1 - \psi_1 = 0,$$

$$\frac{\partial L}{\partial a_1^U} = -0.8\tau_1 - \omega_1 = 0,$$

$$\frac{\partial L}{\partial a_2^L} = -0.4\tau_2 - \psi_2 = 0,$$

$$\frac{\partial L}{\partial a_2^U} = -0.6\tau_2 - \omega_2 = 0,$$

$$\frac{\partial L}{\partial b_1^L} = 0.3\eta_1 - \phi_1 = 0,$$

$$\frac{\partial L}{\partial b_1^U} = 0.7\eta_1 - \epsilon_1 = 0,$$

$$\frac{\partial L}{\partial b_2^L} = 0.2\eta_2 - \phi_2 = 0,$$

$$\frac{\partial L}{\partial b_2^U} = 0.8\eta_2 - \epsilon_2 = 0,$$

$$\frac{\partial L}{\partial D_1^-} = 3.067 + \xi_1 + v_1 - \zeta_1 = 0,$$

$$\frac{\partial L}{\partial D_1^+} = -\xi_1 - \pi_1 = 0,$$

$$\frac{\partial L}{\partial D_2^-} = 0.839 + \xi_2 + v_2 - \zeta_2 = 0,$$

$$\frac{\partial L}{\partial D_2^+} = -\zeta_2 - \pi_2 = 0,$$

$$v_1\left[-x_{11} - 3x_{12} - x_{21} - 2x_{22} + D_1^- - 2\right] = 0, \text{ i.e., } v_1 = 0,$$

$$v_2\left[-x_{11} - 2x_{12} - 3x_{21} - x_{22} + D_2^- - 4\right] = 0, \text{ i.e., } v_2 = 0,$$

$$\tau_1\left[x_{11} + x_{12} - 0.2a_1^L - 0.8a_1^U\right] = 0, \text{ i.e., } \tau_1 = 0,$$

$$\tau_2\left[x_{21} + x_{22} - 0.4a_2^L - 0.6a_2^U\right] = 0, \text{ i.e., } \tau_2 \geq 0,$$

$$\eta_1\left[-x_{11} - x_{21} + 0.3b_1^L + 0.7b_1^U\right] = 0, \text{ i.e., } \eta_1 = 0,$$

$$\eta_2\left[-x_{12} - x_{22} + 0.2b_2^L + 0.8b_2^U\right] = 0, \text{ i.e., } \eta_2 = 0,$$

$$\varphi_1[-x_{11}] = 0, \text{ i.e., } \varphi_1 \geq 0,$$

$$\varphi_2[-x_{12}] = 0, \text{ i.e., } \varphi_2 = 0,$$

$$\varphi_3[-x_{21}] = 0, \text{ i.e., } \varphi_3 = 0,$$

$$\varphi_4[-x_{22}] = 0, \text{ i.e., } \varphi_4 = 0,$$

$$\psi_1\left[-a_1^L\right] = 0, \text{ i.e., } \psi_1 = 0,$$

$$\psi_2\left[-a_2^L\right] = 0, \text{ i.e., } \psi_2 = 0,$$

$$\phi_1\left[-b_1^L\right] = 0, \text{ i.e., } \phi_1 = 0,$$

$$\phi_2\left[-b_2^L\right] = 0, \text{ i.e., } \phi_2 = 0,$$

$$\varpi_1\left[-a_1^U\right] = 0, \text{ i.e., } \varpi_1 = 0,$$

$$\varpi_2\left[-a_2^U\right] = 0, \text{ i.e., } \varpi_2 = 0,$$

$$\epsilon_1\left[-b_1^U\right] = 0, \text{ i.e., } \epsilon_1 = 0,$$

$$\epsilon_2\left[-b_2^U\right] = 0, \text{ i.e., } \epsilon_2 = 0,$$

$$\zeta_1[-D_1^-] = 0, \text{ i.e., } \zeta_1 \geq 0,$$

$$\zeta_2[-D_2^-] = 0, \text{ i.e., } \zeta_2 = 0,$$

$$\pi_1[-D_1^+] = 0, \text{ i.e., } \pi_1 \geq 0,$$

$$\pi_2[-D_2^+] = 0, \text{ i.e., } \pi_2 \geq 0,$$

$$-x_{11} - 3x_{12} - x_{21} - 2x_{22} + D_1^- \leq 2,$$

$$-x_{11} - 2x_{12} - 3x_{21} - x_{22} + D_2^- \leq 4,$$

$$x_{11} + x_{12} \leq 0.2a_1^L + 0.8a_1^U,$$

$$x_{21} + x_{22} \leq 0.4a_2^L + 0.6a_2^U,$$

$$x_{11} + x_{21} \geq 0.3b_1^L + 0.7b_1^U,$$

$$x_{12} + x_{22} \geq 0.2b_2^L + 0.8b_2^U,$$

$$x_{11}, x_{12}, x_{21}, x_{22}, a_1^L, a_1^U, a_2^L, a_2^U, b_1^L, b_1^U, b_2^L, b_2^U, D_1^-, D_1^+, D_2^-, D_2^+ \geq 0; \vartheta \in R$$

Solving the above system of Equation. we get: $v_1 = v_2 = \tau_1 = \tau_2 = \eta_1 = \eta_2 = \varphi_2 = \varphi_3 = \varphi_4 = \psi_1 = \psi_2 = \phi_1 = \phi_2 = \varpi_1 = \varpi_2 = \epsilon_1 = \epsilon_2 = \zeta_2 = 0$, and $\varphi_1, \zeta_1, \pi_1, \pi_2 \geq 0$.

Also, $\zeta_2 = -\pi_2 = -0.839$, $\zeta_1 = -\pi_1$. The above system of Equation is reduced to the following:

$$(-8.518 + 3.067\vartheta)\zeta_1 - 2.863\zeta_2 - \varphi_1 = 0,$$

$$(-19.42 + 3.067\vartheta)\zeta_1 + (-6.564 + 0.839\vartheta)\zeta_2 = 0,$$

$$(0.682 + 6.133\vartheta)\zeta_1 + (-10.266 + 1.677\vartheta)\zeta_2 = 0,$$

$$1.364\zeta_1 + (-0.347 + 0.839\vartheta)\zeta_2 = 0,$$

Therefore, the SSFK for the parametric IF-MOFTP is given by:

$$= \left\{ \begin{array}{l} \vartheta \in R, \\ \alpha, \beta \in [0,1] \end{array} \left| \begin{array}{l} S_1(0, 165.88, 76.39, 163.61, 0, 0, 726.78, 0) \\ 12.948\, \zeta_1 + [-1.41 + 6.133\zeta_1]\vartheta + 5.799 - \varphi_1 = 0, \\ \zeta_1 = \zeta_1 - 3.67;\ \zeta_1 = -\pi_1;\ \zeta_2 = -\pi_2 = -0.839, \\ \zeta_1, \varphi_1, \pi_1, \pi_2 \geq 0;\ \zeta_1, \zeta_2 \in R \end{array} \right. \right\}$$

After applying the KKT optimality conditions we obtain a large system of algebraic equations. By reducing and solving the algebraic system of equations the SSFK is obtained. The SSFK introduces the values and relations between different parameters which generate the same solution of the PIF-MOFTP as indicated by set S_1. To test the obtained results of the SSFK, different values of $\alpha, \beta \in [0,1]$ will be taken and the solution will remain the same.

8. Conclusions

The SSFK for the PIF-MOFTP was investigated in this study. Also, we characterized definitions of the set of feasible parameters and the solvability set for PIF-MOFTP. First, the concept of (α, β)-cut methodology was applied to get the parametric model. Moreover, the FGP approach was applied to find a (α, β)-Pareto optimal solution for PIF-MOFTP which has not been published in the literature to date. To obtain the SSFK for the novel model of PIF-MOFTP, the KKT necessary optimality conditions are applied. After applying the KKT optimality conditions, we obtained a large system of algebraic equations. By reducing and solving the algebraic system of equations, the SSFK was obtained. A detailed procedure that determines the SSFK for the PIF-MOFTP was exhibited. A numerical example was given to ensure the applicability and efficiency of the proposed PIF-MOFTP.

The major limitation of the proposed PIF-MOFTP is that a specific (α, β)-level is adopted in the proposed methods to represent the confidence level on DMs' subjective uncertainty to specify parameter values in the PIF-MOFTP. For simplification, the (α, β)-level for all parameters of the supply and demand in the solution process are assumed to be the same. However, these may be limitations in practical applications. The determination of (α, β)-levels for various DMs' subjective uncertainties could be different in the real world due to DMs' different consideration of the real transportation data. Thus, this will be addressed in future studies.

Several remaining areas of research in the topic of parametric MOFTP include the following:

1. The parametric study of multi-choice MOTP should be addressed.
2. Real-world PIF-MOFTP is a vital field in the future research.
3. Rough parametric MOFTP is a vital topic to be investigated.

Author Contributions: Conceptualization, M.A.E.S., M.A.E.-S. and F.A.F.; Methodology, M.A.E.S., F.A.F. and M.A.E.; Investigation, M.A.E.S., M.A.E.-S., A.F.F. and M.A.E.; writing—review and editing, M.A.E.S., M.A.E.-S., A.F.F., M.A.E. and F.A.F. All authors have read and agreed to the published version of the manuscript.

Funding: This research received no external funding.

Institutional Review Board Statement: Not applicable.

Informed Consent Statement: Not applicable.

Data Availability Statement: Not applicable.

Conflicts of Interest: The authors declare no conflict of interest.

References

1. Ammar, E.E.; Youness, E.A. Study on multi-objective transportation problem with fuzzy numbers. *Appl. Math. Comput.* **2005**, *166*, 241–253.
2. Bit, A.K.; Biswal, M.P.; Alam, S.S. Fuzzy programming approach to multi-criteria decision-making transportation problem. *Fuzzy Sets Syst.* **1992**, *50*, 135–141. [CrossRef]
3. Cetin, N.; Tiryaki, F. A Fuzzy Approach Using Generalized Dinkelbach's Algorithm for Multiobjective Linear Fractional Transportation Problem. *Math. Probl. Eng.* **2014**, *2014*, 702319. [CrossRef]
4. Chanas, S.; Kuchta, D. A concept of the optimal solution of the transportation problem with fuzzy cost coefficients. *Fuzzy Sets Syst.* **1996**, *82*, 299–305. [CrossRef]
5. Charnes, A.; Cooper, W.W. The steppingstone method for explaining linear programming calculation in transportation problem. *Manag. Sci.* **1954**, *1*, 49–69. [CrossRef]
6. Diaz, J.A. Finding a complete description of all efficient solutions to a multi-objective transportation problem. *Ekon.-Mat. Obz.* **1979**, *15*, 62–73.
7. El Sayed, M.; Abo-Sinna, M.A. A novel Approach for Fully Intuitionistic Fuzzy Multi-Objective Fractional Transportation Problem. *Alex. Eng. J.* **2020**, *60*, 1447–1463. [CrossRef]
8. Beaula, T.; Priyadharsini, M. A new algorithm for finding a fuzzy optimal solution for intuitionistic fuzzy transportation problems. *Int. J. Appl. Fuzzy Sets Artif. Intell.* **2015**, *5*, 183–192.
9. Arya, R.; Singh, P.; Kumari, S.; Obaidat, M.S. An approach for solving fully fuzzy multi-objective linear fractional optimization problems. *Soft Comput.* **2020**, *24*, 9105–9119. [CrossRef]
10. Ebrahimnejad, A.; Verdegay, J.L. A new approach for solving fully intuitionistic fuzzy transportation problems. *Fuzzy Optim. Decis. Mak.* **2017**, *17*, 447–474. [CrossRef]
11. Liou, J.J.; Hsu, C.-C.; Chen, Y.-S. Improving transportation service quality based on information fusion. *Transp. Res. Part A Policy Pract.* **2014**, *67*, 225–239. [CrossRef]
12. Ergun, O.; Kuyzu, G.; Savelsbergh, M. Reducing truckload transportation through collaboration. *Transp. Sci.* **2007**, *41*, 206–221. [CrossRef]
13. Sheu, J.B.; Chen, Y.J. Transportation and economics of scale in recycling low-value materials. *Transp. Res. Part B Methodol.* **2014**, *65*, 65–76. [CrossRef]
14. Roy, S.K.; Maity, G.; Weber, G.-W. Multi-objective two-stage grey transportation problem using utility function with goals. *Cent. Eur. J. Oper. Res.* **2016**, *25*, 417–439. [CrossRef]
15. Roy, S.K.; Maity, G.; Weber, G.W.; Gök, S.Z.A. Conic scalarization approach to solve multi-choice multi-objective transportation problem with interval goal. *Ann. Oper. Res.* **2017**, *253*, 599–620. [CrossRef]
16. Roy, S.K.; Mahapatra, D.R. Multi-objective interval valued transportation probabilistic problem involving lognormal. *Int. J. Math. Sci. Comput.* **2011**, *1*, 14–21.
17. Roy, S.K. Multi-choice stochastic transportation problem involving Weibull distribution. *Int. J. Oper. Res.* **2014**, *21*, 38. [CrossRef]
18. Maity, G.; Roy, S.K.; Verdegay, J.L. Multi-objective transportation problem with cost reliability under uncertain environment. *Int. J. Comput. Intell. Syst.* **2016**, *9*, 839–849. [CrossRef]
19. Maity, G.; Roy, S.K. Solving a multi-objective transportation problem with nonlinear cost and multi-choice demand. *Int. J. Manag. Sci. Eng. Manag.* **2014**, *11*, 62–70. [CrossRef]
20. Atanassov, K.T. Intuitionistic fuzzy sets. *Fuzzy Sets Syst.* **1986**, *20*, 87–96. [CrossRef]
21. Gong, Z.; Zhangc, N.; Chiclanad, F. The optimization ordering model for intuitionistic fuzzy preference relations with utility functions. *Knowl.-Based Syst.* **2018**, *162*, 174–184. [CrossRef]
22. Jana, B.; Roy, T.K. Multi-objective intuitionistic fuzzy linear programming and its application in transportation model. *Notes Intuit. Fuzzy Sets* **2007**, *13*, 34–51.
23. Verma, R.; Biswal, M.; Biswas, A. Fuzzy programming technique to solve multi-objective transportation problems with some non-linear membership functions. *Fuzzy Sets Syst.* **1997**, *91*, 37–43. [CrossRef]
24. Gourav, G.; Kumari, A. An efficient method for solving intuitionistic fuzzy transportation problem of type-2. *Int. J. Appl. Comput. Math.* **2017**, *3*, 3795–3804.
25. Gupta, A.; Kumar, A. A new method for solving linear multi-objective transportation problems with fuzzy parameters. *Appl. Math. Model.* **2012**, *36*, 1421–1430. [CrossRef]
26. Maity, G.; Roy, S.K. Solving multi-choice multi-objective transportation problem: A utility function approach. *J. Uncertain. Anal. Appl.* **2014**, *2*, 11. [CrossRef]
27. Mahajan, S.; Gupta, S.K. On fully intuitionistic fuzzy multiobjective transportation problems using different membership functions. *Ann. Oper. Res.* **2019**, *296*, 211–241. [CrossRef]

28. El Sayed, M.; Farahat, F. Study of Achievement Stability Set for Parametric Linear FGP Problems. *Ain Shams Eng. J.* **2020**, *11*, 1345–1353. [CrossRef]
29. Veeramani, C.; Edalatpanah, S.A.; Sharanya, S. Solving the Multiobjective Fractional Transportation Problem through the Neutrosophic Goal Programming Approach. *Discret. Dyn. Nat. Soc.* **2021**, *2021*, 7308042. [CrossRef]
30. Pramanik, S.; Banerjee, D. Multi-objective chance constrained capacitated transportation problem based on fuzzy goal programming. *Int. J. Comput. Appl.* **2012**, *44*, 42–46. [CrossRef]
31. Edalatpanah, S.A. A nonlinear approach for neutrosophic linear programming. *J. Appl. Res. Ind. Eng.* **2019**, *6*, 367–373. [CrossRef]
32. Rizk-Allah, R.M.; Hassanien, A.E.; Elhoseny, M. A multi-objective transportation model under neutrosophic environment. *Comput. Electr. Eng.* **2018**, *69*, 705–719. [CrossRef]
33. Veeramani, C.; Sumathi, M. Fuzzy Mathematical Programming approach for Solving Fuzzy Linear Fractional Programming Problem. *RAIRO-Oper. Res.* **2014**, *48*, 109–122. [CrossRef]
34. Emam, O.E. A parametric study on multi-objective integer quadratic programming problems under uncertainty. *Gen. Math. Notes* **2011**, *6*, 49–60.
35. Osman, M.S. Qualitative analysis of basic notions in parametric convex programming, (parameters in the objective function). *Apl. Mat.* **1977**, *22*, 333–348. [CrossRef]
36. Osman, M.S.; Emam, O.E.; El Sayed, M.A. On Parametric Multi-level Multi-objective Fractional Programming Problems with Fuzziness in the Constraints. *Br. J. Math. Comput. Sci.* **2016**, *18*, 1–19. [CrossRef] [PubMed]
37. Hsien-Chung, W. The Karush-Kuhn-Tucker optimality conditions for multi-objective programming problems with fuzzy-valued objective functions. *Fuzzy Optim. Decis. Mak.* **2009**, *8*, 1–28.
38. Saad, O.M. On stability of proper efficient solutions in multi-objective fractional programming problems under fuzziness. *Math. Comput. Model.* **2007**, *45*, 221–231. [CrossRef]
39. Saad, O.M.; Hughes, J.B. Bicriterion integer linear fractional programs with parameters in the objective functions. *J. Inf. Optim. Sci.* **1998**, *19*, 97–108. [CrossRef]
40. Saad, O.M.; Elshafei, M.M.; Sleem, M.M. On some stability notions for fuzzy three-level fractional programming problem. *Math. Sci. Lett.* **2021**, *10*, 23–34.
41. Mahapatra, G.S.; Roy, T.K. Intuitionistic Fuzzy Number and Its Arithmetic Operation with Application on System Failure. *J. Uncertain Syst.* **2013**, *7*, 92–107.
42. Mahmoodirad, A.; Allahviranloo, T.; Niroomand, S. A new effective solution method for fully intuitionistic fuzzy transportation problem. *Soft Comput.* **2019**, *23*, 4521–4530. [CrossRef]
43. Roy, S.K.; Ebrahimnejad, A.; Verdegay, J.L.; Das, S. New approach for solving intuitionistic fuzzy multi-objective transportation problem. *Sadhana* **2018**, *43*, 3. [CrossRef]
44. Mohamed, R.H. The relationship between goal programming and fuzzy programming. *Fuzzy Sets Syst.* **1997**, *89*, 215–222. [CrossRef]
45. Pal, B.B.; Moitra, B.N.; Maulik, U. A goal programming procedure for fuzzy multi-objective linear fractional programming problem. *Fuzzy Sets Syst.* **2003**, *139*, 395–405. [CrossRef]
46. Zangiabadi, M.; Maleki, H.R. Fuzzy goal programming technique to solve multi-objective transportation problems with some non-linear membership functions. *Iran. J. Fuzzy Syst.* **2013**, *10*, 61–74.
47. Osman, M.; Emam, O.; El Sayed, M.A. Interactive Approach for Multi-Level Multi-Objective Fractional Programming Problems with Fuzzy Parameters. *Beni-Suef Univ. J. Basic Appl. Sci.* **2018**, *7*, 139–149. [CrossRef]

fractal and fractional

Article

Hermite–Jensen–Mercer-Type Inequalities via Caputo–Fabrizio Fractional Integral for h-Convex Function

Miguel Vivas-Cortez [1,*,†], **Muhammad Shoaib Saleem** [2,†], **Sana Sajid** [2,†] **and Muhammad Sajid Zahoor** [2,†] **and Artion Kashuri** [3,†]

1. Faculty of Exact and Natural Sciences, School of Physical and Mathematical Sciences, Pontificia Universidad Católica del Ecuador, Av. 12 October 1076, Quito 17-01-2184, Ecuador
2. Department of Mathematics, University of Okara, Okara 56300, Pakistan; shoaib83455@gmail.com (M.S.S.); sanasajid1022@gmail.com (S.S.); sajidzahoor308@gmail.com (M.S.Z.)
3. Department of Mathematics, Faculty of Technical Science, University Ismail Qemali , 9401 Vlore, Albania; artion.kashuri@univlora.edu.al
* Correspondence: mjvivas@puce.edu.ec
† These authors contributed equally to this work.

Abstract: Integral inequalities involving many fractional integral operators are used to solve various fractional differential equations. In the present paper, we will generalize the Hermite–Jensen–Mercer-type inequalities for an h-convex function via a Caputo–Fabrizio fractional integral. We develop some novel Caputo–Fabrizio fractional integral inequalities. We also present Caputo–Fabrizio fractional integral identities for differentiable mapping, and these will be used to give estimates for some fractional Hermite–Jensen–Mercer-type inequalities. Some familiar results are recaptured as special cases of our results.

Keywords: convex function; h-convex function; Hermite–Hadamard inequality; Caputo–Fabrizio fractional integral; Hermite–Hadamard inequality; Jensen inequality; Jensen–Mercer inequality

1. Introduction

Fractional calculus has undergone rapid development in both applied and pure mathematics because of its enormous use in image processing, physics, machine learning, networking, and other branches. For more on fractional calculus identities, see [1–3]. The fractional derivative has received rapid attention among experts from different branches of science. Most of the applied problems can not be modeled by classical derivations. The complications in real-world problems are addressed by fractional differential equations. The famous fractional integral contains Riemann–Liouville [4–6], Hadamard [6,7], Caputo–Fabrizio [8], and Katugampola [6], etc.

In this paper, we will restrict ourselves to the Caputo–Fabrizio fractional integral operator. In the current direction of fractional calculus, numerous analysts are characterizing new operators by various methods to cover most of the real-world problems. Usually, the operators are not the same as each other in terms of singularity and locality of kernels. The main aspect that makes Caputo–Fabrizio different from others is that it has a non-singular kernel, and it is useful to find exact solutions for various issues.

For convex functions, the Hermite–Hadamard inequality is a famous inequality that has been proved in many ways and has several extensions and generalizations in the literature (see [9–19]). The Hermite–Hadamard inequality for the convex function is defined as:

Let $\xi : I \subseteq \mathbb{R} \to \mathbb{R}$ be a convex function. Then

$$\xi\left(\frac{v+\mu}{2}\right) \leq \frac{1}{\mu-v}\int_v^\mu \xi(\chi)d\chi \leq \frac{\xi(v)+\xi(\mu)}{2},$$

holds $\forall\, v, \mu \in I$ and $v < \mu$.

The generalization of the Hermite–Hadamard inequality for h-convex are defined as (see [20]):

Let $\xi : I \subseteq \mathbb{R} \to \mathbb{R}$ be a convex function. Then

$$\frac{1}{2h\left(\frac{1}{2}\right)} \xi\left(\frac{v+\mu}{2}\right) \leq \frac{1}{\mu-v} \int_v^\mu \xi(\chi) d\chi \leq [\xi(v) + \xi(\mu)] \int_0^1 h(\sigma) d\sigma,$$

holds $\forall\, v, \mu \in I$ and $v < \mu$.

In the literature, some more interesting extensions and refinements of the Hermite–Hadamard integral inequality with the help of h-convex functions have been widely studied (see [21–26]).

In the literature, for the Jensen inequality, several interesting studies are given. In [27], for a convex function, a variant of Jensen's inequality is proved by Mercer within the year 2003. Later, Matković et al. presented the Jensen–Mercer inequality for operators with applications in the year 2006 (see [28]).

Vivas-Cortez et al. presented the following variant of the Jensen–Mercer inequality (see [29]).

Theorem 1 ([29]). *Let ξ be a h-convex function defined on interval $[v, \mu]$. Then*

$$\xi\left(v + \mu - \sum_{i=1}^n \chi_i x_i\right) \leq M[\xi(v) + \xi(\mu)] - \sum_{i=1}^n h(\chi_i)\xi(x_i), \tag{1}$$

holds $\forall\, x_i \in [v, \mu]$ and $\chi_i \in [0, 1]$ with $\sum_{i=1}^n \chi_i = 1$, where $M = \sup\{h(\sigma) : \sigma \in (0, 1)\}$.

In 2019, the authors established the Hermite–Hadamard–Mercer-like inequalities for fractional integrals [30]. In [31], Butt et al. presented the Hermite–Jensen–Mercer type inequalities for conformable fractional integrals within the year 2020. Furthermore, they developed the Hermite–Jensen–Mercer-like inequalities for k-fractional integrals, generalized fractional integrals and ψ-Riemann–Liouville k-fractional integrals (see [32–34]). In 2020, several researchers presented Hermite–Jensen–Mercer-like inequalities in the setting of a k-Caputo fractional derivative and Caputo fractional derivative (see [35,36]). In [37], the authors developed the weighted Hermite–Hadamard–Mercer-type inequalities for convex functions within the year 2020. Chu et al. presented the new fractional estimates for Hermite–Hadamard–Mercer inequalities in the year 2020 (see [38]).

The present paper is organized as follows. First, we write definitions and preliminary material associated with our present paper. In Section 2, we will present Hermite–Jensen–Mercer-type inequalities for a Caputo–Fabrizio fractional integral operator with the help of an h-convex function. In Section 3, we will develop new Lemmas and then present some results for an h-convex function via a Caputo–Fabrizio fractional integral operator. In Section 4, some more integral inequalities for h-convex functions are established making use of the Hölder–İşcan integral inequality for an improved power mean integral inequality, and at last, we will write concluding remarks to our present paper.

Throughout the paper, we need the following assumption:

Let $\xi : I = [v, \mu] \to \mathbb{R}$ be a positive function, $0 \leq v < \mu$ and $\xi \in L_1[v, \mu]$. Furthermore, consider $h : (0, 1) \to \mathbb{R}$ is a non-negative function, $h \neq 0$ and $I \subseteq \mathbb{R}$ is an interval.

Now, we begin with definitions and preliminary results, which will be used in this work.

Definition 1. *(Convex function) [39] The function $\xi : [v, \mu] \to \mathbb{R}$ is called convex, if*

$$\xi(\chi x_1 + (1-\chi)x_2) \leq \chi \xi(x_1) + (1-\chi)\xi(x_2),$$

holds $\forall\, x_1, x_2 \in [v, \mu]$ and $\chi \in [0, 1]$.

Definition 2. *(h-Convex function)* [40] *A function* $\xi : [v, \mu] \subseteq \mathbb{R} \to \mathbb{R}$ *is said to be h-convex if*

$$\xi(\chi x_1 + (1 - \chi)x_2) \leq h(\chi)\xi(x_1) + h(1 - \chi)\xi(x_2),$$

holds $\forall\ x_1, x_2 \in [v, \mu]$ *and* $\chi \in [0, 1]$.

Definition 3. *(Superadditive function) A function* $h : [v, \mu] \subseteq \mathbb{R} \to \mathbb{R}$ *is called superadditive function if*

$$h(x_1 + x_2) \geq h(x_1) + h(x_2),$$

holds $\forall\ x_1, x_2 \in [v, \mu]$.

Definition 4 ([8,41,42]). *Let* $\xi \in H^1(x_1, x_2), x_1 < x_2, \theta \in [0, 1]$, *then the definition of the left fractional derivative in the sense of Caputo and Fabrizio is defined as*

$$\left({}^{CFC}_{x_1} D^\theta \xi \right)(t) = \frac{B(\theta)}{1 - \theta} \int_{x_1}^t \xi'(z) e^{\frac{-\theta(t-z)^\theta}{1-\theta}} dz,$$

and the associated fractional integral is

$$\left({}^{CF}_{x_1} I^\theta \xi \right)(t) = \frac{1 - \theta}{B(\theta)} \xi(t) + \frac{\theta}{B(\theta)} \int_{x_1}^t \xi(z) dz,$$

where $B(\theta) > 0$ *is a normalization function satisfying* $B(0) = B(1) = 1$.

The right fractional derivative is defined as

$$\left({}^{CFC} D^\theta_{x_2} \xi \right)(t) = \frac{-B(\theta)}{1 - \theta} \int_t^{x_2} \xi'(z) e^{\frac{-\theta(z-t)^\theta}{1-\theta}} dz,$$

and the associated fractional integral is

$$\left({}^{CF} I^\theta_{I_2} \xi \right)(t) = \frac{1 - \theta}{B(\theta)} \xi(t) + \frac{\theta}{B(\theta)} \int_t^{l_2} \xi(z) dz.$$

In [43,44], the Hölder-İşcan integral inequality and improved power-mean integral inequality is explained as follows.

Theorem 2. *(Hölder–İşcan integral inequality)* [43] *Let* ξ_1 *and* ξ_2 *be real functions defined on* $[x_1, x_2]$ *and if* $|\xi_1|^q$ *and* $|\xi_2|^q$ *are integrable on* $[x_1, x_2]$. *If* $p > 1$ *and* $\frac{1}{p} + \frac{1}{q} = 1$, *then*

$$\int_{x_1}^{x_2} |\xi_1(z)\xi_2(z)| dz \leq \frac{1}{x_2 - x_1} \left\{ \left(\int_{x_1}^{x_2} (x_2 - z)|\xi_1(z)|^p dz \right)^{\frac{1}{p}} \left(\int_{x_1}^{x_2} (x_2 - z)|\xi_2(z)|^q dz \right)^{\frac{1}{q}} \right.$$

$$\left. + \left(\int_{x_1}^{x_2} (z - x_1)|\xi_1(z)|^p dz \right)^{\frac{1}{p}} \left(\int_{x_1}^{x_2} (z - x_1)|\xi_2(z)|^q dz \right)^{\frac{1}{q}} \right\}$$

$$\leq \left(\int_{x_1}^{x_2} |\xi_1(z)|^p dz \right)^{\frac{1}{p}} \left(\int_{x_1}^{x_2} |\xi_2(z)|^q dz \right)^{\frac{1}{q}}.$$

Theorem 3. *(Improved power-mean integral inequality)* [44] *Let* ξ_1 *and* ξ_2 *be real functions defined on* $[x_1, x_2]$ *and if* $|\xi_1|, |\xi_1||\xi_2|^q$ *are integrable functions on* $[x_1, x_2]$. *Let* $q \geq 1$, *then*

$$\int_{x_1}^{x_2} |\xi_1(z)\xi_2(z)| dz \leq \frac{1}{x_2 - x_1} \left\{ \left(\int_{x_1}^{x_2} (x_2 - z)|\xi_1(z)| dz \right)^{1 - \frac{1}{q}} \left(\int_{x_1}^{x_2} (x_2 - z)|\xi_1(z)||\xi_2(z)|^q dz \right)^{\frac{1}{q}} \right.$$

$$\left. + \left(\int_{x_1}^{x_2} (z - x_1)|\xi_1(z)| dz \right)^{1 - \frac{1}{q}} \left(\int_{x_1}^{x_2} (z - x_1)|\xi_1(z)||\xi_2(z)|^q dz \right)^{\frac{1}{q}} \right\}$$

$$\leq \left(\int_{x_1}^{x_2} |\xi_1(z)| dz \right)^{1 - \frac{1}{q}} \left(\int_{x_1}^{x_2} |\xi_1(z)||\xi_2(z)|^q dz \right)^{\frac{1}{q}}.$$

2. Hermite–Jensen–Mercer-Type Inequalities via the Caputo–Fabrizio Fractional Operator

Theorem 4. *Let $\xi : I = [v, \mu] \to \mathbb{R}$ be a h-convex function and $\xi \in L_1[v, \mu]$. If h is a super-additive function and $\theta \in [0, 1]$, then*

$$\frac{1}{2h\left(\frac{1}{2}\right)}\xi\left(v + \mu - \frac{x_1 + x_2}{2}\right) \leq \frac{B(\theta)}{\theta(x_2 - x_1)}$$

$$\times \left[\left(^{CF}_{v+\mu-x_2}I^{\theta}\xi\right)(t) + \left(^{CF}I^{\theta}_{v+\mu-x_1}\xi\right)(t) - \frac{2(1-\theta)}{B(\theta)}\xi(t)\right]$$

$$\leq \int_0^1 h(1)d\chi\left(M[\xi(v) + \xi(\mu)] - \frac{\xi(x_1) + \xi(x_2)}{2}\right), \quad (2)$$

holds for all $x_1, x_2 \in [v, \mu]$, $t \in [v, \mu]$, $B(\theta) > 0$ is a normalization function and $M = \sup\{h(\chi) : \chi \in (0, 1)\}$.

Proof. Since ξ is h-convex function on $[x_1, x_2]$ yields that

$$\xi\left(v + \mu - \frac{x_1 + x_2}{2}\right) = \xi\left(\frac{v + \mu - x_1 + v + \mu - x_2}{2}\right)$$

$$\leq h\left(\frac{1}{2}\right)\left(\xi(v + \mu - x_1) + \xi(v + \mu - x_2)\right)$$

$$= h\left(\frac{1}{2}\right)\Big(\xi(v + \mu - (\chi x_1 + (1 - \chi)x_2))$$

$$+ \xi(v + \mu - ((1 - \chi)x_1 + \chi x_2))\Big),$$

holds for all $x_1, x_2 \in [v, \mu]$.

The above inequality is integrated with respect to χ over $[0, 1]$ and by change of variable technique, we can deduce

$$\frac{1}{h\left(\frac{1}{2}\right)}\xi\left(v + \mu - \frac{x_1 + x_2}{2}\right) \leq \frac{2}{x_2 - x_1}\int_{v+\mu-x_2}^{v+\mu-x_1}\xi(z)dz$$

$$= \frac{2}{x_2 - x_1}\left(\int_{v+\mu-x_2}^{\chi}\xi(z)dz + \int_{\chi}^{v+\mu-x_1}\xi(z)dz\right). \quad (3)$$

Both sides of (3) multiplied by $\frac{\theta(x_2-x_1)}{2B(\theta)}$ and adding $\frac{2(1-\theta)}{B(\theta)}\xi(t)$, we have

$$\frac{2(1-\theta)}{B(\theta)}\xi(t) + \frac{\theta(x_2 - x_1)}{2h\left(\frac{1}{2}\right)B(\theta)}\xi\left(v + \mu - \frac{x_1 + x_2}{2}\right)$$

$$\leq \frac{2(1-\theta)}{B(\theta)}\xi(t) + \frac{\theta}{B(\theta)}\left(\int_{v+\mu-x_2}^{t}\xi(z)dz + \int_{t}^{v+\mu-x_1}\xi(z)dz\right)$$

$$= \left(\frac{(1-\theta)}{B(\theta)}\xi(t) + \frac{\theta}{B(\theta)}\int_{v+\mu-x_2}^{t}\xi(z)dz\right) + \left(\frac{(1-\theta)}{B(\theta)}\xi(t) + \int_{t}^{v+\mu-x_1}\xi(z)dz\right)$$

$$= \left(^{CF}_{v+\mu-x_2}I^{\theta}\xi\right)(t) + \left(^{CF}I^{\theta}_{v+\mu-x_1}\xi\right)(t). \quad (4)$$

Suitable rearrangement of (4) yields the first inequality of (2).

By using h-convexity of ξ, we have

$$\xi(\chi(v + \mu - x_1) + (1 - \chi)(v + \mu - x_2)) \leq h(\chi)\xi(v + \mu - x_1) + h(1 - \chi)\xi(v + \mu - x_2),$$

and

$$\xi((1 - \chi)(v + \mu - x_1) + \chi(v + \mu - x_2)) \leq h(1 - \chi)\xi(v + \mu - x_1) + h(\chi)\xi(v + \mu - x_2).$$

Adding the above two inequalities and then by using the super additivity of function and Jensen–Mercer inequality yields that

$$\xi(\chi(v+\mu-x_1)+(1-\chi)(v+\mu-x_2))+\xi((1-\chi)(v+\mu-x_1)+\chi(v+\mu-x_2))$$
$$\leq h(1)\Big(\xi(v+\mu-x_1)+\xi(v+\mu-x_2)\Big)$$
$$\leq h(1)\Big(2M[\xi(v)+\xi(\mu)]-(\xi(x_1)+\xi(x_2))\Big). \tag{5}$$

Integrating the inequality (5) with respect to χ over $[0,1]$ and by the change of variable technique, we can write

$$\frac{2}{x_2-x_1}\int_{v+\mu-x_2}^{v+\mu-x_1}\xi(z)dz \leq \int_0^1 h(1)d\chi\Big(2M[\xi(v)+\xi(\mu)]-(\xi(x_1)+\xi(x_2))\Big). \tag{6}$$

By making use of the same operations with (3) in (6), we have

$$\left({}^{CF}_{v+\mu-x_2}I^\theta\xi\right)(t)+\left({}^{CF}I^\theta_{v+\mu-x_1}\xi\right)(t)$$
$$\leq \frac{2(1-\theta)}{B(\theta)}\xi(t)+\frac{\theta(x_2-x_1)}{2B(\theta)}\left[\int_0^1 h(1)d\chi\Big(2M[\xi(v)+\xi(\mu)]-(\xi(x_1)+\xi(x_2))\Big)\right]. \tag{7}$$

By suitable rearrangement of (7), we obtain inequality (2). □

Remark 1. By putting $h(\chi)=\chi$, $M=\sup\{h(\chi):\chi\in(0,1)\}=1$, $x_1=v$ and $x_2=\mu$ in Theorem 2, then we obtain Theorem 2 of (see [45]).

Theorem 5. Assume that $\xi:I=[v,\mu]\to\mathbb{R}$ is a h-convex function and $\xi\in L_1[v,\mu]$. If $\theta\in[0,1]$, then

$$\frac{1}{h\left(\frac{1}{2}\right)}\xi\left(v+\mu-\frac{x_1+x_2}{2}\right)\int_0^1 h(\chi)d\chi$$
$$\leq \frac{1}{h\left(\frac{1}{2}\right)}M[\xi(v)+\xi(\mu)]\int_0^1 h(\chi)d\chi$$
$$-\frac{B(\theta)}{\theta(x_2-x_1)}\left[\left({}^{CF}_{x_1}I^\theta\xi\right)(t)+\left({}^{CF}I^\theta_{x_2}\xi\right)(t)-\frac{2(1-\theta)}{B(\theta)}\xi(t)\right]$$
$$\leq \frac{1}{h\left(\frac{1}{2}\right)}M[\xi(v)+\xi(\mu)]\int_0^1 h(\chi)d\chi-\frac{1}{2h\left(\frac{1}{2}\right)}\xi\left(\frac{x_1+x_2}{2}\right), \tag{8}$$

holds $\forall\, x_1,x_2\in[v,\mu]$, $t\in[v,\mu]$, $B(\theta)>0$ is a normalization function and $M=\sup\{h(\chi):\chi\in(0,1)\}$.

Proof. By the Jensen–Mercer inequality, we have

$$\xi\left(v+\mu-\frac{x_1+x_2}{2}\right)\leq M[\xi(v)+\xi(\mu)]-h\left(\frac{1}{2}\right)[\xi(x_1)+\xi(x_2)].$$

Both sides of the above inequality are multiplied by $h(\chi)$ and integrated with respect to χ over $[0,1]$, and we obtain

$$\xi\left(v+\mu-\frac{x_1+x_2}{2}\right)\int_0^1 h(\chi)d\chi$$
$$\leq M[\xi(v)+\xi(\mu)]\int_0^1 h(\chi)d\chi-h\left(\frac{1}{2}\right)[\xi(x_1)+\xi(x_2)]\int_0^1 h(\chi)d\chi,$$

which implies that

$$\frac{1}{h\left(\frac{1}{2}\right)}\xi\left(v+\mu-\frac{x_1+x_2}{2}\right)\int_0^1 h(\chi)d\chi$$
$$\leq \frac{1}{h\left(\frac{1}{2}\right)}M[\xi(v)+\xi(\mu)]\int_0^1 h(\chi)d\chi - [\xi(x_1)+\xi(x_2)]\int_0^1 h(\chi)d\chi.$$

Now, we will use the right-hand side of the Hermite–Hadamard inequality for the h-convex function, and we obtain

$$\frac{1}{h\left(\frac{1}{2}\right)}\xi\left(v+\mu-\frac{x_1+x_2}{2}\right)\int_0^1 h(\chi)d\chi$$
$$\leq \frac{1}{h\left(\frac{1}{2}\right)}M[\xi(v)+\xi(\mu)]\int_0^1 h(\chi)d\chi - \frac{1}{x_2-x_1}\int_{x_1}^{x_2}\xi(z)dz$$
$$= \frac{1}{h\left(\frac{1}{2}\right)}M[\xi(v)+\xi(\mu)]\int_0^1 h(\chi)d\chi - \frac{1}{x_2-x_1}\left(\int_{x_1}^t \xi(z)dz + \int_t^{x_2}\xi(z)dz\right). \quad (9)$$

Both sides of (9) multiplying by $\frac{\theta(x_2-x_1)}{B(\theta)}$ and subtracting $\frac{2(1-\theta)}{B(\theta)}\xi(t)$, we have

$$\frac{\theta(x_2-x_1)}{B(\theta)h\left(\frac{1}{2}\right)}\xi\left(v+\mu-\frac{x_1+x_2}{2}\right)\int_0^1 h(\chi)d\chi - \frac{2(1-\theta)}{B(\theta)}\xi(t)$$
$$\leq \frac{\theta(x_2-x_1)}{B(\theta)h\left(\frac{1}{2}\right)}M[\xi(v)+\xi(\mu)]\int_0^1 h(\chi)d\chi$$

$$-\frac{\theta}{B(\theta)}\left(\int_{x_1}^t \xi(z)dz + \int_t^{x_2}\xi(z)dz\right) - \frac{2(1-\theta)}{B(\theta)}\xi(t)$$
$$= \frac{\theta(x_2-x_1)}{B(\theta)h\left(\frac{1}{2}\right)}M[\xi(v)+\xi(\mu)]\int_0^1 h(\chi)d\chi - \left[\left(\frac{\theta}{B(\theta)}\int_{x_1}^t \xi(z)dz + \frac{(1-\theta)}{B(\theta)}\xi(t)\right)\right.$$
$$\left.+\left(\frac{\theta}{B(\theta)}\int_t^{x_2}\xi(z)dz + \frac{(1-\theta)}{B(\theta)}\xi(t)\right)\right]$$
$$= \frac{\theta(x_2-x_1)}{h\left(\frac{1}{2}\right)B(\theta)}M[\xi(v)+\xi(\mu)]\int_0^1 h(\chi)d\chi - \left[\left(_{x_1}^{CF}I^\theta\xi\right)(t)+\left(^{CF}I_{x_2}^\theta\xi\right)(t)\right]. \quad (10)$$

After suitable rearrangement, (10) yields the first inequality of (8).

For the second part of the inequality of (8), we will use the right-hand side of the Hermite–Hadamard integral inequality for the h-convex function, and we can write

$$-\frac{1}{x_2-x_1}\int_{x_1}^{x_2}\xi(z)dz \leq -\frac{1}{2h\left(\frac{1}{2}\right)}\xi\left(\frac{x_1+x_2}{2}\right). \quad (11)$$

By using the same operations with (9) in (11), we have

$$-\frac{B(\theta)}{\theta(x_2-x_1)}\left[\left(_{x_1}^{CF}I^\theta\xi\right)(t)+\left(^{CF}I_{x_2}^\theta\xi\right)(t)-\frac{2(1-\theta)}{B(\theta)}\xi(t)\right] \leq -\frac{1}{2h\left(\frac{1}{2}\right)}\xi\left(\frac{x_1+x_2}{2}\right). \quad (12)$$

Adding $\frac{1}{h(\frac{1}{2})}M[\xi(v)+\xi(\mu)]\int_0^1 h(\chi)d\chi$ to both sides of (12), we have

$$\frac{1}{h(\frac{1}{2})}M[\xi(v)+\xi(\mu)]\int_0^1 h(\chi)d\chi$$
$$-\frac{B(\theta)}{\theta(x_2-x_1)}\left[\left({}^{CF}_{x_1}I^\theta\xi\right)(t)+\left({}^{CF}I^\theta_{x_2}\xi\right)(t)-\frac{2(1-\theta)}{B(\theta)}\xi(t)\right]$$
$$\leq \frac{1}{h(\frac{1}{2})}M[\xi(v)+\xi(\mu)]\int_0^1 h(\chi)d\chi - \frac{1}{2h(\frac{1}{2})}\xi\left(\frac{x_1+x_2}{2}\right),$$

which completes the proof. □

Theorem 6. *Let $\xi_1,\xi_2 : I \subseteq \mathbb{R} \to \mathbb{R}$ be an h-convex function on I. If $\xi_1\xi_2 \in L[v,\mu]$, then*

$$\frac{2B(\theta)}{\theta(x_2-x_1)}\left[\left({}^{CF}_{v+\mu-x_2}I^\theta\xi_1\xi_2\right)(t)+\left({}^{CF}I^\theta_{v+\mu-x_1}\xi_1\xi_2\right)(t)-\frac{2(1-\theta)}{B(\theta)}\xi_1(t)\xi_2(t)\right]$$
$$\leq 2M^2 B_1(v,\mu) - 2MB_2(v,\mu,x_1)\int_0^1 h(1-\chi)d\chi$$
$$-2MB_3(v,\mu,x_2)]\int_0^1 h(\chi)d\chi + 2B_4(x_1,x_2)\int_0^1 h(\chi)h(1-\chi)d\chi$$
$$+2K_1(x_1)\int_0^1 (h(1-\chi))^2 d\chi + 2K_2(x_2)\int_0^1 (h(\chi))^2 d\chi, \tag{13}$$

where

$B_1(v,\mu) = \xi_1(v)\xi_2(v) + \xi_1(v)\xi_2(\mu) + \xi_1(\mu)\xi_2(v) + \xi_1(\mu)\xi_2(\mu),$
$B_2(v,\mu,x_1) = \xi_1(v)\xi_2(x_1) + \xi_1(\mu)\xi_2(x_1) + \xi_1(x_1)\xi_2(v) + \xi_1(x_1)\xi_2(\mu),$
$B_3(v,\mu,x_2) = \xi_1(v)\xi_2(x_2) + \xi_1(\mu)\xi_2(x_2) + \xi_1(x_2)\xi_2(v) + \xi_1(x_2)\xi_2(\mu),$
$B_4(x_1,x_2) = \xi_1(x_1)\xi_2(x_2) + \xi_1(x_2)\xi_2(x_1),$
$K_1(x_1) = \xi_1(x_1)\xi_2(x_1),$

and

$$K_2(x_2) = \xi_1(x_2)\xi_2(x_2),$$

holds $\forall\ x_1, x_2 \in [v,\mu]$, $M = \sup\{h(\chi) : \chi \in (0,1)\}$, $t \in [v,\mu]$ and $B(\theta) > 0$ is a normalization function.

Proof. Since ξ_1 and ξ_2 are h-convex functions on $[x_1,x_2]$ and making use of the Jensen–Mercer inequality, we have

$\xi_1(v+\mu-((1-\chi)x_1+\chi x_2))$
$\quad \leq M[\xi_1(v)+\xi_1(\mu)]-(h(1-\chi)\xi_1(x_1)+h(\chi)\xi_1(x_2)),\ \forall \chi \in [0,1], x_1, x_2 \in I,$

and

$\xi_2(v+\mu-((1-\chi)x_1+\chi x_2))$
$\quad \leq M[\xi_2(v)+\xi_2(\mu)]-(h(1-\chi)\xi_2(x_1)+h(\chi)\xi_2(x_2)),\ \forall \chi \in [0,1], x_1, x_2 \in I.$

Multiplying both sides of the above inequalities, we can write

$$\xi_1(v+\mu-((1-\chi)x_1+\chi x_2))\xi_2(v+\mu-((1-\chi)x_1+\chi x_2))$$
$$\leq M^2[\xi_1(v)\xi_2(v)+\xi_1(v)\xi_2(\mu)+\xi_1(\mu)\xi_2(v)+\xi_1(\mu)\xi_2(\mu)]$$
$$-Mh(1-\chi)[\xi_1(v)\xi_2(x_1)+\xi_1(\mu)\xi_2(x_1)+\xi_1(x_1)\xi_2(v)+\xi_1(x_1)\xi_2(\mu)]$$
$$-Mh(\chi)[\xi_1(v)\xi_2(x_2)+\xi_1(\mu)\xi_2(x_2)+\xi_1(x_2)\xi_2(v)+\xi_1(x_2)\xi_2(\mu)]$$
$$+h(\chi)h(1-\chi)[\xi_1(x_1)\xi_2(x_2)+\xi_1(x_2)\xi_2(x_1)]+(h(1-\chi))^2[\xi_1(x_1)\xi_2(x_1)]$$
$$+(h(\chi))^2[\xi_1(x_2)\xi_2(x_2)].$$

Integrating the above inequality with respect to χ over [0,1] and then by the change of variable technique, we obtain

$$\frac{1}{x_2-x_1}\int_{v+\mu-x_2}^{v+\mu-x_1}\xi_1(z)\xi_2(z)dz$$
$$\leq M^2[\xi_1(v)\xi_2(v)+\xi_1(v)\xi_2(\mu)+\xi_1(\mu)\xi_2(v)+\xi_1(\mu)\xi_2(\mu)]$$
$$-M[\xi_1(v)\xi_2(x_1)+\xi_1(\mu)\xi_2(x_1)+\xi_1(x_1)\xi_2(v)+\xi_1(x_1)\xi_2(\mu)]\int_0^1 h(1-\chi)d\chi$$
$$-M[\xi_1(v)\xi_2(x_2)+\xi_1(\mu)\xi_2(x_2)+\xi_1(x_2)\xi_2(v)+\xi_1(x_2)\xi_2(\mu)]\int_0^1 h(\chi)d\chi$$
$$+[\xi_1(x_1)\xi_2(x_2)+\xi_1(x_2)\xi_2(x_1)]\int_0^1 h(\chi)h(1-\chi)d\chi$$
$$+[\xi_1(x_1)\xi_2(x_1)]\int_0^1 (h(1-\chi))^2 d\chi+[\xi_1(x_2)\xi_2(x_2)]\int_0^1 (h(\chi))^2 d\chi,$$

which implies

$$\frac{2}{x_2-x_1}\left[\int_{v+\mu-x_2}^{\chi}\xi_1(z)\xi_2(z)dz+\int_{\chi}^{v+\mu-x_1}\xi_1(z)\xi_2(z)dz\right]$$
$$\leq 2M^2 B_1(v,\mu)-2MB_2(v,\mu,x_1)\int_0^1 h(1-\chi)d\chi$$
$$-2MB_3(v,\mu,x_2)]\int_0^1 h(\chi)d\chi+2B_4(x_1,x_2)\int_0^1 h(\chi)h(1-\chi)d\chi$$
$$+2K_1(x_1)\int_0^1 (h(1-\chi))^2 d\chi+2K_2(x_2)\int_0^1 (h(\chi))^2 d\chi.$$

The above inequality is multipled by $\frac{\theta(x_2-x_1)}{2B(\theta)}$, and adding $\frac{2(1-\theta)}{B(\theta)}\xi_1(t)\xi_2(t)$, we have

$$\frac{\theta}{B(\theta)}\left[\int_{v+\mu-x_2}^{t}\xi_1(z)\xi_2(z)dz+\int_{t}^{v+\mu-x_1}\xi_1(z)\xi_2(z)dz\right]+\frac{2(1-\theta)}{B(\theta)}\xi_1(t)\xi_2(t)$$
$$\leq \frac{\theta(x_2-x_1)}{2B(\theta)}\left[2M^2 B_1(v,\mu)-2MB_2(v,\mu,x_1)\int_0^1 h(1-\chi)d\chi\right.$$
$$-2MB_3(v,\mu,x_2)]\int_0^1 h(\chi)d\chi+2B_4(x_1,x_2)\int_0^1 h(\chi)h(1-\chi)d\chi$$
$$\left.+2K_1(x_1)\int_0^1 (h(1-\chi))^2 d\chi+2K_2(x_2)\int_0^1 (h(\chi))^2 d\chi\right]+\frac{2(1-\theta)}{B(\theta)}\xi_1(t)\xi_2(t).$$

Therefore,

$$\left[\frac{(1-\theta)}{B(\theta)}\xi_1(t)\xi_2(t) + \frac{\theta}{B(\theta)}\int_{v+\mu-x_2}^{t}\xi_1(w)\xi_2(w)dw\right] + \left[\frac{(1-\theta)}{B(\theta)}\xi_1(t)\xi_2(t)\right.$$
$$\left. + \frac{\theta}{B(\theta)}\int_{t}^{v+\mu-x_1}\xi_1(w)\xi_2(w)dw\right]$$
$$\leq \frac{\theta(x_2-x_1)}{2B(\theta)}\left[2M^2B_1(v,\mu) - 2MB_2(v,\mu,x_1)\int_0^1 h(1-\chi)d\chi\right.$$
$$- 2MB_3(v,\mu,x_2)]\int_0^1 h(\chi)d\chi + 2B_4(x_1,x_2)\int_0^1 h(\chi)h(1-\chi)d\chi$$
$$\left. + 2K_1(x_1)\int_0^1 (h(1-\chi))^2 d\chi + 2K_2(x_2)\int_0^1 (h(\chi))^2 d\chi\right] + \frac{2(1-\theta)}{B(\theta)}\xi_1(t)\xi_2(t).$$

Thus,

$$\left[\left(_{v+\mu-x_2}^{CF}I^\theta \xi_1\xi_2\right)(t) + \left(^{CF}I_{v+\mu-x_1}^{\theta}\xi_1\xi_2\right)(t)\right]$$
$$\leq \frac{\theta(x_2-x_1)}{2B(\theta)}\left[2M^2B_1(v,\mu) - 2MB_2(v,\mu,x_1)\int_0^1 h(1-\chi)d\chi\right.$$
$$- 2MB_3(v,\mu,x_2)]\int_0^1 h(\chi)d\chi + 2B_4(x_1,x_2)\int_0^1 h(\chi(1-\chi))$$
$$\left. + 2K_1(x_1)\int_0^1 h((1-\chi)^2)d\chi + 2K_2(x_2)\int_0^1 h(\chi^2)d\chi\right] + \frac{2(1-\theta)}{B(\theta)}\xi_1(t)\xi_2(t). \quad (14)$$

By suitable rearrangement, (14) yields required inequality (13). □

Remark 2. *By putting $h(\chi) = \chi$, $M = \sup\{h(\chi) : \chi \in (0,1)\} = 1$, $x_1 = v$ and $x_2 = \mu$ in Theorem 2, then we obtain Theorem 3 of [45].*

3. Some Novel Results Related to the Caputo–Fabrizio Fractional Operator

In this section, we will present some new Lemmas, and then we develop some novel results for an h-convex function with the help of the Caputo–Fabrizio fractional integral operator.

Lemma 1. *Let $\xi : I = [v,\mu] \to \mathbb{R}$ be a differentiable mapping on I°, where $v, \mu \in I$ with $v < \mu$. If $\xi' \in L_1[v,\mu]$, then*

$$\frac{\xi(v+\mu-x_1) + \xi(v+\mu-x_2)}{2} - \frac{1}{x_2-x_1}\int_{v+\mu-x_2}^{v+\mu-x_1}\xi(z)dz$$
$$= \frac{x_2-x_1}{2}\int_0^1 (1-2\chi)\xi'(v+\mu-((1-\chi)x_1+\chi x_2))d\chi, \quad (15)$$

holds for all $x_1, x_2 \in [v,\mu]$.

Proof. Note that

$$I = \int_0^1 (1-2\chi)\xi'(v+\mu-((1-\chi)x_1+\chi x_2))d\chi$$
$$= \left.\frac{\xi(v+\mu-((1-\chi)x_1+\chi x_2))}{x_1-x_2}(1-2\chi)\right|_0^1 + 2\int_0^1 \frac{\xi(v+\mu-((1-\chi)x_1+\chi x_2))}{x_1-x_2}d\chi$$
$$= \frac{\xi(v+\mu-x_1)+\xi(v+\mu-x_2)}{x_2-x_1} - \frac{2}{x_2-x_1}\cdot\frac{1}{x_2-x_1}\int_{v+\mu-x_2}^{v+\mu-x_1}\xi(z)dz.$$

After suitable rearrangements, we obtain the required inequality (15). □

Remark 3. For $x_1 = v$ and $x_2 = \mu$ in Lemma 3, we obtain Lemma 2.1 of (see [46]).

Lemma 2. *Suppose that $\xi : I = [v, \mu] \to \mathbb{R}$ is a differentiable mapping on $I°$, $v, \mu \in I$ with $v < \mu$. If $\xi' \in L_1[v, \mu]$ and take $\theta \in [0, 1]$, then*

$$\frac{x_2 - x_1}{2} \int_0^1 (1 - 2\chi) \xi'(v + \mu - ((1 - \chi)x_1 + \chi x_2)) d\chi - \frac{2(1 - \theta)}{\theta(x_2 - x_1)} \xi(t)$$
$$= \frac{\xi(v + \mu - x_1) + \xi(v + \mu - x_2)}{2} - \frac{B(\theta)}{\theta(x_2 - x_1)} \left[\left({}_{v+\mu-x_2}^{CF} I^\theta \xi \right)(t) + \left({}^{CF} I_{v+\mu-x_1}^\theta \xi \right)(t) \right],$$

holds for all $x_1, x_2 \in [v, \mu]$, where $t \in [v, \mu]$ and $B(\theta) > 0$ is a normalization function.

Proof. It is easy to see that

$$\int_0^1 (1 - 2\chi) \xi'(v + \mu - ((1 - \chi)x_1 + \chi x_2)) d\chi$$
$$= \frac{\xi(v + \mu - x_1) + \xi(v + \mu - x_2)}{x_2 - x_1} - \frac{2}{(x_2 - x_1)^2} \left(\int_{v+\mu-d}^t \xi(z) dz + \int_t^{v+\mu-x_1} \xi(z) dz \right).$$

With both sides of the above inequality multiplied by $\frac{\theta(x_2 - x_1)^2}{2B(\theta)}$ and subtracting $\frac{2(1-\theta)}{B(\theta)} \xi(t)$, we have

$$\frac{\theta(x_2 - x_1)^2}{2B(\theta)} \int_0^1 (1 - 2\chi) \xi'(v + \mu - ((1 - \chi)x_1 + \chi x_2)) d\chi - \frac{2(1 - \theta)}{B(\theta)} \xi(t)$$
$$= \frac{\theta(x_2 - x_1)(\xi(v + \mu - x_1) + \xi(v + \mu - x_2))}{2B(\theta)} - \frac{2(1 - \theta)}{B(\theta)} \xi(t)$$
$$- \frac{\theta}{B(\theta)} \left(\int_{v+\mu-x_2}^t \xi(z) dz + \int_t^{v+\mu-x_1} \xi(z) dz \right)$$
$$= \frac{\theta(x_2 - x_1)(\xi(v + \mu - x_1) + \xi(v + \mu - x_2))}{2B(\theta)} - \left(\frac{(1 - \theta)}{B(\theta)} \xi(t) + \frac{\theta}{B(\theta)} \int_{v+\mu-x_2}^t \xi(z) dz \right)$$
$$- \left(\frac{(1 - \theta)}{B(\theta)} \xi(t) + \frac{\theta}{B(\theta)} \int_t^{v+\mu-x_1} \xi(z) dz \right)$$
$$= \frac{\theta(x_2 - x_1)(\xi(v + \mu - x_1) + \xi(v + \mu - x_2))}{2B(\theta)} - \left[\left({}_{v+\mu-x_2}^{CF} I^\theta \xi \right)(t) + \left({}^{CF} I_{v+\mu-x_1}^\theta \xi \right)(t) \right].$$

After suitable rearrangements, we obtain the desired result. □

Remark 4. For $x_1 = v$ and $x_2 = \mu$ in Lemma 3, then we obtain Lemma 2 of (see [45]).

Theorem 7. *Let $\xi : I \to \mathbb{R}$ be a positive differentiable function on $I°$. If $|\xi'|$ is a h-convex function on $[v, \mu]$ where $x_1, x_2 \in I$ with $v < \mu$, $\xi' \in L_1[v, \mu]$ and $\theta \in [0, 1]$, then*

$$\left| \frac{\xi(v + \mu - x_1) + \xi(v + \mu - x_2)}{2} - \frac{B(\theta)}{\theta(x_2 - x_1)} \left[\left({}_{v+\mu-x_2}^{CF} I^\theta \xi \right)(t) + \left({}^{CF} I_{v+\mu-x_1}^\theta \xi \right)(t) \right] \right.$$
$$\left. + \frac{2(1 - \theta)}{\theta(x_2 - x_1)} \xi(t) \right|$$
$$\leq \frac{x_2 - x_1}{2} \left[\frac{1}{2} M \left(|\xi'(v)| + |\xi'(\mu)| \right) - \left\{ B_h(1 - \chi) |\xi'(x_1)| + B_h(\chi) |\xi'(x_2)| \right\} \right], \quad (16)$$

where

$$B_h(1-\chi) = \int_0^{\frac{1}{2}} (1-2\chi)h(1-\chi)d\chi + \int_{\frac{1}{2}}^1 (2\chi-1)h(1-\chi)d\chi,$$

$$B_h(\chi) = \int_0^{\frac{1}{2}} (1-2\chi)h(\chi)d\chi + \int_{\frac{1}{2}}^1 (2\chi-1)h(\chi)d\chi,$$

holds $\forall\ x_1, x_2 \in [v,\mu],\ t \in [v,\mu],\ B(\theta) > 0$ is a normalization function and $M = \sup\{h(\chi): \chi \in (0,1)\}$.

Proof. By making use of Lemma 3, the properties of the absolute value, the h-convexity of $|\xi'|$ and the Jensen–Mercer inequality yields

$$\left| \frac{\xi(v+\mu-x_1) + \xi(v+\mu-x_2)}{2} - \frac{B(\theta)}{\theta(x_2-x_1)}\left[\left({}^{CF}_{v+\mu-x_2}I^\theta \xi\right)(t) + \left({}^{CF}I^\theta_{v+\mu-x_1}\xi\right)(t)\right]\right.$$
$$\left. + \frac{2(1-\theta)}{\theta(x_2-x_1)}\xi(t) \right|$$
$$\leq \frac{x_2-x_1}{2}\int_0^1 |1-2\chi|\left|\xi'(v+\mu-((1-\chi)x_1+\chi x_2))\right|d\chi$$
$$\leq \frac{x_2-x_1}{2}\int_0^1 |1-2\chi|\left(M\left[|\xi'(v)|+|\xi'(\mu)|\right] - \left(h(1-\chi)|\xi'(x_1)| + h(\chi)|\xi'(x_2)|\right)\right)d\chi$$
$$\leq \frac{x_2-x_1}{2}\left(\int_0^{\frac{1}{2}}(1-2\chi)\left(M\left[|\xi'(v)|+|\xi'(\mu)|\right] - \left(h(1-\chi)|\xi'(x_1)| + h(\chi)|\xi'(x_2)|\right)\right)d\chi\right.$$
$$\left. + \int_{\frac{1}{2}}^1 (2\chi-1)\left(M\left[|\xi'(v)|+|\xi'(\mu)|\right] - \left(h(1-\chi)|\xi'(x_1)| + h(\chi)|\xi'(x_2)|\right)\right)d\chi\right)$$
$$\leq \frac{x_2-x_1}{2}\left[\frac{1}{2}M\left(|\xi'(v)|+|\xi'(\mu)|\right)\right.$$
$$- \left\{|\xi'(x_1)|\left(\int_0^{\frac{1}{2}}(1-2\chi)h(1-\chi)d\chi + \int_{\frac{1}{2}}^1 (2\chi-1)h(1-\chi)d\chi\right)\right.$$
$$\left.\left. + |\xi'(x_2)|\left(\int_0^{\frac{1}{2}}(1-2\chi)h(\chi)d\chi + \int_{\frac{1}{2}}^1 (2\chi-1)h(\chi)d\chi\right)\right\}\right]$$
$$\leq \frac{x_2-x_1}{2}\left[\frac{1}{2}M\left(|\xi'(v)|+|\xi'(\mu)|\right) - \left\{B_h(1-\chi)|\xi'(x_1)| + B_h(\chi)|\xi'(x_2)|\right\}\right].$$

This completes the proof. □

Remark 5. By putting $h(\chi) = \chi$, $M = \sup\{h(\chi) : \chi \in (0,1)\} = 1$, $x_1 = v$ and $x_2 = \mu$ in Theorem 3, we obtain Theorem 5 of [45].

Theorem 8. Suppose that $\xi : I \to \mathbb{R}$ is a positive differentiable function on I° and $|\xi'|^q$ is a h-convex function on $[v,\mu]$, $v, \mu \in I^\circ$ with $v < \mu$ for $p, q > 1$ with $\frac{1}{p} + \frac{1}{q} = 1$, where $v, \mu \in I$ with $v < \mu$. If $\xi' \in L_1[v,\mu]$ and $\theta \in [0,1]$, then

$$\left| \frac{\xi(v+\mu-x_1) + \xi(v+\mu-x_2)}{2} - \frac{B(\theta)}{\theta(x_2-x_1)}\left[\left({}^{CF}_{v+\mu-x_2}I^\theta \xi\right)(t) + \left({}^{CF}I^\theta_{v+\mu-x_1}\xi\right)(t)\right]\right.$$
$$\left. + \frac{2(1-\theta)}{\theta(x_2-x_1)}\xi(t) \right|$$
$$\leq \frac{x_2-x_1}{2}\left(\frac{1}{p+1}\right)^{\frac{1}{p}}\left(M\left[|\xi'(v)|^q + |\xi'(\mu)|^q\right]\right.$$
$$\left. - \left(|\xi'(x_1)|^q \int_0^1 h(1-\chi)d\chi + |\xi'(x_2)|^q \int_0^1 h(\chi)d\chi\right)\right)^{\frac{1}{q}}, \tag{17}$$

holds $\forall\ x_1, x_2 \in [v,\mu],\ t \in [v,\mu],\ B(\theta) > 0$ is a normalization function and $M = \sup\{h(\chi): \chi \in (0,1)\}$.

Proof. From Lemma 3, Hölder's integral inequality, the h-convexity of $|\xi'|^q$ and the Jensen–Mercer inequality yields that

$$\left| \frac{\xi(v+\mu-x_1)+\xi(v+\mu-x_2)}{2} - \frac{B(\theta)}{\theta(x_2-x_1)}\left[\left(^{CF}_{v+\mu-x_2}I^\theta \xi\right)(t) + \left(^{CF}I^\theta_{v+\mu-x_1}\xi\right)(t)\right] \right.$$

$$\left. + \frac{2(1-\theta)}{\theta(x_2-x_1)}\xi(t) \right|$$

$$\leq \frac{x_2-x_1}{2}\int_0^1 |1-2\chi| \left|\xi'(v+\mu-((1-\chi)x_1+\chi x_2))\right| d\chi$$

$$\leq \frac{x_2-x_1}{2}\left(\int_0^1 |1-2\chi|^p d\chi\right)^{\frac{1}{p}}\left(\int_0^1 \left|\xi'(v+\mu-((1-\chi)x_1+\chi x_2))\right|^q d\chi\right)^{\frac{1}{q}}$$

$$\leq \frac{x_2-x_1}{2}\left(\int_0^1 |1-2\chi|^p d\chi\right)^{\frac{1}{p}}\left(\int_0^1 \left(M\left[|\xi'(v)|^q + |\xi'(\mu)|^q\right]\right.\right.$$

$$\left.\left. - \left(h(1-\chi)|\xi'(x_1)|^q + h(\chi)|\xi'(x_2)|^q\right)\right)d\chi\right)^{\frac{1}{q}}$$

$$\leq \frac{x_2-x_1}{2}\left(\frac{1}{p+1}\right)^{\frac{1}{p}}\left(M\left[|\xi'(v)|^q + |\xi'(\mu)|^q\right]\right.$$

$$\left. - \left(|\xi'(x_1)|^q \int_0^1 h(1-\chi)d\chi + |\xi'(x_2)|^q \int_0^1 h(\chi)d\chi\right)\right)^{\frac{1}{q}}.$$

This completes the proof. □

Remark 6. *By putting $h(\chi) = \chi$, $M = \sup\{h(\chi) : \chi \in (0,1)\} = 1$, $x_1 = v$ and $x_2 = \mu$ in Theorem 3, we obtain Theorem 6 of [45].*

Next, we will prove the following theorems using the Hölder–İscan integral inequality and for improved power mean integral inequality, respectively.

Theorem 9. *Assume that $\xi : I \to \mathbb{R}$ is a positive differentiable mapping on I° and $|\xi'|^q$ is a h-convex function on $[v,\mu]$, $v,\mu \in I^\circ$ with $v < \mu$ for $q \geq 1$, where $v,\mu \in I$ with $v < \mu$. If $\xi' \in L_1[v,\mu]$ and $\theta \in [0,1]$, then*

$$\left| \frac{\xi(v+\mu-x_1)+\xi(v+\mu-x_2)}{2} - \frac{B(\theta)}{\theta(x_2-x_1)}\left[\left(^{CF}_{v+\mu-x_2}I^\theta \xi\right)(t) + \left(^{CF}I^\theta_{v+\mu-x_1}\xi\right)(t)\right] \right.$$

$$\left. + \frac{2(1-\theta)}{\theta(x_2-x_1)}\xi(t) \right|$$

$$\leq \frac{x_2-x_1}{2}\left(\frac{1}{2}\right)^{1-\frac{1}{q}}\left(\frac{1}{2}M\left[|\xi'(v)|^q + |\xi'(\mu)|^q\right]\right.$$

$$\left. - \left(|\xi'(x_1)|^q \int_0^1 |1-2\chi|h(1-\chi)d\chi + |\xi'(x_2)|^q \int_0^1 |1-2\chi|h(\chi)d\chi\right)\right)^{\frac{1}{q}}, \quad (18)$$

holds $\forall\, x_1, x_2 \in [v,\mu]$, $t \in [v,\mu]$, $B(\theta) > 0$ is a normalization function and $M = \sup\{h(\chi) : \chi \in (0,1)\}$.

Proof. Take $q > 1$, by using Lemma 3, the power mean inequality, the h-convexity of $|\xi'|^q$ and the Jensen–Mercer inequality, and we have

$$\left| \frac{\xi(v+\mu-x_1) + \xi(v+\mu-x_2)}{2} - \frac{B(\theta)}{\theta(x_2-x_1)} \left[\left(^{CF}_{v+\mu-x_2} I^\theta \xi\right)(t) + \left(^{CF} I^\theta_{v+\mu-x_1} \xi\right)(t) \right] \right.$$

$$\left. + \frac{2(1-\theta)}{\theta(x_2-x_1)} \xi(t) \right|$$

$$\leq \frac{x_2-x_1}{2} \int_0^1 |1-2\chi| \left|\xi'(v+\mu-((1-\chi)x_1+\chi x_2))\right| d\chi$$

$$\leq \frac{x_2-x_1}{2} \left(\int_0^1 |1-2\chi| d\chi \right)^{1-\frac{1}{q}}$$

$$\times \left(\int_0^1 |1-2\chi| \left|\xi'(v+\mu-((1-\chi)x_1+\chi x_2))\right|^q d\chi \right)^{\frac{1}{q}}$$

$$\leq \frac{x_2-x_1}{2} \left(\frac{1}{2}\right)^{1-\frac{1}{q}} \left(\int_0^1 |1-2\chi| \right.$$

$$\times \left. \left(M\left[|\xi'(v)|^q + |\xi'(\mu)|^q\right] - \left(h(1-\chi)|\xi'(x_1)|^q + h(\chi)|\xi'(x_2)|^q\right) \right) d\chi \right)^{\frac{1}{q}}$$

$$\leq \frac{x_2-x_1}{2} \left(\frac{1}{2}\right)^{1-\frac{1}{q}} \left(\frac{1}{2} M\left[|\xi'(v)|^q + |\xi'(\mu)|^q\right] \right.$$

$$\left. - \left(|\xi'(x_1)|^q \int_0^1 |1-2\chi|h(1-\chi)d\chi + |\xi'(x_2)|^q \int_0^1 |1-2\chi|h(\chi)d\chi \right) \right)^{\frac{1}{q}}. \tag{19}$$

This completes the proof. □

4. Some Results in Improved Hölder Setting

In this section, we will present some results for the h-convex function in the setting of the Hölder–İşcan integral inequality and improved power mean integral inequality via the Caputo–Fabrizio fractional integral operator.

Theorem 10. Let $\xi : I \to \mathbb{R}$ be a positive differentiable mapping on I° and $|\xi'|^q$ be a h-convex function on $[v,\mu]$, $v,\mu \in I^\circ$ with $v < \mu$ for $p, q > 1$ with $\frac{1}{p} + \frac{1}{q} = 1$, where $v, \mu \in I$ with $v < \mu$. If $\xi' \in L_1[v,\mu]$ and $\theta \in [0,1]$, then

$$\left| \frac{\xi(v+\mu-x_1) + \xi(v+\mu-x_2)}{2} - \frac{B(\theta)}{\theta(x_2-x_1)} \left[\left(^{CF}_{v+\mu-x_2} I^\theta \xi\right)(t) + \left(^{CF} I^\theta_{v+\mu-x_1} \xi\right)(t) \right] \right.$$

$$\left. + \frac{2(1-\theta)}{\theta(x_2-x_1)} \xi(t) \right|$$

$$\leq \frac{x_2-x_1}{2} \left[\left(\frac{1}{2(p+1)}\right)^{\frac{1}{p}} \left(\frac{1}{2}M\left(|\xi'(v)|^q + |\xi'(\mu)|^q\right) \right. \right.$$

$$\left. - \left(|\xi'(x_1)|^q \int_0^1 (1-\chi)h(1-\chi)d\chi + |\xi'(x_2)|^q \int_0^1 (1-\chi)h(\chi)d\chi \right) \right)^{\frac{1}{q}}$$

$$+ \left(\frac{1}{2(p+1)}\right)^{\frac{1}{p}} \left(\frac{1}{2}M\left(|\xi'(v)|^q + |\xi'(\mu)|^q\right) \right.$$

$$\left. \left. - \left(|\xi'(x_1)|^q \int_0^1 \chi h(1-\chi)d\chi + |\xi'(x_2)|^q \int_0^1 \chi h(\chi)d\chi \right) \right)^{\frac{1}{q}} \right], \tag{20}$$

holds $\forall x_1, x_2 \in [v,\mu]$, $t \in [v,\mu]$, $B(\theta) > 0$ is a normalization function and $M = \sup \{h(\chi) : \chi \in (0,1)\}$.

Proof. From Lemma 3, using the Hölder–İscan integral inequality, the h-convexity of $|\zeta'|^q$ and the Jensen–Mercer inequality yields

$$\left| \frac{\zeta(v+\mu-x_1)+\zeta(v+\mu-x_2)}{2} - \frac{B(\theta)}{\theta(x_2-x_1)}\left[\left(^{CF}_{v+\mu-x_2}I^\theta\zeta\right)(t) + \left(^{CF}I^\theta_{v+\mu-x_1}\zeta\right)(t)\right] \right.$$
$$\left. + \frac{2(1-\theta)}{\theta(x_2-x_1)}\zeta(t) \right|$$

$$\leq \frac{x_2-x_1}{2}\int_0^1 |1-2\chi|\left|\zeta'(v+\mu-((1-\chi)x_1+\chi x_2))\right|d\chi$$

$$\leq \frac{x_2-x_1}{2}\left[\left(\int_0^1 (1-\chi)|1-2\chi|^p d\chi\right)^{\frac{1}{p}}\right.$$
$$\times \left(\int_0^1 (1-\chi)\left|\zeta'(v+\mu-((1-\chi)x_1+\chi x_2))\right|^q d\chi\right)^{\frac{1}{q}}$$
$$\left. + \left(\int_0^1 \chi|1-2\chi|^p d\chi\right)^{\frac{1}{p}}\left(\int_0^1 \chi\left|\zeta'(v+\mu-((1-\chi)x_1+\chi x_2))\right|^q d\chi\right)^{\frac{1}{q}}\right]$$

$$\leq \frac{x_2-x_1}{2}\left[\left(\frac{1}{2(p+1)}\right)^{\frac{1}{p}}\left(\int_0^1 (1-\chi)\left|\zeta'(v+\mu-((1-\chi)x_1+\chi x_2))\right|^q d\chi\right)^{\frac{1}{q}}\right.$$
$$\left. + \left(\frac{1}{2(p+1)}\right)^{\frac{1}{p}}\left(\int_0^1 \chi\left|\zeta'(v+\mu-((1-\chi)x_1+\chi x_2))\right|^q d\chi\right)^{\frac{1}{q}}\right]$$

$$\leq \frac{x_2-x_1}{2}\left[\left(\frac{1}{2(p+1)}\right)^{\frac{1}{p}}\left(\int_0^1 (1-\chi)\right.\right.$$
$$\times \left(M\left[|\zeta'(v)|^q+|\zeta'(\mu)|^q\right] - \left(h(1-\chi)|\zeta'(x_1)|^q+h(\chi)|\zeta'(x_2)|^q\right)\right)d\chi\right)^{\frac{1}{q}}$$
$$+ \left(\frac{1}{2(p+1)}\right)^{\frac{1}{p}}\left(\int_0^1 \chi\right.$$
$$\left.\left.\times \left(M\left[|\zeta'(v)|^q+|\zeta'(\mu)|^q\right] - \left(h(1-\chi)|\zeta'(x_1)|^q+h(\chi)|\zeta'(x_2)|^q\right)\right)d\chi\right)^{\frac{1}{q}}\right]$$

$$\leq \frac{x_2-x_1}{2}\left[\left(\frac{1}{2(p+1)}\right)^{\frac{1}{p}}\left(\frac{1}{2}M\left(|\zeta'(v)|^q+|\zeta'(\mu)|^q\right)\right.\right.$$
$$\left. - \left(|\zeta'(x_1)|^q \int_0^1 (1-\chi)h(1-\chi)d\chi + |\zeta'(x_2)|^q\int_0^1 (1-\chi)h(\chi)d\chi\right)\right)^{\frac{1}{q}}$$
$$+ \left(\frac{1}{2(p+1)}\right)^{\frac{1}{p}}\left(\frac{1}{2}M\left(|\zeta'(v)|^q+|\zeta'(\mu)|^q\right)\right.$$
$$\left.\left. - \left(|\zeta'(x_1)|^q \int_0^1 \chi h(1-\chi)d\chi + |\zeta'(x_2)|^q\int_0^1 \chi h(\chi)d\chi\right)\right)^{\frac{1}{q}}\right].$$

This completes the proof. □

Theorem 11. *Let $\zeta : I \to \mathbb{R}$ be a positive differentiable mapping on I° and $|\zeta'|^q$ be a h-convex function on $[v,\mu]$, $v, \mu \in I^\circ$ with $v < \mu$ for $q \geq 1$, where $v, \mu \in I$ with $v < \mu$. If $\zeta' \in L_1[v,\mu]$ and $\theta \in [0,1]$, then*

$$\left| \frac{\zeta(v+\mu-x_1)+\zeta(v+\mu-x_2)}{2} - \frac{B(\theta)}{\theta(x_2-x_1)}\left[\left(^{CF}_{v+\mu-x_2}I^\theta\zeta\right)(t) + \left(^{CF}I^\theta_{v+\mu-x_1}\zeta\right)(t)\right] \right.$$
$$\left. + \frac{2(1-\theta)}{\theta(x_2-x_1)}\zeta(t) \right|$$

$$\leq \frac{x_2 - x_1}{2}\left[\left(\frac{1}{4}\right)^{1-\frac{1}{q}}\left(\frac{1}{4}M\left(\left|\xi'(v)\right|^q + \left|\xi'(\mu)\right|^q\right)\right.\right.$$
$$\left.- \left(\left|\xi'(x_1)\right|^q \int_0^1 (1-\chi)|1 - 2\chi| h(1-\chi) d\chi + \left|\xi'(x_2)\right|^q \int_0^1 (1-\chi)|1 - 2\chi| h(\chi) d\chi\right)\right)^{\frac{1}{q}}$$
$$+ \left(\frac{1}{4}\right)^{1-\frac{1}{q}}\left(\frac{1}{4}M\left(\left|\xi'(v)\right|^q + \left|\xi'(\mu)\right|^q\right)\right.$$
$$\left.\left.- \left(\left|\xi'(x_1)\right|^q \int_0^1 \chi|1 - 2\chi| h(1-\chi) d\chi + \left|\xi'(x_2)\right|^q \int_0^1 \chi|1 - 2\chi| h(\chi) d\chi\right)\right)^{\frac{1}{q}}\right], \tag{21}$$

holds $\forall\, x_1, x_2 \in [v, \mu]$, $t \in [v, \mu]$, $B(\theta) > 0$ is a normalization function and $M = \sup\{h(\chi): \chi \in (0,1)\}$.

Proof. Take $q > 1$, from Lemma 3, and using the improved power-mean integral inequality, the definition of the h-convexity of $|\xi'|^q$, and the Jensen–Mercer inequality, we have

$$\left|\frac{\xi(v+\mu-x_1)+\xi(v+\mu-x_2)}{2} - \frac{B(\theta)}{\theta(x_2-x_1)}\left[\left({}^{CF}I^{\theta}_{v+\mu-x_2}\xi\right)(t) + \left({}^{CF}I^{\theta}_{v+\mu-x_1}\xi\right)(t)\right]\right|$$

$$\leq \frac{x_2-x_1}{2}\int_0^1 |1-2\chi|\left|\xi'(v+\mu-((1-\chi)x_1+\chi x_2))\right|d\chi$$

$$\leq \frac{x_2-x_1}{2}\left[\left(\int_0^1 (1-\chi)|1-2\chi|d\chi\right)^{1-\frac{1}{q}}\right.$$
$$\times \left(\int_0^1 (1-\chi)|1-2\chi|\left|\xi'(v+\mu-((1-\chi)x_1+\chi x_2))\right|^q d\chi\right)^{\frac{1}{q}}$$
$$+ \left(\int_0^1 \chi|1-2\chi|d\chi\right)^{1-\frac{1}{q}}$$
$$\left.\times \left(\int_0^1 \chi|1-2\chi|\left|\xi'(v+\mu-((1-\chi)x_1+\chi x_2))\right|^q d\chi\right)^{\frac{1}{q}}\right]$$

$$\leq \frac{x_2-x_1}{2}\left[\left(\frac{1}{4}\right)^{1-\frac{1}{q}}\left(\int_0^1 (1-\chi)|1-2\chi|\left|\xi'(v+\mu-((1-\chi)x_1+\chi x_2))\right|^q d\chi\right)^{\frac{1}{q}}\right.$$
$$\left.+ \left(\frac{1}{4}\right)^{1-\frac{1}{q}}\left(\int_0^1 \chi|1-2\chi|\left|\xi'(v+\mu-((1-\chi)x_1+\chi x_2))\right|^q d\chi\right)^{\frac{1}{q}}\right]$$

$$\leq \frac{x_2-x_1}{2}\left[\left(\frac{1}{4}\right)^{1-\frac{1}{q}}\left(\int_0^1 (1-\chi)|1-2\chi|\right.\right.$$
$$\left.\times \left(M\left[\left|\xi'(v)\right|^q + \left|\xi'(\mu)\right|^q\right] - \left(h(1-\chi)\left|\xi'(x_1)\right|^q + h(\chi)\left|\xi'(x_2)\right|^q\right)\right)d\chi\right)^{\frac{1}{q}}$$
$$+ \left(\frac{1}{4}\right)^{1-\frac{1}{q}}\left(\int_0^1 \chi|1-2\chi|\right.$$
$$\left.\left.\times \left(M\left[\left|\xi'(v)\right|^q + \left|\xi'(\mu)\right|^q\right] - \left(h(1-\chi)\left|\xi'(x_1)\right|^q + h(\chi)\left|\xi'(x_2)\right|^q\right)\right)d\chi\right)^{\frac{1}{q}}\right]$$

$$\leq \frac{x_2-x_1}{2}\left[\left(\frac{1}{4}\right)^{1-\frac{1}{q}}\left(\frac{1}{4}M\left(\left|\xi'(v)\right|^q + \left|\xi'(\mu)\right|^q\right)\right.\right.$$
$$\left.- \left(\left|\xi'(x_1)\right|^q \int_0^1 (1-\chi)|1-2\chi|h(1-\chi)d\chi + \left|\xi'(x_2)\right|^q \int_0^1 (1-\chi)|1-2\chi|h(\chi)d\chi\right)\right)^{\frac{1}{q}}$$
$$+ \left(\frac{1}{4}\right)^{1-\frac{1}{q}}\left(\frac{1}{4}M\left(\left|\xi'(v)\right|^q + \left|\xi'(\mu)\right|^q\right)\right.$$
$$\left.\left.- \left(\left|\xi'(x_1)\right|^q \int_0^1 \chi|1-2\chi|h(1-\chi)d\chi + \left|\xi'(x_2)\right|^q \int_0^1 \chi|1-2\chi|h(\chi)d\chi\right)\right)^{\frac{1}{q}}\right].$$

This completes the proof. □

5. Conclusions

In this note, we established the Hermite–Jensen–Mercer-type inequalities for an h-convex function in the Caputo–Fabrizio setting, and various Caputo–Fabrizio fractional integral inequalities are provided as well. We expect that this work will lead to the novel fractional integral research for Hermite–Hadamard inequalities. The remarks at the end of the results verify the generalization of the results. These results are new and set various interesting directions. In the future, we will prove the inequalities (2) and (8) by using any other method.

Author Contributions: Conceptualization, M.V.-C., M.S.S., S.S., M.S.Z. and A.K.; Funding acquisition, M.V.-C.; Investigation, M.V.-C., M.S.S., S.S., M.S.Z. and A.K.; Methodology, M.V.-C. and M.S.Z.; Writing—original draft, M.V.-C., M.S.S., S.S., M.S.Z. and A.K.; Writing—review and editing, M.V.-C., M.S.S., S.S., M.S.Z. and A.K. All authors have read and agreed to the published version of the manuscript.

Funding: This research received external funding from Dirección de Investigción in Ponticial Catholic University of Ecuador.

Acknowledgments: The authors thank Ponticial Catholic University of Ecuador for the technical support given for this project.

Conflicts of Interest: The authors declare no conflict of interest.

References

1. Miller, K.S.; Ross, B. *An Introduction to the Fractional Calculus and Fractional Differential Equations*; Wiley: New York, NY, USA, 1993.
2. Ross, B. Fractional Calculus. *Math. Mag.* **1977**, *50*, 115–122. [CrossRef]
3. Baleanu, D.; Diethelm, K.; Scalas, E.; Trujillo, J.J. *Fractional Calculus: Models and Numerical Methods*, 2nd ed.; World Scientific: Singapore, 2016.
4. Kilbas, A.A.; Srivastava, H.M.; Trujillo, J.J. *Theory and Applications of Fractional Differential Equations*; Elsevier: Amsterdam, The Netherlands, 2006.
5. Sarikaya, M.Z.; Set, E.; Yaldiz, H.; Başak, N. Hermite-Hadamard's inequalities for fractional integrals and related fractional inequalities. *Math. Comp. Modell.* **2013**, *57*, 2403–2407. [CrossRef]
6. Katugampola, U.N. New approach to a genaralized fractional integral. *Appl. Math. Comp.* **2011**, *218*, 860–865. [CrossRef]
7. Samko, S.G.; Kilbas, A.A.; Marichev, O.I. *Fractional Integrals and Derivatives: Theory and Applications*; Gordon and Breach: Yverdon, Switzerland, 1993.
8. Abdeljawad, T.; Baleanu, D. On fractional derivatives with exponential kernel and their discrete versions. *Rep. Math. Phys.* **2017**, *80*, 11–27. [CrossRef]
9. Adil Khan, M.; Mohammad, N.; Nwaeze, E.R. Quantum Hermite-Hadamard inequality by means of a Green function. *Adv. Differ. Equ.* **2020**, *2020*, 99. [CrossRef]
10. Awan, M.U.; Talib, S.; Chu, Y.-M.; Noor, M.A.; Noor, K.I. Some new refinements of Hermite-Hadamard-type inequalities involving Ψ_k-Riemann-Liouville fractional integrals and applications. *Math. Probl. Eng.* **2020**, *2020*, 3051920. [CrossRef]
11. Yang, X.-Z.; Farid, G.; Nazeer, W.; Chu, Y.-M.; Dong, C.F. Fractional generalized Hadamard and Fejér-Hadamard inequalities for m-convex function. *AIMS Math.* **2020**, *5*, 6325–6340. [CrossRef]
12. Guo, S.Y.; Chu, Y.M.; Farid, G.; Mehmood, S.; Nazeer, W. Fractional Hadamard and Fejér-Hadamard inequalities associated with exponentially (s,m)-convex functions. *J. Funct. Spaces* **2020**, *2020*, 2410385. [CrossRef]
13. Iqbal, A.; Khan, M.A.; Ullah, S.; Chu, Y.M. Some new Hermite-Hadamard-type inequalities associated with conformable fractional integrals and their applications. *J. Funct. Spaces* **2020**, *2020*, 9845407. [CrossRef]
14. Khurshid, Y.; Khan, M.A.; Chu, Y.M. Conformable integral version of Hermite-Hadamard-Fejér inequalities via η-convex functions. *AIMS Math.* **2020**, *5*, 5106–5120. [CrossRef]
15. Vivas-Cortez, M.; Kashuri, A.; Liko, R.; Hernández Hernández, J.E. Some New q-Integral Inequalities Using Generalized Quantum Montgomery Identity via Preinvex Functions. *Symmetry* **2020**, *12*, 553. [CrossRef]
16. Vivas-Cortez, M.J.; Kashuri, A.; Liko, A.; Hernández Hernández, J.E. Quantum Trapezium-Type Inequalities Using Generalized f-Convex Functions. *Axioms* **2020**, *9*, 12. [CrossRef]
17. Vivas-Cortez, M.; Kashuri, A.; Liko, R.; Hernández Hernández, J.E. Trapezium-Type Inequalities for an Extension of Riemann–Liouville Fractional Integrals Using Raina's Special Function and Generalized Coordinate Convex Functions. *Axioms* **2020**, *9*, 117. [CrossRef]
18. Vivas-Cortez, M.; Kashuri, A.; Hernández Hernández, J.E. Trapezium-Type Inequalities for Raina's Fractional Integrals Operator Using Generalized Convex Functions. *Symmetry* **2020**, *12*, 1034. [CrossRef]

19. Vivas-Cortez, M.; Hernández Hernández, J.E.; Turhan, S. On exponentially (h_1, h_2)-convex functions and fractional integral inequalities related. *Math. Moravica.* **2020**, *24*, 45–62. [CrossRef]
20. Sarikaya, M.Z.; Saglam, A.; Yildirim, H. On some Hadamard-type inequalities for h-convex functions. *J. Math. Inequal.* **2008**, *2*, 335–341. [CrossRef]
21. Sarikaya, M.Z.; Set, E.; Özdemir, M.E. On some new inequalities of Hadamard-type involving h-convex functions. *Acta Math. Univ. Comen.* **2010**, *79*, 265–272.
22. Noor, M.A.; Noor, K.I.; Awan, M.U.; Costache, S. Some integral inequalities for harmonically h-convex functions. *Sci. Bull. Politeh. Univ. Buchar. Ser. A Appl. Math. Phys.* **2015**, *77*, 5–16.
23. Noor, M.A.; Noor, K.I.; Awan, M.U. A new Hermite-Hadamard type inequality for h-convex functions. *Creat. Math. Inform.* **2015**, *24*, 191–197.
24. Matłoka, M. On Hadamard's inequality for h-convex function on a disk. *Appl. Math. Comput.* **2014**, *235*, 118–123. [CrossRef]
25. Dragomir, S.S. Inequalities of Hermite-Hadamard type for h-convex functions on linear spaces. *Proyecciones* **2015**, *34*, 323–341. [CrossRef]
26. Bombardelli, M.; Varošanec, S. Properties of h-convex functions related to the Hermite-Hadamard-Fejér inequalities. *Comput. Math. Appl.* **2009**, *58*, 1869–1877. [CrossRef]
27. Mercer, A.M. A variant of Jensen's inequality. *JIPAM J. Inequal. Pure Appl. Math.* **2003**, *4*, 1–15
28. Matković, A.; Pečarixcx, J.; Perixcx, I. A variant of Jensen's inequality of Mercer's type for operators with applications. *Linear Algebra Appl.* **2006**, *418*, 551–564. [CrossRef]
29. Vivas-Cortez, M.J.; Hernández, J.E. A variant of Jensen-Mercer Inequality for h-convex functions and Operator h-convex functions. *Matua Rev. Mat. Univ. Atl.* **2017**, *4*, 62–76.
30. Öğülmüs, H.; Sarikaya, M.Z. Hermite-Hadamard-Mercer Type Inequalities for Fractional Integrals. Available online: https://www.researchgate.net/publication/337682540_HERMITE-HADAMARD-MERCER_TYPE_INEQUALITIES_FOR_FRACTIONAL_INTEGRALS?channel=doi&linkId=5de5677092851c83645ce741&showFulltext=true (accessed on 1 December 2019).
31. Butt, S.I.; Nadeem, M.; Qaisar, S.; Akdemir, A.O.; Abdeljawad, T. Hermite-Jensen-Mercer type inequalities for conformable integrals and related results. *Adv. Differ. Equ.* **2020**, *501*. [CrossRef]
32. Butt, S.I.; Umar, M.; Rashid, S. New Hermite-Jensen-Mercer-type inequalities via k-fractional integrals. *Adv. Differ. Equ.* **2020**, *2020*, 635 [CrossRef]
33. Kang, Q.; Butt, S.I.; Nazeer, W.; Nadeem, M.; Nasir, J.; Yang, H. New variant of Hermite–Jensen–Mercer inequalities via Riemann–Liouville fractional integral operators. *J. Math.* **2020**, *2020*, 4303727.
34. Butt, S.I.; Akdemir, A.O.; Nasir, J.; Jarad, F. Some Hermite-Jensen-Mercer like inequalities for convex functions through a certain generalized fractional integrals and related results. *Miskolc Math. Notes* **2020**, *21*, 689–715. [CrossRef]
35. Zhao, S.; Butt, S.I.; Nazeer, W.; Nasir, J.; Umar, M.; Liu, Y. Some Hermite-Jensen-Mercer type inequalities for k-Caputo-fractional derivatives and related results. *Adv. Differ. Equ.* **2020**, *2020*, 262. [CrossRef]
36. Zhao, J.; Butt, S.I.; Nasir, J.; Wang, Z.; Tlili, I. Hermite-Jensen-Mercer type inequalities for Caputo fractional derivatives. *J. Funct. Spaces* **2020**, *2020*, 7061549. [CrossRef]
37. Işcan, I. Weighted Hermite–Hadamard–Mercer type inequalities for convex functions. *Numer. Methods Partial Differ. Equ.* **2020**, *37*, 118–130. [CrossRef]
38. Chu, H.-H.; Rashid, S.; Hammouch, Z.; Chu, Y.-M. New fractional estimates for Hermite-Hadamard-Merccer's type inequalities. *Alex. Eng. J.* **2020**, *59*, 3079–3089. [CrossRef]
39. Hadamard, J. E'tude sur les prope'rtie's des fonctions entieres et en particular dune fonction considere'e par Riemann. *J. Math. Pures Appl.* **1893**, *58*, 171–215.
40. Varošanec, S. On h-convexity. *J. Math. Anal. Appl.* **2007**, *326*, 303–311 [CrossRef]
41. Temur, Z. Kalanov Vector calculus and Maxwells equations: Logic errors in mathematics and electrodynamics. *Open J. Math. Sci.* **2020**, *4*, 343–355.
42. Abdeljawad, T. Fractional operators with exponential kernels and a Lyapunov type inequality. *Adv. Differ. Equ.* **2017**, *2017*, 313–323. [CrossRef]
43. Işcan, I. New refinements for integral and sum forms of Hölder inequality. *J. Inequal. Appl.* **2019**, *2019*, 304. [CrossRef]
44. Kadakal, M.; Işcan, I.; Kadakal, H.; Bekar, K. On improvements of some integral inequalities. *Honam Math. J.* **2021**, *43*, 441–452.
45. Gurbuz, M.; Akdemir, A.O.; Rashid, S.; Set, E. Hermite-Hadamard inequality for fractional integrals of Caputo-Fabrizio type and related inequalities. *J. Inequal. Appl.* **2020**, *2020*, 172–189. [CrossRef]
46. Dragomir, S.S.; Agarwal, R.P. Two inequalities for differentiable mappings and applications to special means of real numbers and to trapezoidal formula. *Appl. Math. Lett.* **1998**, *11*, 91–95. [CrossRef]

fractal and fractional

Article

Semidefinite Multiobjective Mathematical Programming Problems with Vanishing Constraints Using Convexificators

Kin Keung Lai [1,*,†], Mohd Hassan [2,†], Sanjeev Kumar Singh [2,†], Jitendra Kumar Maurya [3,†] and Shashi Kant Mishra [2,†]

1. International Business School, Shaanxi Normal University, Xi'an 710119, China
2. Department of Mathematics, Institute of Science, Banaras Hindu University, Varanasi 221005, India; mohd.hassan10@bhu.ac.in (M.H.); sanjeevk.singh1@bhu.ac.in (S.K.S.); shashikant.mishra@bhu.ac.in (S.K.M.)
3. Kashi Naresh Government Postgraduate College, Gyanpur, Bhadohi 221304, India; jitendrak.maurya1@bhu.ac.in
* Correspondence: mskklai@outlook.com
† These authors contributed equally to this work.

Abstract: In this paper, we establish Fritz John stationary conditions for nonsmooth, nonlinear, semidefinite, multiobjective programs with vanishing constraints in terms of convexificator and introduce generalized Cottle type and generalized Guignard type constraints qualification to achieve strong $S-$stationary conditions from Fritz John stationary conditions. Further, we establish strong $S-$stationary necessary and sufficient conditions, independently from Fritz John conditions. The optimality results for multiobjective semidefinite optimization problem in this paper is related to two recent articles by Treanta in 2021. Treanta in 2021 discussed duality theorems for special class of quasiinvex multiobjective optimization problems for interval-valued components. The study in our article can also be seen and extended for the interval-valued optimization motivated by Treanta (2021). Some examples are provided to validate our established results.

Keywords: multiobjective programs with vanishing constraints; semidefinite programming; convexificators; nonsmooth analysis; constraint qualifications

1. Introduction

Nonlinear semidefinite programming problems (SDP) include several classes of optimization problems, such as linear programming, quadratic programming, second order cone programming [1], and semidefinite programming [2]. The nonlinear semidefinite programming problem has broad applications in system control [3], truss topology optimization [4], and other several fields. It has been at the center point of optimization research for the last two decades. For instance, in the release of library COMPleib [5], where 168 test examples on nonlinear semidefinite programs from various fields, such as control system design, academia, and many real-life based problems are collected.

In this paper, we consider the following semidefinite multiobjective mathematical programs with vanishing constraints $(S - MMPVC)$,

$$\min \mathfrak{f}(A) = (\mathfrak{f}_i(A), ..., \mathfrak{f}_p(A)) \qquad (1)$$
$$\text{subject to } A \in M = \{A \in \mathbb{M}_+^n : \mathscr{H}_i(A) \geqq 0, \mathscr{G}_i(A)\mathscr{H}_i(A) \leqq 0\},$$

where $\mathbb{M}_+^n$ is set of $n \times n$ positive semidefinite matrix, $\mathfrak{f}_i : \mathbb{M}_+^n \to \mathbb{R} \cup \{+\infty\}$ $(i = 1, ...p)$ and $\mathscr{G}_i, \mathscr{H}_i : \mathbb{M}_+^n \to \mathbb{R} \cup \{+\infty\}(i = 1, ..., m)$ are extended real-valued locally Lipschitz functions.

Nonlinear semidefinite programming problems consist of the nonlinear problems where vector variables are replaced by symmetric positive semidefinite matrices. Nonlinear SDPs have been studied extensively due to a wide range of applications, see for instance, [6,7]. Shapiro [6] established first and second order necessary and sufficient optimality conditions under the convexity assumptions. Forsgren [8] extended those results for nonconvex semidefinite programming. Further, Sun et al. [7] and Sun [9] discussed the algorithmic approaches to solve nonlinear semidefinite programming problems. Yamashita and Yabe [10] introduced some numerical methods to solve nonlinear SDP and studied the algorithmic consequences. Recently, Golestani and Nobakhtian [11] proposed the generalized Abadie constraint qualification $(GACQ)$ and established necessary and sufficient optimality conditions for nonlinear semidefinite programming problems using convexificators.

Mathematical programs with vanishing constraints($MPVC$) has many applications in truss topology optimization [12], pathfinding problem with logic communication constraints in robot motion planning [13], mixed integer nonlinear optimal control problems [14], scheduling problems with disjoint feasible regions in power generation dispatch [15] and many more fields of the current research [16–18]. Initially, mathematical programs with vanishing constraints (MPVC) was introduced by Achtziger and Kanzow in 2008. MPVC is closely related to an optimization problem known as mathematical programs with equilibrium constraints (MPEC), for more details on MPEC, we refer, [19–28].

Due to the constraints $\mathscr{G}_i(z)\mathscr{H}_i(z) \leq 0$, the feasible set may not be convex even disconnected, most of the basic constraint qualifications such as linearly independent constraint qualification and Mangasarian–Fromovitz constraint qualification do not hold, therefore, standard Karush–Kuhn–Tucker conditions are of no use in such cases. Several constraint qualifications and necessary optimality conditions have been established in [12] for mathematical programs with vanishing constraints. First order sufficient optimality conditions, as well as second order necessary and sufficient optimality conditions, have been discussed in [29] using generalized convexity for mathematical programs with vanishing constraints. In [30], various stationary conditions under weaker assumptions of constraint qualifications were derived. Further, Hoheisel and Kanzow [31] investigated necessary and sufficient optimality conditions through Abadie and Guignard type constraint qualifications for mathematical programs with vanishing constraints. For more details on the MPVC, we refer to [16,32,33] and the references therein.

Multiobjective optimization problems (MOP) plays a vital role in science, technology, business, economics, and many others field of daily demand, where optimal decisions need to be taken among many conflicting objectives and all objective functions to be optimized simultaneously. Effect of conflict on objectives leads to some change in the solution of (MOP) compared to the optimal solution of single-objective optimization problems. Therefore, weak efficient point (weak Pareto optimal solution), efficient point (Pareto optimal solution) like terms are coined for the solutions of (MOP). Initially, the concept of Pareto optimal solutions was given by Italian civil engineer and economist Vilfredo Pareto and was applied in the studies of economic efficiency and income distribution. Basic concept and literature on the solution of multiobjective optimization problems can be found [34,35]. Maeda [36] studied the strong KKT optimality conditions and differentiable functions. Preda and Chitescu [37] extends these results for semidifferentiable functions. Further, Li [38] discussed these results for the nonsmooth case. Recently, Lai et al. [39] proposed saddle point necessary and sufficient Pareto optimality conditions for multiobjective convex optimization problems. Treanta [40] established dual pair of multiobjective interval-valued variational control problems. Further, Treanta [41] discussed duality theorems for special class of quasiinvex multiobjective optimization problems for interval-valued components.

Since nonsmoothness in optimization is naturally generated from the mathematical formulation of real-world problems, therefore, proper effective way for solving these problems should be discovered. Even the solution of some smooth problems, sometimes requires the use of nonsmooth optimization techniques, in order to either make it easy or simplify its form. Thus, the field of nonsmooth optimization is an important branch of mathematical programming that is based on classical concepts of variational analysis and generalized derivatives. In recent years, research in nonsmooth analysis has focused on the growth of generalized subdifferentials that give sharp results and good calculus rules for nonsmooth functions. It is convexificators [42], that has been used to extend, unify, and sharpen the results in various aspects of optimization. Jeyakumar and Luc [43] provided a more sophisticated version of convexificators by introducing the new notion of convexificators which are the closed set but not necessarily bounded or convex. The new version of convexificators consists only finitely many points so it is advantageous for application point of view. We have used the convexificator due to Jeyakumar and Luc [43] in our study.

Recently, Dorsch et al. [44] established a new result for nonlinear semidefinite programming (NLSDP) where almost all linear perturbations of a given NLSDP are shown to be nondegenerate. Semidefinite programming is a powerful framework from convex optimization that has striking potential for data science applications [45]. Sequential optimality conditions have played a vital role in unifying and extending global convergence results for several classes of algorithms for general nonlinear optimization, Andreani et al. [46] extended these concepts for nonlinear semidefinite programming. Andreani et al. [47] discussed simple extensions of constant rank-type constraint qualifications to semidefinite programming, which are based on the Approximate Karush–Kuhn–Tucker necessary optimality condition and on the application of the reduction approach.

Motivated by the above mentioned work, we propose some new constraints qualification to establish necessary and sufficient type optimality conditions for nonsmooth, nonlinear, semidefinite, multiobjective mathematical programs with vanishing constraints. The organization of this article is as follows: In Section 2, we recall some needful preliminaries and fundamental results. In Section 3, we establish Fritz John necessary optimality conditions and propose generalized Cottle and generalized Guignard type constraint qualification to establish strong Karush–Kuhn–Tucker necessary optimality conditions. Further, sufficient optimality conditions are also established under generalized convexity. Section 4, presents the conclusion of the paper, as well as some possible views towards future work.

2. Preliminaries

This section recalls needful notation, definitions, and preliminaries that will be used throughout the paper. $\mathbb{M}^n$ is denoted as the space of $n \times n$ symmetric matrices. The notation $A \succeq 0 (A \succ 0)$ means that A is a positive semidefinite matrix (positive definite matrix) and we denote by $\mathbb{M}^n_+ (\mathbb{M}^n_{++})$ the set of all positive semidefinite matrices (positive definite matrices). The inner product of the symmetric matrices $P, Q \in \mathbb{M}^n$ is denoted by $\langle P, Q \rangle$ and defined by $\langle P, Q \rangle = tr(PQ)$ where $tr(.)$ denotes the summation of the diagonal elements of a square matrix. The inner product of $x = (x_1, ..., x_n)$, $y = (y_1, ..., y_n) \in \mathbb{R}^n$ is denoted and defined by $x^T y = \sum_{i=1}^{n} x_i y_i$. The norm associated with matrix inner product is called the Frobenius norm $||P||_F = tr(PP)^{\frac{1}{2}} = (\sum_{i,j=1}^{n} a_{ij}^2)^{\frac{1}{2}}$. The vector space $\mathbb{M}^n$ with this norm is a Hilbert space and $\mathbb{M}^n_+$ is a closed convex cone in $\mathbb{M}^n$. The interior of the positive semidefinite matrices is the positive definite matrices, for more basics on matrices see [48,49]. For $y, \mathfrak{z} \in \mathbb{R}^n$,

$$y \leqq \mathfrak{z} \iff y_i \leqq \mathfrak{z}_i, \ i = 1, ..., n,$$
$$y \leq \mathfrak{z} \iff y \leqq \mathfrak{z}, \ y \neq \mathfrak{z},$$
$$y < \mathfrak{z} \iff y_i < \mathfrak{z}_i, \ i = 1, ..., n.$$

Some index sets are as follows

$$M = \{A \in \mathbb{M}^n_+ : \mathcal{H}_i(A) \geqq 0, \ \mathcal{G}_i(A)\mathcal{H}_i(A) \leqq 0\}, \ \theta_i(A) = \mathcal{G}_i(A)\mathcal{H}_i(A),$$
$$\mathbb{J}_\mathfrak{f} = \{1, ..., p\}, \ \mathbb{J}_\mathfrak{f}^k = \{1, ..., p\} \setminus \{k\}, \ \mathbb{J}_{\mathcal{G}\mathcal{H}} := \{1, ..., m\},$$
$$Q = \{A \in \mathbb{M}^n_+ : \mathfrak{f}_i(A) \leqq \mathfrak{f}_i(\bar{A}) \ (i \in \mathbb{J}_\mathfrak{f}), \ \mathcal{H}_i(A) \geqq 0, \ \mathcal{G}_i(A)\mathcal{H}_i(A) \leqq 0\},$$
$$Q^k = \{A \in \mathbb{M}^n_+ : \mathfrak{f}_i(A) \leqq \mathfrak{f}_i(\bar{A}) \ (i \in \mathbb{J}_\mathfrak{f}^k), \ \mathcal{H}_i(A) \geqq 0, \ \mathcal{G}_i(A)\mathcal{H}_i(A) \leqq 0\}, \text{ where } \bar{A} \in M,$$
$$\mathbb{R}^n_+ = \{x \in \mathbb{R}^n : x \geqq 0\}, \ \mathbb{R}^n_{++} = \{x \in \mathbb{R}^n : x > 0\},$$

$$\mathtt{J}_0 = \mathtt{J}_0(\bar{A}) := \{i \in \mathtt{J}_{\mathscr{GH}} : \mathscr{H}_i(\bar{A}) = 0\}, \mathtt{J}_+ = \mathtt{J}_+(\bar{A}) := \{i \in \mathtt{J}_{\mathscr{GH}} : \mathscr{H}_i(\bar{A}) > 0\},$$
$$\mathtt{J}_{0+} = \mathtt{J}_{0+}(\bar{A}) := \{i \in \mathtt{J}_{\mathscr{GH}} : \mathscr{H}_i(\bar{A}) = 0, \mathscr{G}_i(\bar{A}) > 0\},$$
$$\mathtt{J}_{00} = \mathtt{J}_{00}(\bar{A}) := \{i \in \mathtt{J}_{\mathscr{GH}} : \mathscr{H}_i(\bar{A}) = 0, \mathscr{G}_i(\bar{A}) = 0\},$$
$$\mathtt{J}_{0-} = \mathtt{J}_{0-}(\bar{A}) := \{i \in \mathtt{J}_{\mathscr{GH}} : \mathscr{H}_i(\bar{A}) = 0, \mathscr{G}_i(\bar{A}) < 0\},$$
$$\mathtt{J}_{+0} = \mathtt{J}_{+0}(\bar{A}) := \{i \in \mathtt{J}_{\mathscr{GH}} : \mathscr{H}_i(\bar{A}) > 0, \mathscr{G}_i(\bar{A}) = 0\},$$
$$\mathtt{J}_{+-} = \mathtt{J}_{+-}(\bar{A}) := \{i \in \mathtt{J}_{\mathscr{GH}} : \mathscr{H}_i(\bar{A}) > 0, \mathscr{G}_i(\bar{A}) < 0\}.$$

We discuss the solution concepts of $S - MMPVC$ motivated by Miettinen [34].

Definition 1. *A feasible point $\bar{A}$ is said to be a weak efficient solution of $S - MMPVC$ if there is no any $A \in M$, such that*

$$\mathfrak{f}_i(A) < \mathfrak{f}_i(\bar{A}), \; \forall \, i \in \mathtt{J}_\mathfrak{f}.$$

Definition 2. *A feasible point $\bar{A}$ is said to be a local weak efficient solution of $S - MMPVC$ if there exist a neighborhood $\mathscr{N}(\bar{A})$ of $\bar{A}$, such that there is no any $A \in M \cap \mathscr{N}(\bar{A})$, for which*

$$\mathfrak{f}_i(A) < \mathfrak{f}_i(\bar{A}), \; \forall \, i \in \mathtt{J}_\mathfrak{f},$$

holds.

Given a nonempty subset M of $\mathbb{M}^n$, the closure, the convex hull and the convex cone (including the origin) generated by M are denoted by clM, coM, and $coneM$, respectively. The negative and the strictly negative polar cone of M are defined respectively by

$$M^- := \{V \in \mathbb{M}^n : \langle V, \mathscr{W} \rangle \leq 0, \; \forall \, \mathscr{W} \in M\}, \; M^s := \{V \in \mathbb{M}^n : \langle V, \mathscr{W} \rangle < 0, \; \forall \, \mathscr{W} \in M\}.$$

Contingent cone $T(M, A)$ to M at point $A \in clM$ are defined by

$$T(M, A) := \{V \in \mathbb{M}^n : \exists \, t_n \downarrow 0, \; V_n \to V \text{ such that } A + t_n V_n \in M \; \forall \, n\}.$$

The notion of semi-regular convexificators [43] will be used here. It is observed that for locally Lipschitz function many generalized subdifferential like Clarke subdifferential [50], Michel-Penot subdifferential [51], Mordukhovich subdifferential [52], and Treiman subdifferential [53] are examples of upper semi-regular convexificators.

Let $\mathfrak{f} : \mathbb{M}^n \to \mathbb{R} \cup \{+\infty\}$ be an extended real-valued function and let $A \in \mathbb{M}^n$ at which $\mathfrak{f}$ is finite. The lower and upper Dini derivatives of $\mathfrak{f}$ at A in the direction $V \in \mathbb{M}^n$ are defined, respectively, by

$$\mathfrak{f}^-(A; V) := \liminf_{t \downarrow 0} \frac{\mathfrak{f}(A + tV) - \mathfrak{f}(A)}{t},$$

$$\mathfrak{f}^+(A; V) := \limsup_{t \downarrow 0} \frac{\mathfrak{f}(A + tV) - \mathfrak{f}(A)}{t}.$$

Now, we recall the definition of upper and lower semi-regular convexificators from [42,43].

Definition 3. *Let $\mathfrak{f} : \mathbb{M}^n \to \mathbb{R} \cup \{+\infty\}$ be an extended real-valued function and let $A \in \mathbb{M}^n$ at which $\mathfrak{f}$ is finite. The function $\mathfrak{f}$ is said to admit an upper semi-regular convexificator $\partial^* \mathfrak{f}(A) \subset \mathbb{M}^n$ at A if $\partial^* \mathfrak{f}(A)$ is closed and for each $V \in \mathbb{M}^n$,*

$$\mathfrak{f}^+(A; V) \leq \sup_{\xi \in \partial^* \mathfrak{f}(A)} \langle \xi, V \rangle.$$

The function $\mathfrak{f}$ is said to admit a lower semi-regular convexificator $\partial_* \mathfrak{f}(A) \subset \mathbb{M}^n$ at A if $\partial_* \mathfrak{f}(A)$ is closed and for each $V \in \mathbb{M}^n$

$$\mathfrak{f}^-(A;V) \geq \inf_{\xi \in \partial^*\mathfrak{f}(A)} \langle \xi, V \rangle.$$

Definition 4. *Set $\partial\mathfrak{f}(A)$ is said to be semi-regular convexificators if it satisfy both upper semi-regular convexificators, as well as lower semi-regular convexificators.*

Definition 5. *Let $\mathfrak{f} : \mathbb{M}^n \to \mathbb{R} \cup \{+\infty\}$ be an extended real-valued function. Suppose that $A \in \mathbb{M}^n, \mathfrak{f}(A)$ is finite and admits a convexificator $\partial^*\mathfrak{f}(A)$ at A.*

- $\mathfrak{f}$ *is said to be ∂^*—convex at A if, and only if, for all $B \in \mathbb{M}^n$,*

$$\mathfrak{f}(B) - \mathfrak{f}(A) \geq \langle \xi, B - A \rangle, \ \forall \ \xi \in \partial^*\mathfrak{f}(A).$$

- $\mathfrak{f}$ *is said to be strictly ∂^*—convex at A if, and only if, for all $B \in \mathbb{M}^n$,*

$$\mathfrak{f}(B) - \mathfrak{f}(A) > \langle \xi, B - A \rangle, \ \forall \ \xi \in \partial^*\mathfrak{f}(A).$$

- $\mathfrak{f}$ *is said to be ∂^*-pseudoconvex at A if, and only if, for all $B \in \mathbb{M}^n$,*

$$\mathfrak{f}(B) < \mathfrak{f}(A) \implies \langle \xi, B - A \rangle < 0, \ \forall \ \xi \in \partial^*\mathfrak{f}(A).$$

- $\mathfrak{f}$ *is said to be strictly ∂^*-pseudoconvex at A if, and only if, for all $B(\neq A) \in \mathbb{M}^n$,*

$$\langle \xi, B - A \rangle \geq 0 \implies \mathfrak{f}(B) > \mathfrak{f}(A) \ \forall \ \xi \in \partial^*\mathfrak{f}(A).$$

- $\mathfrak{f}$ *is said to be ∂^*—quasiconvex at A if, and only if, for all $B \in \mathbb{M}^n$,*

$$\mathfrak{f}(B) \leq \mathfrak{f}(A) \implies \langle \xi, B - A \rangle \leq 0, \ \forall \ \xi \in \partial^*\mathfrak{f}(A).$$

Now, we recall generalized version of Farkas' lemma [54], which will play the vital role in the derivation of main result of this paper.

Lemma 1. *(Farkas' Lemma) Let $\mathfrak{h} : \mathbb{M}^n \to \mathbb{R}^m$ be convex functions. Then, the following system:*

$$\begin{cases} \mathfrak{h}(A) < 0, \\ A \in M_{++}^n. \end{cases}$$

has no solution if, and only if, there exists $(\lambda, \mathscr{W}) \in \mathbb{R}^m \times \mathbb{M}^n$ with $\lambda \geqq 0, \mathscr{W} \preceq 0$ and $(\lambda, \mathscr{W}) \neq (0,0)$, such that

$$\lambda^T \mathfrak{h}(A) + \langle \mathscr{W}, A \rangle \geqq 0, \ \forall \ A \in \mathbb{M}^n.$$

3. Optimality Conditions

In this section, we deal with the traditional Fritz John necessary optimality conditions and propose some constraint qualifications to establish strong Karush–Kuhn–Tucker necessary optimality conditions, as well as sufficient optimality conditions for semidefinite multiobjective mathematical programs with vanishing constraints in terms of convexificators.

Theorem 1. *(Fritz–John necessary optimality conditions) Let $\bar{A}$ be a local weak efficient solution for $(S - MMPVC)$. Suppose that $\mathfrak{f}_i \ (i \in \mathtt{I}_\mathfrak{f})$ and $\mathscr{H}_i \ (i \in \mathtt{I}_0)$, $\mathscr{G}_i \ (i \in \mathtt{I}_{+0})$, admit bounded upper semi-regular convexificators and for each $\mathscr{H}_i \ (i \in \mathtt{I}_+)$, $\mathscr{G}_i \ (i \in \mathtt{I}_0 \cup \mathtt{I}_{+-})$, is continuous. Then, there exist $\bar{\lambda}_i^\mathfrak{f} \geqq 0 \ (i \in \mathtt{I}_\mathfrak{f})$, $\bar{\lambda}_i^\mathscr{H} \geqq 0 \ (i \in \mathtt{I}_{0-} \cup \mathtt{I}_{00})$, $\bar{\lambda}_i^\mathscr{H}$ free $(i \in \mathtt{I}_{0+})$, $\bar{\lambda}_i^\mathscr{G} \geqq 0 \ (i \in \mathtt{I}_{+0})$, $\bar{\lambda}_i^\mathscr{G} = 0 \ (i \in \mathtt{I}_0 \cup \mathtt{I}_{+-})$, $\mathscr{W} \in \mathbb{M}_+^n$ and not all multipliers along with $\mathscr{W}$ can be simultaneously zero, such that*

$$0 \in \sum_{i=1}^p \bar{\lambda}_i^\mathfrak{f} co\partial^*\mathfrak{f}_i(\bar{A}) + \sum_{i=1}^m [\bar{\lambda}_i^\mathscr{G} co\partial^*\mathscr{G}_i(\bar{A}) - \bar{\lambda}_i^\mathscr{H} co\partial^*\mathscr{H}_i(\bar{A})] - \mathscr{W}, \ \langle \bar{A}, \mathscr{W} \rangle = 0.$$

Proof. We have to show that

$$\left(\left(\bigcup_{i\in J_f}\partial^*f_i(\bar{A})\right)^s+\bar{A}\right)\cap\left(\left(\bigcup_{i\in J_{0+}\cup J_{00}\cup J_{0-}}-\partial^*\mathscr{H}_i(\bar{A})\right)^s+\bar{A}\right)$$
$$\cap\left(\left(\bigcup_{i\in J_{0+}\cup J_{00}\cup J_{0-}\cup J_{+0}}\partial^*\theta_i(\bar{A})\right)^s+\bar{A}\right)\cap M_{++}^n=\varnothing. \quad (2)$$

Suppose, on the contrary,

$$A\in\left(\left(\bigcup_{i\in J_f}\partial^*f_i(\bar{A})\right)^s+\bar{A}\right)\cap\left(\left(\bigcup_{i\in J_{0+}\cup J_{00}\cup J_{0-}}-\partial^*\mathscr{H}_i(\bar{A})\right)^s+\bar{A}\right)$$
$$\cap\left(\left(\bigcup_{i\in J_{0+}\cup J_{00}\cup J_{0-}\cup J_{+0}}\partial^*\theta_i(\bar{A})\right)^s+\bar{A}\right)\cap M_{++}^n. \quad (3)$$

As, f_i ($i\in J_f$), $\mathscr{H}_i$ ($i\in J_{0+}\cup J_{00}\cup J_{0-}$) and θ_i ($i\in J_{+0}\cup J_{00}\cup J_{0-}\cup J_{0+}$), admit bounded upper semi-regular convexificators, we deduce that

$$f_i^+(\bar{A},A-\bar{A})<0, i\in J_f,$$
$$-\mathscr{H}_i^+(\bar{A},A-\bar{A})<0, i\in J_{0+}\cup J_{00}\cup J_{0-},$$
$$\theta_i^+(\bar{A},A-\bar{A})<0, i\in J_{0+}\cup J_{00}\cup J_{0-}\cup J_{+0}.$$

Therefore, there exists $\tau>0$ and $t\in(0,\tau)$ such that

$$f_i(\bar{A}+t(A-\bar{A}))<f_i(\bar{A}), i\in J_f, \quad (4)$$
$$-\mathscr{H}_i(\bar{A}+t(A-\bar{A}))<-\mathscr{H}_i(\bar{A}), i\in J_{0+}\cup J_{00}\cup J_{0-}, \quad (5)$$
$$\theta_i(\bar{A}+t(A-\bar{A}))<\theta_i(\bar{A}), i\in J_{0+}\cup J_{00}\cup J_{0-}\cup J_{+0}. \quad (6)$$

The continuity of $\mathscr{H}_i$ ($i\in J_{+-}\cup J_{+0}$) and θ_i ($i\in J_{+-}$) implies there exists $\tau>0$, such that $\forall\, t\in(0,\tau)$,

$$-\mathscr{H}_i(\bar{A}+t(A-\bar{A}))<0\ (i\in J_{+-}\cup J_{+0}),\ \theta_i(\bar{A}+t(A-\bar{A}))<0\ (i\in J_{+-}). \quad (7)$$

From (4)–(7) and the convexity of M_+^n we find the contradiction with the local weak efficient point of $\bar{A}$. Consider

$$\phi_i(A)=\sup_{\xi_i\in\partial^*f_i(\bar{A})}\langle\xi_i,A-\bar{A}\rangle, i\in J_f,$$
$$\psi_i(A)=\sup_{\eta_i\in-\partial^*\mathscr{H}_i(\bar{A})}\langle\eta_i,A-\bar{A}\rangle, i\in J_{0+}\cup J_{00}\cup J_{0-},$$
$$\varphi_i(A)=\sup_{\zeta_i\in\partial^*\theta_i(\bar{A})}\langle\zeta_i,A-\bar{A}\rangle, i\in J_{0+}\cup J_{00}\cup J_{0-}\cup J_{+0}.$$

Easily, we can seen that $\phi_i(\cdot)$, $\psi_i(\cdot)$ and $\varphi_i(\cdot)$ are convex functions. From (2), it follows that the following system has no solution

$$K=\begin{cases}\phi_i(A)<0 & \text{if } i\in J_f,\\ \psi_i(A)<0 & \text{if } i\in J_{0+}\cup J_{00}\cup J_{0-},\\ \varphi_i(A)<0 & \text{if } i\in J_{0+}\cup J_{00}\cup J_{0-}\cup J_{+0},\\ M_{++}^n.\end{cases}$$

Farkas' Lemma 1 implies that there exist $\bar{\lambda}_i^f \geqq 0$ ($i \in \mathbb{J}_f$), $\lambda_i^{\mathcal{H}} \geqq 0$ ($i \in \mathbb{J}_{0+} \cup \mathbb{J}_{00} \cup \mathbb{J}_{0-}$), $\lambda_i^\theta \geqq 0$ ($i \in \mathbb{J}_{0+} \cup \mathbb{J}_{00} \cup \mathbb{J}_{0-} \cup \mathbb{J}_{+0}$) and $\bar{\mathscr{W}} \in \mathbb{M}_+^n$ and not all multipliers along with $\bar{\mathscr{W}}$ can be simultaneously zero, such that

$$\sum_{i \in \mathbb{J}_f} \bar{\lambda}_i^f \phi_i(A) + \sum_{i \in \mathbb{J}_{0+} \cup \mathbb{J}_{00} \cup \mathbb{J}_{0-}} \lambda_i^{\mathcal{H}} \psi_i(A) + \sum_{i \in \mathbb{J}_{0+} \cup \mathbb{J}_{00} \cup \mathbb{J}_{0-} \cup \mathbb{J}_{+0}} \lambda_i^\theta \varphi_i(A) - \langle \bar{\mathscr{W}}, A \rangle \geqq 0, \ \forall \ A \in \mathbb{M}^n. \quad (8)$$

The above inequality (8) implies that $\langle \bar{\mathscr{W}}, \bar{A} \rangle \leqq 0$. Differently, $\bar{\mathscr{W}}$ and $\bar{A}$ are two elements in $\mathbb{M}_+^n$, hence $\langle \bar{\mathscr{W}}, \bar{A} \rangle = 0$. Therefore,

$$\nu(A) = \sum_{i \in \mathbb{J}_f} \bar{\lambda}_i^f \phi_i(A) + \sum_{i \in \mathbb{J}_{0+} \cup \mathbb{J}_{00} \cup \mathbb{J}_{0-}} \lambda_i^{\mathcal{H}} \psi_i(A) + \sum_{i \in \mathbb{J}_{0+} \cup \mathbb{J}_{00} \cup \mathbb{J}_{0-} \cup \mathbb{J}_{+0}} \lambda_i^\theta \varphi_i(A) - \langle \bar{\mathscr{W}}, A \rangle,$$

is a convex function and $\nu(\bar{A}) = 0$. This implies $0 \in \partial \nu(\bar{A})$, where $\partial \nu(\bar{A})$ is the subdifferential set for ν. Hence,

$$0 \in \sum_{i \in \mathbb{J}_f} \bar{\lambda}_i^f \partial \phi_i(\bar{A}) + \sum_{i \in \mathbb{J}_{0+} \cup \mathbb{J}_{00} \cup \mathbb{J}_{0-}} \lambda_i^{\mathcal{H}} \partial \psi_i(\bar{A}) + \sum_{i \in \mathbb{J}_{0+} \cup \mathbb{J}_{00} \cup \mathbb{J}_{0-} \cup \mathbb{J}_{+0}} \lambda_i^\theta \partial \varphi_i(\bar{A}) - \bar{\mathscr{W}}.$$

This implies,

$$0 \in \sum_{i \in \mathbb{J}_f} \bar{\lambda}_i^f \partial^* \phi_i(\bar{A}) + \sum_{i \in \mathbb{J}_{0+} \cup \mathbb{J}_{00} \cup \mathbb{J}_{0-}} \lambda_i^{\mathcal{H}} \partial^* \psi_i(\bar{A}) + \sum_{i \in \mathbb{J}_{0+} \cup \mathbb{J}_{00} \cup \mathbb{J}_{0-} \cup \mathbb{J}_{+0}} \lambda_i^\theta \partial^* \varphi_i(\bar{A}) - \bar{\mathscr{W}}.$$

$$0 \in \sum_{i \in \mathbb{J}_f} \bar{\lambda}_i^f \partial^* \mathfrak{f}_i(\bar{A}) - \sum_{i \in \mathbb{J}_{0+} \cup \mathbb{J}_{00} \cup \mathbb{J}_{0-}} \lambda_i^{\mathcal{H}} \partial^* \mathscr{H}_i(\bar{A}) + \sum_{i \in \mathbb{J}_{0+} \cup \mathbb{J}_{00} \cup \mathbb{J}_{0-} \cup \mathbb{J}_{+0}} \lambda_i^\theta \partial^* \theta_i(\bar{A}) - \bar{\mathscr{W}},$$

$$0 \in \sum_{i \in \mathbb{J}_f} \bar{\lambda}_i^f \partial^* \mathfrak{f}_i(\bar{A}) - \sum_{i \in \mathbb{J}_{0+} \cup \mathbb{J}_{00} \cup \mathbb{J}_{0-}} \lambda_i^{\mathcal{H}} \partial^* \mathscr{H}_i(\bar{A})$$
$$+ \sum_{i \in \mathbb{J}_{0+} \cup \mathbb{J}_{00} \cup \mathbb{J}_{0-} \cup \mathbb{J}_{+0}} \lambda_i^\theta [\mathscr{H}_i(\bar{A}) \partial^* \mathscr{G}_i(\bar{A}) + \mathscr{G}_i(\bar{A}) \partial^* \mathscr{H}_i(\bar{A})] - \bar{\mathscr{W}}. \quad (9)$$

For $\lambda_i^{\mathcal{H}} = 0$ ($i \in \mathbb{J}_{+-} \cup \mathbb{J}_{+0}$), $\lambda_i^\theta = 0$ ($i \in \mathbb{J}_{+-}$), we obtain from (9)

$$0 \in \sum_{i \in \mathbb{J}_f} \bar{\lambda}_i^f co\partial^* \mathfrak{f}_i(\bar{A}) + \sum_{i=1}^{m} [\bar{\lambda}_i^{\mathscr{G}} co\partial^* \mathscr{G}_i(\bar{A}) - \bar{\lambda}_i^{\mathcal{H}} co\partial^* \mathscr{H}_i(\bar{A})] - \bar{\mathscr{W}},$$

where $\bar{\lambda}_i^{\mathcal{H}} = \lambda_i^{\mathcal{H}} - \lambda_i^\theta \mathscr{G}_i(\bar{A})$ ($i \in \mathbb{J}_{0+} \cup \mathbb{J}_{0-} \cup \mathbb{J}_{00} \cup \mathbb{J}_{+0}$),
$\bar{\lambda}_i^{\mathcal{H}} = \lambda_i^\theta = 0$ ($i \in \mathbb{J}_{+-}$), $\bar{\lambda}_i^{\mathscr{G}} = \lambda^\theta \mathscr{H}_i(\bar{A})$ ($i \in \mathbb{J}_{0+} \cup \mathbb{J}_{0-} \cup \mathbb{J}_{00} \cup \mathbb{J}_{+0}$),
$\bar{\lambda}_i^{\mathscr{G}} = \lambda_i^\theta = 0$ ($i \in \mathbb{J}_{+-}$).

Thus, we have

$$0 \in \sum_{i \in \mathbb{J}_f} \bar{\lambda}_i^f co\partial^* \mathfrak{f}_i(\bar{A}) + \sum_{i=1}^{m} [\bar{\lambda}_i^{\mathscr{G}} co\partial^* \mathscr{G}_i(\bar{A}) - \bar{\lambda}_i^{\mathcal{H}} co\partial^* \mathscr{H}_i(\bar{A})] - \bar{\mathscr{W}},$$

$\bar{\lambda}_i^f \geqq 0$ ($i \in \mathbb{J}_f$), $\langle \bar{\mathscr{W}}, \bar{A} \rangle = 0$, $\bar{\lambda}_i^{\mathcal{H}} = 0$ ($i \in \mathbb{J}_{+0} \cup \mathbb{J}_{+-}$), $\bar{\lambda}_i^{\mathcal{H}} \geqq 0$ ($i \in \mathbb{J}_{0-} \cup \mathbb{J}_{00}$), $\bar{\lambda}_i^{\mathcal{H}}$ free ($i \in \mathbb{J}_{0+}$),
$\bar{\lambda}_i^{\mathscr{G}} = 0$ ($i \in \mathbb{J}_{0+} \cup \mathbb{J}_{0-} \cup \mathbb{J}_{00} \cup \mathbb{J}_{+-}$), $\bar{\lambda}_i^{\mathscr{G}} \geqq 0$ ($i \in \mathbb{J}_{+0}$).

□

Definition 6. *The generalized Cottle constraint qualification (GCCQ) is said to satisfy at $\bar{A}$ if*

$$\left(\bigcup_{i \in \mathbb{J}_f^k} co\partial^* \mathfrak{f}_i(\bar{A}) \right)^s \cap \left(\bigcup_{i \in \mathbb{J}_{0+}} co\partial^* \mathscr{H}_i(\bar{A}) \bigcup_{i \in \mathbb{J}_{0+}} -co\partial^* \mathscr{H}_i(\bar{A}) \right.$$

$$\left(\bigcup_{i\in\mathbb{I}_{0-}\cup\mathbb{I}_{00}} -co\partial^*\mathcal{H}_i(\bar{A}) \bigcup_{i\in\mathbb{I}_{+0}} co\partial^*\mathcal{G}_i(\bar{A})\right)^s \bigcap M^n_+ \neq \emptyset, \ \forall k \in \mathbb{I}_\mathfrak{f}. \tag{10}$$

Theorem 2. *Let $\bar{A}$ be a local weak efficient solution for (S − MMPVC). Suppose that $\mathfrak{f}_i$ ($i \in \mathbb{I}_\mathfrak{f}$), $\mathcal{H}_i$ ($i \in \mathbb{I}_0$) and $\mathcal{G}_i$ ($i \in \mathbb{I}_{+0}$) admit bounded upper semi-regular convexificators and $\mathcal{H}_i$ ($i \in \mathbb{I}_+$), $\mathcal{G}_i$ ($i \in \mathbb{I}_0 \cup \mathbb{I}_{+-}$) are continuous. If (GCCQ) holds at $\bar{A}$ then there exist $\bar{\lambda}_i^\mathfrak{f} > 0$ ($i \in \mathbb{I}_\mathfrak{f}$), $\bar{\lambda}_i^\mathcal{H}, \bar{\lambda}_i^\mathcal{G} \in \mathbb{R}^m$, $\bar{\mathscr{W}} \in \mathbb{M}^n_+$, such that*

$$0 \in \sum_{i\in\mathbb{I}_\mathfrak{f}} \bar{\lambda}_i^\mathfrak{f} co\partial^*\mathfrak{f}_i(\bar{A}) + \sum_{i=1}^m [\bar{\lambda}_i^\mathcal{G} co\partial^*\mathcal{G}_i(\bar{A}) - \bar{\lambda}_i^\mathcal{H} co\partial^*\mathcal{H}_i(\bar{A})] - \bar{\mathscr{W}},$$

$$\langle \bar{\mathscr{W}}, \bar{A}\rangle = 0,\ \bar{\lambda}_i^\mathcal{H} = 0\ (i \in \mathbb{I}_{+0} \cup \mathbb{I}_{+-}),\ \bar{\lambda}_i^\mathcal{H} \geqq 0\ (i \in \mathbb{I}_{0-} \cup \mathbb{I}_{00}),\ \bar{\lambda}_i^\mathcal{H}\ free\ (i \in \mathbb{I}_{0+}),$$

$$\bar{\lambda}_i^\mathcal{G} = 0\ (i \in \mathbb{I}_{0+} \cup \mathbb{I}_{0-} \cup \mathbb{I}_{00} \cup \mathbb{I}_{+-}),\ \bar{\lambda}_i^\mathcal{G} \geqq 0\ (i \in \mathbb{I}_{+0}).$$

Proof. Since $\bar{A}$ is a local weak efficient solution, Theorem 1 implies that there exist $\bar{\lambda}_i^\mathfrak{f} \geqq 0$ ($i \in \mathbb{I}_\mathfrak{f}$), $\bar{\lambda}_i^\mathcal{H} \geqq 0$, $\bar{\lambda}_i^\mathcal{G} \geqq 0$ and $\bar{\mathscr{W}} \in \mathbb{M}^n_+$, such that

$$0 \in \sum_{i\in\mathbb{I}_\mathfrak{f}} \bar{\lambda}_i^\mathfrak{f} co\partial^*\mathfrak{f}_i(\bar{A}) + \sum_{i=1}^m [\bar{\lambda}_i^\mathcal{G} co\partial^*\mathcal{G}_i(\bar{A}) - \bar{\lambda}_i^\mathcal{H} co\partial^*\mathcal{H}_i(\bar{A})] - \bar{\mathscr{W}},$$

$$\langle \bar{\mathscr{W}}, \bar{A}\rangle = 0,\ \bar{\lambda}_i^\mathcal{H} = 0\ (i \in \mathbb{I}_{+0} \cup \mathbb{I}_{+-}),\ \bar{\lambda}_i^\mathcal{H} \geqq 0\ (i \in \mathbb{I}_{0-} \cup \mathbb{I}_{00}),\ \bar{\lambda}_i^\mathcal{H}\ free\ (i \in \mathbb{I}_{0+}),$$

$$\bar{\lambda}_i^\mathcal{G} = 0\ (i \in \mathbb{I}_{0+} \cup \mathbb{I}_{0-} \cup \mathbb{I}_{00} \cup \mathbb{I}_{+-}),\ \bar{\lambda}_i^\mathcal{G} \geqq 0\ (i \in \mathbb{I}_{+0}). \tag{11}$$

Without loss of generality, assume that $\lambda_1 = 0$, then there exist $\xi_i \in co\partial\mathfrak{f}_i(\bar{A})$ ($i \in \mathbb{I}_\mathfrak{f}^1$), $\eta_i \in co\partial\mathcal{H}_i(\bar{A})$, $\zeta_i \in co\partial\mathcal{G}_i(\bar{A})$, such that Equation (11) becomes

$$0 = \sum_{i\in\mathbb{I}_\mathfrak{f}^1} \bar{\lambda}_i^\mathfrak{f} \xi_i + \sum_{i=1}^m [\bar{\lambda}_i^\mathcal{G} \zeta_i - \bar{\lambda}_i^\mathcal{H} \eta_i] - \bar{\mathscr{W}}.$$

it follows from (GCCQ), there exists $A \in \mathbb{M}^n_+$ such that

$$0 > \sum_{i\in\mathbb{I}_\mathfrak{f}^1} \bar{\lambda}_i^\mathfrak{f} \langle \xi_i, A\rangle + \sum_{i=1}^m [\bar{\lambda}_i^\mathcal{G} \langle \zeta_i, A\rangle - \bar{\lambda}_i^\mathcal{H} \langle \eta_i, A\rangle] - \langle \bar{\mathscr{W}}, A\rangle$$

$$= \left\langle \sum_{i\in\mathbb{I}_\mathfrak{f}^1} \bar{\lambda}_i^\mathfrak{f} \xi_i + \sum_{i=1}^m [\bar{\lambda}_i^\mathcal{G} \zeta_i - \bar{\lambda}_i^\mathcal{H} \eta_i] - \bar{\mathscr{W}}, A \right\rangle = 0.$$

This contradicts the assumption. Thus, we obtain $\lambda_1^\mathfrak{f} > 0$. Repeating the above process for each $k \in \mathbb{I}_\mathfrak{f}$ we find the required result. □

Now, we introduce more relaxed constraint qualifications than (GCCQ).

Definition 7. *The generalized Guignard constraint qualification (GGCQ) is said to be hold at $\bar{A}$ if*

$$C = cone\ co\left(\bigcup_{i\in\mathbb{I}_{0+}} co\partial^*\mathcal{H}_i(\bar{A}) \bigcup_{i\in\mathbb{I}_{0+}} -co\partial^*\mathcal{H}_i(\bar{A})\right.$$

$$\left.\bigcup_{i\in\mathbb{I}_{0-}\cup\mathbb{I}_{00}} -co\partial^*\mathcal{H}_i(\bar{A}) \bigcup_{i\in\mathbb{I}_{+0}} co\partial^*\mathcal{G}_i(\bar{A})\right) - \mathbb{M}^n_+\ is\ closed\ set\ and$$

$$\left(\bigcup_{i\in\mathbb{I}_\mathfrak{f}} co\partial^*\mathfrak{f}_i(\bar{A})\right)^- \bigcap \left(\bigcup_{i\in\mathbb{I}_{0+}} co\partial^*\mathcal{H}_i(\bar{A}) \bigcup_{i\in\mathbb{I}_{0+}} -co\partial^*\mathcal{H}_i(\bar{A})\right.$$

$$\left(\bigcup_{i\in \mathbb{J}_{0-}\cup \mathbb{J}_{00}} -co\partial^*\mathscr{H}_i(\bar{A}) \bigcup_{i\in \mathbb{J}_{+0}} co\partial^*\mathscr{G}_i(\bar{A})\right)^{-} \bigcap M_+^n \subset \bigcap_{i=1}^{p} coT(Q^i, \bar{A}).$$

Lemma 2. Let $\bar{A}$ be any feasible solution to problem $(S - MMPVC)$. Suppose that $\mathfrak{f}_i$ $(i \in \mathbb{J}_\mathfrak{f})$, $\mathscr{H}_i$ $(i \in \mathbb{J}_0)$, $\mathscr{G}_i$ $(i \in \mathbb{J}_{+0})$, admit bounded upper semi-regular convexificators and for each $\mathscr{H}_i$ $(i \in \mathbb{J}_+)$, $\mathscr{G}_i$ $(i \in \mathbb{J}_0 \cup \mathbb{J}_{+-})$, are continuous. If C is closed and GCCQ holds at $\bar{A}$, then GGCQ holds at $\bar{A}$.

Proof. Without loss of generality, we assume that A satisfies GCCQ for $k = 1$.

$$A \in \left(\bigcup_{i\in \mathbb{J}_\mathfrak{f}^1} co\partial^*\mathfrak{f}_i(\bar{A})\right)^{-} \bigcap \left(\bigcup_{i\in \mathbb{J}_{0+}} co\partial^*\mathscr{H}_i(\bar{A}) \bigcup_{i\in \mathbb{J}_{0+}} -co\partial^*\mathscr{H}_i(\bar{A})\right.$$
$$\left.\bigcup_{i\in \mathbb{J}_{0-}\cup \mathbb{J}_{00}} -co\partial^*\mathscr{H}_i(\bar{A}) \bigcup_{i\in \mathbb{J}_{+0}} co\partial^*\mathscr{G}_i(\bar{A})\right)^{-} \bigcap M_+^n \neq \emptyset. \quad (12)$$

Since all $\mathfrak{f}_i$ $(i \in \mathbb{J}_\mathfrak{f})$, $\mathscr{H}_i$ $(i \in \mathbb{J}_0)$, $\mathscr{G}_i$ $(i \in \mathbb{J}_{+0})$, admit bounded upper semi-regular convexificators, so we have

$$\mathfrak{f}_i^+(\bar{A}; A) < 0, \forall i \in \mathbb{J}_\mathfrak{f}^1,$$
$$-\mathscr{H}_i^+(\bar{A}; A) < 0, \forall i \in \mathbb{J}_0,$$
$$\mathscr{G}_i^+(\bar{A}; A) < 0, \forall i \in \mathbb{J}_{+0}.$$

Since $\mathbb{M}_+^n$ is a convex cone, there exists $\tau > 0$, such that

$$\mathfrak{f}_i(\bar{A}+tA) < \mathfrak{f}_i(\bar{A}) \ (i \in \mathbb{J}_\mathfrak{f}^1), \ -\mathscr{H}_i(\bar{A}+tA) < 0, \forall i \in \mathbb{J}_0, \ \mathscr{G}_i(\bar{A}+tA) < 0, \forall i \in \mathbb{J}_{+0},$$
$$\bar{A}+tA \in \mathbb{M}_+^n \ \forall t \in (0, \tau). \quad (13)$$

On the other hand $\mathscr{H}_i$ $(i \in \mathbb{J}_+)$, $\mathscr{G}_i$ $(i \in \mathbb{J}_0 \cup \mathbb{J}_{+-})$ are a continuous. Therefore, there exists $\tau > 0$, such that

$$-\mathscr{H}_i(\bar{A}+tA) < 0 \ (i \in \mathbb{J}_+), \ \mathscr{G}_i(\bar{A}+tA) < 0 \ (i \in \mathbb{J}_0 \cup \mathbb{J}_{+-}) \ \bar{A}+tA \in \mathbb{M}_+^n, \ t \in (0, \tau).$$

Thus, $A \in T(Q^1, \bar{A})$. Therefore, we have

$$\mathscr{A} = \left(\bigcup_{i\in \mathbb{J}_\mathfrak{f}} co\partial^*\mathfrak{f}_i(\bar{A})\right)^{-} \bigcap \left(\bigcup_{i\in \mathbb{J}_{0+}} co\partial^*\mathscr{H}_i(\bar{A}) \bigcup_{i\in \mathbb{J}_{0+}} -co\partial^*\mathscr{H}_i(\bar{A})\right.$$
$$\left.\bigcup_{i\in \mathbb{J}_{0-}\cup \mathbb{J}_{00}} -co\partial^*\mathscr{H}_i(\bar{A}) \bigcup_{i\in \mathbb{J}_{+0}} co\partial^*\mathscr{G}_i(\bar{A})\right)^{-} \bigcap M_+^n$$
$$= cl\left(\left(\bigcup_{i\in \mathbb{J}_\mathfrak{f}} co\partial^*\mathfrak{f}_i(\bar{A})\right)^s \bigcap \left(\bigcup_{i\in \mathbb{J}_{0+}} co\partial^*\mathscr{H}_i(\bar{A}) \bigcup_{i\in \mathbb{J}_{0+}} -co\partial^*\mathscr{H}_i(\bar{A})\right.\right.$$
$$\left.\left.\bigcup_{i\in \mathbb{J}_{0-}\cup \mathbb{J}_{00}} -co\partial^*\mathscr{H}_i(\bar{A}) \bigcup_{i\in \mathbb{J}_{+0}} co\partial^*\mathscr{G}_i(\bar{A})\right)^s \bigcap M_{++}^n\right)$$
$$\subset cl\left(\left(\bigcup_{i\in \mathbb{J}_\mathfrak{f}^1} co\partial^*\mathfrak{f}_i(\bar{A})\right)^s \bigcap \left(\bigcup_{i\in \mathbb{J}_{0+}} co\partial^*\mathscr{H}_i(\bar{A}) \bigcup_{i\in \mathbb{J}_{0+}} -co\partial^*\mathscr{H}_i(\bar{A})\right.\right.$$
$$\left.\left.\bigcup_{i\in \mathbb{J}_{0-}\cup \mathbb{J}_{00}} -co\partial^*\mathscr{H}_i(\bar{A}) \bigcup_{i\in \mathbb{J}_{+0}} co\partial^*\mathscr{G}_i(\bar{A})\right)^s \bigcap M_{++}^n\right)$$

$$\subset clcoT(Q^1, \bar{A}) = coT(Q^1, \bar{A}).$$

Similarly, it can be proved that $\mathscr{A} \subset coT(Q^i, \bar{A}), \forall i \in \mathbb{J}_f$. Therefore

$$\left(\bigcup_{i \in \mathbb{J}_f} co\partial^* \mathfrak{f}_i(\bar{A})\right)^- \cap \left(\bigcup_{i \in \mathbb{J}_{0+}} co\partial^* \mathscr{H}_i(\bar{A}) \bigcup_{i \in \mathbb{J}_{0+}} -co\partial^* \mathscr{H}_i(\bar{A})\right.$$

$$\left.\bigcup_{i \in \mathbb{J}_{0-} \cup \mathbb{J}_{00}} -co\partial^* \mathscr{H}_i(\bar{A}) \bigcup_{i \in \mathbb{J}_{+0}} co\partial^* \mathscr{G}_i(\bar{A})\right)^- \cap M_+^n \subset \bigcap_{i=1}^p coT(Q^i, \bar{A}).$$

□

We present an example to show that converse of the above Lemma (2) does not hold.

Example 1. *Consider the problem*

$$\min (\mathfrak{f}_1(A), \mathfrak{f}_2(A)), \text{ subject to } \mathscr{H}(A) = x_1 \geqq 0, \mathscr{G}(A)\mathscr{H}(A) = x_3.x_1 \leqq 0,$$

$$A = \begin{bmatrix} x_1 & x_2 \\ x_2 & x_3 \end{bmatrix} \in \mathbb{M}_+^2, \text{ where } \mathfrak{f}_1(A) = |x_1|, \mathfrak{f}_2(A) = |x_3|.$$

Feasible set $M = \left\{\begin{bmatrix} x_1 & x_2 \\ x_2 & x_3 \end{bmatrix} \in \mathbb{M}_+^2 : x_1 \geqq 0, x_1 x_3 \leqq 0\right\}$. Since $\bar{A} = \begin{bmatrix} 0 & 0 \\ 0 & 0 \end{bmatrix}$, is weak efficient solution for the considered problem. Now, we can find upper semi-regular convexificator of each functions at point $\bar{A}$ as follows:

$$\partial^* \mathfrak{f}_1(\bar{A}) = \left\{\begin{bmatrix} -1 & 0 \\ 0 & 0 \end{bmatrix}, \begin{bmatrix} 1 & 0 \\ 0 & 0 \end{bmatrix}\right\}, \partial^* \mathfrak{f}_2(\bar{A}) = \left\{\begin{bmatrix} 0 & 0 \\ 0 & -1 \end{bmatrix}, \begin{bmatrix} 0 & 0 \\ 0 & 1 \end{bmatrix}\right\},$$

$$\partial^* \mathscr{H}(\bar{A}) = \left\{\begin{bmatrix} 1 & 0 \\ 0 & 0 \end{bmatrix}\right\}, \partial^* \mathscr{G}(\bar{A}) = \left\{\begin{bmatrix} 0 & 0 \\ 0 & 1 \end{bmatrix}\right\}.$$

$$Q^1 = \left\{\begin{bmatrix} x_1 & x_2 \\ x_2 & x_3 \end{bmatrix} \in \mathbb{M}_+^2 : x_1 \geqq 0, x_2 = 0, x_3 = 0\right\},$$

$$Q^2 = \left\{\begin{bmatrix} x_1 & x_2 \\ x_2 & x_3 \end{bmatrix} \in \mathbb{M}_+^2 : x_1 = 0, x_2 = 0, x_3 \in \mathbb{R}\right\}.$$

So, we conclude that

$$\begin{bmatrix} 0 & 0 \\ 0 & 0 \end{bmatrix} \in \bigcap_{i=1}^2 coT(Q^i, \bar{A}) \text{ and } \bigcup_{i=1}^2 co\partial^* \mathfrak{f}_i(\bar{A}) = \left\{\begin{bmatrix} t & 0 \\ 0 & 0 \end{bmatrix}, \begin{bmatrix} 0 & 0 \\ 0 & s \end{bmatrix} : t, s \in [-1, 1]\right\},$$

thus, we have

$$\left(\bigcup_{i=1}^2 co\partial^* \mathfrak{f}_i(\bar{A})\right)^- = \left\{\begin{bmatrix} 0 & x_2 \\ x_2 & 0 \end{bmatrix} : x_2 \in \mathbb{R}\right\}.$$

Since,

$$co\partial^* \mathscr{H}(\bar{A}) = \left\{\begin{bmatrix} 1 & 0 \\ 0 & 0 \end{bmatrix}\right\}, \text{ then } \left(-co\partial^* \mathscr{H}(\bar{A})\right)^- = \left\{\begin{bmatrix} x_1 & x_2 \\ x_2 & x_3 \end{bmatrix} : x_1 \geqq 0\right\}.$$

Consequently, we have

$$\left(\bigcup_{i=1}^{2} co\partial^* f_i(\bar{A})\right)^{-} \cap \left(-co\partial^* \mathcal{H}(\bar{A})\right)^{-} \cap \mathbb{M}_{+}^{2} = \left\{\begin{bmatrix} 0 & 0 \\ 0 & 0 \end{bmatrix}\right\} \subset \bigcap_{i=1}^{2} coT(Q^i, \bar{A}).$$

Obviously, $C = cone\ co\partial^* \mathcal{H}(\bar{A}) - \mathbb{M}_{+}^{2}$ is closed set. Hence, (GGCQ) satisfied at $\bar{A}$.
Now,

$$\left(\bigcup_{i \in \mathbb{J}_f^1} co\partial^* f_i(\bar{A})\right)^s = \left(co\partial^* f_2(\bar{A})\right)^s = \varnothing, \left(\bigcup_{i \in \mathbb{J}_f^2} co\partial^* f_i(\bar{A})\right)^s = \left(co\partial^* f_1(\bar{A})\right)^s = \varnothing,$$

which implies that

$$\left(\bigcup_{i \in \mathbb{J}_f^k} co\partial^* f_i(\bar{A})\right)^s \cap \left(\bigcup_{i \in \mathbb{J}_{0+}} co\partial^* \mathcal{H}_i(\bar{A}) \bigcup_{i \in \mathbb{J}_{0+}} -co\partial^* \mathcal{H}_i(\bar{A})\right.$$

$$\left.\bigcup_{i \in \mathbb{J}_{0-} \cup \mathbb{J}_{00}} -co\partial^* \mathcal{H}_i(\bar{A}) \bigcup_{i \in \mathbb{J}_{+0}} co\partial^* \mathcal{G}_i(\bar{A})\right)^s \cap \mathbb{M}_{+}^{n} = \varnothing, \forall k \in \mathbb{J}_f.$$

Hence, GCCQ not satisfied.

Applying the generalized Guignard constraint qualification, we derive the Karush–Kuhn–Tucker type necessary optimality conditions for $(S - MMPVC)$.

Theorem 3. *Suppose $\bar{A}$ is a local weak efficient solution for $(S - MMPVC)$. Assume that f_i, $\mathcal{H}_i$, $\mathcal{G}_i$ admits bounded upper semi-regular convexificator $\partial^* f_i(\bar{A})$ $(i \in \mathbb{J}_f)$, $\partial^* \mathcal{H}_i(\bar{A})$ $(i \in \mathbb{J}_0)$, $\partial^* \mathcal{G}_i(\bar{A})$ $(i \in \mathbb{J}_{+0})$, respectively, at $\bar{A}$. If (GGCQ) holds at $\bar{A}$ then there exists $\bar{\lambda}_i^f > 0$ $(i \in \mathbb{J}_f)$, $\bar{\lambda}^{\mathcal{G}} \in \mathbb{R}^m$, $\bar{\lambda}^{\mathcal{H}} \in \mathbb{R}^m$ and $\bar{\mathcal{W}} \in \mathbb{M}_{+}^{n}$ such that*

$$0 \in \sum_{i \in \mathbb{J}_f} \bar{\lambda}_i^f co\partial^* f_i(\bar{A}) + \sum_{i=1}^{m} [\bar{\lambda}_i^{\mathcal{G}} co\partial^* \mathcal{G}_i(\bar{A}) - \bar{\lambda}_i^{\mathcal{H}} co\partial^* \mathcal{H}_i(\bar{A})] - \bar{\mathcal{W}},$$

$$\langle \bar{\mathcal{W}}, \bar{A} \rangle = 0, \bar{\lambda}_i^{\mathcal{H}} = 0\ (i \in \mathbb{J}_{+0} \cup \mathbb{J}_{+-}), \bar{\lambda}_i^{\mathcal{H}} \geqq 0\ (i \in \mathbb{J}_{0-} \cup \mathbb{J}_{00}), \bar{\lambda}_i^{\mathcal{H}}\ free\ (i \in \mathbb{J}_{0+}),$$

$$\bar{\lambda}_i^{\mathcal{G}} = 0\ (i \in \mathbb{J}_{0+} \cup \mathbb{J}_{0-} \cup \mathbb{J}_{00} \cup \mathbb{J}_{+-}), \bar{\lambda}_i^{\mathcal{G}} \geqq 0\ (i \in \mathbb{J}_{+0}).$$

Proof. For the claim of the theorem, it suffices to show that,

$$0 \in \sum_{i=1}^{p} \lambda_i^f co\partial^* f_i(\bar{A}) + C, \lambda^f > 0. \tag{14}$$

Suppose, on the contrary, assume that

$$0 \notin \sum_{i=1}^{p} \lambda_i^f co\partial^* f_i(\bar{A}) + C, \lambda^f > 0. \tag{15}$$

As f_i $(i \in \mathbb{J}_f)$ admits an upper semi-regular convexificator, this implies that the right side in (14) is a closed convex set in $\mathbb{M}^n$. The classical separation theorem implies that there exists $A \in \mathbb{M}^n$, such that

$$\langle \tau, A \rangle < 0, \forall \tau \in \sum_{i=1}^{p} \lambda_i^f co\partial^* f_i(\bar{A}) + C, \lambda^f > 0. \tag{16}$$

Consequently,

$$\langle \xi_i, A \rangle < 0, \forall \xi_i \in co\partial^* f_i(\bar{A})\ (i \in \mathbb{J}_f), \tag{17}$$

$$-\langle \eta_i, A \rangle \leqq 0, \ \forall \ \eta_i \in co\partial^* \mathcal{H}_i(\bar{A}) \ (i \in \mathbb{J}_{0-} \cup \mathbb{J}_{00}), \tag{18}$$

$$-\langle \eta_i, A \rangle \leqq 0, \ \forall \ \eta_i \in co\partial^* \mathcal{H}_i(\bar{A}) \ (i \in \mathbb{J}_{0+}), \tag{19}$$

$$\langle \eta_i, A \rangle \leqq 0, \ \forall \ \eta_i \in co\partial^* \mathcal{H}_i(\bar{A}) \ (i \in \mathbb{J}_{0+}), \tag{20}$$

$$\langle \zeta_i, A \rangle \leqq 0, \ \forall \ \zeta_i \in co\partial^* \mathcal{G}_i(\bar{A}) \ (i \in \mathbb{J}_{+0}), \tag{21}$$

$$-\langle \mathcal{W}, A \rangle \leqq 0, \ \forall \ \mathcal{W} \in \mathbb{M}_+^n. \tag{22}$$

Inequalities (17)–(22) and (GGCQ) implies that

$$A \in \left(\bigcup_{i \in \mathbb{J}_f} co\partial^* f_i(\bar{A}) \right)^- \cap \left(\bigcup_{i \in \mathbb{J}_{0+}} co\partial^* \mathcal{H}_i(\bar{A}) \bigcup_{i \in \mathbb{J}_{0+}} -co\partial^* \mathcal{H}_i(\bar{A}) \right.$$

$$\left. \bigcup_{i \in \mathbb{J}_{0-} \cup \mathbb{J}_{00}} -co\partial^* \mathcal{H}_i(\bar{A}) \bigcup_{i \in \mathbb{J}_{+0}} co\partial^* \mathcal{G}_i(\bar{A}) \right)^- \cap \mathbb{M}_+^n \subset \bigcap_{i=1}^p coT(Q^i, \bar{A}).$$

Hence, $A \in \bigcap_{i=1}^p coT(Q^i, \bar{A})$, which implies that, there exist $t_n \downarrow 0$, such that $\bar{A} + t_n A \in M$. Therefore, from (17), we obtain

$$f_i(\bar{A} + tA) < f_i(\bar{A}), \ \forall \ i \in \mathbb{J}_f.$$

Thus, we obtain the contradiction that the feasible point $\bar{A}$ is a local weak efficient solution for $(S - MMPVC)$. Hence, the result. □

Motivated by Achtziger and Kanzow [12] and Sadeghieh et al. [55], we define S-stationary point for S-MMPVC.

Definition 8. *A feasible point $\bar{A}$ is said to be weak S-stationary point for $(S - MMPVC)$ if there exist $\lambda^f \in \mathbb{R}^p$, $\lambda^{\mathcal{H}} \in \mathbb{R}^m$, $\lambda^{\mathcal{G}} \in \mathbb{R}^M$, $\mathcal{W} \in \mathbb{M}_+^n$, and not all multipliers along with $\mathcal{W}$ can be simultaneously zero, such that*

$$0 \in \sum_{i \in \mathbb{J}_f} \lambda_i^f co\partial^* f_i(\bar{A}) + \sum_{i=1}^m [\lambda_i^{\mathcal{G}} co\partial^* \mathcal{G}_i(\bar{A}) - \lambda_i^{\mathcal{H}} co\partial^* \mathcal{H}_i(\bar{A})] - \mathcal{W},$$

$\lambda_i^f \geqq 0 \ (i \in \mathbb{J}_f)$, $\langle \mathcal{W}, \bar{A} \rangle = 0$, $\lambda_i^{\mathcal{H}} = 0 \ (i \in \mathbb{J}_{+0} \cup \mathbb{J}_{+-})$, $\lambda_i^{\mathcal{H}} \geqq 0 \ (i \in \mathbb{J}_{0-} \cup \mathbb{J}_{00})$, $\lambda_i^{\mathcal{H}} \text{ free } (i \in \mathbb{J}_{0+})$, $\lambda_i^{\mathcal{G}} = 0 \ (i \in \mathbb{J}_{0+} \cup \mathbb{J}_{0-} \cup \mathbb{J}_{00} \cup \mathbb{J}_{+-})$, $\lambda_i^{\mathcal{G}} \geqq 0 \ (i \in \mathbb{J}_{+0})$.

Definition 9. *A feasible point $\bar{A}$ is said to be strong S-stationary point for $(S - MMPVC)$ if there exist $\lambda^f \in \mathbb{R}^p$, $\lambda^{\mathcal{H}} \in \mathbb{R}^m$, $\lambda^{\mathcal{G}} \in \mathbb{R}^M$ and $\mathcal{W} \in \mathbb{M}_+^n$, such that*

$$0 \in \sum_{i \in \mathbb{J}_f} \lambda_i^f co\partial^* f_i(\bar{A}) + \sum_{i=1}^m [\lambda_i^{\mathcal{G}} co\partial^* \mathcal{G}_i(\bar{A}) - \lambda_i^{\mathcal{H}} co\partial^* \mathcal{H}_i(\bar{A})] - \mathcal{W},$$

$\lambda_i^f > 0 \ (i \in \mathbb{J}_f)$, $\langle \mathcal{W}, \bar{A} \rangle = 0$, $\lambda_i^{\mathcal{H}} = 0 \ (i \in \mathbb{J}_{+0} \cup \mathbb{J}_{+-})$, $\lambda_i^{\mathcal{H}} \geqq 0 \ (i \in \mathbb{J}_{0-} \cup \mathbb{J}_{00})$, $\lambda_i^{\mathcal{H}} \text{ free } (i \in \mathbb{J}_{0+})$, $\lambda_i^{\mathcal{G}} = 0 \ (i \in \mathbb{J}_{0+} \cup \mathbb{J}_{0-} \cup \mathbb{J}_{00} \cup \mathbb{J}_{+-})$, $\lambda_i^{\mathcal{G}} \geqq 0 \ (i \in \mathbb{J}_{+0})$.

Note that, if multipliers of gradients of objective functions are strictly greater than zero, then it is considered as strong S-stationary conditions.

Example 2. *Consider following optimization problem*

$$\min\ (\mathfrak{f}_1(A), \mathfrak{f}_2(A)),\ \text{subject to}\ \mathscr{H}(A) = x_1 \geqq 0,\ \mathscr{G}(A)\mathscr{H}(A) = x_3.x_1 \leqq 0,$$

$$A = \begin{bmatrix} x_1 & x_2 \\ x_2 & x_3 \end{bmatrix} \in \mathbb{M}_+^2,\ \text{where}\ \mathfrak{f}_1(A) = |x_1 - 1|,\ \mathfrak{f}_2(A) = |x_3|.$$

Feasible set $M = \left\{ \begin{bmatrix} x_1 & x_2 \\ x_2 & x_3 \end{bmatrix} \in \mathbb{M}_+^2 : x_1 \geqq 0,\ x_1 x_3 \leqq 0 \right\}$. Since $\bar{A} = \begin{bmatrix} 1 & 0 \\ 0 & 0 \end{bmatrix}$ is weak efficient solution for the considered problem. Now, we can find upper semi-regular convexificator of each functions at point $\bar{A}$ as follows:

$$\partial^* \mathfrak{f}_1(\bar{A}) = \left\{ \begin{bmatrix} -1 & 0 \\ 0 & 0 \end{bmatrix}, \begin{bmatrix} 1 & 0 \\ 0 & 0 \end{bmatrix} \right\},\ \partial^* \mathfrak{f}_2(\bar{A}) = \left\{ \begin{bmatrix} 0 & 0 \\ 0 & -1 \end{bmatrix}, \begin{bmatrix} 0 & 0 \\ 0 & 1 \end{bmatrix} \right\},$$

$$\partial^* \mathscr{H}(\bar{A}) = \left\{ \begin{bmatrix} 1 & 0 \\ 0 & 0 \end{bmatrix} \right\},\ \partial^* \mathscr{G}(\bar{A}) = \left\{ \begin{bmatrix} 0 & 0 \\ 0 & 1 \end{bmatrix} \right\}.$$

$$Q^1 = \left\{ \begin{bmatrix} x_1 & x_2 \\ x_2 & x_3 \end{bmatrix} \in \mathbb{M}_+^2 : x_1 \geqq 0,\ x_2 = 0,\ x_3 = 0 \right\},\ Q^2 = \left\{ \begin{bmatrix} x_1 & x_2 \\ x_2 & x_3 \end{bmatrix} \in \mathbb{M}_+^2 : x_1 = 1,\ x_2 = 0,\ x_3 = 0 \right\}.$$

So, we conclude that

$$\begin{bmatrix} 0 & 0 \\ 0 & 0 \end{bmatrix} \in \bigcap_{i=1}^{2} coT(Q^i, \bar{A})\ \text{and}\ \bigcup_{i=1}^{2} co\partial^* \mathfrak{f}_i(\bar{A}) = \left\{ \begin{bmatrix} t & 0 \\ 0 & 0 \end{bmatrix}, \begin{bmatrix} 0 & 0 \\ 0 & s \end{bmatrix} : t, s \in [-1, 1] \right\},$$

thus, we have

$$\left(\bigcup_{i=1}^{2} co\partial^* \mathfrak{f}_i(\bar{A}) \right)^- = \left\{ \begin{bmatrix} x_1 & x_2 \\ x_2 & x_3 \end{bmatrix} : x_1 = 0,\ x_2 = 0,\ x_3 = 0 \right\}.$$

Since,

$$co\partial^* \mathscr{H}(\bar{A}) = \left\{ \begin{bmatrix} 1 & 0 \\ 0 & 0 \end{bmatrix} \right\},\ \text{then}\ \left(-co\partial^* \mathscr{H}(\bar{A}) \right)^- = \left\{ \begin{bmatrix} x_1 & x_2 \\ x_2 & x_3 \end{bmatrix} : x_1 \geqq 0 \right\}.$$

Consequently, we have

$$\left(\bigcup_{i=1}^{2} co\partial^* \mathfrak{f}_i(\bar{A}) \right)^- \cap \left(-co\partial^* \mathscr{H}(\bar{A}) \right)^- \cap \mathbb{M}_+^2 = \left\{ \begin{bmatrix} 0 & 0 \\ 0 & 0 \end{bmatrix} \right\} \subset \bigcap_{i=1}^{2} coT(Q^i, \bar{A}).$$

Obviously, $C = \text{cone}\ co\partial^* \mathscr{H}(\bar{A}) - \mathbb{M}_+^2$ is closed set. Hence, (GGCQ) satisfied at $\bar{A}$.
Now, for $\lambda_1^{\mathfrak{f}} = 1,\ \lambda_2^{\mathfrak{f}} = 1,\ \lambda^{\mathscr{H}} = 0,\ \bar{\mathscr{W}} = \begin{bmatrix} 0 & 0 \\ 0 & 1 \end{bmatrix},\ \xi_1 = \begin{bmatrix} 0 & 0 \\ 0 & 0 \end{bmatrix} \in co\partial^* \mathfrak{f}_1(\bar{A})$,
$\xi_2 = \begin{bmatrix} 0 & 0 \\ 0 & 1 \end{bmatrix} \in co\partial^* \mathfrak{f}_2(\bar{A})$, and $\eta = \begin{bmatrix} 1 & 0 \\ 0 & 0 \end{bmatrix} \in co\partial^* \mathscr{H}(\bar{A})$, we have

$$0 = \lambda_1^{\mathfrak{f}} \xi_1 + \lambda_2^{\mathfrak{f}} \xi_2 - \lambda^{\mathscr{H}} \eta - \bar{\mathscr{W}} = 1\begin{bmatrix} 0 & 0 \\ 0 & 0 \end{bmatrix} + 1\begin{bmatrix} 0 & 0 \\ 0 & 1 \end{bmatrix} - 0\begin{bmatrix} 1 & 0 \\ 0 & 0 \end{bmatrix} - \begin{bmatrix} 0 & 0 \\ 0 & 1 \end{bmatrix}$$

$$\in \lambda_1^{\mathfrak{f}} co\partial^* \mathfrak{f}_1(\bar{A}) + \lambda_2^{\mathfrak{f}} co\partial^* \mathfrak{f}_2(\bar{A}) - \lambda^{\mathscr{H}} co\partial^* \mathscr{H}(\bar{A}) - \bar{\mathscr{W}},$$

and $\langle \bar{A}, \bar{\mathscr{W}} \rangle = Tr\left(\begin{bmatrix} 1 & 0 \\ 0 & 0 \end{bmatrix} \begin{bmatrix} 0 & 0 \\ 0 & 1 \end{bmatrix} \right) = 0$. Hence, strong S−stationary conditions satisfied at weak efficient point $\bar{A}$.

Corollary 1. Let $\bar{A}$ be a local weak efficient solution for $(S - MMPVC)$. Suppose that $\mathfrak{f}_i$, $\mathscr{H}_i$, $\mathscr{G}_i$ admits bounded upper semi-regular convexificator $\partial^* \mathfrak{f}_i(\bar{A})$ $(i \in \mathbf{J}_\mathfrak{f})$, $\partial^* \mathscr{H}_i(\bar{A})$ $(i \in \mathbf{J}_0)$, $\partial^* \mathscr{G}_i(\bar{A})$ $(i \in \mathbf{J}_{+0})$, respectively, at $\bar{A}$. If (GGCQ) holds at $\bar{A}$ then there exists $\bar{\lambda}_i^\mathfrak{f} > 0$ $(i \in \mathbf{J}_\mathfrak{f})$, $\bar{\lambda}^\mathscr{G} \in \mathbb{R}^m$, $\bar{\lambda}^\mathscr{H} \in \mathbb{R}^m$ and $\bar{\mathscr{W}} \in \mathbb{M}_+^n$ such that

$$0 \in \sum_{i \in \mathbf{J}_\mathfrak{f}} \bar{\lambda}_i^\mathfrak{f} co \partial^* \mathfrak{f}_i(\bar{A}) + \sum_{i=1}^m [\bar{\lambda}_i^\mathscr{G} co \partial^* \mathscr{G}_i(\bar{A}) - \bar{\lambda}_i^\mathscr{H} co \partial^* \mathscr{H}_i(\bar{A})] - \bar{\mathscr{W}},$$

$\langle \bar{\mathscr{W}}, \bar{A} \rangle = 0$, $\bar{\lambda}_i^\mathscr{H} = 0$ $(i \in \mathbf{J}_{+0} \cup \mathbf{J}_{+-})$, $\bar{\lambda}_i^\mathscr{H} \geqq 0$ $(i \in \mathbf{J}_{0-} \cup \mathbf{J}_{00})$, $\bar{\lambda}_i^\mathscr{H}$ free $(i \in \mathbf{J}_{0+})$,

$$\sum_{i=1}^p \bar{\lambda}_i^\mathfrak{f} = 1, \quad \bar{\lambda}_i^\mathscr{G} = 0 \ (i \in \mathbf{J}_{0+} \cup \mathbf{J}_{0-} \cup \mathbf{J}_{00} \cup \mathbf{J}_{+-}), \quad \bar{\lambda}_i^\mathscr{G} \geqq 0 \ (i \in \mathbf{J}_{+0}).$$

Proof. Since, all conditions of Theorem 3 are satisfying for some $\lambda^\mathfrak{f} > 0$, $\lambda^\mathscr{H}$, $\lambda^\mathscr{G} \in \mathbb{R}^m$, and $\mathscr{W}$ as follows:

$$0 \in \sum_{i \in \mathbf{J}_\mathfrak{f}} \lambda_i^\mathfrak{f} co \partial^* \mathfrak{f}_i(\bar{A}) + \sum_{i=1}^m [\lambda_i^\mathscr{G} co \partial^* \mathscr{G}_i(\bar{A}) - \lambda_i^\mathscr{H} co \partial^* \mathscr{H}_i(\bar{A})] - \mathscr{W}, \quad (23)$$

$\langle \mathscr{W}, \bar{A} \rangle = 0$, $\lambda_i^\mathscr{H} = 0$ $(i \in \mathbf{J}_{+0} \cup \mathbf{J}_{+-})$, $\lambda_i^\mathscr{H} \geqq 0$ $(i \in \mathbf{J}_{0-} \cup \mathbf{J}_{00})$, $\lambda_i^\mathscr{H}$ free $(i \in \mathbf{J}_{0+})$, $\lambda_i^\mathscr{G} = 0$ $(i \in \mathbf{J}_{0+} \cup \mathbf{J}_{0-} \cup \mathbf{J}_{00} \cup \mathbf{J}_{+-})$, $\lambda_i^\mathscr{G} \geqq 0$ $(i \in \mathbf{J}_{+0})$.

Now, dividing (23) by $\sum_{i=1}^p \lambda_i^\mathfrak{f}$ and taking

$$\bar{\lambda}_i^\mathfrak{f} = \frac{\lambda_i^\mathfrak{f}}{\sum_{i=1}^p \lambda_i}, \quad \bar{\lambda}_i^\mathscr{H} = \frac{\lambda_i^\mathscr{H}}{\sum_{i=1}^p \lambda_i^\mathfrak{f}}, \quad \bar{\lambda}_i^\mathscr{G} = \frac{\lambda_i^\mathscr{G}}{\sum_{i=1}^p \lambda_i^\mathfrak{f}}, \quad \bar{\mathscr{W}} = \frac{\mathscr{W}}{\sum_{i=1}^p \lambda_i^\mathfrak{f}},$$

we obtain the required result. □

Now, we propose some index sets to show sufficient optimality conditions for S-MMPVC:

$$\mathbf{J}_{00}^+ := \{i \in \mathbf{J}_{00} : \lambda_i^\mathscr{H} > 0\},$$
$$\mathbf{J}_{00}^0 := \{i \in \mathbf{J}_{00} : \lambda_i^\mathscr{H} = 0\},$$
$$\mathbf{J}_{0-}^+ := \{i \in \mathbf{J}_{0-} : \lambda_i^\mathscr{H} > 0\},$$
$$\mathbf{J}_{0-}^0 := \{i \in \mathbf{J}_{0-} : \lambda_i^\mathscr{H} = 0\},$$
$$\mathbf{J}_{0+}^+ := \{i \in \mathbf{J}_{0+} : \lambda_i^\mathscr{H} > 0\},$$
$$\mathbf{J}_{0+}^- := \{i \in \mathbf{J}_{0+} : \lambda_i^\mathscr{H} < 0\},$$
$$\mathbf{J}_{0+}^0 := \{i \in \mathbf{J}_{0+} : \lambda_i^\mathscr{H} = 0\},$$
$$\mathbf{J}_{+0}^{0+} := \{i \in \mathbf{J}_{+0} : \lambda_i^\mathscr{H} = 0, \lambda_i^\mathscr{G} > 0\},$$
$$\mathbf{J}_{+0}^{00} := \{i \in \mathbf{J}_{+0} : \lambda_i^\mathscr{H} = 0, \lambda_i^\mathscr{G} = 0\}.$$

Following result is motivated by Sadeghieh et al. ([55], Theorem 9).

Theorem 4. (Sufficient conditions) Suppose $\mathfrak{f}_i$ $(i \in \mathbf{J}_\mathfrak{f})$, $\mathscr{H}_i$ $(i \in \mathbf{J}_{0+} \cup \mathbf{J}_{00} \cup \mathbf{J}_{0-})$, $\mathscr{G}_i$ $(i \in \mathbf{J}_{+0})$ admit bounded upper semi-regular convexificators at $\bar{A}$. Assume that feasible point $\bar{A}$ satisfies weak S–stationary conditions under suitable choice of multipliers $\lambda^\mathfrak{f} \in \mathbb{R}^p, \lambda^\mathscr{H} \in \mathbb{R}^m$, $\lambda^\mathscr{G} \in \mathbb{R}^m, \mathscr{W} \in \mathbb{M}_+^n$ for $S - MMPVC$. If $\mathscr{H}_i$ $(i \in \mathbf{J}_{0+}^-)$, $-\mathscr{H}_i$ $(i \in \mathbf{J}_{0+}^+ \cup \mathbf{J}_{00}^+ \cup \mathbf{J}_{0-}^-)$, $\mathscr{G}_i$ $(i \in \mathbf{J}_{+0}^{0+})$, are ∂^*–quasiconvex and $\mathfrak{f}_i$ $(i \in \mathbf{J}_\mathfrak{f})$ are ∂^*–pseudoconvex at $\bar{A}$ and at least one $\lambda_i^\mathfrak{f} > 0$. Then,

(i) $\bar{A}$ is a local weak efficient solution for $S - MMPVC$;

(ii) In addition to that if $\mathbb{I}_{0+}^{-} \cup \mathbb{I}_{+0}^{0+} = \emptyset$, then $\bar{A}$ is a weak efficient solution for $S - MMPVC$.

Proof. (i) From continuity of $\mathscr{G}_i (i \in \mathbb{I}_{0+})$ and $\mathscr{H}_i (i \in \mathbb{I}_{+0})$ there exist neighborhoods $\mathcal{N}$ and $\mathcal{M}$ for $\bar{A}$, such that

$$\mathscr{H}_i(A) = 0, \mathscr{G}_i(A) > 0, \forall A \in M \cap \mathcal{N} \; \forall i \in \mathbb{I}_{0+}, \tag{24}$$

$$\mathscr{H}_i(A) > 0, \mathscr{G}_i(A) \leq 0, \forall A \in M \cap \mathcal{M} \; \forall i \in \mathbb{I}_{+0}. \tag{25}$$

Since $\bar{A}$ is a weak S-stationary point, so there exist $\lambda^{\mathfrak{f}} \in \mathbb{R}^p, \lambda^{\mathscr{H}} \in \mathbb{R}^m, \lambda^{\mathscr{G}} \in \mathbb{R}^m, \mathscr{W}$ and not all multipliers along with $\mathscr{W}$ can be simultaneously zero, such that satisfies weak S-stationary conditions. Thus, there exist $\xi_i \in co\partial^* \mathfrak{f}_i(\bar{A}) \; (i \in \mathbb{I}_{\mathfrak{f}})$, $\eta_i \in co\partial^* \mathscr{H}_i(\bar{A}) \; (i \in \mathbb{I}_0)$, $\zeta_i \in co\partial^* \mathscr{G}_i(\bar{A}) \; (i \in \mathbb{I}_{+0})$, such that

$$\sum_{i \in \mathbb{I}_{\mathfrak{f}}} \lambda_i^{\mathfrak{f}} \xi_i + \sum_{i \in \mathbb{I}_{+0}} \lambda_i^{\mathscr{G}} \zeta_i - \sum_{i \in \mathbb{I}_0} \lambda_i^{\mathscr{H}} \eta_i - \mathscr{W} = 0. \tag{26}$$

Suppose, on contrary $\bar{A}$ is not local weak efficient solution for $S - MMPVC$. Then, there exists $B \in M \cap \mathcal{N} \cap \mathcal{M}$, such that

$$\mathfrak{f}_i(B) < \mathfrak{f}_i(\bar{A}), \forall \, i \in \mathbb{I}_{\mathfrak{f}}. \tag{27}$$

By the ∂^*-pseudoconvexity of $\mathfrak{f}_i \; (i \in \mathbb{I}_{\mathfrak{f}})$ and (27), we obtain

$$\langle \xi_i, B - \bar{A} \rangle < 0, \forall \, i \in \mathbb{I}_{\mathfrak{f}}. \tag{28}$$

By the ∂^*-quasiconvexity of functions $\mathscr{G}_i \; (i \in \mathbb{I}_{+0}^{0+})$, $\mathscr{H}_i \; (i \in \mathbb{I}_{0+}^{-})$ and (24) and (25), we obtain

$$\mathscr{G}_i(B) \leq 0 = \mathscr{G}_i(\bar{A}) \implies \langle \zeta_i, B - \bar{A} \rangle \leq 0, \forall \, i \in \mathbb{I}_{+0}^{0+}. \tag{29}$$

$$\mathscr{H}_i(B) = 0 \leq \mathscr{H}_i(\bar{A}) \implies \langle \eta_i, B - \bar{A} \rangle \leq 0, \forall \, i \in \mathbb{I}_{0+}^{-}. \tag{30}$$

On the other hand, $\forall \, i \in \mathbb{I}_{0+}^{+} \cup \mathbb{I}_{0-}^{+} \cup \mathbb{I}_{00}^{+}$,

$$-\mathscr{H}_i(B) \leq 0 = -\mathscr{H}_i(\bar{A}) \implies \langle -\eta_i, B - \bar{A} \rangle \leq 0, \forall \, -\eta_i \in -co\partial^* \mathscr{H}_i(\bar{A}). \tag{31}$$

Since $\mathscr{W}, B \in \mathbb{M}_+^n$, so we have

$$-\langle \mathscr{W}, B \rangle + \langle \mathscr{W}, \bar{A} \rangle = -\langle \mathscr{W}, B - \bar{A} \rangle \leq 0. \tag{32}$$

Multiplying their corresponding multiplier in (29) to (32) and adding, we obtain contradictions to (26). Hence, the result.

(ii) We proceed similar to (i) and using $\mathbb{I}_{0+}^{+0} \cup \mathbb{I}_{+0}^{-} = \emptyset$, therefore without making use of neighborhood $\mathcal{N}$ and $\mathcal{M}$, we obtain the required result. □

To validate the sufficient optimality conditions we present following example.

Example 3. *Consider following optimization problem*

$$\min \, (\mathfrak{f}_1(A), \mathfrak{f}_2(A)), \text{ subject to } \mathscr{H}_1(A) = -x_2 \geq 0, \mathscr{G}_1(A)\mathscr{H}_1(A) = -|x_3|x_2 \leq 0,$$

$$A = \begin{bmatrix} x_1 & x_2 \\ x_2 & x_3 \end{bmatrix} \in \mathbb{M}_+^2, \text{where } \mathfrak{f}_1(A) = x_2, \mathfrak{f}_2(A) = x_3.$$

Feasible set,

$$M = \left\{ \begin{bmatrix} x_1 & x_2 \\ x_2 & x_3 \end{bmatrix} \in \mathbb{M}_+^2 : x_2 \leqq 0, |x_3|x_2 \geqq 0 \right\},$$

$$= \left\{ \begin{bmatrix} x_1 & x_2 \\ x_2 & x_3 \end{bmatrix} : x_1 \geqq 0, x_1 x_3 - x_2^2 \geqq 0, x_2 \leqq 0, |x_3|x_2 \geqq 0 \right\}.$$

Consider at feasible point $\bar{A} = \begin{bmatrix} 0 & 0 \\ 0 & 0 \end{bmatrix}$. *We observe that* $\mathfrak{f}_1$, $\mathfrak{f}_2$ *are* ∂^*−*pseudoconvex,* −$\mathscr{H}_1$ *is* ∂^*−*quasiconvex at* $\bar{A}$ *and* $\mathscr{H}_i$ ($i = 1 \in \mathbf{J}_{00}$), $\mathscr{G}_i$ ($i = 1 \notin \mathbf{J}_{+0}$) *also* $\mathbf{J}_{0+}^{+0} \cup \mathbf{J}_{+0}^{-} = \emptyset$. *Now, we can find upper semi-regular convexificator of each functions at point* $\bar{A}$ *as follows:*

$$\partial^* \mathfrak{f}_1(\bar{A}) = \left\{ \begin{bmatrix} 0 & \frac{1}{2} \\ \frac{1}{2} & 0 \end{bmatrix} \right\}, \ \partial^* \mathfrak{f}_2(\bar{A}) = \left\{ \begin{bmatrix} 0 & 0 \\ 0 & 1 \end{bmatrix} \right\}, \ \partial^* \mathscr{H}_1(\bar{A}) = \left\{ \begin{bmatrix} 0 & -\frac{1}{2} \\ -\frac{1}{2} & 0 \end{bmatrix} \right\}.$$

Thus, for $\lambda_1^\mathfrak{f} = 0$, $\lambda_2^\mathfrak{f} > 0$, $\lambda_1^\mathscr{H} = 0$, *and* $\mathscr{W} = \begin{bmatrix} 0 & 0 \\ 0 & \lambda_2^\mathfrak{f} \end{bmatrix}$, *we have*

$$\lambda_1^\mathfrak{f} co \partial^* \mathfrak{f}_1(\bar{A}) + \lambda_2^\mathfrak{f} co \partial^* \mathfrak{f}_2(\bar{A}) - \lambda_1^\mathscr{H} co \partial^* \mathscr{H}_1(\bar{A}) - \mathscr{W} = 0.$$

That is, $\bar{A}$ *satisfying weak* S−*stationary conditions. Hence,* $\bar{A}$ *is weak efficient solution, which is true by simple observations.*

4. Conclusions and Future Remarks

Golestani and Nobakhtian [11] established optimality conditions for nonsmooth semidefinite single optimization problems. We have established the optimality conditions for a more interesting class of nonlinear optimization namely, mathematical programming problems with vanishing constraints (MPVC), which is more applicable in topology optimization and many real-life problems. We have further extended the single objective semidefinite optimization problems to multiobjective semidefinite optimization problems. We established Fritz John stationary conditions for nonsmooth, nonlinear, semidefinite, multiobjective programs with vanishing constraints using convexificator and generalized Cottle type and generalized Guignard type constraints qualification have been introduced to achieve strong S−stationary conditions from Fritz John stationary conditions. Sufficient conditions are also established under generalized convexity assumptions and through an example, we validate our established results. We have used the constraint qualifications technique motivated by Li [38] and provided some generalized constraint qualifications for semidefinite optimization problems. We have also used the linearization technique inspired by Kanzow et al. [56]. Recently, Treanta [41] discussed duality theorems for a special class of quasiinvex multiobjective optimization problems for interval-valued components. Further, Treanta established dual pair of multiobjective interval-valued variational control problems. We can extend the results on multiobjective semidefinite optimization problems to variational control problems and interval-valued optimization problems motivated by [40,41,57–61] for the application point of view.

Author Contributions: Writing-original draft preparation, K.K.L., M.H., S.K.S., J.K.M. and S.K.M.; writing-review and editing, K.K.L., M.H., S.K.S., J.K.M. and S.K.M.; funding acquisition, K.K.L. All authors have read and agreed to the published version of the manuscript.

Funding: The second author is financially supported by CSIR-UGC JRF, New Delhi, India, through Reference no.: 1009/(CSIR-UGC NET JUNE 2018). The third author is financially supported by CSIR-UGC JRF, New Delhi, India, through Reference no.: 1272/(CSIR-UGC NET DEC.2016). The fifth author is financially supported by "Research Grant for Faculty" (IoE Scheme) under Dev. Scheme NO. 6031 and Department of Science and Technology, SERB, New Delhi, India, through grant no.: MTR/2018/000121.

Institutional Review Board Statement: Not Applicable.

Informed Consent Statement: Not Applicable.

Data Availability Statement: No data were used to support this study.

Acknowledgments: The authors are indebted to the anonymous reviewers for their valuable comments and remarks that helped to improve the presentation and quality of the manuscript.

Conflicts of Interest: The authors declare no conflict of interest.

References

1. Lobo, M.S.; Vandenberghe, L.; Boyd, S.; Lebret, H. Applications of second-order cone programming. *Linear Algebra Its Appl.* **1998**, *284*, 193–228. [CrossRef]
2. Vandenberghe, L.; Boyd, S. Semidefinite programming. *SIAM Rev.* **1996**, *38*, 49–95. [CrossRef]
3. Fares, B.; Noll, D.; Apkarian, P. Robust control via sequential semidefinite programming. *SIAM J. Control Optim.* **2002**, *4*, 1791–1820. [CrossRef]
4. Ben-Tal, A.; Jarre, F.; Kočvara, M.; Nemirovski, A.; Zowe, J. Optimal design of trusses under a nonconvex global buckling constraint. *Optim. Eng.* **2000**, *1*, 189–213. [CrossRef]
5. Leibfritz, F. *COMPleib 1.1: COnstraint Matrix-Optimization Problem Library—A Collection of Test Examples for Nonlinear Semidefinite Programs, Control System Design and Related Problems*; Technical Report; Department of Mathematics, University of Trier: Trier, Germany, 2005.
6. Shapiro, A. First and second order analysis of nonlinear semidefinite programs. *Math. Program.* **1997**, *77*, 301–320. [CrossRef]
7. Sun, D.; Sun, J.; Zhang, L. The rate of convergence of the augmented Lagrangian method for nonlinear semidefinite programming. *Math. Program. Ser. A* **2008**, *114*, 349–391. [CrossRef]
8. Forsgren, A. Optimality conditions for nonconvex semidefinite programming. *Math. Program. Ser. A* **2000**, *88*, 105–128. [CrossRef]
9. Sun, J. On method for solving nonlinear semidefinite optimization problems. *Numer. Algebr. Control Optim.* **2011**, *1*, 1–14. [CrossRef]
10. Yamashita, H.; Yabe, H. A survey of numerical methods doe nonlinear semidefinite programming. *J. Oper. Res. Soc. Jpn.* **2015**, *58*, 24–60.
11. Golestani, M.; Nobakhtian, S. Optimality conditions for nonsmooth semidefinite programming via convexificators. *Positivity* **2015**, *19*, 221–236. doi:10.1007/s11117-014-0292-6. [CrossRef]
12. Achtziger, W.; Kanzow, C. Mathematical programs with vanishing constraints: optimality conditions and constraint qualifications. *Math. Program. Ser. A* **2008**, *114*, 69–99. [CrossRef]
13. Kirches, C.; Potschka, A.; Bock, H.G.; Sager, S. A parametric active-set method for QPS with vanishing constraints arising in a robot motion planning problem. *Pac. J. Optim.* **2013**, *9*, 275–299.
14. Jung, M.N.; Kirches, C.; Sager, S. *On Perspective Functions and Vanishing Constraints in Mixed-Integer Nonlinear Optimal Control*; Springer: Berlin/Heidelberg, Germany, 2013; pp. 387–417.
15. Jabr, R. Solution to economic dispatching with disjoint feasible regions via semidefinite programming. *IEEE Trans. Power Syst.* **2012**, *27*, 572–573. [CrossRef]
16. Guu, S.M.; Singh, Y.; Mishra, S.K. On strong KKT type sufficient optimality conditions for multiobjective semi-infinite programming problems with vanishing constraints. *J. Inequalities Appl.* **2017**, *2017*, 1–9. [CrossRef] [PubMed]
17. Mishra, S.K.; Singh, V.; Laha, V. On duality for mathematical programs with vanishing constraints. *Ann. Oper. Res.* **2016**, *243*, 249–272. [CrossRef]
18. Mishra, S.K.; Singh, V.; Laha, V.; Mohapatra, R.N. On constraint qualifications for multiobjective optimization problems with vanishing constraints. In *Optimization Methods, Theory and Applications*; Springer: Berlin/Heidelberg, Germany, 2015; pp. 95–135.
19. Luo, Z.Q.; Pang, J.S.; Ralph, D. *Mathematical Programs with Equilibrium Constraints*; Cambridge University Press: Cambridge, UK, 1996.
20. Outrata, J.; Kocvara, M.; Zowe, J. Nonsmooth approach to optimization problems with equilibrium constraints. In *Nonconvex Optimization and Its Applications*; Kluwer Academic Publishers: Dordrecht, The Netherlands, 1998.
21. Mishra, S.K.; Jaiswal, M. Optimality conditions and duality for semi-infinite mathematical programming problem with equilibrium constraints. *Numer. Funct. Anal. Optim.* **2015**, *36*, 460–480. [CrossRef]
22. Pandey, Y.; Mishra, S.K. Optimality conditions and duality for semi-infinite mathematical programming problems with equilibrium constraints, using convexificators. *Ann. Oper. Res.* **2018**, *269*, 549–564. [CrossRef]
23. Pandey, Y.; Mishra, S.K. On strong KKT type sufficient optimality conditions for nonsmooth multiobjective semi-infinite mathematical programming problems with equilibrium constraints. *Oper. Res. Lett.* **2016**, *44*, 148–151. [CrossRef]
24. Singh, Y.; Pandey, Y.; Mishra, S.K. Saddle point optimality criteria for mathematical programming problems with equilibrium constraints. *Oper. Res. Lett.* **2017**, *45*, 254–258. [CrossRef]
25. Guu, S.M.; Mishra, S.K.; Pandey, Y. Duality for nonsmooth mathematical programming problems with equilibrium constraints. *J. Inequalities Appl.* **2016**, *2016*, 28. [CrossRef]

26. Pandey, Y.; Mishra, S.K. Duality of mathematical programming problems with equilibrium constraints. *Pac. J. Optim.* **2017**, *13*, 105–122.
27. Mishra, S.K.; Pandey, Y. On sufficiency for mathematical programming problems with equilibrium constraints. *Yugosl. J. Oper. Res.* **2013**, *23*, 173–182.
28. Pandey, Y.; Mishra, S.K. Duality for nonsmooth optimization problems with equilibrium constraints, using convexificators. *J. Optim. Theory Appl.* **2016**, *171*, 694–707. [CrossRef]
29. Hoheisel, T.; Kanzow, C. First- and second-order optimality conditions for mathematical programs with vanishing constraints. *Appl. Math.* **2007**, *52*, 495–514. [CrossRef]
30. Hoheisel, T.; Kanzow, C. Stationary conditions for mathematical programs with vanishing constraints using weak constraint qualifications. *J. Math. Anal. Appl.* **2008**, *337*, 292–310. [CrossRef]
31. Hoheisel, T.; Kanzow, C. On the Abadie and Guignard constraint qualifications for mathematical programmes with vanishing constraints. *Optimization* **2009**, *58*, 431–448. [CrossRef]
32. Hoheisel, T.; Kanzow, C.; Outrata, J.V. Exact penalty results for mathematical programs with vanishing constraints. *Nonlinear Anal.* **2010**, *72*, 2514–2526. [CrossRef]
33. Izmailov, A.F.; Pogosyan, A.L. Optimality conditions and newton-type methods for mathematical programs with vanishing constraints. *Comput. Math. Math. Phys.* **2009**, *49*, 1128–1140. [CrossRef]
34. Miettinen, K. *Nonlinear Multiobjective Optimization*; Kluwer Academic Publishers: Boston, MA, USA, 1999.
35. Wendell, R.E.; Lee, D.N. Efficiency in multiple objective optimization problems. *Math. Program* **1977**, *12*, 406–414. [CrossRef]
36. Maeda, T. Constraint qualifications in multiobjective optimization problems: Differentiable case. *J. Optim. Theory Appl.* **1994**, *80*, 483–500. [CrossRef]
37. Preda, V.; Chiţescu, I. On constraint qualification in multiobjective optimization problems: Semidifferentiable case. *J. Optim. Theory Appl.* **1999**, *100*, 417–433. [CrossRef]
38. Li, X.F. Constraint qualifications in nonsmooth multiobjective optimization. *J. Optim. Theory Appl.* **2000**, *106*, 373–398. [CrossRef]
39. Lai, K.K.; Hassan, M.; Maurya, J.K.; Singh, S.K.; Mishra, S.K. Multiobjective Convex Optimization in Real Banach Space. *Symmetry* **2021**, *13*, 2148. [CrossRef]
40. Treanţă, S. On a Dual Pair of Multiobjective Interval-Valued Variational Control Problems. *Mathematics* **2021**, *9*, 893. [CrossRef]
41. Treanta, S. Duality Theorems for (ρ, ψ, d)—Quasiinvex multiobjective optimization problems with interval-valued components. *Mathematics* **2021**, *9*, 894. [CrossRef]
42. Demyanov, V.F.; Jeyakumar, V. Hunting for a smaller convex subdifferential. *J. Glob. Optim.* **1997**, *10*, 305–326. [CrossRef]
43. Jeyakumar, V.; Luc, D.T. Nonsmooth calculus, minimality, and monotonicity of convexificators. *J. Optim. Theory Appl.* **1999**, *101*, 599–621. [CrossRef]
44. Dorsch, D.; Gómez, W.; Shikhman, V. Sufficient optimality conditions hold for almost all nonlinear semidefinite programs. *Math. Program.* **2016**, *158*, 77–97. [CrossRef]
45. Yurtsever, A.; Tropp, J.A.; Fercoq, O.; Udell, M.; Cevherk, V. Scalable Semidefinite Programming. *SIAM J. Math. Data Sci.* **2021**, *3*, 171–200. [CrossRef]
46. Andreani, R.; Haeser, G.; Viana, D.S. Optimality conditions and global convergence for nonlinear semidefinite programming. *Math. Program. Ser. A* **2020**, *180*, 203–235. [CrossRef]
47. Andreani, R.; Haeser, G.; Mito, L.M.; Ramírez, H.; Santos, D.O.; Silveira, T.P. Naive constant rank-type constraint qualifications for multifold second-order cone programming and semidefinite programming. *Optim. Lett.* **2021**. **2021**, 1–22. doi: 10.1007/s11590-021-01737-w. [CrossRef]
48. Berman, A.; Shaked-Monderer, N. *Completely Positive Matrices*; World Scientific Publishing Co., Inc.: River Edge, NJ, USA, 2003.
49. Boyd, S.; El Ghaoui, L.; Feron, E.; Balakrishnan, V. Linear Matrix Inequalities in System and Control Theory. In *SIAM Studies in Applied Mathematics*; Society for Industrial and Applied Mathematics (SIAM): Philadelphia, PA, USA, 1994; Volume 15.
50. Clarke, F.H. *Optimization and Nonsmooth Analysis, Classics in Applied Mathematics*, 2nd ed.; Society for Industrial and Applied Mathematics (SIAM): Philadelphia, PA, USA, 1990; Volume 5.
51. Michel, P.; Penot, J.P. A generalized derivative for calm and stable functions. *Differ. Integr. Equ.* **1992**, *5*, 433–454.
52. Mordukhovich, B.S.; Shao, Y.H. On nonconvex subdifferential calculus in Banach spaces. *J. Convex Anal.* **1995**, *2*, 211–227.
53. Treiman, J.S. The linear nonconvex generalized gradient and Lagrange multipliers. *SIAM J. Optim.* **1995**, *5*, 670–680. [CrossRef]
54. Fan, J. Generalized separation theorems and the Farkas' lemma. *Appl. Math. Lett.* **2005**, *18*, 791–796. [CrossRef]
55. Sadeghieh, A.; Kanzi, N.; Caristi, G.; Barilla, D. On stationarity for nonsmooth multiobjective problems with vanishing constraints. *J. Glob. Optim.* **2021**, *2021*, 1–21. doi: 10.1007/s10898-021-01030-1. [CrossRef]
56. Kanzow, C.; Nagel, C.; Kato, H.; Fukushima, M. Successive linearization methods for nonlinear semidefinite programs. *Comput. Optim. Appl.* **2005**, *31*, 251–273. [CrossRef]
57. Treanţă, S. On locally and globally optimal solutions in scalar variational control problems. *Mathematics* **2019**, *7*, 829. [CrossRef]
58. Treanta, S.; Udriste, C. On efficiency conditions for multiobjective variational problems involving higher order derivatives. In Proceedings of the 15th International Conference on Automatic Control, Modelling & Simulation (ACMOS'13), Brasov, Romania, 1–3 June 2013; pp. 157–164.
59. Treanţă, S. Well Posedness of New Optimization Problems with Variational Inequality Constraints. *Fractal Fract.* **2021**, *5*, 123. [CrossRef]

60. Mititelu, Ş.; Treanţă, S. Efficiency conditions in vector control problems governed by multiple integrals. *J. Appl. Math. Comput.* **2018**, *57*, 647–665. [CrossRef]
61. Treanţă, S. On modified interval-valued variational control problems with first-order PDE constraints. *Symmetry* **2020**, *12*, 472. [CrossRef]

fractal and fractional

Article

Some Hadamard–Fejér Type Inequalities for LR-Convex Interval-Valued Functions

Muhammad Bilal Khan [1], Savin Treanță [2], Mohamed S. Soliman [3], Kamsing Nonlaopon [4,*] and Hatim Ghazi Zaini [5]

1 Department of Mathematics, COMSATS University Islamabad, Islamabad 44000, Pakistan; bilal42742@gmail.com
2 Department of Applied Mathematics, University Politehnica of Bucharest, 060042 Bucharest, Romania; savin.treanta@upb.ro
3 Department of Electrical Engineering, College of Engineering, Taif University, P.O. Box 11099, Taif 21944, Saudi Arabia; soliman@tu.edu.sa
4 Department of Mathematics, Faculty of Science, Khon Kaen University, Khon Kaen 40002, Thailand
5 Department of Computer Science, College of Computers and Information Technology, Taif University, P.O. Box 11099, Taif 21944, Saudi Arabia; h.zaini@tu.edu.sa
* Correspondence: nkamsi@kku.ac.th; Tel.: +66-8-6642-1582

Abstract: The purpose of this study is to introduce the new class of Hermite–Hadamard inequality for LR-convex interval-valued functions known as LR-interval Hermite–Hadamard inequality, by means of pseudo-order relation ($\leq_p$). This order relation is defined on interval space. We have proved that if the interval-valued function is LR-convex then the inclusion relation " $\subseteq$ " coincident to pseudo-order relation " $\leq_p$ " under some suitable conditions. Moreover, the interval Hermite–Hadamard–Fejér inequality is also derived for LR-convex interval-valued functions. These inequalities also generalize some new and known results. Useful examples that verify the applicability of the theory developed in this study are presented. The concepts and techniques of this paper may be a starting point for further research in this area.

Keywords: interval-valued function; Riemann integral; LR-convex interval-valued function; interval Hermite–Hadamard inequality; interval Hermite–Hadamard–Fejér inequality

1. Introduction

In the development of pure and applied mathematics [1,2] convexity has played a key role. Due to their resilience, convex sets and convex functions have been refined and expanded in many mathematical fields; see [3–8]. Convexity theory may be used to generate numerous inequalities in the literature. Integral inequalities [9] have uses in linear programming, combinatory, orthogonal polynomials, quantum theory, number theory, optimization theory, dynamics, and the theory of relativity. Researchers have given this problem a lot of attention [10–14], and it is now regarded an integrative topic involving economics, mathematics, physics, and statistics [15,16]. The Hermite–Hadamard inequality (HH-inequality) is, to the best of my knowledge, a well-known, ultimate, and broadly applied inequality. Other classical inequalities, such as the Oslen and Gagliardo–Nirenberg, Oslen, Opial, Hardy, Young, Linger, Ostrowski, levison, Arithmetic's-Geometric, Ky-fan, Minkowski, Beckenbach–Dresher, and Holer inequality, are closely linked to the classical HH-inequality [17–20], and it can be put in the following manner.

Let $\mathfrak{S}: K \to \mathbb{R}$ be a convex function on a convex set K and $t,\ v \in K$ with $t \leq v$. Then,

$$\mathfrak{S}\left(\frac{t+v}{2}\right) \leq \frac{1}{v-t} \int_t^v \mathfrak{S}(\omega) d\omega \leq \frac{\mathfrak{S}(t)+\mathfrak{S}(v)}{2}. \qquad (1)$$

In [21], Fejér looked at the key extensions of HH-inequality, dubbed Hermite–Hadamard–Fejér inequality (HH-Fejér inequality).

Let $\mathfrak{S} : K \to \mathbb{R}$ be a convex function on a convex set K and t, $v \in K$ with t $\leq v$. Then,

$$\mathfrak{S}\left(\frac{t+v}{2}\right) \leq \frac{1}{\int_t^v \mathfrak{D}(\omega)d\omega} \int_t^v \mathfrak{S}(\omega)\mathfrak{D}(\omega)d\omega \leq \frac{\mathfrak{S}(t)+\mathfrak{S}(v)}{2} \int_t^v \mathfrak{D}(\omega))d\omega. \quad (2)$$

If $\mathfrak{D}(\omega) = 1$ then, we obtain (1) from (2). Many classical inequalities may be derived by specific convex functions with the help of inequality (1). Furthermore, in both pure and industrial mathematics, these inequalities play a crucial role for convex functions. We encourage readers to go more into the literature on generalized convex functions and HH-integral inequalities, particularly [22–29] and the references therein.

Interval analysis, on the other hand, was mostly forgotten for a long time due to a lack of applicability in other fields. Moore [30] and Kulish and W. Miranker [31] introduced and researched the notion of interval analysis. It is the first time in numerical analysis that it is utilized to calculate the error boundaries of numerical solutions of a finite state machine. Since then, a number of analysts have focused on and studied interval analysis and interval-valued functions (I.V-Fs) in both mathematics and applications. As a result, various writers looked into the literature and applications of neural network output optimization, automatic error analysis, computational physics, robotics, computer graphics, and a variety of other well-known scientific and technology fields. We encourage readers to conduct more research into essential aspects and applications in the literature (see [32–40] and the references therein).

The theory of fuzzy sets and systems has progressed in a number of ways from its introduction five decades ago, as seen in [41]. As a result, it is useful in the study of a variety of issues in pure mathematics and applied sciences, such as operation research, computer science, management sciences, artificial intelligence, control engineering, and decision sciences. Convex analysis has contributed significantly to the advancement of several sectors of practical and pure research. Similarly, the concepts of convexity and non-convexity are important in fuzzy optimization because we obtain fuzzy variational inequalities when we characterize the optimality condition of convexity, so variational inequality theory and fuzzy complementary problem theory established powerful mechanisms of mathematical problems and have a friendly relationship. Costa [42], Costa and Roman-Flores [43], Flores-Franulic et al. [44], Roman-Flores et al. [45,46], and Chalco-Cano et al. [47,48] have recently generalized several classical discrete and integral inequalities not only to the environment of the I.V-Fs and fuzzy I.V-Fs, but also to more general set valued maps by Nikodem et al. Zhang et al. [49] used a pseudo order relation to establish a novel version of Jensen's inequalities for set-valued and fuzzy set-valued functions, proving that these Jensen's inequalities are an expanded form of Costa Jensen's inequalities [42]. Zhao et al. [50], inspired by the literature, introduced $\hbar$-convex I.V-Fs and established that the HH-inequality for $\hbar$-convex I.V-Fs. Yanrong An et al. [51] took a step forward by introducing the class of $(\hbar_1, \hbar_2)$ $\hbar$-convex I.V-Fs and establishing the interval HH-inequality for $(\hbar_1, \hbar_2)$-convex I.V-Fs.

This research is structured as follows: preliminary and novel notions and results in interval space and interval-valued convex analysis are presented in Section 2. Section 3 uses LR-convex I.V-Fs to generate LR-interval HH-inequalities and HH-Fejér inequalities. In addition, several intriguing cases are provided to support our findings. Conclusions and future plans are presented in Section 4.

2. Preliminaries

Let $\mathcal{K}_C$ be the collection of all closed and bounded intervals of $\mathbb{R}$ that is $\mathcal{K}_C = \{[\mathcal{Z}_*, \mathcal{Z}^*] : \mathcal{Z}_*, \mathcal{Z}^* \in \mathbb{R} \text{ and } \mathcal{Z}_* \leq \mathcal{Z}^*\}$. If $\mathcal{Z}_* \geq 0$, then $[\mathcal{Z}_*, \mathcal{Z}^*]$ is named as positive interval. The set of all positive interval is denoted by $\mathcal{K}_C^+$ and defined as $\mathcal{K}_C^+ = \{[\mathcal{Z}_*, \mathcal{Z}^*] : \mathcal{Z}_*, \mathcal{Z}^* \in \mathcal{K}_C \text{ and } \mathcal{Z}_* \geq 0\}$.

If $[\mathfrak{A}_*, \mathfrak{A}^*], [\mathcal{Z}_*, \mathcal{Z}^*] \in \mathcal{K}_C$ and $s \in \mathbb{R}$, then arithmetic operations are defined by

$$[\mathfrak{A}_*, \mathfrak{A}^*] + [\mathcal{Z}_*, \mathcal{Z}^*] = [\mathfrak{A}_* + \mathcal{Z}_*, \mathfrak{A}^* + \mathcal{Z}^*],$$

$$[\mathfrak{A}_*, \mathfrak{A}^*] \times [\mathcal{Z}_*, \mathcal{Z}^*] = [\min\{\mathfrak{A}_*\mathcal{Z}_*, \mathfrak{A}^*\mathcal{Z}_*, \mathfrak{A}_*\mathcal{Z}^*, \mathfrak{A}^*\mathcal{Z}^*\}, \max\{\mathfrak{A}_*\mathcal{Z}_*, \mathfrak{A}^*\mathcal{Z}_*, \mathfrak{A}_*\mathcal{Z}^*, \mathfrak{A}^*\mathcal{Z}^*\}],$$

$$s.[\mathfrak{A}_*, \mathfrak{A}^*] = \begin{cases} [s\mathfrak{A}_*, s\mathfrak{A}^*] & \text{if } s > 0 \\ \{0\} & \text{if } s = 0, \\ [s\mathfrak{A}^*, s\mathfrak{A}_*] & \text{if } s < 0. \end{cases}$$

For $[\mathfrak{A}_*, \mathfrak{A}^*]$, $[\mathcal{Z}_*, \mathcal{Z}^*] \in \mathcal{K}_C$, the inclusion "$\subseteq$" is defined by

$$[\mathfrak{A}_*, \mathfrak{A}^*] \subseteq [\mathcal{Z}_*, \mathcal{Z}^*], \text{ if and only if } \mathcal{Z}_* \leq \mathfrak{A}_*, \mathfrak{A}^* \leq \mathcal{Z}^*.$$

Remark 1. [49]. *(i) The relation "$\leq_p$" defined on $\mathcal{K}_C$ by $[\mathfrak{A}_*, \mathfrak{A}^*] \leq_p [\mathcal{Z}_*, \mathcal{Z}^*]$ if and only if $\mathfrak{A}_* \leq \mathcal{Z}_*, \mathfrak{A}^* \leq \mathcal{Z}^*$, for all $[\mathfrak{A}_*, \mathfrak{A}^*]$, $[\mathcal{Z}_*, \mathcal{Z}^*] \in \mathcal{K}_C$, it is a pseudo-order relation. The relation $[\mathfrak{A}_*, \mathfrak{A}^*] \leq_p [\mathcal{Z}_*, \mathcal{Z}^*]$ coincident to $[\mathfrak{A}_*, \mathfrak{A}^*] \leq [\mathcal{Z}_*, \mathcal{Z}^*]$ on $\mathcal{K}_C$.*

(ii) It can be easily seen that "$\leq_p$" looks similar to "left and right" on the real line $\mathbb{R}$, so we call "$\leq_p$" is "left and right" (or "LR" order, in short).

The concept of Riemann integral for I.V-F first introduced by Moore [30] is defined as follow:

Theorem 1. [30]. *If $\mathfrak{S} : [t, v] \subset \mathbb{R} \to \mathcal{K}_C$ is an I.V-F on such that $\mathfrak{S}(\omega) = [\mathfrak{S}_*(\omega), \mathfrak{S}^*(\omega)]$. Then $\mathfrak{S}$ is Riemann integrable over $[t, v]$ if and only if, $\mathfrak{S}_*$ and $\mathfrak{S}^*$ both are Riemann integrable over $[t, v]$ such that*

$$(IR)\int_t^v \mathfrak{S}(\omega)d\omega = [(R)\int_t^v \mathfrak{S}_*(\omega)d\omega, (R)\int_t^v \mathfrak{S}^*(\omega)d\omega].$$

The collection of all Riemann integrable real valued functions and Riemann integrable I.V-F is denoted by $\mathcal{R}_{[t, v]}$ and $\mathcal{IR}_{[t, v]}$, respectively.

Definition 1. *The real mapping $\mathfrak{S} : [t, v] \to \mathbb{R}$ is named as convex function if for all $\omega, y \in [t, v]$ and $\varsigma \in [0, 1]$ we have*

$$\mathfrak{S}(\varsigma\omega + (1 - \varsigma)y) \leq \varsigma\mathfrak{S}(\omega) + (1 - \varsigma)\mathfrak{S}(y), \qquad (3)$$

If inequality (3) is reversed, then $\mathfrak{S}$ is named as concave on $[t, v]$. A function $\mathfrak{S}$ is named as affine if $\mathfrak{S}$ is both convex and cocave function. The set of all convex (concave) functions is denoted by

$$SX([t, v],)(SV([t, v], \mathbb{R}^+), SA([t, v], \mathbb{R}^+)).$$

Definition 2. [50]. *The I.V-F $\mathfrak{S} : [t, v] \to \mathbb{R}_I^+$ is named as convex I.V-F if for all $\omega, y \in [t, v]$ and $\varsigma \in [0, 1]$, the coming inequality*

$$\mathfrak{S}(\varsigma\omega + (1 - \varsigma)y) \supseteq \hbar(\varsigma)\mathfrak{S}(\omega) + \hbar(1 - \varsigma)\mathfrak{S}(y), \qquad (4)$$

is valid. If inequality (4) is reversed, then $\mathfrak{S}$ is named as concave on $[t, v]$. A I.V-F $\mathfrak{S}$ is named as affine if $\mathfrak{S}$ is both convex and cocave I.V-F. The set of all convex (concave, affine) I.V-Fs is denoted by

$$SX([t, v], \mathcal{K}_C^+) \quad (SV([t, v], \mathcal{K}_C^+), SA([t, v], \mathcal{K}_C^+)).$$

Definition 3. [49]. *The I.V-F $\mathfrak{S} : [t, v] \to \mathcal{K}_C^+$ is named as LR-convex I.V-F if for all $\omega, y \in [t, v]$ and $\varsigma \in [0, 1]$, the coming inequality*

$$\mathfrak{S}(\varsigma\omega + (1 - \varsigma)y) \leq_p \varsigma\mathfrak{S}(\omega) + (1 - \varsigma)\mathfrak{S}(y), \qquad (5)$$

is valid. If inequality (5) is reversed, then $\mathfrak{S}$ is named as LR-concave on $[t, v]$. A I.V-F $\mathfrak{S}$ is named as LR-affine if $\mathfrak{S}$ is both LR-convex and LR-cocave I.V-F. The set of all LR-convex (LR-concave) I.V-Fs is denoted by

$$LRSX([t, v], \mathcal{K}_C^+)(LRSV([t, v], \mathcal{K}_C^+), LRSA([t, v], \mathcal{K}_C^+)).$$

Theorem 2. [49]. *Let $\mathfrak{S} : [t, v] \to \mathcal{K}_C^+$ be an I.V-F defined by $\mathfrak{S}(\omega) = [\mathfrak{S}_*(\omega), \mathfrak{S}^*(\omega)]$, for all $\omega \in [t, v]$. Then $\mathfrak{S} \in LRSX([t, v], \mathcal{K}_C^+)$ if and only if, $\mathfrak{S}_*, \mathfrak{S}^* \in SX([t, v])$.*

Example 1. *We consider the I.V-F $\mathfrak{S} : [1, 4] \to \mathcal{K}_C^+$ defined by $\mathfrak{S}(\omega) = [2\omega, 2\omega^2]$. Since end point functions $\mathfrak{S}_*(\omega)$ and $\mathfrak{S}^*(\omega)$ are convex functions. Hence $\mathfrak{S}(\omega)$ is LR-convex I.V-F.*

Remark 2. *By using our Definition 3 and Example 1, it can be easily observed that the concept of set inclusion " $\supseteq$ " coincident to relation " $\leq_p$ " (or " $\leq_p$ " coincident to " $\supseteq$ ") when one of the end point function $\mathfrak{S}_*$ or $\mathfrak{S}^*$ is affine function such that "If $\mathfrak{S} \in SX([t, v], \mathcal{K}_C^+)$ then $\mathfrak{S} \in LRSV([t, v], \mathcal{K}_C^+)$ if and only if $\mathfrak{S}_* \in SA([t, v], \mathbb{R}^+)$ and $\mathfrak{S}^* \in SX([t, v], \mathbb{R}^+)$ ". Similarly, "If $\mathfrak{S} \in SV([t, v], \mathcal{K}_C^+)$ then $\mathfrak{S} \in LRSX([t, v], \mathcal{K}_C^+)$, if and only if $\mathfrak{S}_* \in SV([t, v], \mathbb{R}^+)$ and $\mathfrak{S}^* \in SA([t, v], \mathbb{R}^+)$ ".*

Remark 3. *From Theorem 2, it can be easily seen that if $\mathfrak{S}_*(\omega) = \mathfrak{S}^*(\omega)$ then, LR-convex I.V-Fs becomes classical convex functions.*

Example 2. *We consider the I.V-F $\mathfrak{S} : [1, 4] \to \mathcal{K}_C^+$ defined by $\mathfrak{S}(\omega) = [2\omega^2, 2\omega^2]$. Since end point functions $\mathfrak{S}_*(\omega), \mathfrak{S}^*(\omega)$, are equal and convex functions. Hence, $\mathfrak{S}(\omega)$ is a convex function.*

3. Interval Inequalities

In this section, we present two classes of *HH*-inequalities and discuss some related results, and verify with the help of use examples. First of all, we derive *HH*-inequality for LR-convex I.V-F.

Theorem 3. *Let $\mathfrak{S} : [t, v] \to \mathcal{K}_C^+$ be an I.V-F such that $\mathfrak{S}(\omega) = [\mathfrak{S}_*(\omega), \mathfrak{S}^*(\omega)]$ for all $\omega \in [t, v]$ and $\mathfrak{S} \in \mathcal{IR}_{([t, v])}$. If $\mathfrak{S} \in LRSX([t, v], \mathcal{K}_C^+)$, then*

$$\mathfrak{S}\left(\frac{t+v}{2}\right) \leq_p \frac{1}{v-t} (IR) \int_t^v \mathfrak{S}(\omega) d\omega \leq_p \frac{\mathfrak{S}(t) + \mathfrak{S}(v)}{2}. \tag{6}$$

If $\mathfrak{S} \in LRSV([t, v], \mathcal{K}_C^+)$, then

$$\mathfrak{S}\left(\frac{t+v}{2}\right) \geq_p \frac{1}{v-t} (IR) \int_t^v \mathfrak{S}(\omega) d\omega \geq_p \frac{\mathfrak{S}(t) + \mathfrak{S}(v)}{2}.$$

Proof. Let $\mathfrak{S} \in LRSX([t, v], \mathcal{K}_C^+)$ convex I.V-F. Then, by hypothesis, we have

$$2\mathfrak{S}_*\left(\frac{t+v}{2}\right) \leq \mathfrak{S}_*(\varsigma t + (1-\varsigma)v) + \mathfrak{S}_*((1-\varsigma)t + \varsigma v),$$
$$2\mathfrak{S}^*\left(\frac{t+v}{2}\right) \leq \mathfrak{S}^*(\varsigma t + (1-\varsigma)v) + \mathfrak{S}^*((1-\varsigma)t + \varsigma v).$$

Then

$$2\int_0^1 \mathfrak{S}_*\left(\frac{t+v}{2}\right) d\varsigma \leq \int_0^1 \mathfrak{S}_*(\varsigma t + (1-\varsigma)v) d\varsigma + \int_0^1 \mathfrak{S}_*((1-\varsigma)t + \varsigma v) d\varsigma,$$
$$2\int_0^1 \mathfrak{S}^*\left(\frac{t+v}{2}\right) d\varsigma \leq \int_0^1 \mathfrak{S}^*(\varsigma t + (1-\varsigma)v) d\varsigma + \int_0^1 \mathfrak{S}^*((1-\varsigma)t + \varsigma v) d\varsigma.$$

It follows that

$$\mathfrak{S}_*\left(\frac{t+v}{2}\right) \leq \frac{1}{v-t} \int_t^v \mathfrak{S}_*(\omega) d\omega,$$
$$\mathfrak{S}^*\left(\frac{t+v}{2}\right) \leq \frac{1}{v-t} \int_t^v \mathfrak{S}^*(\omega) d\omega.$$

That is

$$\left[\mathfrak{S}_*\left(\frac{t+v}{2}\right), \mathfrak{S}^*\left(\frac{t+v}{2}\right)\right] \leq_p \frac{1}{v-t}\left[\int_t^v \mathfrak{S}_*(\omega) d\omega, \int_t^v \mathfrak{S}^*(\omega) d\omega\right].$$

Thus,

$$\mathfrak{S}\left(\frac{t+v}{2}\right) \leq_p \frac{1}{v-t} (IR) \int_t^v \mathfrak{S}(\omega) d\omega. \tag{7}$$

In a similar way as above, we have

$$\frac{1}{v-t} (IR) \int_t^v \mathfrak{S}(\omega)d\omega \leq_p \frac{\mathfrak{S}(t)+\mathfrak{S}(v)}{2}. \tag{8}$$

Combining (7) and (8), we have

$$\mathfrak{S}\left(\frac{t+v}{2}\right) \leq_p \frac{1}{v-t} (IR) \int_t^v \mathfrak{S}(\omega)d\omega \leq_p \frac{\mathfrak{S}(t)+\mathfrak{S}(v)}{2}.$$

Hence, the required result. □

Remark 4. *If $\mathfrak{S}_*(\omega) = \mathfrak{S}^*(\omega)$, then Theorem 3, reduces to the result for convex function:*

$$\mathfrak{S}\left(\frac{t+v}{2}\right) \leq \frac{1}{v-t} (R) \int_t^v \mathfrak{S}(\omega)d\omega \leq \frac{\mathfrak{S}(t)+\mathfrak{S}(v)}{2}.$$

It is easy to see that due to the convexity of end point functions $\mathfrak{S}_*(\omega)$ and $\mathfrak{S}^*(\omega)$ have following two possibilities to satisfy (1) either both are convex or affine convex functions. However, in the case of interval inclusion both functions $\mathfrak{S}_*(\omega)$ and $\mathfrak{S}^*(\omega)$ has only one possibility to satisfy (1) such that both end point functions should be affine convex because in interval inclusion $\mathfrak{S}_*(\omega)$ is convex and $\mathfrak{S}^*(\omega)$ is concave, see [50].

Example 3. *We consider the function $\mathfrak{S} : [t, v] = [0, 2] \to \mathcal{K}_\mathcal{C}^+$ defined by, $\mathfrak{S}(\omega) = [\omega^2, 2\omega^2]$. Since end point functions $\mathfrak{S}_*(\omega) = \omega^2$, $\mathfrak{S}^*(\omega) = 2\omega^2$ LR-convex functions. Hence $\mathfrak{S}(\omega)$ is LR-convex I.V-F. We now compute the following*

$$\mathfrak{S}_*\left(\frac{t+v}{2}\right) \leq \frac{1}{v-t} \int_t^v \mathfrak{S}_*(\omega)d\omega \leq \frac{\mathfrak{S}_*(t)+\mathfrak{S}_*(v)}{2}.$$

$$\mathfrak{S}_*\left(\frac{t+v}{2}\right) = \mathfrak{S}_*(1) = 1,$$

$$\frac{1}{v-t} \int_t^v \mathfrak{S}_*(\omega)d\omega = \frac{1}{2}\int_0^2 \omega^2 d\omega = \frac{4}{3},$$

$$\frac{\mathfrak{S}_*(t)+\mathfrak{S}_*(v)}{2} = 2.$$

That means

$$1 \leq \frac{4}{3} \leq 2.$$

Similarly, it can be easily show that

$$\mathfrak{S}^*\left(\frac{t+v}{2}\right) \leq \frac{1}{v-t} \int_t^v \mathfrak{S}^*(\omega)d\omega \leq \frac{\mathfrak{S}^*(t)+\mathfrak{S}^*(v)}{2}.$$

such that

$$\mathfrak{S}^*\left(\frac{t+v}{2}\right) = \mathfrak{S}_*(1) = 2,$$

$$\frac{1}{v-t} \int_t^v \mathfrak{S}^*(\omega)d\omega = \frac{1}{2}\int_0^2 2\omega^2 d\omega = \frac{8}{3},$$

$$\frac{\mathfrak{S}^*(t)+\mathfrak{S}^*(v)}{2} = 4,$$

from which, it follows that

$$2 \leq \frac{8}{3} \leq 4,$$

that is

$$[1, 2] \leq \left[\frac{4}{3}, \frac{8}{3}\right] \leq [2, 4].$$

Hence,

$$\mathfrak{S}\left(\frac{t+v}{2}\right) \leq_p \frac{1}{v-t} (IR) \int_t^v \mathfrak{S}(\omega) d\omega \leq_p \frac{\mathfrak{S}(t) + \mathfrak{S}(v)}{2}.$$

Theorem 4. *Let $\mathfrak{S} : [t, v] \to \mathcal{K}_C^+$ be an I.V-F such that $\mathfrak{S}(\omega) = [\mathfrak{S}_*(\omega), \mathfrak{S}^*(\omega)]$ for all $\omega \in [t, v]$ and $\mathfrak{S} \in \mathcal{IR}_{([t, v])}$. If $\mathfrak{S} \in LRSX([t, v], \mathcal{K}_C^+)$, then*

$$\mathfrak{S}\left(\frac{t+v}{2}\right) \leq_p \triangleright_2 \leq_p \frac{1}{v-t} (IR) \int_t^v \mathfrak{S}(\omega) d\omega \leq_p \triangleright_1 \leq_p \frac{\mathfrak{S}(t) + \mathfrak{S}(v)}{2},$$

where

$$\triangleright_1 = \frac{\frac{\mathfrak{S}(t) + \mathfrak{S}(v)}{2} + \mathfrak{S}\left(\frac{t+v}{2}\right)}{2}, \triangleright_2 = \frac{\mathfrak{S}\left(\frac{3t+v}{4}\right) + \mathfrak{S}\left(\frac{t+3v}{4}\right)}{2}$$

and $\triangleright_1 = [\triangleright_{1*}, \triangleright_1^*]$, $\triangleright_2 = [\triangleright_{2*}, \triangleright_2^*]$.

Proof. Take $\left[t, \frac{t+v}{2}\right]$, we have

$$2\mathfrak{S}\left(\frac{\varsigma t + (1-\varsigma)\frac{t+v}{2}}{2} + \frac{(1-\varsigma)t + \varsigma\frac{t+v}{2}}{2}\right) \leq_p \mathfrak{S}\left(\varsigma t + (1-\varsigma)\frac{t+v}{2}\right) + \mathfrak{S}\left((1-\varsigma)t + \varsigma\frac{t+v}{2}\right).$$

From which, we have

$$2\mathfrak{S}_*\left(\frac{\varsigma t + (1-\varsigma)\frac{t+v}{2}}{2} + \frac{(1-\varsigma)t + \varsigma\frac{t+v}{2}}{2}\right) \leq \mathfrak{S}_*\left(\varsigma t + (1-\varsigma)\frac{t+v}{2}\right) + \mathfrak{S}_*\left((1-\varsigma)t + \varsigma\frac{t+v}{2}\right),$$

$$2\mathfrak{S}^*\left(\frac{\varsigma t + (1-\varsigma)\frac{t+v}{2}}{2} + \frac{(1-\varsigma)t + \varsigma\frac{t+v}{2}}{2}\right) \leq \mathfrak{S}^*\left(\varsigma t + (1-\varsigma)\frac{t+v}{2}\right) + \mathfrak{S}^*\left((1-\varsigma)t + \varsigma\frac{t+v}{2}\right).$$

In consequence, we obtain

$$\frac{\mathfrak{S}_*\left(\frac{3t+v}{4}\right)}{2} \leq \frac{1}{v-t} \int_t^{\frac{t+v}{2}} \mathfrak{S}_*(\omega) d\omega,$$

$$\frac{\mathfrak{S}^*\left(\frac{3t+v}{4}\right)}{2} \leq \frac{1}{v-t} \int_t^{\frac{t+v}{2}} \mathfrak{S}^*(\omega) d\omega.$$

That is

$$\frac{\left[\mathfrak{S}_*\left(\frac{3t+v}{4}\right), \mathfrak{S}^*\left(\frac{3t+v}{4}\right)\right]}{2} \leq \frac{1}{v-t}\left[\int_t^{\frac{t+v}{2}} \mathfrak{S}_*(\omega) d\omega, \int_t^{\frac{t+v}{2}} \mathfrak{S}^*(\omega) d\omega\right].$$

It follows that

$$\frac{\mathfrak{S}\left(\frac{3t+v}{4}\right)}{2} \leq_p \frac{1}{v-t} (IR) \int_t^{\frac{t+v}{2}} \mathfrak{S}(\omega) d\omega. \tag{9}$$

In a similar way as above, we have

$$\frac{\mathfrak{S}\left(\frac{t+3v}{4}\right)}{2} \leq_p \frac{1}{v-t} (IR) \int_{\frac{t+v}{2}}^v \mathfrak{S}(\omega) d\omega. \tag{10}$$

Combining (9) and (10), we have

$$\frac{\left[\mathfrak{S}\left(\frac{3t+v}{4}\right) + \mathfrak{S}\left(\frac{t+3v}{4}\right)\right]}{2} \leq_p \frac{1}{v-t} (IR) \int_t^v \mathfrak{S}(\omega) d\omega.$$

By using Theorem 3, we have

$$\mathfrak{S}\left(\frac{t+v}{2}\right) = \mathfrak{S}\left(\frac{1}{2} \cdot \frac{3t+v}{4} + \frac{1}{2} \cdot \frac{t+3v}{4}\right).$$

From which, we have

$$\mathfrak{S}_*\left(\tfrac{t+v}{2}\right) = \mathfrak{S}_*\left(\tfrac{1}{2}\cdot\tfrac{3t+v}{4} + \tfrac{1}{2}\cdot\tfrac{t+3v}{4}\right),$$
$$\mathfrak{S}^*\left(\tfrac{t+v}{2}\right) = \mathfrak{S}^*\left(\tfrac{1}{2}\cdot\tfrac{3t+v}{4} + \tfrac{1}{2}\cdot\tfrac{t+3v}{4}\right),$$
$$\leq \left[\tfrac{1}{2}\mathfrak{S}_*\left(\tfrac{3t+v}{4}\right) + \tfrac{1}{2}\mathfrak{S}_*\left(\tfrac{t+3v}{4}\right)\right],$$
$$\leq \left[\tfrac{1}{2}\mathfrak{S}^*\left(\tfrac{3t+v}{4}\right) + \tfrac{1}{2}\mathfrak{S}^*\left(\tfrac{t+3v}{4}\right)\right],$$
$$= \triangleright_{2*},$$
$$= \triangleright_2^*,$$
$$\leq \tfrac{1}{v-t}\int_t^v \mathfrak{S}_*(\omega)d\omega,$$
$$\leq \tfrac{1}{v-t}\int_t^v \mathfrak{S}^*(\omega)d\omega,$$
$$\leq \tfrac{1}{2}\left[\tfrac{\mathfrak{S}_*(t)+\mathfrak{S}_*(v)}{2} + \mathfrak{S}_*\left(\tfrac{t+v}{2}\right)\right],$$
$$\leq \tfrac{1}{2}\left[\tfrac{\mathfrak{S}^*(t)+\mathfrak{S}^*(v)}{2} + \mathfrak{S}^*\left(\tfrac{t+v}{2}\right)\right],$$
$$= \triangleright_{1*},$$
$$= \triangleright_1^*,$$
$$\leq \tfrac{1}{2}\left[\tfrac{\mathfrak{S}_*(t)+\mathfrak{S}_*(v)}{2} + \tfrac{\mathfrak{S}_*(t)+\mathfrak{S}_*(v)}{2}\right],$$
$$\leq \tfrac{1}{2}\left[\tfrac{\mathfrak{S}^*(t)+\mathfrak{S}^*(v)}{2} + \tfrac{\mathfrak{S}_*(t)+\mathfrak{S}_*(v)}{2}\right],$$
$$= \tfrac{\mathfrak{S}_*(t)+\mathfrak{S}_*(v)}{2},$$
$$= \tfrac{\mathfrak{S}^*(t)+\mathfrak{S}^*(v)}{2},$$

that is

$$\mathfrak{S}\left(\tfrac{t+v}{2}\right) \leq_p \triangleright_2 \leq_p \tfrac{1}{v-t}(IR)\int_t^v \mathfrak{S}(\omega)d\omega \leq_p \triangleright_1 \leq_p \tfrac{\mathfrak{S}(t)+\mathfrak{S}(v)}{2},$$

hence, the result follows. □

Example 4. *We consider the function* $\mathfrak{S}: [t, v] = [0, 2] \to \mathcal{K}_C^+$ *defined by,* $\mathfrak{S}(\omega) = [\omega^2, 2\omega^2]$, *as in Example 3, then* $\mathfrak{S}(\omega)$ *is LR-convex I.V-F and satisfying (10). We have* $\mathfrak{S}_*(\omega) = \omega^2$ *and* $\mathfrak{S}^*(\omega) = 2\omega^2$. *We now compute the following*

$$\tfrac{\mathfrak{S}_*(t)+\mathfrak{S}_*(v)}{2} = 2,$$
$$\tfrac{\mathfrak{S}^*(t)+\mathfrak{S}^*(v)}{2} = 4,$$
$$\triangleright_{1*} = \tfrac{\tfrac{\mathfrak{S}_*(t)+\mathfrak{S}_*(v)}{2} + \mathfrak{S}_*\left(\tfrac{t+v}{2}\right)}{2} = \tfrac{3}{2},$$
$$\triangleright_1^* = \tfrac{\tfrac{\mathfrak{S}^*(t)+\mathfrak{S}^*(v)}{2} + \mathfrak{S}^*\left(\tfrac{t+v}{2}\right)}{2} = 3,$$
$$\triangleright_{2*} = \tfrac{\mathfrak{S}_*\left(\tfrac{3t+v}{4}\right) + \mathfrak{S}_*\left(\tfrac{t+3v}{4}\right)}{2} = \tfrac{5}{4},$$
$$\triangleright_2^* = \tfrac{\mathfrak{S}^*\left(\tfrac{3t+v}{4}\right) + \mathfrak{S}^*\left(\tfrac{t+3v}{4}\right)}{2} = \tfrac{5}{2},$$

Then we obtain that

$$1 \leq \tfrac{5}{4} \leq \tfrac{4}{3} \leq \tfrac{3}{2} \leq 2,$$
$$2 \leq \tfrac{5}{2} \leq \tfrac{8}{3} \leq 3 \leq 4,$$

Hence, Theorem 4 is verified.

Theorem 5. Let $\mathfrak{S}, g : [t, v] \to \mathcal{K}_C^+$ be two I.V-F such that $\mathfrak{S}(\omega) = [\mathfrak{S}_*(\omega), \mathfrak{S}^*(\omega)]$ and $g(\omega) = [g_*(\omega), g^*(\omega)]$ for all $\omega \in [t, v]$ and $\mathfrak{S}g \in \mathcal{IR}_{([t, v])}$. If $\mathfrak{S}, g \in LRSX([t, v], \mathcal{K}_C^+)$, then

$$\frac{1}{v-t}(IR)\int_t^v \mathfrak{S}(\omega)g(\omega)d\omega \leq_p \frac{\mathfrak{B}(t,v)}{3} + \frac{\mathfrak{C}(t,v)}{6},$$

where $\mathfrak{B}(t,v) = \mathfrak{S}(t)g(t) + \mathfrak{S}(v)g(v)$, $\mathfrak{C}(t,v) = \mathfrak{S}(t)g(v) + \mathfrak{S}(v)g(t)$, and $\mathfrak{B}(t,v) = [\mathfrak{B}_*((t,v)), \mathfrak{B}^*((t,v))]$ and $\mathfrak{C}(t,v) = [\mathfrak{C}_*((t,v)), \mathfrak{C}^*((t,v))]$.

Proof. Since $\mathfrak{S}, g \in \mathcal{IR}_{([t,v])}$, then we have

$$\mathfrak{S}_*(\varsigma t + (1-\varsigma)v) \leq \varsigma \mathfrak{S}_*(t) + (1-\varsigma)\mathfrak{S}_*(v),$$
$$\mathfrak{S}^*(\varsigma t + (1-\varsigma)v) \leq \varsigma \mathfrak{S}^*(t) + (1-\varsigma)\mathfrak{S}^*(v).$$

And
$$g_*(\varsigma t + (1-\varsigma)v) \leq \varsigma g_*(t) + (1-\varsigma)g_*(v),$$
$$g^*(\varsigma t + (1-\varsigma)v) \leq \varsigma g^*(t) + (1-\varsigma)g^*(v).$$

From the definition of LR-convex I.V-Fs it follows that $0 \leq_p \mathfrak{S}(\omega)$ and $0 \leq_p g(\omega)$, so

$$\mathfrak{S}_*(\varsigma t + (1-\varsigma)v)g_*(\varsigma t + (1-\varsigma)v)$$
$$\leq \left(\varsigma\mathfrak{S}_*(t) + (1-\varsigma)\mathfrak{S}_*(v)\right)\left(\varsigma g_*(t) + (1-\varsigma)g_*(v)\right)$$
$$= \mathfrak{S}_*(t)g_*(t)\varsigma^2 + \mathfrak{S}_*(v)g_*(v)\varsigma^2 + \mathfrak{S}_*(t)g_*(v)\varsigma(1-\varsigma) + \mathfrak{S}_*(v)g_*(t)\varsigma(1-\varsigma)$$
$$\mathfrak{S}^*(\varsigma t + (1-\varsigma)v)g^*(\varsigma t + (1-\varsigma)v)$$
$$\leq \left(\varsigma\mathfrak{S}^*(t) + (1-\varsigma)\mathfrak{S}^*(v)\right)\left(\varsigma g^*(t) + (1-\varsigma)g^*(v)\right)$$
$$= \mathfrak{S}^*(t)g^*(t)\varsigma^2 + \mathfrak{S}^*(v)g^*(v)\varsigma^2 + \mathfrak{S}^*(t)g^*(v)\varsigma(1-\varsigma) + \mathfrak{S}^*(v)g^*(t)\varsigma(1-\varsigma),$$

Integrating both sides of above inequality over $[0,1]$ we obtain

$$\int_0^1 \mathfrak{S}_*(\varsigma t + (1-\varsigma)v)g_*(\varsigma t + (1-\varsigma)v) = \frac{1}{v-t}\int_t^v \mathfrak{S}_*(\omega)g_*(\omega)d\omega$$
$$\leq (\mathfrak{S}_*(t)g_*(t) + \mathfrak{S}_*(v)g_*(v))\int_0^1 \varsigma^2 d\varsigma$$
$$+ (\mathfrak{S}_*(t)g_*(v) + \mathfrak{S}_*(v)g_*(t))\int_0^1 \varsigma(1-\varsigma)d\varsigma,$$
$$\int_0^1 \mathfrak{S}^*(\varsigma t + (1-\varsigma)v)g^*(\varsigma t + (1-\varsigma)v) = \frac{1}{v-t}\int_t^v \mathfrak{S}^*(\omega)g^*(\omega)d\omega$$
$$\leq (\mathfrak{S}^*(t)g^*(t) + \mathfrak{S}^*(v)g^*(v))\int_0^1 \varsigma^2 d\varsigma$$
$$+ (\mathfrak{S}^*(t)g^*(v) + \mathfrak{S}^*(v)g^*(t))\int_0^1 \varsigma(1-\varsigma)d\varsigma.$$

It follows that,

$$\frac{1}{v-t}\int_t^v \mathfrak{S}_*(\omega)g_*(\omega)d\omega \leq \mathfrak{B}_*((t,v))\int_0^1 \varsigma^2 d\varsigma + \mathfrak{C}_*((t,v))\int_0^1 \varsigma(1-\varsigma)d\varsigma,$$
$$\frac{1}{v-t}\int_t^v \mathfrak{S}^*(\omega)g^*(\omega)d\omega \leq \mathfrak{B}^*((t,v))\int_0^1 \varsigma^2 d\varsigma + \mathfrak{C}^*((t,v))\int_0^1 \varsigma(1-\varsigma)d\varsigma,$$

that is

$$\frac{1}{v-t}\left[\int_t^v \mathfrak{S}_*(\omega)g_*(\omega)d\omega, \int_t^v \mathfrak{S}^*(\omega)g^*(\omega)d\omega\right] \leq_p \left[\frac{\mathfrak{B}_*((t,v))}{3}, \frac{\mathfrak{B}^*((t,v))}{3}\right] + \left[\frac{\mathfrak{C}_*((t,v))}{6}, \frac{\mathfrak{C}^*((t,v))}{6}\right].$$

Thus,
$$\frac{1}{v-t}(IR)\int_t^v \mathfrak{S}(\omega)g(\omega)d\omega \leq_p \frac{\mathfrak{B}(t,v)}{3} + \frac{\mathfrak{C}(t,v)}{6},$$

and the theorem has been established. □

Theorem 6. Let $\mathfrak{S}, g : [t, v] \to \mathcal{K}_C^+$ be two I.V-Fs such that $\mathfrak{S}(\omega) = [\mathfrak{S}_*(\omega), \mathfrak{S}^*(\omega)]$ and $g(\omega) = [g_*(\omega), g^*(\omega)]$ for all $\omega \in [t, v]$ and $\mathfrak{S}g \in \mathcal{IR}_{([t, v])}$. If $\mathfrak{S}, g \in LRSX([t, v], \mathcal{K}_C^+)$, then

$$2\mathfrak{S}\left(\frac{t+v}{2}\right)g\left(\frac{t+v}{2}\right) \leq_p \frac{1}{v-t}(IR)\int_t^v \mathfrak{S}(\omega)g(\omega)d\omega + \frac{\mathfrak{B}(t,v)}{6} + \frac{\mathfrak{C}(t,v)}{3},$$

where $\mathfrak{B}(t, v) = \mathfrak{S}(t)g(t) + \mathfrak{S}(v)g(v)$, $\mathfrak{C}(t, v) = \mathfrak{S}(t)g(v) + \mathfrak{S}(v)g(t)$, and $\mathfrak{B}(t, v) = [\mathfrak{B}_*((t, v)), \mathfrak{B}^*((t, v))]$ and $\mathfrak{C}(t, v) = [\mathfrak{C}_*((t, v)), \mathfrak{C}^*((t, v))]$.

Proof. By hypothesis, we have

$$\mathfrak{S}_*\left(\tfrac{t+v}{2}\right)g_*\left(\tfrac{t+v}{2}\right)$$
$$\mathfrak{S}^*\left(\tfrac{t+v}{2}\right)g^*\left(\tfrac{t+v}{2}\right)$$

$$\leq \frac{1}{4}\begin{bmatrix} \mathfrak{S}_*(\varsigma t + (1-\varsigma)v)g_*(\varsigma t + (1-\varsigma)v) \\ +\mathfrak{S}_*(\varsigma t + (1-\varsigma)v)g_*((1-\varsigma)t + \varsigma v) \end{bmatrix}$$

$$+\frac{1}{4}\begin{bmatrix} \mathfrak{S}_*((1-\varsigma)t + \varsigma v)g_*(\varsigma t + (1-\varsigma)v) \\ +\mathfrak{S}_*((1-\varsigma)t + \varsigma v)g_*((1-\varsigma)t + \varsigma v) \end{bmatrix},$$

$$\leq \frac{1}{4}\begin{bmatrix} \mathfrak{S}^*(\varsigma t + (1-\varsigma)v)g^*(\varsigma t + (1-\varsigma)v) \\ +\mathfrak{S}^*(\varsigma t + (1-\varsigma)v)g^*((1-\varsigma)t + \varsigma v) \end{bmatrix}$$

$$+\frac{1}{4}\begin{bmatrix} \mathfrak{S}^*((1-\varsigma)t + \varsigma v)g^*(\varsigma t + (1-\varsigma)v) \\ +\mathfrak{S}^*((1-\varsigma)t + \varsigma v)g^*((1-\varsigma)t + \varsigma v) \end{bmatrix},$$

$$\leq \frac{1}{4}\begin{bmatrix} \mathfrak{S}_*(\varsigma t + (1-\varsigma)v)g_*(\varsigma t + (1-\varsigma)v) \\ +\mathfrak{S}_*((1-\varsigma)t + \varsigma v)g_*((1-\varsigma)t + \varsigma v) \end{bmatrix}$$

$$+\frac{1}{4}\begin{bmatrix} (\varsigma\mathfrak{S}_*(t) + (1-\varsigma)\mathfrak{S}_*(v)) \\ ((1-\varsigma)g_*(t) + \varsigma g_*(v)) \\ +((1-\varsigma)\mathfrak{S}_*(t) + \varsigma\mathfrak{S}_*(v)) \\ (\varsigma g_*(t) + (1-\varsigma)g_*(v)) \end{bmatrix},$$

$$\leq \frac{1}{4}\begin{bmatrix} \mathfrak{S}_*(\varsigma t + (1-\varsigma)v)g_*(\varsigma t + (1-\varsigma)v) \\ +\mathfrak{S}_*((1-\varsigma)t + \varsigma v)g_*((1-\varsigma)t + \varsigma v) \end{bmatrix}$$

$$+\frac{1}{4}\begin{bmatrix} (\varsigma\mathfrak{S}^*(t) + (1-\varsigma)\mathfrak{S}^*(v)) \\ ((1-\varsigma)g^*(t) + \varsigma g^*(v)) \\ +((1-\varsigma)\mathfrak{S}^*(t) + \varsigma\mathfrak{S}^*(v)) \\ (\varsigma g^*(t) + (1-\varsigma)g^*(v)) \end{bmatrix},$$

$$= \frac{1}{4}\begin{bmatrix} \mathfrak{S}_*(\varsigma t + (1-\varsigma)v)g_*(\varsigma t + (1-\varsigma)v) \\ +\mathfrak{S}_*((1-\varsigma)t + \varsigma v)g_*((1-\varsigma)t + \varsigma v) \end{bmatrix}$$

$$+\frac{1}{2}\begin{bmatrix} \{\varsigma^2 + (1-\varsigma)^2\}\mathfrak{C}_*((t,v)) \\ +\{\varsigma(1-\varsigma) + (1-\varsigma)\varsigma\}\mathfrak{B}_*((t,v)) \end{bmatrix},$$

$$= \frac{1}{4}\begin{bmatrix} \mathfrak{S}^*(\varsigma t + (1-\varsigma)v)g^*(\varsigma t + (1-\varsigma)v) \\ +\mathfrak{S}^*((1-\varsigma)t + \varsigma v)g^*((1-\varsigma)t + \varsigma v) \end{bmatrix}$$

$$+\frac{1}{2}\begin{bmatrix} \{\varsigma^2 + (1-\varsigma)^2\}\mathfrak{C}^*((t,v)) \\ +\{\varsigma(1-\varsigma) + (1-\varsigma)\varsigma\}\mathfrak{B}^*((t,v)) \end{bmatrix}.$$

$\mathcal{IR}$-Integrating over $[0, 1]$, we have

$$2\mathfrak{S}_*\left(\tfrac{t+v}{2}\right)g_*\left(\tfrac{t+v}{2}\right) \leq \tfrac{1}{v-t}\int_t^v \mathfrak{S}_*(\omega)g_*(\omega)d\omega + \tfrac{\mathfrak{B}_*((t,v))}{6} + \tfrac{\mathfrak{C}_*((t,v))}{3},$$
$$2\mathfrak{S}^*\left(\tfrac{t+v}{2}\right)g^*\left(\tfrac{t+v}{2}\right) \leq \tfrac{1}{v-t}\int_t^v \mathfrak{S}^*(\omega)g^*(\omega)d\omega + \tfrac{\mathfrak{B}^*((t,v))}{6} + \tfrac{\mathfrak{C}^*((t,v))}{3},$$

that is

$$2\mathfrak{S}\left(\frac{t+v}{2}\right)g\left(\frac{t+v}{2}\right) \leq_p \frac{1}{v-t}\,(IR)\int_t^v \mathfrak{S}(\omega)g(\omega)d\omega + \frac{\mathfrak{B}(t,v)}{6} + \frac{\mathfrak{C}(t,v)}{3}.$$

Hence, the required result. □

Example 5. *We consider the I.V-Fs $\mathfrak{S}$, $g : [t, v] = [0, 1] \to \mathcal{K}_C^+$ defined by $\mathfrak{S}(\omega) = [2\omega^2, 4\omega^2]$ and $g(\omega) = [\omega, 2\omega]$. Since end point functions $\mathfrak{S}_*(\omega) = 2\omega^2$, $\mathfrak{S}^*(\omega) = 4\omega^2$ and $g_*(\omega) = \omega$, $g^*(\omega) = 2\omega$ are convex functions. Hence $\mathfrak{S}$, g both are LR-convex I.V-Fs. We now compute the following*

$$\tfrac{1}{v-t}\int_t^v \mathfrak{S}_*(\omega)g_*(\omega)d\omega = \tfrac{1}{2},$$
$$\tfrac{1}{v-t}\int_t^v \mathfrak{S}^*(\omega)g^*(\omega)d\omega = 2,$$
$$\tfrac{\mathfrak{B}_*((t,v))}{3} = \tfrac{2}{3},$$
$$\tfrac{\mathfrak{B}^*((t,v))}{3} = \tfrac{8}{3},$$
$$\tfrac{\mathfrak{C}_*((t,v))}{6} = 0,$$
$$\tfrac{\mathfrak{C}^*((t,v))}{6} = 0,$$

that means

$$\tfrac{1}{2} \leq \tfrac{2}{3} + 0 = \tfrac{2}{3},$$
$$2 \leq \tfrac{8}{3} + 0 = \tfrac{8}{3},$$

Consequently, Theorem 5 is verified.

For Theorem 6, we have

$$2\mathfrak{S}_*\left(\tfrac{t+v}{2}\right)g_*\left(\tfrac{t+v}{2}\right) = \tfrac{1}{2},$$
$$2\mathfrak{S}^*\left(\tfrac{t+v}{2}\right)g^*\left(\tfrac{t+v}{2}\right) = 2,$$
$$\tfrac{1}{v-t}\int_t^v \mathfrak{S}_*(\omega)g_*(\omega)d\omega = \tfrac{1}{2},$$
$$\tfrac{1}{v-t}\int_t^v \mathfrak{S}^*(\omega)g^*(\omega)d\omega = 2,$$
$$\tfrac{\mathfrak{B}_*((t,v))}{6} = \tfrac{1}{3},$$
$$\tfrac{\mathfrak{B}^*((t,v))}{6} = \tfrac{4}{3},$$
$$\tfrac{\mathfrak{C}_*((t,v))}{3} = 0,$$
$$\tfrac{\mathfrak{C}^*((t,v))}{3} = 0,$$

From which, we have

$$\tfrac{1}{2} \leq \tfrac{1}{2} + 0 + \tfrac{1}{3} = \tfrac{5}{6},$$
$$2 \leq 2 + 0 + \tfrac{4}{3} = \tfrac{10}{3},$$

Consequently, Theorem 6 is demonstrated.

We now give *HH*-Fejér inequalities for LR-convex *I.V-Fs*. Firstly, we obtain the second *HH*-Fejér inequality for LR-convex *I.V-F*.

Theorem 7. Let $\mathfrak{S} : [t, v] \to \mathcal{K}_\mathcal{C}^+$ be an I.V-F with $t < v$, such that $\mathfrak{S}(\omega) = [\mathfrak{S}_*(\omega), \mathfrak{S}^*(\omega)]$ for all $\omega \in [t, v]$ and $\mathfrak{S} \in \mathcal{IR}_{([t, v])}$. If $\mathfrak{S} \in LRSX([t, v], \mathcal{K}_\mathcal{C}^+)$, then $\mathfrak{D} : [t, v] \to \mathbb{R}$, $\mathfrak{D}(\omega) \geq 0$, symmetric with respect to $\frac{t+v}{2}$, then

$$\frac{1}{v-t} (IR) \int_t^v \mathfrak{S}(\omega) \mathfrak{D}(\omega) d\omega \leq_p [\mathfrak{S}(t) + \mathfrak{S}(v)] \int_0^1 \varsigma \mathfrak{D}((1-\varsigma)t + \varsigma v) d\varsigma. \quad (11)$$

Proof. Let $\mathfrak{S} \in LRSX([t, v], \mathcal{K}_\mathcal{C}^+)$. Then we have

$$\mathfrak{S}_*(\varsigma t + (1-\varsigma)v) D(\varsigma t + (1-\varsigma)v)$$
$$\leq (\varsigma \mathfrak{S}_*(t) + (1-\varsigma)\mathfrak{S}_*(v)) D(\varsigma t + (1-\varsigma)v),$$
$$\mathfrak{S}^*(\varsigma t + (1-\varsigma)v) D(\varsigma t + (1-\varsigma)v) \quad (12)$$
$$\leq (\varsigma \mathfrak{S}^*(t) + (1-\varsigma)\mathfrak{S}^*(v)) D(\varsigma t + (1-\varsigma)v).$$

And

$$\mathfrak{S}_*((1-\varsigma)t + \varsigma v) D((1-\varsigma)t + \varsigma v) \leq ((1-\varsigma)\mathfrak{S}_*(t) + \varsigma \mathfrak{S}_*(v)) D((1-\varsigma)t + \varsigma v), \quad (13)$$
$$\mathfrak{S}^*((1-\varsigma)t + \varsigma v) D((1-\varsigma)t + \varsigma v) \leq ((1-\varsigma)\mathfrak{S}^*(t) + \varsigma \mathfrak{S}^*(v)) D((1-\varsigma)t + \varsigma v).$$

After adding (12) and (13), and integrating over $[0, 1]$, we obtain

$$\int_0^1 \mathfrak{S}_*(\varsigma t + (1-\varsigma)v) \mathfrak{D}(\varsigma t + (1-\varsigma)v) d\varsigma + \int_0^1 \mathfrak{S}_*((1-\varsigma)t + \varsigma v) \mathfrak{D}((1-\varsigma)t + \varsigma v) d\varsigma$$
$$\leq \int_0^1 \left[\begin{array}{l} \mathfrak{S}_*(t)\{\varsigma \mathfrak{D}(\varsigma t + (1-\varsigma)v) + (1-\varsigma)\mathfrak{D}((1-\varsigma)t + \varsigma v)\} \\ +\mathfrak{S}_*(v)\{(1-\varsigma)\mathfrak{D}(\varsigma t + (1-\varsigma)v) + \varsigma \mathfrak{D}((1-\varsigma)t + \varsigma v)\} \end{array} \right] d\varsigma,$$
$$\int_0^1 \mathfrak{S}^*((1-\varsigma)t + \varsigma v) \mathfrak{D}((1-\varsigma)t + \varsigma v) d\varsigma + \int_0^1 \mathfrak{S}^*(\varsigma t + (1-\varsigma)v) \mathfrak{D}(\varsigma t + (1-\varsigma)v) d\varsigma$$
$$\leq \int_0^1 \left[\begin{array}{l} \mathfrak{S}^*(t)\{\varsigma \mathfrak{D}(\varsigma t + (1-\varsigma)v) + (1-\varsigma)\mathfrak{D}((1-\varsigma)t + \varsigma v)\} \\ +\mathfrak{S}^*(v)\{(1-\varsigma)\mathfrak{D}(\varsigma t + (1-\varsigma)v) + \varsigma \mathfrak{D}((1-\varsigma)t + \varsigma v)\} \end{array} \right] d\varsigma.$$
$$= 2\mathfrak{S}_*(t) \int_0^1 \varsigma D(\varsigma t + (1-\varsigma)v) d\varsigma + 2\mathfrak{S}_*(v) \int_0^1 \varsigma D((1-\varsigma)t + \varsigma v) d\varsigma,$$
$$= 2\mathfrak{S}^*(t) \int_0^1 \varsigma D(\varsigma t + (1-\varsigma)v) d\varsigma + 2\mathfrak{S}^*(v) \int_0^1 \varsigma D((1-\varsigma)t + \varsigma v) d\varsigma.$$

Since $\mathfrak{D}$ is symmetric, then

$$= 2[\mathfrak{S}_*(t) + \mathfrak{S}_*(v)] \int_0^1 \varsigma D((1-\varsigma)t + \varsigma v) d\varsigma, \quad (14)$$
$$= 2[\mathfrak{S}^*(t) + \mathfrak{S}^*(v)] \int_0^1 \varsigma D((1-\varsigma)t + \varsigma v) d\varsigma.$$

Since

$$\int_0^1 \mathfrak{S}_*(\varsigma t + (1-\varsigma)v) \mathfrak{D}(\varsigma t + (1-\varsigma)v) d\varsigma$$
$$= \int_0^1 \mathfrak{S}_*((1-\varsigma)t + \varsigma v) \mathfrak{D}((1-\varsigma)t + \varsigma v) d\varsigma = \frac{1}{v-t} \int_t^v \mathfrak{S}_*(\omega) \mathfrak{D}(\omega) d\omega$$
$$\int_0^1 \mathfrak{S}^*((1-\varsigma)t + \varsigma v) \mathfrak{D}((1-\varsigma)t + \varsigma v) d\varsigma$$
$$= \int_0^1 \mathfrak{S}^*(\varsigma t + (1-\varsigma)v) \mathfrak{D}(\varsigma t + (1-\varsigma)v) d\varsigma = \frac{1}{v-t} \int_t^v \mathfrak{S}^*(\omega) \mathfrak{D}(\omega) d\omega \quad (15)$$

From (15), we have

$$\frac{1}{v-t} \int_t^v \mathfrak{S}_*(\omega) \mathfrak{D}(\omega) d\omega \leq [\mathfrak{S}_*(t) + \mathfrak{S}_*(v)] \int_0^1 \varsigma D((1-\varsigma)t + \varsigma v) d\varsigma,$$
$$\frac{1}{v-t} \int_t^v \mathfrak{S}^*(\omega) \mathfrak{D}(\omega) d\omega \leq [\mathfrak{S}^*(t) + \mathfrak{S}^*(v)] \int_0^1 \varsigma D((1-\varsigma)t + \varsigma v) d\varsigma,$$

that is

$$\left[\frac{1}{v-t} \int_t^v \mathfrak{S}_*(\omega) \mathfrak{D}(\omega) d\omega, \frac{1}{v-t} \int_t^v \mathfrak{S}^*(\omega) \mathfrak{D}(\omega) d\omega \right]$$
$$\leq_p [\mathfrak{S}_*(t) + \mathfrak{S}_*(v), \mathfrak{S}^*(t) + \mathfrak{S}^*(v)] \int_0^1 \varsigma D((1-\varsigma)t + \varsigma v) d\varsigma$$

hence

$$\frac{1}{v-t}(IR)\int_t^v \mathfrak{S}(\omega)\mathfrak{D}(\omega)d\omega \leq_p [\mathfrak{S}(t)+\mathfrak{S}(v)]\int_0^1 \varsigma\mathfrak{D}((1-\varsigma)t+\varsigma v)d\varsigma.$$

Next, we construct first *HH*-Fejér inequality for LR-convex *I.V-F*, which generalizes first *HH*-Fejér inequalities for convex function, see [21]. □

Theorem 8. *Let* $\mathfrak{S}:[t,v]\to\mathcal{K}_\mathcal{C}^+$ *be an I.V-F with* $t<v$, *such that* $\mathfrak{S}(\omega)=[\mathfrak{S}_*(\omega),\mathfrak{S}^*(\omega)]$ *for all* $\omega\in[t,v]$ *and* $\mathfrak{S}\in\mathcal{IR}_{([t,v])}$. *If* $\mathfrak{S}\in LRSX([t,v],\mathcal{K}_\mathcal{C}^+)$ *and* $\mathfrak{D}:[t,v]\to\mathbb{R}$, $\mathfrak{D}(\omega)\geq 0$, *symmetric with respect to* $\frac{t+v}{2}$, *and* $\int_t^v \mathfrak{D}(\omega)d\omega>0$, *then*

$$\mathfrak{S}\left(\frac{t+v}{2}\right)\leq_p \frac{1}{\int_t^v \mathfrak{D}(\omega)d\omega}(IR)\int_t^v \mathfrak{S}(\omega)\mathfrak{D}(\omega)d\omega. \quad (16)$$

Proof. Since $\mathfrak{S}\in LRSX([t,v],\mathcal{K}_\mathcal{C}^+)$, then we have

$$\begin{aligned}\mathfrak{S}_*\left(\tfrac{t+v}{2}\right) &\leq \tfrac{1}{2}(\mathfrak{S}_*(\varsigma t+(1-\varsigma)v)+\mathfrak{S}_*((1-\varsigma)t+\varsigma v)),\\ \mathfrak{S}^*\left(\tfrac{t+v}{2}\right) &\leq \tfrac{1}{2}(\mathfrak{S}^*(\varsigma t+(1-\varsigma)v)+\mathfrak{S}^*((1-\varsigma)t+\varsigma v)),\end{aligned} \quad (17)$$

By multiplying (17) by $\mathfrak{D}(\varsigma t+(1-\varsigma)v)=\mathfrak{D}((1-\varsigma)t+\varsigma v)$ and integrate it by ς over $[0,1]$, we obtain

$$\begin{aligned}\mathfrak{S}_*\left(\tfrac{t+v}{2}\right)&\int_0^1 \mathfrak{D}((1-\varsigma)t+\varsigma v)d\varsigma\\ &\leq \tfrac{1}{2}\left(\begin{array}{l}\int_0^1 \mathfrak{S}_*(\varsigma t+(1-\varsigma)v)\mathfrak{D}(\varsigma t+(1-\varsigma)v)d\varsigma\\ +\int_0^1 \mathfrak{S}_*((1-\varsigma)t+\varsigma v)\mathfrak{D}((1-\varsigma)t+\varsigma v)d\varsigma\end{array}\right),\\ \mathfrak{S}^*\left(\tfrac{t+v}{2}\right)&\int_0^1 \mathfrak{D}((1-\varsigma)t+\varsigma v)d\varsigma\\ &\leq \tfrac{1}{2}\left(\begin{array}{l}\int_0^1 \mathfrak{S}^*(\varsigma t+(1-\varsigma)v)\mathfrak{D}(\varsigma t+(1-\varsigma)v)d\varsigma\\ +\int_0^1 \mathfrak{S}^*((1-\varsigma)t+\varsigma v)\mathfrak{D}((1-\varsigma)t+\varsigma v)d\varsigma\end{array}\right),\end{aligned} \quad (18)$$

Since

$$\begin{aligned}\int_0^1 \mathfrak{S}_*(\varsigma t&+(1-\varsigma)v)\mathfrak{D}(\varsigma t+(1-\varsigma)v)d\varsigma\\ &=\int_0^1 \mathfrak{S}_*((1-\varsigma)t+\varsigma v)\mathfrak{D}((1-\varsigma)t+\varsigma v)d\varsigma\\ &=\tfrac{1}{v-t}\int_t^v \mathfrak{S}_*(\omega)\mathfrak{D}(\omega)d\omega\\ \int_0^1 \mathfrak{S}^*((1-\varsigma)&t+\varsigma v)\mathfrak{D}((1-\varsigma)t+\varsigma v)d\varsigma\\ &=\int_0^1 \mathfrak{S}^*(\varsigma t+(1-\varsigma)v)\mathfrak{D}(\varsigma t+(1-\varsigma)v)d\varsigma\\ &=\tfrac{1}{v-t}\int_t^v \mathfrak{S}^*(\omega)\mathfrak{D}(\omega)d\omega\end{aligned} \quad (19)$$

From (19), we have

$$\begin{aligned}\mathfrak{S}_*\left(\tfrac{t+v}{2}\right) &\leq \tfrac{1}{\int_t^v \mathfrak{D}(\omega)d\omega}\int_t^v \mathfrak{S}_*(\omega)\mathfrak{D}(\omega)d\omega,\\ \mathfrak{S}^*\left(\tfrac{t+v}{2}\right) &\leq \tfrac{1}{\int_t^v \mathfrak{D}(\omega)d\omega}\int_t^v \mathfrak{S}^*(\omega)\mathfrak{D}(\omega)d\omega,\end{aligned}$$

From which, we have

$$\begin{aligned}\left[\mathfrak{S}_*\left(\tfrac{t+v}{2}\right),\ \mathfrak{S}^*\left(\tfrac{t+v}{2}\right)\right]\\ \leq_p \tfrac{1}{\int_t^v \mathfrak{D}(\omega)d\omega}\left[\int_t^v \mathfrak{S}_*(\omega)\mathfrak{D}(\omega)d\omega,\ \int_t^v \mathfrak{S}^*(\omega)\mathfrak{D}(\omega)d\omega\right],\end{aligned}$$

that is

$$\mathfrak{S}\left(\frac{t+v}{2}\right)\leq_p \frac{1}{\int_t^v \mathfrak{D}(\omega)d\omega}(IR)\int_t^v \mathfrak{S}(\omega)\mathfrak{D}(\omega)d\omega.$$

This completes the proof. □

Remark 5. If $\mathfrak{D}(\omega) = 1$ then, combining Theorems 7 and 8, we obtain Theorem 3.

If $\mathfrak{S}_*(t) = \mathfrak{S}^*(t)$ then, Theorems 7 and 8 reduces to classical first and second HH-Fejér inequality for convex function, see [21].

If $\mathfrak{S}_*(t) = \mathfrak{S}^*(t)$ with $\mathfrak{D}(\omega) = 1$ then, Theorems 7 and 8 reduces to classical first and second HH-Fejér inequality for convex function, see [17,18].

Example 6. We consider the I.V-F $\mathfrak{S} : [t, v] = \left[\frac{\pi}{4}, \frac{\pi}{2}\right] \to \mathcal{K}_C^+$ defined by,

$$\mathfrak{S}(\omega) = [\exp(\sin(\omega)), 2\exp(\sin(\omega))]$$

Since end point functions $\mathfrak{S}_*(\omega) = \exp(\sin(\omega))$, $\mathfrak{S}^*(\omega) = 2\exp(\sin(\omega))$ convex functions then, by Theorem 2, $\mathfrak{S}(\omega)$ is LR-convex I.V-F. If

$$\mathfrak{D}(\omega) = \begin{cases} \omega - \frac{\pi}{4}, & S \in \left[\frac{\pi}{4}, \frac{3\pi}{8}\right], \\ \frac{\pi}{2} - \omega, & S \in \left(\frac{3\pi}{8}, \frac{\pi}{2}\right]. \end{cases}$$

then, we have

$\frac{1}{v-t}\int_t^v [\mathfrak{S}_*(\omega)]\mathfrak{D}(\omega)d\omega = \frac{4}{\pi}\int_{\frac{\pi}{4}}^{\frac{\pi}{2}} [\mathfrak{S}_*(\omega)]\mathfrak{D}(\omega)d\omega = \frac{4}{\pi}\int_{\frac{\pi}{4}}^{\frac{3\pi}{8}} [\mathfrak{S}_*(\omega)]\mathfrak{D}(\omega)d\omega + \frac{4}{\pi}\int_{\frac{3\pi}{8}}^{\frac{\pi}{2}} \mathfrak{S}_*(\omega)\mathfrak{D}(\omega)d\omega,$

$\frac{1}{v-t}\int_t^v [\mathfrak{S}^*(\omega)]\mathfrak{D}(\omega)d\omega = \frac{4}{\pi}\int_{\frac{\pi}{4}}^{\frac{\pi}{2}} [\mathfrak{S}^*(\omega)]\mathfrak{D}(\omega)d\omega = \frac{4}{\pi}\int_{\frac{\pi}{4}}^{\frac{3\pi}{8}} [\mathfrak{S}^*(\omega)]\mathfrak{D}(\omega)d\omega + \frac{4}{\pi}\int_{\frac{3\pi}{8}}^{\frac{\pi}{2}} \mathfrak{S}^*(\omega)\mathfrak{D}(\omega)d\omega,$

$$= \frac{4}{\pi}\int_{\frac{\pi}{4}}^{\frac{3\pi}{8}} [\exp(\sin(\omega))]\left(\omega - \frac{\pi}{4}\right)d\omega + \frac{4}{\pi}\int_{\frac{3\pi}{8}}^{\frac{\pi}{2}} \exp(\sin(\omega))\left(\frac{\pi}{2} - \omega\right)d\omega \approx \frac{63}{100\pi}, \tag{20}$$

$$= \frac{8}{\pi}\int_{\frac{\pi}{4}}^{\frac{3\pi}{8}} \exp(\sin(\omega))\left(\omega - \frac{\pi}{4}\right)d\omega + \frac{8}{\pi}\int_{\frac{3\pi}{8}}^{\frac{\pi}{2}} \exp(\sin(\omega))\left(\frac{\pi}{2} - \omega\right)d\omega \approx \frac{63}{50\pi},$$

and

$$[\mathfrak{S}_*(t) + \mathfrak{S}_*(v)]\int_0^1 \varsigma D(t + \varsigma\partial(v, t))\, d\varsigma$$

$$[\mathfrak{S}^*(t) + \mathfrak{S}^*(v)]\int_0^1 \varsigma D(t + \varsigma\partial(v, t))d\varsigma$$

$$= \frac{\pi}{2}\left[\int_0^{\frac{1}{2}} \varsigma^2 d\varsigma + \int_{\frac{1}{2}}^1 \varsigma(1+\varsigma)d\varsigma\right] = \frac{17\pi}{48}, \tag{21}$$

$$= \pi\left[\int_0^{\frac{1}{2}} \varsigma^2 d\varsigma + \int_{\frac{1}{2}}^1 \varsigma(1+\varsigma)d\varsigma\right] = \frac{17\pi}{24}.$$

From (20) and (21), we have

$$\left[\frac{63}{100\pi}, \frac{63}{50\pi}\right] \leq_p \left[\frac{17\pi}{48}, \frac{17\pi}{24}\right].$$

Hence, Theorem 7 is verified.

For Theorem 8, we have

$$\mathfrak{S}_*\left(\frac{t+v}{2}\right) = \mathfrak{S}_*\left(\frac{3\pi}{8}\right) \approx 1,$$
$$\mathfrak{S}^*\left(\frac{t+v}{2}\right) = \mathfrak{S}^*\left(\frac{3\pi}{8}\right) \approx 2, \tag{22}$$

$$\int_t^v \mathfrak{D}(\omega)d\omega = \int_{\frac{\pi}{4}}^{\frac{3\pi}{8}} \left(\omega - \frac{\pi}{4}\right)d\omega + \int_{\frac{3\pi}{8}}^{\frac{\pi}{2}} \left(\frac{\pi}{2} - \omega\right)d\omega \approx \frac{4}{25},$$

$$\frac{1}{\int_t^v \mathfrak{D}(\omega)d\omega}\int_t^v \mathfrak{S}_*(\omega)\mathfrak{D}(\omega)d\omega \approx 1.1$$
$$\frac{1}{\int_t^v \mathfrak{D}(\omega)d\omega}\int_t^v \mathfrak{S}^*(\omega)\mathfrak{D}(\omega)d\omega \approx 2.1. \tag{23}$$

From (22) and (23), we have

$$[1, 2] \leq {}_p[1.1, 2.1].$$

Hence, Theorem 8 is verified.

4. Results and Discussion

For LR-convex I.V-Fs, we find Hermite–Hadamard type inequalities. Our findings not only improve on Zhao's work, but they also investigate some of the findings of Sarikaya et al. We have not looked into inequalities using interval derivatives since there are not any "interval derivatives" with desirable characteristics.

5. Conclusions

In this paper, *HH*-inequalities have been investigated for the concept of LR-convex I.V-Fs. The most important thing in this study is that we have proved that both concepts LR-convex I.V-F and convex I.V-Fs coincide under some mild conditions when these conditions are defined on the endpoint functions. As for future research, we try to explore this concept for generalized LR-convex I.V-Fs and some applications in interval nonlinear programing. This is an open problem for the readers and anyone can investigate this concept, "the optimality conditions of LR-convex I.V-Fs can be obtained through variational inequalities". We hope that this concept will be helpful for other authors to play their roles in different fields of sciences. Moreover, in future, we will also start exploring this concept and their generalizations by using different fractional integral operators.

Author Contributions: Conceptualization, M.B.K.; methodology, M.B.K.; validation, S.T., M.S.S. and H.G.Z.; formal analysis, K.N.; investigation, M.S.S.; resources, S.T.; data curation, H.G.Z.; writing—original draft preparation, M.B.K., K.N. and H.G.Z.; writing—review and editing, M.B.K. and S.T.; visualization, H.G.Z.; supervision, M.B.K. and M.S.S.; project administration, M.B.K.; funding acquisition, K.N., M.S.S. and H.G.Z. All authors have read and agreed to the published version of the manuscript.

Funding: This research received no external funding.

Institutional Review Board Statement: Not applicable.

Informed Consent Statement: Not applicable.

Data Availability Statement: Not applicable.

Acknowledgments: The authors wish to thank the Rector, COMSATS University Islamabad, Islamabad, Pakistan, for providing excellent research and this work was funded by the Taif University Researchers Supporting Project (Number TURSP-2020/345), Taif University, Taif, Saudi Arabia. Moreover, this research has also received funding support from the National Science, Research and Innovation Fund (NSRF), Thailand.

Conflicts of Interest: The authors declare no conflict of interest.

References

1. Chang, S.S. *Variational Inequality and Complementarity Problems Theory and Applications*; Shanghai Scientific and Technological Literature Publishing House: Shanghai, China, 1991.
2. Hudzik, H.; Maligranda, L. Some remarks on s-convex functions. *Aequ. Math.* **1994**, *48*, 100–111. [CrossRef]
3. Alomari, M.; Darus, M.; Dragomir, S.S.; Cerone, P. Ostrowski type inequalities for functions whose derivatives are s-convex in the second sense. *Appl. Math. Lett.* **2010**, *23*, 1071–1076. [CrossRef]
4. Bede, B. Studies in Fuzziness and Soft Computing. In *Mathematics of Fuzzy Sets and Fuzzy Logic*; Springer: Berlin/Heidelberg, Germany, 2013; Volume 295.
5. Khan, M.B.; Noor, M.A.; Noor, K.L.; Chu, Y.M. New Hermite–Hadamard Type Inequalities for -Convex Fuzzy-Interval-Valued Functions. *Adv. Differ. Equ.* **2021**, *2021*, 1–20.
6. Anderson, G.D.; Vamanamurthy, M.K.; Vuorinen, M. Generalized convexity and inequalities. *J. Math. Anal. Appl.* **2007**, *335*, 1294–1308. [CrossRef]

7. Avci, M.; Kavurmaci, H.; Ozdemir, M.E. New inequalities of Hermite–Hadamard type via s-convex functions in the second sense with applications. *Appl. Math. Comput.* **2011**, *217*, 5171–5176. [CrossRef]
8. Awan, M.U.; Noor, M.A.; Noor, K.I. Hermite–Hadamard inequalities for exponentially convex functions. *Appl. Math. Inf. Sci.* **2018**, *12*, 405–409. [CrossRef]
9. Iscan, I. Hermite–Hadamard type inequalities for p-convex functions. *Int. J. Anal. Appl.* **2016**, *11*, 137–145.
10. Matkowski, J.; Nikodem, K. An integral Jensen inequality for convex multifunctions. *Results Math.* **1994**, *26*, 348–353. [CrossRef]
11. Mihai, M.V.; Noor, M.A.; Noor, K.I.; Awan, M.U. Some integral inequalities for harmonic h-convex functions involving hypergeometric functions. *Appl. Math. Comput.* **2015**, *252*, 257–262. [CrossRef]
12. Iscan, I. Hermite–Hadamard type inequalities for harmonically convex functions. *Hacet. J. Math. Stat.* **2014**, *43*, 935–942. [CrossRef]
13. Nanda, S.; Kar, K. Convex fuzzy mappings. *Fuzzy Sets Syst.* **1992**, *48*, 129–132. [CrossRef]
14. Nikodem, K.; Snchez, J.L.; Snchez, L. Jensen and Hermite–Hadamard inequalities for strongly convex set-valued maps. *Math. Aeterna* **2014**, *4*, 979–987.
15. Chen, F.; Wu, S. Integral inequalities of Hermite–Hadamard type for products of two h-convex functions. *Abstr. Appl. Anal.* **2014**, *5*, 1–6.
16. Bede, B.; Gal, S.G. Generalizations of the differentiability of fuzzy-number-valued functions with applications to fuzzy differential equations. *Fuzzy Sets Syst.* **2005**, *151*, 581–599. [CrossRef]
17. Hadamard, J. Étude sur les propriétés des fonctions entières et en particulier d'une fonction considérée par Riemann. *J. Math. Pures Appl.* **1893**, *58*, 171–215.
18. Hermite, C. Sur deux limites d'une intégrale définie. *Mathesis* **1883**, *3*, 1–82.
19. Noor, M.A. Hermite–Hadamard integral inequalities for log-preinvex functions. *J. Math. Anal. Approx. Theory* **2007**, *5*, 126–131.
20. Pachpatte, B.G. On some inequalities for convex functions. *RGMIA Res. Rep. Coll.* **2003**, *6*, 1–9.
21. Fejer, L. Uberdie Fourierreihen II. *Math. Naturwise. Anz. Ungar. Akad. Wiss.* **1906**, *24*, 369–390.
22. Niculescu, P.C. The Hermite–Hadamard inequality for log convex functions. *Nonlinear Anal.* **2012**, *75*, 662–669. [CrossRef]
23. Noor, M.A. Fuzzy preinvex functions. *Fuzzy Sets Syst.* **1994**, *64*, 95–104. [CrossRef]
24. Yan, H.; Xu, J. A class convex fuzzy mappings. *Fuzzy Sets Syst.* **2002**, *129*, 47–56. [CrossRef]
25. Hussain, S.; Khalid, J.; Chu, Y.M. Some generalized fractional integral Simpson's type inequalities with applications. *AIMS Math.* **2020**, *5*, 5859–5883. [CrossRef]
26. Sarikaya, M.Z.; Saglam, A.; Yildrim, H. On some Hadamard-type inequalities for h-convex functions. *J. Math. Inequalities* **2008**, *2*, 335–341. [CrossRef]
27. Xu, L.; Chu, Y.M.; Rashid, S.; El-Deeb, A.A.; Nisar, K.S. On new unified bounds for a family of functions with fractional q-calculus theory. *J. Funct. Spaces* **2020**, *2020*, 1–9.
28. Costa, T.M.; Román-Flores, H.; Chalco-Cano, Y. Opial-type inequalities for interval-valued functions. *Fuzzy Sets Syst.* **2019**, *358*, 48–63. [CrossRef]
29. Fang, Z.B.; Shi, R. On the, (p, h)-convex function and some integral inequalities. *J. Inequalities Appl.* **2014**, *45*, 1–16. [CrossRef]
30. Moore, R.E. *Interval Analysis*; Prentice Hall: Englewood Cliffs, NJ, USA, 1966.
31. Kulish, U.; Miranker, W. *Computer Arithmetic in Theory and Practice*; Academic Press: New York, NY, USA, 2014.
32. Moore, R.E.; Kearfott, R.B.; Cloud, M.J. *Introduction to Interval Analysis*; SIAM: Philadelphia, PA, USA, 2009.
33. Rothwell, E.J.; Cloud, M.J. Automatic error analysis using intervals. *IEEE Trans. Ed.* **2012**, *55*, 9–15. [CrossRef]
34. Snyder, J.M. Interval analysis for computer graphics. *SIGGRAPH Comput. Graph.* **1992**, *26*, 121–130. [CrossRef]
35. de Weerdt, E.; Chu, Q.P.; Mulder, J.A. Neural network output optimization using interval analysis. *IEEE Trans. Neural Netw.* **2009**, *20*, 638–653. [CrossRef]
36. Khan, M.B.; Mohammed, P.O.; Machado, J.A.T.; Guirao, J.L. Integral Inequalities for Generalized Harmonically Convex Functions in Fuzzy-Interval-Valued Settings. *Symmetry* **2021**, *13*, 2352. [CrossRef]
37. Khan, M.B.; Noor, M.A.; Abdeljawad, T.; Mousa, A.A.A.; Abdalla, B.; Alghamdi, S.M. LR-Preinvex Interval-Valued Functions and Riemann–Liouville Fractional Integral Inequalities. *Fractal Fract.* **2021**, *5*, 243. [CrossRef]
38. Khan, M.B.; Srivastava, H.M.; Mohammed, P.O.; Guirao, J.L. Fuzzy mixed variational-like and integral inequalities for strongly preinvex fuzzy mappings. *Symmetry* **2021**, *13*, 1816. [CrossRef]
39. Mohan, S.R.; Neogy, S.K. On invex sets and preinvex functions. *J. Math. Anal. Appl.* **1995**, *189*, 901–908. [CrossRef]
40. Iscan, I. A new generalization of some integral inequalities for, (α, m)-convex functions. *Math. Sci.* **2013**, *7*, 1–8. [CrossRef]
41. Zadeh, L.A. Fuzzy sets. *Inf. Control* **1965**, *8*, 338–353. [CrossRef]
42. Costa, T.M. Jensen's inequality type integral for fuzzy-interval-valued functions. *Fuzzy Sets Syst.* **2017**, *327*, 31–47. [CrossRef]
43. Costa, T.M.; Roman-Flores, H. Some integral inequalities for fuzzy-interval-valued functions. *Inf. Sci.* **2017**, *420*, 110–125. [CrossRef]
44. Flores-Franulic, A.; Chalco-Cano, Y.; Roman-Flores, H. An Ostrowski type inequality for interval-valued functions. In Proceedings of the IFSA World Congress and AFIPS Annual Meeting IEEE, Edmonton, AB, Canada, 24–28 June 2013; Volume 35, pp. 1459–1462.
45. Roman-Flores, H.; Chalco-Cano, Y.; Lodwick, W.A. Some integral inequalities for interval-valued functions. *Comput. Appl. Math.* **2016**, *35*, 1–13. [CrossRef]

46. Roman-Flores, H.; Chalco-Cano, Y.; Silva, G.N. A note on Gronwall type inequality for interval-valued functions. In Proceedings of the IFSA World Congress and NAFIPS Annual Meeting IEEE, Edmonton, AB, Canada, 24–28 June 2013; Volume 35, pp. 1455–1458.
47. Chalco-Cano, Y.; Flores-Franulič, A.; Román-Flores, H. Ostrowski type inequalities for interval-valued functions using generalized Hukuhara derivative. *Comput. Appl. Math.* **2012**, *31*, 457–472.
48. Chalco-Cano, Y.; Lodwick, W.A.; Condori-Equice, W. Ostrowski type inequalities and applications in numerical integration for interval-valued functions. *Soft Comput.* **2015**, *19*, 3293–3300. [CrossRef]
49. Zhang, D.; Guo, C.; Chen, D.; Wang, G. Jensen's inequalities for set-valued and fuzzy set-valued functions. *Fuzzy Sets Syst.* **2020**, *2020*, 1–27. [CrossRef]
50. Zhao, D.F.; An, T.Q.; Ye, G.J.; Liu, W. New Jensen and Hermite–Hadamard type inequalities for h-convex interval-valued functions. *J. Inequalities Appl.* **2018**, *3*, 1–14. [CrossRef]
51. An, Y.; Ye, G.; Zhao, D.; Liu, W. Hermite–hadamard type inequalities for interval, (h1, h2)-convex functions. *Mathematics* **2019**, *7*, 436. [CrossRef]

 *fractal and fractional*

Article

Application of the Pick Function in the Lieb Concavity Theorem for Deformed Exponentials

Guozeng Yang [1], Yonggang Li [2,*], Jing Wang [3] and Huafei Sun [4,5]

1. School of Mathematics and Statistics, Zhengzhou Normal University, Zhengzhou 450000, China; yangguozeng@zznu.edu.cn
2. School of Science, Zhengzhou University of Aeronautics, Zhengzhou 450000, China
3. School of Information, Beijing Wuzi University, Beijing 101149, China; wangjing3@bwu.edu.cn
4. Beijing Institute of Technology, School of Mathematics and Statistics, Beijing 100081, China; huafeisun@bit.edu.cn
5. Yangtze Delta Region Academy of Beijing Institute of Technology, Jiaxing 314000, China
* Correspondence: liyonggang914@126.com

Abstract: The Lieb concavity theorem, successfully solved in the Wigner–Yanase–Dyson conjecture, is an important application of matrix concave functions. Recently, the Thompson–Golden theorem, a corollary of the Lieb concavity theorem, was extended to deformed exponentials. Hence, it is worthwhile to study the Lieb concavity theorem for deformed exponentials. In this paper, the Pick function is used to obtain a generalization of the Lieb concavity theorem for deformed exponentials, and some corollaries associated with exterior algebra are obtained.

Keywords: Lieb concavity theorem; deformed exponential; Pick function; convexity of matrix

MSC: 15A42; 15A16; 47A56

1. Introduction

Matrix theory is widely used in statistics [1], physics [2], computer science [3] and so on. For convenience, $M(n, \mathbb{C})$ is denoted as the set of all $n \times n$ complex matrices ($\mathbb{C}$ is the set of complex numbers) [4]. A is called a Hermitian matrix when $A \in M(n, \mathbb{C})$ satisfies $A^* = A$ (A^* denotes conjugate transposition of A). The Hermitian matrix is frequently used in quadratic forms and their correlation theory [5]. Let H_n denote the set of $n \times n$ Hermitian matrices and H_n^+ denote the $n \times n$ positive semidefinite Hermitian matrix ($\mathbb{C}^n$ is the n dimensional complex Euclidean space).

Set $u_1, u_2, \cdots, u_n$ to be any orthonormal basis of $\mathbb{C}^n$, and then the trace operator Tr is defined as [4]

$$\text{Tr}[A] = \sum_{i=1}^{n} (u_i, Au_i),$$

where $(\cdot, \cdot)$ is the inner product of $\mathbb{C}^n$. It is well known that for any $A = (a_{ij}) \in M(n, \mathbb{C})$, the following equalities hold [6]

$$\text{Tr}[A] = \sum_{j=1}^{n} \lambda_i = \sum_{j=1}^{n} a_{ii},$$

where λ_i is the eigenvalue of A.

From the spectral theorem [5], $A \in H_n^+$ can be decomposed as

$$A = P^* \Lambda_A P,$$

where P is a unitary matrix and $\Lambda_A := \text{diag}\{\lambda_1, ..., \lambda_n\}$ is a diagonal matrix with eigenvalues $\lambda_1, ..., \lambda_n$. Then, matrix function $f(A)$ is defined as

$$f(A) = P^* f(\Lambda_A) P = \sum_{i=1}^n f(\lambda_i) P_i, \quad (1)$$

where $f(\Lambda_A) := \text{diag}\{f(\lambda_1), ..., f(\lambda_n)\}$ and $P_i^2 = P_i$.

Based on the above definition, in 1963, the Wigner–Yanase skew information

$$I_{WY}(\rho) = -\frac{1}{2} \text{Tr}\left[[\sqrt{\rho}, H]^2\right]$$

was introduced by Wigner and Yanase ([7]), where ρ is a density matrix ($\rho \geq 0, \text{tr}\,\rho = 1$) and H is a Hermitian matrix. Then, an open problem was left

$$\text{Tr}[A^s K A^{1-s} K^*], \quad (2)$$

which is concave for any positive semidefinite matrix A.

In 1973, (2) was proven by Lieb for all $0 < s < 1$ [8], and a more generalized result was obtained from the following fact [9]

$$\text{Tr}[A^s K B^{1-s} K^*] = \langle K, B^{1-s} K^* A^s \rangle_{\mathcal{L}(H)}$$
$$= \langle K, \Psi^{-1}(B^{1-s} \otimes A^s) K^* \rangle_{\mathcal{L}(H)}.$$

where $\Psi^{-1}(A) = \sum_j (Ae_j) \otimes e_j^*$. In fact, the Lieb concavity theorem is equivalent to the concavity of $B^{1-s} \otimes A^s$.

A more elegant proof of the Lieb concavity theorem appeared in [10] using

$$\text{Tr}[K^* A^s K B^{1-s}] = \langle K, (A^s \otimes B^{1-s}) K \rangle_{\mathcal{L}(H)},$$

where

$$[(A \otimes B)K]_{i,j} = \sum_{k,l} A_{i,k} B_{j,l} K_{k,l}.$$

In 2009, Effros gave another proof of the Lieb concavity theorem based on the Hansen–Pedersen–Jensen inequality ([11]). Using

$$L_A(K) = AK, R_B(K) = KB,$$

then one obtains

$$\text{Tr}[K^* A^s K B^{1-s}] = \langle K, L_{A^s} R_{B^{1-s}}(K) \rangle_{\mathcal{L}(H)}$$
$$= \langle K, R_B^{\frac{1}{2}} (R_B^{-\frac{1}{2}} L_A R_B^{-\frac{1}{2}})^s R_B^{\frac{1}{2}}(K) \rangle_{\mathcal{L}(H)}.$$

All the above proof of the Lieb concave theorem is equivalent to the joint concavity of commutative operators. In addition, Epstein also obtained the Lieb concave theorem using the theory of Herglotz functions [12].

Recently, Shi and Hansen [13] generalized the Thompson–Golden theorem

$$\text{Tr}[\exp_q(A+B)] \leq \text{Tr}[(\exp_q(A))^{2-q}(A(q-1) + \exp_q(B))]$$

As the Thompson–Golden theorem can be regarded as a special form of the Lieb concave theorem, it is worthwhile to study the Lieb concavity theorem for deformed exponentials. In this paper, we will use the theory of the Pick function to obtain a generalization of the Lieb concavity theorem and some other corollaries. The rest of the paper is organized as follows. In Section 2, some general definitions and important conclusions are introduced.

With these preparations, we obtain some useful results, such as the Lieb concavity theorem, presented in the final Section 3.

2. Preliminary

In this section, some general definitions and some important properties are introduced.

2.1. The q-Logarithm Function and q-Exponential Function

It is well known that the q-logarithm function $\ln_q(x)$ is defined as [13]

$$\ln_q(x) = \left\{ \begin{array}{ll} \frac{x^{q-1}-1}{q-1}, & q \neq 1 \\ \ln x, & q = 1 \end{array} \right\}$$

for any $x > 0$. The deformed exponential function or the q-exponential $\exp_q(x)$ is the inverse function of the q-logarithm and is defined as

$$\exp_q(x) = \left\{ \begin{array}{lll} [(q-1)x+1]^{\frac{1}{q-1}}, & x > \frac{1}{q-1}, & q > 1 \\ [(q-1)x+1]^{\frac{1}{q-1}}, & x < \frac{1}{q-1}, & q < 1 \\ \exp(x), & x \in \mathbb{R}, & q = 1 \end{array} \right\}$$

2.2. Tensor Product and Exterior Algebra

The tensor product, denoted by "$\otimes$", is also called the Kronecker product. It is a generalization of the outer product from vectors to matrices, and the tensor product of matrices is also referred to as the outer product in certain contexts ([9]). For an $m \times n$ matrix A and a $p \times q$ matrix B, the tensor product of A and B is defined by

$$A \otimes B := \begin{pmatrix} a_{11}B & \cdots & a_{1n}B \\ \vdots & \ddots & \vdots \\ a_{m1}B & \cdots & a_{mn}B \end{pmatrix},$$

where $A = (a_{ij})_{1 \leq i \leq m, 1 \leq j \leq n}$.

The tensor product is different from matrix multiplication, and one of the differences is commutativity

$$(I \otimes B)(A \otimes I) = (A \otimes I)(I \otimes B) = A \otimes B.$$

From the above equations, we obtain

$$AC \otimes BD = (AC \otimes I)(I \otimes BD)$$
$$= (A \otimes I)(C \otimes I)(I \otimes B)(I \otimes D)$$
$$= (A \otimes I)(I \otimes B)(C \otimes I)(I \otimes D)$$
$$= (A \otimes B)(C \otimes D).$$

For convenience, we denote

$$\otimes_k A = \underbrace{A \otimes A \otimes \cdots \otimes A}_{k}.$$

In addition to the tensor product, there is another common product called exterior algebra [6]. Exterior algebra, denoted by "$\wedge$", is a binary operation for any $A_i \in H_n^+$, and the definition is

$$\underbrace{(A_1 \wedge A_2 \wedge \cdots \wedge A_k)}_{k}(\xi_{i_1} \wedge \xi_{i_2} \cdots \wedge \xi_{i_k})_{1 \leq i_1 < \cdots < i_k \leq n}$$
$$= (A_1 \xi_{i_1} \wedge A_2 \xi_{i_2} \cdots \wedge A_k \xi_{i_k})_{1 \leq i_1 < \cdots < i_k \leq n},$$

where $\{\xi_j\}_{j=1}^n$ is an orthonormal basis of $\mathbb{C}^n$, and

$$\xi_{i_1} \wedge \xi_{i_2} \cdots \wedge \xi_{i_k} = \frac{1}{\sqrt{n!}} \sum_{\pi \in \sigma_n} (-1)^\pi \xi_{\pi(i_1)} \otimes \xi_{\pi(i_2)} \cdots \otimes \xi_{\pi(i_k)},$$

σ_n is the family of all permutations on $\{1, 2, \cdots, n\}$.

Let $\bigwedge^k \mathbb{C}^n$ be the span of the $\{\xi_{i_1} \wedge \xi_{i_2} \cdots \wedge \xi_{i_k}\}_{1 \leq i_1 < \cdots < i_k \leq n}$, and then a simple calculation shows that

$$\wedge_n A = (\underbrace{A \wedge A \wedge \cdots \wedge A}_{k}) = \det(A)$$

2.3. Pick Function

Let $z = x + iy$ be a complex number where i is the imaginary unit and $f(z) = U(z) + iV(z)$ is analytic where $U(z), V(z)$ are all real functions. $\operatorname{Re} z = x$ denotes the real part of z, and $\operatorname{Im} z = y$ is the imaginary part of z. If $\operatorname{Im} f(z) > 0$ for any $\operatorname{Im} z > 0$, then we call the analytic function $f(z)$ a Pick function [14]. It is equivalent that $f(z)$ is analytic in the upper half-plane with the positive imaginary part.

The Pick functions evidently form a convex cone—for instance, if α and β are positive numbers and $f(z)$ and $g(z)$ are two Pick functions, then the function $\alpha f(z) + \beta g(z)$ is also a Pick function. A simple example is that $\tan(z)$ is a Pick function.

$$\tan(x + iy) = \frac{\tan(x) + \tan(iy)}{1 - \tan(x)\tan(iy)}$$

$$= \frac{\tan(x) + i\tanh(y)}{1 - i\tan(x)\tanh(y)}.$$

Hence, $\operatorname{Im} \tan(z) = \frac{(1+\tan^2(x))\tanh(y)}{1+\tan^2(x)\tanh^2(y)}$, and this implies that $\operatorname{Im} \tan(z) > 0$ when $y > 0$.

It is well known that the Pick function has a integral representation, such as the following lemma [14].

Lemma 1. *Let $f(z)$ be a Pick function. Then, $f(z)$ has a unique canonical representation of the form*

$$f(z) = \alpha + \beta z + \int_R \left(\frac{1}{\lambda - z} - \frac{\lambda}{1 + \lambda^2} \right) d\mu(\lambda),$$

where α is real, $\beta \geq 0$ and $d\mu(\lambda)$ is a positive Borel measure on the real λ-axis that $\int_R (1 + \lambda^2)^{-1} d\mu(\lambda)$ is finite. Conversely, any function of this form is also a Pick function.

Lemma 1 is frequently used for functions that are positive and harmonic in the half-plane.

2.4. The Matrix-Monotone Function

A matrix function f is said to be matrix-monotonic if it satisfies

$$f(A) \geq f(B) \quad \text{for all} \quad A \geq B > 0. \tag{3}$$

where $A \geq B >$ is equivalent to $A - B$ is a positive semidefinite Hermitian matrix.

Since the matrix-monotone function is a special kind of operator monotone function, we have the following general conclusions [14].

Lemma 2. *The following statements for a real valued continuous function f on $(0, +\infty)$ are equivalent:*

(1) $f(z)$ is matrix-monotone;
(2) $f(z)$ admits an analytic continuation to the whole domain $\operatorname{Im} z \neq 0$ and $\operatorname{Im}(z) \operatorname{Im} f(z) \geq 0$.

(3) f admits an integral representation:

$$f(\lambda) = \alpha + \beta\lambda + \int_{-\infty}^{0} (1+\lambda t)(t-\lambda)^{-1} \mathrm{d}\mu(t), \text{ for any } \lambda > 0, \quad (4)$$

where α is a real number, β is non-negative and μ is a finite positive measure on $(-\infty, 0)$.

From Lemmas 1 and 2, we know that a Pick function must be a matrix-monotone function.

2.5. Convexity of Matrix

Suppose that X is a convex set in $\mathbb{R}^n$ and f is a function defined on X. Then, we call f a convex function if

$$f(tx_1 + (1-t)x_2) \leq tf(x_1) + (1-t)f(x_2), \forall x_1, x_2 \in X, \forall t \in [0,1],$$

for all $x_1, x_2 \in X$ and $t \in [0,1]$.

A matrix function f is called convex if [15–17]

$$f(tA + (1-t)B) \leq tf(A) + (1-t)f(B), \quad (5)$$

for any $A, B \in H_n^+$ and any $t \in [0,1]$. Replacing $\leq$ by $<$ in (5), this gives the definition of a strictly matrix convex function. A matrix function f is called (strictly) concave if $-f$ is (strictly) convex. More details can be found in [18].

A matrix convex function must be a convex function; however, the inverse claim is not always true. For instance, the function $f : [0, +\infty) \to \mathbb{R}$ given by $f(x) = x^3$ is a convex function. However, the matrix function $f(A) = A^3$ for any $A \in H_n^+$ is not convex.

Let $f(\cdot, \cdot)$ be a bivariate function defined on $H_n^+ \times H_n^+$. We call $f(\cdot, \cdot)$ jointly convex if

$$f(tA_1 + (1-t)A_2, tB_1 + (1-t)B_2) \leq tf(A_1, B_1) + (1-t)f(A_2, B_2),$$

for all $A_1, A_2, B_1, B_2 \in H_n^+$ and all $t \in [0,1]$.

2.6. Brunn–Minkowski Inequality

Finally, let us review the Brunn–Minkowski inequality [19].

Lemma 3. *for any $A, B > 0$, and then*

$$\{\mathrm{Tr}[\wedge^k(A+B)]\}^{\frac{1}{k}} \geq \{\mathrm{Tr}[\wedge^k A]\}^{\frac{1}{k}} + \{\mathrm{Tr}[\wedge^k B]\}^{\frac{1}{k}}.$$

Proof. Let $\{\xi_i\}_{i=1}^n$ be the eigenvectors of $A+B$ with the eigenvalue $\{\lambda_i\}_{i=1}^n$, then

$$\{\mathrm{Tr}[\wedge^k(A+B)]\}^{\frac{1}{k}} = \left[\sum_{1 \leq \xi_{i_1} < \cdots < \xi_{i_k} \leq n} \lambda_{i_1} \cdots \lambda_{i_k}\right]^{\frac{1}{k}}$$

$$= \left[\sum_{1 \leq \xi_{i_1} < \cdots < \xi_{i_k} \leq n} \left(\det\left|P^*_{i_1,\cdots,i_k}(A+B)P_{i_1,\cdots,i_k}\right|\right)\right]^{\frac{1}{k}}$$

$$\geq \left[\sum_{1 \leq \xi_{i_1} < \cdots < \xi_{i_k} \leq n} \left(\det\left|P^*_{i_1,\cdots,i_k}AP_{i_1,\cdots,i_k}\right| + \det\left|P^*_{i_1,\cdots,i_k}BP_{i_1,\cdots,i_k}\right|\right)\right]^{\frac{1}{k}}$$

where $P_{i_1,\cdots,i_k} = (\xi_{i_1}, \cdots, \xi_{i_k})$ and $\geq$ holds due to $\det(A+B) \geq \det(A) + \det(B)$.

As $S_k = \left[\sum_{1 \leq \zeta_{i_1} < \cdots < \zeta_{i_k} \leq n} x_{i_1} \cdots x_{i_k}\right]^{\frac{1}{k}}$ is concave [20], we have

$$\{\text{Tr}[\wedge^k(A+B)]\}^{\frac{1}{k}} \geq \left[\sum_{1 \leq \zeta_{i_1} < \cdots < \zeta_{i_k} \leq n} \det\left|P^*_{i_1,\cdots,i_k} A P_{i_1,\cdots,i_k}\right|\right]^{\frac{1}{k}}$$

$$+ \left[\sum_{1 \leq \zeta_{i_1} < \cdots < \zeta_{i_k} \leq n} \det\left|P^*_{i_1,\cdots,i_k} B P_{i_1,\cdots,i_k}\right|\right]^{\frac{1}{k}}$$

$$= \left[\sum_{1 \leq \zeta_{i_1} < \cdots < \zeta_{i_k} \leq n} \left(\xi_{i_1} \wedge \cdots \wedge \xi_{i_k}, A\xi_{i_1} \wedge \cdots \wedge A\xi_{i_k}\right)\right]^{\frac{1}{k}}$$

$$+ \left[\sum_{1 \leq \zeta_{i_1} < \cdots < \zeta_{i_k} \leq n} \left(\xi_{i_1} \wedge \cdots \wedge \xi_{i_k}, B\xi_{i_1} \wedge \cdots \wedge B\xi_{i_k}\right)\right]^{\frac{1}{k}}$$

$$= \{\text{Tr}[\wedge^k A]\}^{\frac{1}{k}} + \{\text{Tr}[\wedge^k B]\}^{\frac{1}{k}}.$$

□

3. Lieb Concavity Theorem for Deformed Exponential

In this section, we obtain some useful conclusions, and some simple and straightforward computations are omitted. Recently, by using the Young inequality,

$$\text{Tr}[Y] = \max_{X \geq 0}\{\text{Tr}[X] - \text{Tr}[X^{2-q}(\ln_q X - \ln_q Y)]\},$$

Shi and Hansen obtained that $F(A) = \text{Tr}\left[\exp_q^{\frac{1}{p}}(K^* \ln_q(A^p)K)\right]$ is concave for any $1 \leq q \leq 2$ where $K^*K \leq I$ (I is the identity matrix of $M(n, \mathbb{C})$) [13], namely, the following theorem.

Theorem 1. *For $0 < p \leq 1, 1 < q \leq 2$ and $K^*K \leq I$, the function*

$$F(A) = \text{Tr}\left[\exp_q^{\frac{1}{p}}(K^* \ln_q(A^p)K)\right] \tag{6}$$

is concave for the strictly positive $A \in H_n^+$.

Proof. (The first proof of Theorem 1)
Since [21]

$$Df(A)(B) = \sum_i \sum_j \frac{f(\lambda_i) - f(\lambda_j)}{\lambda_i - \lambda_i} P_i B P_j,$$

we obtain

$$\frac{d(\text{Tr}[f(A+tB) - f(A)])}{dt} = \text{Tr}\left[\sum_i \sum_j \frac{f(\lambda_i) - f(\lambda_j)}{\lambda_i - \lambda_i} P_i B P_j\right]$$

$$= \text{Tr}\left[\sum_i P_j \sum_j \frac{f(\lambda_i) - f(\lambda_j)}{\lambda_i - \lambda_i} P_i B\right]$$

$$= \text{Tr}\left[\sum_i f'(\lambda_i) P_i B\right] = \text{Tr}[f'(A)B],$$

where λ_i are eigenvalues of A. When $f(x)$ is a convex function, we obtain

$$\text{Tr}[f(A+tB) - f(A)] \geq \text{Tr}[f'(A)tB]$$

for any t. This implies that

$$\text{Tr}[f(C)] = \max\{\text{Tr}[f(D) + f'(D)(C-D)] : D > 0\}.$$

Therefore, we obtain

$$\text{Tr}[(K^* A^{pq-p} K + I - K^* K)^{\frac{1}{pq-p}}]$$
$$= \max\{\text{Tr}[D^{\frac{1}{pq-p}} + \frac{D^{\frac{1}{pq-p}-1}(K^* A^{pq-p} K + I - K^* K - D)}{pq-p}] : D > 0\}$$
$$= \max\{\text{Tr}[C + \frac{C^{1-pq+p}(K^* A^{pq-p} K + I - K^* K - C^{pq-p})}{pq-p}] : C = D^{\frac{1}{pq-p}} > 0\}$$
$$= \max\{\text{Tr}[C(1 - \frac{1}{pq-p}) + \frac{C^{1-pq+p} K^* A^{pq-p} K}{pq-p} + C^{1-pq+p}(I - K^* K)] : C > 0\}$$

Thus, the concavity of $F(A)$ is equivalent to the jointly concavity of $\text{Tr}[\frac{C^{1-pq+p} K^* A^{pq-p} K}{pq-p}]$ for the strictly positive A and C, which is the Lieb concavity theorem [22,23]. □

Unfortunately, Theorem 1 cannot be obtained using Epstein's theorem. Hence, we require a more general generalization of Epstein's theorem. First, for any $\text{Im}(z) > 0$, we know that $A + zB$ is invertible and $x^*(A + zB)x$ is a Pick function for any $x \in \mathbb{C}^n$ [14]. For any $A \in M(n, \mathbb{C})$, we know $f(A)$ is defined as [12]

$$f(A) = \frac{1}{2\pi} \oint_C \frac{f(z)}{z - A} dz,$$

where $f(z)$ is a complex holomorphic function in an open set of the complex plane containing $\text{Sp}(A)$ (the set of all eigenvalues of A). Then, we have the following lemma.

Lemma 4. Let $A, B \in H_n^+$ and $0 < \alpha \leq 1$, then

$$x^*(A + zB)^\alpha x$$

is a Pick function for any $x \in \mathbb{C}^n$ and $0 < \arg(x^*(A+zB)^\alpha x) < \alpha\pi$ if $0 < \arg(z) = \theta < \pi$, such as $\text{Sp}((A+zB)^\alpha) \subseteq (\text{Sp}(A+zB)^\alpha)$. Generally, we can find that

$$x^* f(A + zB) x$$

is a Pick function when f is a Pick function.

Proof. Setting $z = \rho e^{i\theta}$, we have

$$(A + zB)^\alpha = \int_0^{+\infty} (\frac{A + zB}{t + A + zB}) d\mu(t)$$
$$= \int_0^{+\infty} (\frac{1}{\frac{t}{A+zB} + 1}) d\mu(t),$$

where $d\mu(t) = \frac{t^{\alpha-1} \pi}{\sin \alpha \pi}$.

Since $\mathrm{Im}\, z > 0$, we see that $A + zB$ is invertible. Hence, we have

$$x^*(A+zB)^\alpha x = \int_0^{+\infty} x^*(\frac{1}{\frac{t}{A+zB}+1})x\, d\mu(t)$$

$$= \int_0^{+\infty} y^*(\frac{t}{A+z^*B}+1)y\, d\mu(t),\ y = (\frac{t}{A+zB}+1)^{-1}x$$

$$= \int_0^{+\infty} y^*y + tw^*(A+zB)w\, d\mu(t),\ w = (A+z^*B)^{-1}y$$

$$= \int_0^{+\infty} y^*y + tw^*Aw\, d\mu(t) + z\int_0^{+\infty} tw^*Bw\, d\mu(t).$$

This implies that

$$\mathrm{Im}\, x^*(A+zB)^\alpha x = \mathrm{Im}(z)\cdot \int_0^{+\infty} tw^*Bw\, d\mu(t) > 0;$$

hence, $0 < \arg(x^*(A+zB)^\alpha x)$ when $0 < \arg(z) = \theta < \pi$.

In the same way, we can obtain

$$\mathrm{Im}\, w^*[(-A-z^*B)^{-\alpha}]w = \mathrm{Im}(e^{-i\alpha\pi}z^*)\cdot \int_0^{+\infty} tv^*Bv\, d\mu(t) < 0,\ v = (t(A+z^*B)+1)^{-1}w.$$

In particular, letting $w = (A+z^*B)^\alpha x$, we have

$$\mathrm{Im}(e^{-i\alpha\pi}x^*(A+zB)^\alpha x) < 0.$$

This is equivalent to $\arg(x^*(A+zB)^\alpha x) < \alpha\pi$.

To prove $\mathrm{Sp}((A+zB)^\alpha) \subseteq (\mathrm{Sp}(A+zB)^\alpha)$, let $(A+zB)\xi = \lambda\xi$, we find

$$\xi^*(A+zB)^\alpha\xi = [\xi^*(A+zB)\xi]^\alpha = [\xi^*A\xi + z\xi^*B\xi]^\alpha = \rho^\alpha e^{i\alpha\theta},$$

where $\tan\theta = \frac{\xi^*B\xi\, \mathrm{Im}(z)}{\xi^*A\xi + \xi^*B\xi\, \mathrm{Re}(z)} \leq \tan\arg(z)$.

When $f(z)$ is a Pick function, using the integral represented of $f(z)$, in a similar way, we can obtain that

$$x^*f(A+zB)x$$

is a Pick function for any $x \in \mathbb{C}^n$. □

Using Lemma 4, another proof of Theorem 1 can be obtained.

Theorem 2. *For $0 < p \leq 1, 1 < q \leq 2$ and $K^*K \leq I$, the function*

$$F(A) = \mathrm{Tr}\left[\exp_q^{\frac{1}{p}}(K^*\ln_q(A^p)K)\right]$$

is concave for the strictly positive $A \in H_n^+$.

Proof. (The second proof of Theorem 1)

First, setting $f(z) = \text{Tr}[(A(z) + iB(z))^{\frac{1}{pq-p}}]$ where $A(z) = \text{Re}(K^*(A+zB)^{pq-p}K + I - K^*K)$ and $B(z) = \text{Im}(K^*(A+zB)^{pq-p}K + I - K^*K) \in H_n^+$. As

$$\text{Im}\left[\text{Tr}[(A(z)+iB(z))^{\frac{1}{pq-p}}]\right] = \text{Im}\left[\text{Tr}[\int_0^{+\infty}(\frac{A(z)+iB(z)}{t+A(z)+iB(z)})\,d\mu(t)]\right]$$

$$= \text{Im}\left[\int_0^{+\infty}\text{Tr}[(\frac{\Lambda_{A(z)+iB(z)}}{t+\Lambda_{A(z)+iB(z)}})\,d\mu(t)]\right]$$

$$= \text{Im}\left[\int_0^{+\infty}\sum_{i=1}^n[(\frac{\lambda_i(A(z)+iB(z))}{t+\lambda_i(A(z)+iB(z))})\,d\mu(t)]\right]$$

$$= \text{Im}\left[\sum_{i=1}^n[(\lambda_i(A(z)+iB(z)))^{\frac{1}{pq-p}}]\right],$$

when $\arg(z) \in (0, \pi)$ and $K^*K \leq I$, then

$$\arg(\lambda_i(A(z)+iB(z)))$$
$$= \arg(x_i^*(A(z)+iB(z))x_i)$$
$$= \arg(x_i^* K^*(A+zB)^{pq-p}Kx_i + x_i^*(I-K^*K)x_i) \in (0, (pq-p)\pi),$$

where $x_i \in \mathbb{C}^n$ are the eigenvectors of $K^*(A+zB)^{pq-p}K + I - K^*K$.

Hence,

$$\text{Im}\left[\text{Tr}[(A(z)+iB(z))^{\frac{1}{pq-p}}]\right] = \text{Im}\left[\sum_{i=1}^n z_i\right],$$

where z_i is the i eigenvalue of $(A(z)+iB(z))^{\frac{1}{pq-p}}$ and $\arg(z_i) \in (0,\pi)$.

Thus, $f(z) = \text{Tr}[(A(z)+iB(z))^{\frac{1}{pq-p}}]$ is a Pick function, and this implies that $F(A)$ is concave. □

Using a similar method, we can obtain the following corollary.

Corollary 1. *For $0 < p \leq 1$ and $1 < q \leq 2$, the function*

$$E(A) = \text{Tr}\left[\exp_q^{\frac{1}{p}}[B + \ln_q(A^p)]\right] \tag{7}$$

is concave for the strictly positive $A \in H_n^+$.

Since the Thompson–Golden theorem can be seen as a corollary of the Lieb concavity theorem, we discuss the Lieb concavity theorem for deformed exponentials. Setting $\text{SP}(A) \subset \{z = \rho e^{i\theta} : 0 < \rho, 0 < \theta < \alpha\}$ and $\text{SP}(B) \subset \{z = \rho e^{i\theta} : 0 < \rho, 0 < \theta < \beta\}$, then for any $A_1, B_1 \in H_n$, $A_2, B_2 \in H_n^+$ and $A = A_1 + iA_2$, $B = B_1 + iB_2$, we have [12]

$$\text{SP}(AB) \subset \{z = \rho e^{i\theta} : 0 < \rho, 0 < \theta < \alpha + \beta\}. \tag{8}$$

and then the following theorem can be obtained.

Theorem 3. *For $0 < p \leq 1$, $1 < q \leq 2$ and $P^*P \leq I$, the following function*

$$L(A) = \text{Tr}[\exp_q(P^* \ln_q(K^* A^p K)P) \exp_q(P^* \ln_q A^{1-p}P)] \tag{9}$$

is concave for any $A \in H_n^+$.

Proof. Set $L_{A,B}(z) = \text{Tr}[\exp_q(P^* \ln_q(K^*(A+zB)^pK)P) \exp_q(P^* \ln_q(A+zB)^{1-p}P)]$.

When $x_i \in \mathbb{C}^n$ is a eigenvector of $P^*(A+zB)^{pq-p}P + I - P^*P$ and $P^*P \leq I$,

$$\arg(x_i^* P^* K^*(A+zB)^{pq-p}KPx_i + x_i^*(I-P^*P)x_i) \in (0, (pq-p)\pi),$$

if $\arg(z) \in (0, \pi)$. This implies

$$SP(P^*K^*(A+zB)^{pq-p}KP + I - P^*P) \subset \{z = \rho e^{i\theta} : 0 < \rho, 0 < \theta < (pq-p)\pi\},$$

such as

$$SP(\exp_q(P^* \ln_q(K^*(A+zB)^p K)P)) \subset \{z = \rho e^{i\theta} : 0 < \rho, 0 < \theta < p\pi\}.$$

Similarly, we can also obtain

$$SP(\exp_q(P^* \ln_q(A+zB)^{1-p}P)) \subset \{z = \rho e^{i\theta} : 0 < \rho, 0 < \theta < (1-p)\pi\}.$$

Hence, using (8), we see that

$$SP[\exp_q(P^* \ln_q(K^*(A+zB)^p K)P) \exp_q(P^* \ln_q(A+zB)^{1-p}P)]$$
$$\subset \{z = \rho e^{i\theta} : 0 < \rho, 0 < \theta < \pi\}.$$

Thus, we know $\arg(L_{A,B}(z)) \in (0, \pi)$, which implies that $L_{A,B}(z)$ is a Pick function. Hence, L(A) is concave. □

In fact, Theorem 3 is a generalization of the Lieb concavity theorem setting $P = I$, $K = \begin{pmatrix} 0 & 0 \\ H & 0 \end{pmatrix}$ and $A = \begin{pmatrix} Z & 0 \\ 0 & B \end{pmatrix}$. Moreover, we can obtain the following theorem.

Theorem 4. *For $0 < p, s \leq 1, 1 < q \leq 2$ and $P^*P \leq I$, the functions*

$$\mathrm{Tr}\left[\left[\exp_q(P^* \ln_q A^{\frac{ps}{2}} P) \exp_q(P^* \ln_q(K^* A^{s-sp} K)P) \exp_q(P^* \ln_q A^{\frac{ps}{2}} P)\right]^{\frac{1}{s}}\right] \quad (10)$$

and

$$\left[\mathrm{Tr}\exp_q(P^* \ln_q A^{\frac{ps}{2}} P) \exp_q(P^* \ln_q(K^* A^{s-sp} K)P) \exp_q(P^* \ln_q A^{\frac{ps}{2}} P)\right]^{\frac{1}{s}} \quad (11)$$

are jointly concave for any $A \in H_n^+$.

The proof of Theorem 4 is similar to Theorem 3; here, we do not repeat the proof. In [19], Huang used exterior algebra to find that

$$\left[\mathrm{Tr} \wedge^k [\exp(K^* \ln(A) K)]\right]^{\frac{1}{k}}$$

is a concave function for any $A \in H_n^+$, $K^*K \leq I$ and $k \leq n$. Associated with Theorem 1, we can obtain a generalization as the following theorem.

Theorem 5. *For $0 < p \leq 1, 1 < q \leq 2$ and $K^*K \leq I$, the function*

$$\left[\mathrm{Tr} \wedge^k \left[\exp_q^{\frac{1}{p}}(K^* \ln_q(A^p) K)\right]\right]^{\frac{1}{k}} \quad (12)$$

is concave for the strictly positive $A \in H_n^+$ and $k \leq n$.

Proof. In fact, we can prove that

$$\left[\mathrm{Tr} \wedge^k \left[(H^* A^s H + B)^{\frac{1}{s}}\right]\right]^{\frac{1}{k}}$$

is a concave function for any $A \in H_n^+$ where $0 < s \leq 1$ and $B \in H_n^+$.
Using Theorem 1, we know that

$$\text{Tr}\left[(H^* A^p H + C)^{\frac{1}{p}}\right] \tag{13}$$

is a concave function for any $A \in H_n^+$ where $0 < p \leq 1$ and $C \in H_n^+$.
Then, for any $A_1, A_2 \in H_n^+$, we have

$$\left[\text{Tr} \wedge^k \left[(H^*(\frac{A_1 + A_2}{2})^s H + B)^{\frac{1}{s}}\right]\right]^{\frac{1}{k}}$$

$$= \left[\text{Tr}\left[(\tilde{H}^*(\frac{A_1 \wedge^{k-1} I + A_2 \wedge^k I}{2})^s \tilde{H} + \tilde{B})^{\frac{1}{s}}\right]\right]^{\frac{1}{k}}$$

$$\geq \left[\text{Tr}\left[\frac{(\tilde{H}^*(A_1 \wedge^{k-1} I)^s \tilde{H} + \tilde{B})^{\frac{1}{s}} + (\tilde{H}^*(A_2 \wedge^{k-1} I)^s \tilde{H} + \tilde{B})^{\frac{1}{s}}}{2}\right]\right]^{\frac{1}{k}}$$

$$= \left[\text{Tr}\left[\left(\frac{(H^* A_1^s H + B)^{\frac{1}{s}} + (H^* A_2^s H + B)^{\frac{1}{s}}}{2}\right) \wedge^{k-1} (H^*(\frac{A_1 + A_2}{2})^s H + B)^{\frac{1}{s}}\right]\right]^{\frac{1}{k}},$$

where $\tilde{H} = H \wedge^{k-1} (H^*(\frac{A_1+A_2}{2})^s H + B)^{\frac{1}{s}}$ and $\tilde{B} = B \wedge^{k-1} (H^*(\frac{A_1+A_2}{2})^s H + B)^{\frac{1}{s}}$. Analogously, we can obtain

$$\left[\text{Tr} \wedge^k \left[(H^*(\frac{A_1 + A_2}{2})^s H + B)^{\frac{1}{s}}\right]\right]^{\frac{1}{k}}$$

$$\geq \left[\text{Tr}\left[\wedge^k \left(\frac{(H^* A_1^s H + B)^{\frac{1}{s}} + (H^* A_2^s H + B)^{\frac{1}{s}}}{2}\right)\right]\right]^{\frac{1}{k}}.$$

Using lemma 3, we obtain

$$\left[\text{Tr} \wedge^k \left[(H^*(\frac{A_1 + A_2}{2})^s H + B)^{\frac{1}{s}}\right]\right]^{\frac{1}{k}}$$

$$\geq \frac{\left[\text{Tr}\left[\wedge^k \left((H^* A_1^s H + B)^{\frac{1}{s}}\right)\right]\right]^{\frac{1}{k}} + \left[\text{Tr}\left[\wedge^k \left((H^* A_2^s H + B)^{\frac{1}{s}}\right)\right]\right]^{\frac{1}{k}}}{2}.$$

□

Clearly, the proof of Theorem 5 is in the application of exterior algebra and the Brunn–Minkowski inequality. Hence, other theorems, such as the Thompson–Golden theorem in a deformed exponential, can be generalized to a more general form, but we do not discuss this here.

4. Conclusions

In this paper, we used the Pick function to obtain a generalization of the Lieb concavity theorem and some corollaries. The advantage of using the Pick function is that it avoids discussing the commutativity of the matrix and variational method. Generally, we obtain that the following two functions are concave for $0 < p, s \leq 1$, $1 < q \leq 2$ and $P^* P \leq I$

$$\left[\text{Tr} \wedge^k \left[\exp_q(P^* \ln_q A^{\frac{ps}{2}} P) \exp_q(P^* \ln_q(K^* A^{s-sp} K) P) \exp_q(P^* \ln_q A^{\frac{ps}{2}} P)\right]^{\frac{1}{s}}\right]^{\frac{1}{k}} \tag{14}$$

and

$$\left[\text{Tr} \wedge^k [\exp_q(P^* \ln_q A^{\frac{ps}{2}} P) \exp_q(P^* \ln_q (K^* A^{s-sp} K) P) \exp_q(P^* \ln_q A^{\frac{ps}{2}} P)]\right]^{\frac{1}{ks}}, \quad (15)$$

where $A \in H_n^+$ and $k \leq n$, and this provides work for the future.

Author Contributions: Conceptualization, H.S.; writing—original draft, G.Y. and Y.L.; writing—review and editing, J.W.; funding acquisition, J.W. All authors have read and agreed to the published version of the manuscript.

Funding: The present research was supported by the General Project of Science and Technology Plan of Beijing Municipal Education Commission (Grant No. KM202010037003).

Institutional Review Board Statement: Not applicable.

Informed Consent Statement: Not applicable.

Data Availability Statement: Not applicable.

Acknowledgments: The authors thank the referees for detailed reading and comments that were both helpful and insightful.

Conflicts of Interest: The authors declare no conflict of interest.

References

1. Tropp, J.A. An introduction to matrix concentration inequalities. *arXiv* **2015**, arXiv:1501.01571.
2. Petz, D. *Quantum Information Theory and Quantum Statistics*; Springer Science and Business Media: Berlin/Heidelberg, Germany, 2007.
3. Bellman, R. *Introduction to Matrix Analysis*; McGraw-Hill: New York, NY, USA, 1960.
4. Carlen, E. Trace inequalities and quantum entropy: An introductory course. *Contemp. Math.* **2010**, *529*, 73–140.
5. Zhang, F. *Matrix Theory: Basic Results and Techniques*; Springer: Berlin/Heidelberg, Germany, 1999.
6. Simon, B. Trace Ideals and Their Applications. In *Mathematical Surveys and Monographs*, 2nd ed.; American Mathematical Society: Providence, Rhode Island, 2005.
7. Wigner, E.P.; Yanase, M.M. On the positive definite nature of certain matrix expressions. *Cunud. J. Math.* **1964**, *16*, 397–406. [CrossRef]
8. Lieb, E. Convex trace functions and the Wigner–Yanase–Dyson conjecture. *Adv. Math.* **1973**, *11*, 267–288. [CrossRef]
9. Bhatia, R. *Positive Definite Matrices*; Princeton University Press: Princeton, NJ, USA, 2007.
10. Ando, T. Concavity of certain maps on positive definite matrices and applications to Hadamard products. *Linear Algebra Appl.* **1979**, *26*, 203–241. [CrossRef]
11. Effros, E.G. A matrix convexity approach to some celebrated quantum inequalities. *Proc. Natl. Acad. Sci. USA* **2009**, *106*, 1006–1008. [CrossRef] [PubMed]
12. Epstein, H. Remarks on two theorems of E. Lieb. *Comm. Math. Phys.* **1973**, *31*, 317–325. [CrossRef]
13. Shi, G.H.; Hansen, F. Variational representations related to Tsallis relative entropy. *Lett. Math. Phys.* **2020**, *110*, 2203–2220. [CrossRef]
14. Donogue, W. *Monotone Matrix Functions and Analytic Continuution*; Springer: New York, NY, USA, 1974.
15. Davis, C. Notions generalizing convexity for functions defined on spaces of matrices. In *Convexity: Proceedings of the Seventh Symposium in Pure Mathematics of the American Mathematical Society*; American Mathematical Society: Providence, Rhode Island, 1963; pp. 187–201.
16. Choi, M.D. A Schwarz inequality for positive linear maps on C^*-algebras. *Illinois J. Math.* **1974**, *18*, 565–574. [CrossRef]
17. Bendat, J.; Sherman, S. Monotone and convex operator functions. *Trans. Am. Math. Soc.* **1955**, *79*, 58–71. [CrossRef]
18. Hansen, F. Trace Functions with Applications in Quantum Physics. *J. Stat. Phys.* **2014**, *154*, 807–818. [CrossRef]
19. Huang, D. A generalized Lieb's theorem and its applications to spectrum estimates for a sum of random matrices. *Linear Algebra Appl.* **2019**, *579*, 419–448. [CrossRef]
20. Marshall, A.W.; Olkin, I.; Arnold, B.C. *Inequalities: Theory of Majorization and Its Applications*; Springer: New York, NY, USA, 2011.
21. Bhatia, R. *Matrix Analysis*; Springer: Berlin/Heidelberg, Germany, 1997.
22. Aujla, J.S. A simple proof of Lieb concavity theorem. *J. Math. Phys.* **2011**, *52*, 043505. [CrossRef]
23. Nikoufar, I.; Ebadian, A.; Gordji, M.E. The simplest proof of Lieb concavity theorem. *Adv. Math.* **2013**, *248*, 531–533. [CrossRef]

Article

The Role of the Discount Policy of Prepayment on Environmentally Friendly Inventory Management

Shirin Sultana [1], Abu Hashan Md Mashud [2], Yosef Daryanto [3], Sujan Miah [4], Adel Alrasheedi [5] and Ibrahim M. Hezam [5,*]

1 Department of General Educational Development, Daffodil International University, Dhaka 1341, Bangladesh; shirin.ged@daffodilvarsity.edu.bd
2 School of Engineering and IT, University of New South Wales (UNSW), Campbell, Canberra, ACT 2612, Australia; a.mashud@adfa.edu.au
3 Department of Industrial Engineering, Universitas Atma Jaya Yogyakarta, Yogyakarta 55281, Indonesia; yosef.daryanto@uajy.ac.id
4 Department of Mathematics, Hajee Mohammad Danesh Science and Technology University, Dinajpur 5200, Bangladesh; sujan1608217@std.hstu.ac.bd
5 Department of Statistics & Operations Research, College of Sciences, King Saud University, Riyadh 11451, Saudi Arabia; aalrasheedi@ksu.edu.sa
* Correspondence: ialmishnanah@ksu.edu.sa

Abstract: Nowadays, more and more consumers consider environmentally friendly products in their purchasing decisions. Companies need to adapt to these changes while paying attention to standard business systems such as payment terms. The purpose of this study is to optimize the entire profit function of a retailer and to find the optimal selling price and replenishment cycle when the demand rate depends on the price and carbon emission reduction level. This study investigates an economic order quantity model that has a demand function with a positive impact of carbon emission reduction besides the selling price. In this model, the supplier requests payment in advance on the purchased cost while offering a discount according to the payment in the advanced decision. Three different types of payment-in-advance cases are applied: (1) payment in advance with equal numbers of instalments, (2) payment in advance with a single instalment, and (3) the absence of payment in advance. Numerical examples and sensitivity analysis illustrate the proposed model. Here, the total profit increases for all three cases with higher values of carbon emission reduction level. Further, the study finds that the profit becomes maximum for case 2, whereas the selling price and cycle length become minimum. This study considers the sustainable inventory model with payment-in-advance settings when the demand rate depends on the price and carbon emission reduction level. From the literature review, no researcher has undergone this kind of study in the authors' knowledge.

Keywords: low carbon inventory; discount; payment in advance; price-sensitive demand; emission reduction

1. Introduction

Customer preferences have always been a concern in industries in terms of their effect on business growth. Customer preferences are affected by many factors and are reflected in the customers' willingness to buy. The level of consumer demand is usually sensitive to product prices. However, in today's setting, more and more consumers consider the environmental performance of the producer and the green level of product in their purchasing decisions [1–3]. This trend is expanding globally along with the increasing consumer awareness of the importance of environmental conservation in the midst of climate change issues. Hence, many producers and retailers innovate green products and promote green operations to attract these customers [4,5]. Moreover, regulations become another driver for this eco-innovation. Companies try to reduce carbon emissions from production, logistics, and transportation activities and apply green technology to meet new regulations and pressure from these customers.

The inventory decisions on supply chain operations already incorporate environmental parameters with certain intentions such as reducing carbon emission levels. Previous supply chain studies also consider customers' awareness of low carbon emissions [6–8], the green quality of the product [9,10], and the amount of carbon emissions [11], which affect the demand level. Customer awareness and green quality level have positive impacts on the demand function of green products, while the amount of carbon emissions has the opposite effect. Green marketing becomes a powerful strategy for businesses through various green advertising, branding, and eco-labeling. This strategy has been adopted to promote many international and local brands and products in both developed and developing countries [12–16]. Xia et al. [7] incorporated the positive effect of emission reduction and the promotion of this environmental benefit into the demand function. Recently, Dong et al. [17] considered the manufacturer's reduction in carbon emission levels, which shows the company's initiative for greener operations. The positive effect of carbon emission reduction on customer demand was combined with the negative effect of the selling price. The study also examined the payment issue by analyzing the effect of trade credit and bank loans. Based on Dong et al. [17], we study a sustainable inventory model considering a prepayment mechanism, another common payment term in business, in order to consider a real situation.

The payment term in the transaction between a supplier and a buyer is an important issue in supply chain collaboration. The term should be agreed upon by both parties so that it is clear and beneficial for all. The classic economic order quantity (EOQ) model assumes a payment immediately after product delivery. However, in many cases, payment in advance is applied, in which the buyer should pay the purchase cost before the product delivery. The buyer may have to pay all the purchase cost in advance [18–20] or pay only a percentage of it [19,21]. Further, the prepayment can be done in several time intervals [22,23]. While the payment in advance will give the supplier an advantage by mitigating the risk of cancellation, the supplier can offer some discounts to the buyer so that they benefit as well. Our study considers discount offers similar to Mashud et al. [23].

The increase in customers' awareness of green issues, together with the trend in producers' concern on carbon emission reduction and the common practices of payment in advance, has motivated this study to contribute to the development of a sustainable inventory model. This paper presents a profit maximization study of a retailer inventory system to respond to customers' increasing low carbon preferences. When customer demand depends on the selling price and the retailer must pay the purchase cost in advance, the proposed model suggests the optimum replenishment time and selling price. The study aims at providing managerial insights by answering the following questions:

a. How do payment-in-advance models affect pricing and replenishment decisions as well as the total profit when customer demand is sensitive to the selling price and environmental performance of the producer?
b. Will discount policy impact retailers' choice of payment in advance settings and total profit?
c. How do the customer preference for carbon emission reduction levels and various emission costs impact the retailer's total profit?

In this paper, we first study the retailer's optimal selling price and replenishment cycle when payment in advance is fulfilled with some equal numbers of instalments and a discount is offered by the supplier. Then, we address the same issue when the purchase cost is completed in a single payment, and in return, the customer will get a different discount rate. Next, we study the situation in which no payment in advance is considered, hence no discount is offered.

2. Literature Review

Traditional inventory management focuses on the economic benefit from a business point of view. For example, the classic inventory model works under some basic assumptions such as infinite replenishment and planning horizon that aim at optimizing financial profit. In recent years, considerable attention has shifted to economic and environmental aspects of the emergence of the sustainable inventory model terminology. The focus has broadened to include minimizing environmental impacts through carbon emission reductions, energy efficiency, and the adoption of green technologies [24,25]. Most sustainable inventory models seek to reduce supply chain emission levels by considering emissions from production, transportation, and inventory storage activities. Carbon tax systems are widely used to include the cost of carbon emissions in the objective function [26–31]. Other carbon regulations such as carbon cap-and-trade and strict carbon limits are also used, depending on the regulations imposed by the government [32,33]. Datta [34] developed an inventory model with investment in emission reduction technology focusing on emissions from production activities. The study considered carbon emission reduction under a carbon tax policy and optimization of the investment amount. Under a similar carbon tax policy, Mashud et al. [35] considered emission reduction from transportation activities. Simultaneous investments for emission reduction and deterioration rate were studied by Mishra et al. [36]. Lou et al. [37] optimized the green technology investment considering a subsidy from the government. Optimum investment level was considered important because customer demand was assumed to be sensitive to emission reduction level. The study recommended an active role from the government such as providing technology investment subsidies and controlling the emission trading price.

In addition to government regulations, efforts to reduce carbon emission levels are also driven by the increasing number of consumers who consider environmental aspects in their purchasing decisions [9]. Hence, the environmental performance of the product was added to the demand function. Pang et al. [6] and Gao et al. [38] considered the customers' environmental awareness and set a linear demand function in addition to the effect of selling price on demand. Hovelaque and Bironneau [11] set a demand function that depends on price and the amount of carbon emissions. They found two order quantities, one that will maximize the total profit and another one that will minimize the emission level. Lou et al. [37] also set a demand function with a linear effect of price and emission reduction. The percentage of emission reduction per unit product was optimized together with the selling price. Xia et al. [7] incorporated a promotion strategy to support the emission reduction program to gain customer attention. Zhang et al. [3] analyzed a manufacturer's decision to introduce a new green product to customers with high environmental awareness. The study found a conflict if the manufacturer also sells an ordinary product because the products will compete with each other. Recently, Dong et al. [17] considered the positive effect of carbon emission reduction and the negative effect of the selling price on the customer demand rate. For a single-supplier, single-buyer supply chain model, they also studied the effect of financial facilities such as trade credit and bank loans on the manufacturer's decision regarding the level of emission reduction. Using a similar demand function, we study an economic order quantity (EOQ) model of a retailer when payment in advance is requested by the supplier. The retailer's optimal selling price and replenishment cycle are the decision variables.

The inventory model with payment in advance is another research stream in this paper. The practices of payment in advance may be introduced by a powerful supplier to prevent order cancellation, especially for customized and expensive products [19]. The payment may help the supplier to finish the product. Maiti et al. [21] are among the first researchers who incorporated payment in advance into an inventory model. The proposed model says that the buyer gets some price discount based on the amount of the payment in advance. Further, due to the payment in advance, the retailer may need cash aid from any financial institution, which means an additional cost of interest. The buyer may have to pay all the purchase costs in advance or pay only a percentage of it [18,19,39,40].

During the payment-in-advance period, multiple instalments may be applied to reduce the retailer burden [22,23]. Hence, our study examines the effect of three different payment-in-advance settings—(1) payment in advance with equal numbers of instalments, (2) payment in advance with single instalment, and (3) the absence of payment in advance—on the optimum price and replenishment cycle that optimizes the profit.

3. Mathematical Model Formulation for Inventory Model

The proposed mathematical model is based on the following assumptions:

a. Inventory of a single product is considered with a limitless planning horizon.
b. The replenishment rate is boundless.
c. The lead time is constant and the shortages are overlooked.
d. The retailer has to make a payment in advance to the supplier [23].
e. The supplier offers a discount on the purchase cost of the products according to the number of instalment decisions [23].
f. The demand function $D_L = \psi - \gamma p + \eta R_C$ depends on price and carbon emission reduction level [17,41,42]. Here,

ψ is the market potential
γ is the price sensitivity coefficient ($\gamma > 0$)
p is the unit selling price
η is the low carbon preference coefficient ($\eta > 0$)
R_C is the manufacturer's carbon emission reduction level

In addition, the following nomenclatures (Table 1) are used.

Table 1. Notation description.

Notations	Units	Description
D_L	units	Demand function
F_C	$/trip	Fixed cost of transportation
N_t	unit	Number of trips
c_e	$/km	Carbon emission cost per unit distance
c_k	$/unit/km	Carbon emission cost per unit item per unit distance
c_p	liter/ton	Fuel consumption per ton of payload
f_e	liter	Empty vehicle fuel consumption
$\hbar$	$/unit	Holding cost per unit
ℓ	km	One way distance
n	unit	Number of instalments
w_p	kg	Product weight
α	$/liter	Price of fuel
χ	$	Carbon emission reduction investment
δ	constant	Payment in advance portion of purchase cost
ζ	$/cycle	Ordering cost per cycle
κ	months	Lead time
ϕ	constant	Interest rate due to instalment based payment in advance
ϕ_L	constant	Interest rate on loan amount
Ω	units	Order quantity
ω	$/unit	Purchase cost per unit
Decision Variables		
p	$/unit	Selling price
T_C	months	Replenishment time.

An inventory model is developed under consideration of the above assumptions. The model is divided into three cases considering the payment-in-advance cost. In Case I, payment in advance is fulfilled with some equal numbers of instalments, and a discount is offered for the retailer as a benefit according to the number of instalments. Case II considers that the payment is completed with a single instalment and, in return, the retailer will get a different discount from the supplier. Finally, in Case III, no payment-in-advance cost is considered, and hence no discount is offered to the retailer.

3.1. Case I: With Advanced Payment and a Discount for Instalment Based Payment

A retailer runs its business with an initial stock of Ω units. Depending on the demand function, the stock decreases and becomes zero at time $t = T_C$. Thus, one cycle ends, and the process repeats so that the business continues. To get this stock, the retailer pays a percentage of the purchase cost (δ) in n equal number of instalments before the products are delivered. The amount of each prepayment is $\frac{\delta \omega \Omega}{n}$. At the moment of delivery, the retailer needs to pay for the remaining $(1-\delta)\Omega$ quantity. Figure 1 outlines the above facts and the pattern of the level of inventory.

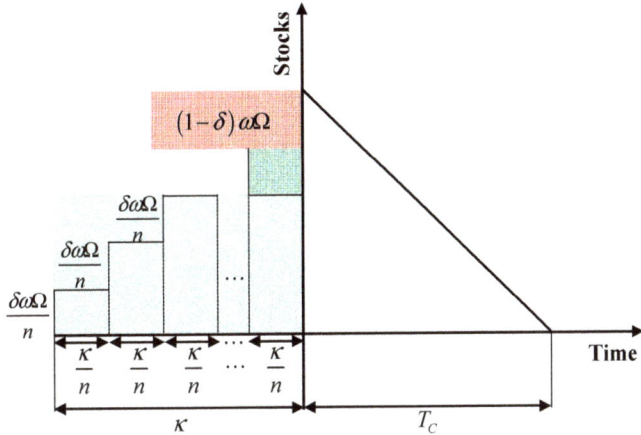

Figure 1. Graphical presentation of the inventory system for Case I.

The inventory system is described by the following differential equation considering the demand $D_L = \psi - \gamma p + \eta R_C$.

$$\frac{dI_S(t)}{dt} = -D_L,\ 0 \leq t \leq T_C \tag{1}$$

With the help of boundary conditions $I_S(0) = \Omega$ and $I_S(T_C) = 0$, by solving Equation (1) we get,

$$I_S(t) = D_L(T_C - t) \tag{2}$$

and

$$\Omega = D_L T_C \tag{3}$$

3.1.1. Total Cost per Unit Time

(a) Ordering cost per cycle:

$$OC = \zeta \tag{4}$$

(b) The inventory holding cost per cycle:

$$HC = \hbar \int_0^{T_C} I_S(t)\ dt = \frac{1}{2}\hbar D_L T_C^2 \tag{5}$$

(c) The purchase cost per cycle:

$$PC = \Omega \omega \tag{6}$$

(d) Transportation cost per cycle:

Three major costs are considered to estimate the transportation cost: fixed cost (F_C), variable cost, and carbon emission cost. However, the variable cost and the carbon emission cost are different for an empty vehicle (truck) and a loaded vehicle. The total travel distance is 2ℓ as the vehicle has to travel a distance ℓ to ship the goods and has to travel another ℓ distance to return with an empty load. For the vehicle only, the variable cost is the total fuel consumption ($2\ell f_e$) times the fuel price (α). An additional variable cost is estimated for one-way distance ℓ based on the vehicle load, that is, a one-way distance (ℓ) multiplied by the fuel consumption per ton of payload (c_p), product weight (w_p), order quantity per trip $\left(\frac{\Omega}{N_t}\right)$, and fuel price ($\alpha$). Similarly, the carbon emission cost for the vehicle is 2ℓ times the cost of carbon emission per unit distance of delivery (c_e), and the carbon emission cost based on the load is ℓ times the cost of carbon emission per unit item per unit distance of delivery (c_k) times $\left(\frac{\Omega}{N_t}\right)$. Thus, the total transportation cost per cycle is

$$TC = N_t\left[F_C + \left(2\ell f_e \alpha + \frac{\ell c_p w_p \Omega \alpha}{N_t}\right) + \left(2\ell c_e + \frac{\ell c_k \Omega}{N_t}\right)\right]$$

$$TC = F_C N_t + 2\ell f_e \alpha N_t + \ell c_p w_p \Omega \alpha + 2\ell c_e N_t + \ell c_k \Omega \tag{7}$$

(e) Instalment capital cost:

The instalment capital cost is estimated following the procedure described by [35,43]:

$$\begin{aligned}IC &= \left(\tfrac{\phi \delta \omega}{n}\Omega \times n \times \tfrac{\kappa}{n}\right) + \left(\tfrac{\phi \delta \omega}{n}\Omega \times (n-1) \times \tfrac{\kappa}{n}\right) + \cdots + \left(\tfrac{\phi \delta \omega}{n}\Omega \times (n-(n-1)) \times \tfrac{\kappa}{n}\right) \\ &= \left(\tfrac{\phi \delta \omega}{n}\Omega \times \tfrac{\kappa}{n}\right)(n + (n-1) + \cdots + 2 + 1) = \left(\tfrac{\phi \delta \omega}{n}\Omega \times \tfrac{\kappa}{n}\right)\tfrac{n(n+1)}{2} \\ &= \tfrac{(n+1)}{2n}\phi \delta \omega \Omega \kappa\end{aligned} \tag{8}$$

(f) Discount on purchase cost:

For the retailer's advanced payment on the purchase cost, the supplier provides $v\%$ discount. The discount rate ξ depends on the number of instalments n; that is, the supplier offers a lower discount rate for more installments as follows:

$$\xi = \frac{v}{n}, \quad 0 \leq v \leq 100. \tag{9}$$

Hence, the total discount is

$$DC = \Omega \omega \xi = \frac{\Omega \omega v}{n}. \tag{10}$$

(g) Carbon emission reduction cost:

The effort of carbon emission reduction needs an investment. The higher emission reduction needs an increasingly accelerated emission reduction cost. This cost is estimated according to Swami and Shah [42] in Equation (11).

$$RC = \chi R_C^2. \tag{11}$$

(h) Sales revenue per cycle:

$$SR = p\int_0^{T_C} D_L \, dt = pD_L T_C \tag{12}$$

3.1.2. Total Profit per Unit Time

Now, for the total profit per unit time, one can write:

$$\tau(p, T_C) = \frac{1}{T_C}(SR - OC - HC - PC - TC - IC - RC + DC)$$

$$\tau(p, T_C) = \frac{1}{T_C}\left(pD_L T_C - \zeta - \frac{1}{2}\hbar D_L T_C^2 - \Omega\omega - \begin{pmatrix} F_C N_t + 2\ell f_e \alpha N_t + \ell c_p w_p \Omega\alpha \\ +2\ell c_e N_t + \ell c_k \Omega \end{pmatrix} \right.$$
$$\left. - \frac{(n+1)}{2n}\phi\delta\omega\Omega\kappa - \chi R_C^2 + \frac{\Omega\omega v}{n} \right)$$

$$\tau(p, T_C) = \frac{1}{T_C}\left(pD_L T_C - \frac{1}{2}\hbar D_L T_C^2 - \begin{pmatrix} \zeta + \Omega\omega + \frac{(n+1)}{2n}\phi\delta\omega\Omega\kappa + F_C N_t + 2\ell f_e \alpha N_t \\ +\ell c_p w_p \Omega\alpha + 2\ell c_e N_t + \ell c_k \Omega + \chi R_C^2 - \frac{\Omega\omega v}{n} \end{pmatrix} \right)$$

$$\tau(p, T_C) = \frac{1}{T_C}\left(p(\psi - \gamma p + \eta R_C)T_C - \frac{1}{2}\hbar(\psi - \gamma p + \eta R_C)T_C^2 - \lambda_1 \right) \quad (13)$$

where

$$\lambda_1 = \zeta + \Omega\omega + \frac{(n+1)}{2n}\phi\delta\omega\Omega\kappa + F_C N_t + 2\ell f_e \alpha N_t + \ell c_p w_p \Omega\alpha + 2\ell c_e N_t + \ell c_k \Omega + \chi R_C^2 - \frac{\Omega\omega v}{n} \quad (14)$$

3.2. Case II: With Advanced Payment and a Discount for Single Time Payment

In this case, the retailer has to pay in advance in a single payment. The payment amount means the whole purchase cost. The scenario is described in Figure 2, which is a modified version of Figure 1.

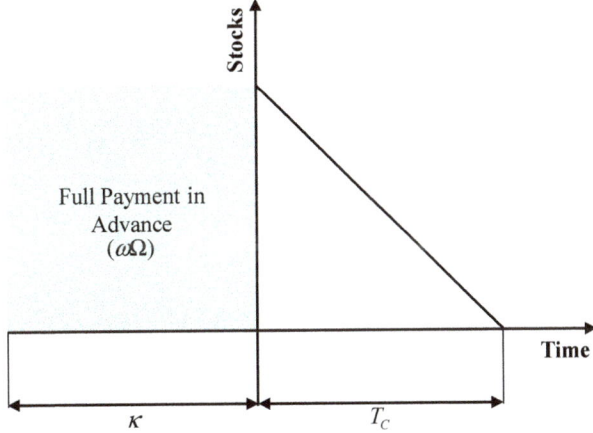

Figure 2. Graphical presentation of the inventory system for Case II.

The supplier offers a $v\%$ discount to the retailer for a single-time prepayment as a benefit. In this situation, the retailer may have a crisis of capital during time κ; in that case, a loan with some interest of $\phi_L\%$ from any financial institutes or other funds can be a suitable option to manage the required capital.

The discount for purchase cost is

$$PC = (1-v)\Omega\omega. \quad (15)$$

The associated cost of taking a loan is

$$LC = \phi_L \kappa (1-v)\Omega\omega. \quad (16)$$

Hence, the total profit per unit time can be written as:

$$\tau_f(p, T_C) = \frac{1}{T_C}(SR - OC - HC - PC - TC - RC - LC)$$

$$\tau_f(p, T_C) = \frac{1}{T_C}\begin{pmatrix} pD_L T_C - \zeta - \frac{1}{2}\hbar D_L T_C^2 - (1-v)\Omega\omega \\ -\begin{pmatrix} F_C N_t + 2\ell f_e \alpha N_t + \ell c_p w_p \Omega\alpha \\ +2\ell c_e N_t + \ell c_k \Omega \end{pmatrix} - \chi R_C^2 - \phi_L \kappa(1-v)\Omega\omega \end{pmatrix}$$

$$\tau_f(p, T_C) = \frac{1}{T_C}\left(p(\psi - \gamma p + \eta R_C)T_C - \frac{1}{2}\hbar(\psi - \gamma p + \eta R_C)T_C^2 - \lambda_2\right) \quad (17)$$

where

$$\lambda_2 = \zeta + (1 + \phi_L \kappa)(1 - v)\Omega\omega + F_C N_t + 2\ell f_e \alpha N_t + \ell c_p w_p \Omega\alpha + 2\ell c_e N_t + \ell c_k \Omega + \chi R_C^2 \quad (18)$$

3.3. Case III: Without Advanced Payment

In this case, the retailer does not pay in advance. If the retailer does not pay any payment in advance, then there is no instalment cost and discount. Thus, there is a necessity to modify Figure 1 into Figure 3. The retailer must pay the full payment during purchase product shipment.

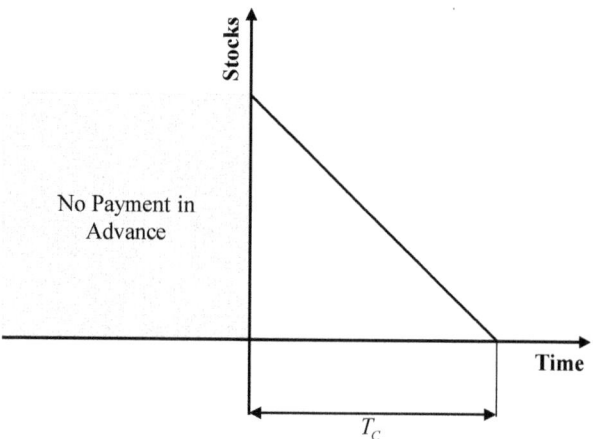

Figure 3. Graphical presentation of the inventory system for Case III.

Then, the total profit per unit time can be written as:

$$\tau_N(p, T_C) = \frac{1}{T_C}(SR - OC - HC - PC - TC - RC)$$

$$\tau_N(p, T_C) = \frac{1}{T_C}\left(pD_L T_C - \zeta - \frac{1}{2}\hbar D_L T_C^2 - \Omega\omega - \begin{pmatrix} F_C N_t + 2\ell f_e \alpha N_t + \ell c_p w_p \Omega\alpha \\ +2\ell c_e N_t + \ell c_k \Omega \end{pmatrix} - \chi R_C^2\right)$$

$$\tau_N(p, T_C) = \frac{1}{T_C}\left(p(\psi - \gamma p + \eta R_C)T_C - \frac{1}{2}\hbar(\psi - \gamma p + \eta R_C)T_C^2 - \lambda_3\right) \quad (19)$$

where

$$\lambda_3 = \zeta + \Omega\omega + F_C N_t + 2\ell f_e \alpha N_t + \ell c_p w_p \Omega\alpha + 2\ell c_e N_t + \ell c_k \Omega + \chi R_C^2 \quad (20)$$

4. Theoretical Development

Here, the concavity of the profit function is analyzed to show the existence of an optimal solution for each case.

4.1. Case I (with Advanced Payment and a Discount for Instalment Based Payment)

It is now important to investigate the concavity nature of the profit function $\tau(p, T_C)$ in Equation (13) for Case I. For this purpose, the priority is to determine the critical points, and one needs to differentiate Equation (13) with respect to the two decision variables p, T_C as follows:

$$\frac{\partial \tau}{\partial T_C} = -\frac{1}{T_C^2}\left[\frac{1}{2}\hbar(\psi - \gamma p + \eta R_C)T_C^2 - \lambda_1\right] \quad (21)$$

$$\frac{\partial \tau}{\partial p} = \psi - 2\gamma p + \eta R_C + \frac{1}{2}\hbar\gamma T_C \quad (22)$$

The critical points can be determined by setting Equations (21) and (22) to zero and doing some manipulations.

$$T_C^* = \sqrt{\frac{2\lambda_1}{\hbar(\psi - \gamma p + \eta R_C)}} \quad (23)$$

$$p^* = \frac{1}{2\gamma}\left(\psi + \eta R_C + \frac{1}{2}\hbar\gamma T_C\right) \quad (24)$$

The concavity of the profit function is next discussed with some conditions.

Proposition 1. *The profit function $\tau(p, T_C)$ in Equation (13) is concave regarding the replenishment time T_C if the selling price p remains fixed, and hence it provides a unique optimal T_C^*.*

Proof. One needs to determine the associated critical points as well as prove the sufficient condition to confirm the concavity of the profit function. The critical point is associated with Equation (23).

$$T_C^* = \sqrt{\frac{2\lambda_1}{\hbar(\psi - \gamma p + \eta R_C)}}$$

Then, differentiating the profit function as in Equation (13) with respect to T_C, one can find:

$$\frac{\partial^2 \tau}{\partial T_C^2} = -\frac{2\lambda_1}{T_C^3} \quad (25)$$

Since $\lambda_1 > 0$ and replenishment time T_C must be positive, $\frac{\partial^2 \tau}{\partial T_C^2} < 0$. Thus, we confirm the concave nature of the profit function regarding T_C, and the critical point T_C becomes the unique optimal point T_C^*. □

Proposition 2. *The profit function $\tau(p, T_C)$ in Equation (13) is concave regarding the replenishment time p if the selling price p remains fixed, and hence it provides a unique optimal p^*.*

Proof. The solution system is akin to the proposed system of Proposition 1; thus, to avoid redundancy the proof is removed. □

Proposition 3. *The profit function $\tau(p, T_C)$ of selling price p and replenishment time T_C in Equation (13) is a strictly pseudo-concave function at a unique optimal investment (p^*, T_C^*).*

Proof. The Hessian matrix of $\tau(p, T_C)$ is of order 2×2.

$$\Delta = \begin{bmatrix} \frac{\partial^2 \tau(p, T_C)}{\partial T_C^2} & \frac{\partial^2 \tau(p, T_C)}{\partial T_C \partial p} \\ \frac{\partial^2 \tau(p, T_C)}{\partial p \partial T_C} & \frac{\partial^2 \tau(p, T_C)}{\partial p^2} \end{bmatrix} \quad (26)$$

To prove that $\tau(p, T_C)$ is a strictly pseudo-concave function, it is essential to confirm that the Hessian matrix Δ is negative definite. Thus, it is necessary to show that the leading principal minors, $(-1)^k \Delta_k > 0$, $1 \leq k \leq 2$ means the first principal minor Δ_1 is negative, and the second principle minor Δ_2 is positive.

$$\Delta_1 = \left| \frac{\partial^2 \tau(p, T_C)}{\partial T_C^2} \right| = \frac{\partial^2 \tau(p, T_C)}{\partial T_C^2} \tag{27}$$

and

$$\Delta_2 = \begin{vmatrix} \frac{\partial^2 \tau(p,T_C)}{\partial T_C^2} & \frac{\partial^2 \tau(p,T_C)}{\partial T_C \partial p} \\ \frac{\partial^2 \tau(p,T_C)}{\partial p \partial T_C} & \frac{\partial^2 \tau(p,T_C)}{\partial p^2} \end{vmatrix} = \frac{\partial^2 \tau(p,T_C)}{\partial T_C^2} \frac{\partial^2 \tau(p,T_C)}{\partial p^2} - \frac{\partial^2 \tau(p,T_C)}{\partial p \partial T_C} \frac{\partial^2 \tau(p,T_C)}{\partial T_C \partial p} \tag{28}$$

Taking the second order partial derivatives of the profit function $\tau(p, T_C)$ in Equation (13) with respect to p and T_C, one gets

$$\frac{\partial^2 \tau}{\partial T_C^2} = -\frac{2\lambda_1}{T_C^3} \tag{29}$$

$$\frac{\partial^2 \tau}{\partial p^2} = -2\gamma \tag{30}$$

$$\frac{\partial^2 \tau}{\partial T_C \partial p} = \frac{1}{2} \hbar \gamma \tag{31}$$

Proposition 1 ensures that the first principal minor Δ_1 is negative at the optimal point $p = p^*$ and $T_C = T_C^*$. Now, the only target should be to prove that the second principal minor Δ_2 is positive and to aim it, after manipulations, one can write:

$$\Delta_2 = \frac{4\gamma\lambda_1}{T_C^3} - \frac{1}{4}\hbar^2 \gamma^2. \tag{32}$$

At the optimal point $p = p^*$ and $T_C = T_C^*$,

$$\Delta_2 = \frac{4\gamma\lambda_1}{T_C^{*3}} - \frac{1}{4}\hbar^2 \gamma^2. \tag{33}$$

Later, Lemma 1 confirms the fact that $\Delta_2 > 0$.

Thus, the proof of Proposition 3 is complete such that $\tau(p, T_C)$ is a strictly pseudo-concave function at a unique optimal investment (p^*, T_C^*). Hence, the profit function affirms the global maximum solution at (p^*, T_C^*). □

Lemma 1. *If replenishment time $T_C < \left(\frac{16\lambda_1}{\hbar^2 \gamma} \right)^{\frac{1}{3}}$, then Equation (32) provides positive results, which consequently shows that Proposition 3 is valid.*

Proof. Replenishment time $T_C < \left(\frac{16\lambda_1}{\hbar^2 \gamma} \right)^{\frac{1}{3}}$

$$\hbar^2 \gamma < \frac{16\lambda_1}{T_C^3}$$

$$\hbar^2 \gamma \frac{\gamma}{4} < \frac{16\lambda_1}{T_C^3} \frac{\gamma}{4} \quad [\text{since } \gamma > 0]$$

$$\frac{1}{4}\hbar^2\gamma^2 < \frac{4\gamma\lambda_1}{T_C^3}$$

$$\frac{4\gamma\lambda_1}{T_C^{*3}} - \frac{1}{4}\hbar^2\gamma^2 > 0$$

Thus, $\Delta_2 > 0$. □

4.2. Case II: With Advanced Payment and a Discount for Single Time Payment

The concavity test for Case II is similar to Case I, so the proof for Case II is not shown to avoid redundancy. From Equations (13) and (17) one has:

$$\tau(p, T_C) = \frac{1}{T_C}\left(p(\psi - \gamma p + \eta R_C)T_C - \frac{1}{2}\hbar(\psi - \gamma p + \eta R_C)T_C^2 - \lambda_1\right)$$

$$\tau_f(p, T_C) = \frac{1}{T_C}\left(p(\psi - \gamma p + \eta R_C)T_C - \frac{1}{2}\hbar(\psi - \gamma p + \eta R_C)T_C^2 - \lambda_2\right)$$

where

$$\lambda_1 = \zeta + \Omega\omega + \frac{(n+1)}{2n}\phi\delta\omega\Omega\kappa + F_C N_t + 2\ell f_e \alpha N_t + \ell c_p w_p \Omega\alpha + 2\ell c_e N_t + \ell c_k \Omega + \chi R_C^2 - \frac{\Omega\omega v}{n}$$

$$\lambda_2 = \zeta + (1 + \phi_L\kappa)(1 - v)\Omega\omega + F_C N_t + 2\ell f_e \alpha N_t + \ell c_p w_p \Omega\alpha + 2\ell c_e N_t + \ell c_k \Omega + \chi R_C^2$$

From $\tau(p, T_C)$ and $\tau_f(p, T_C)$ one can easily notice that parts (λ_1 and λ_2) are the only difference between these two profit functions. Moreover, these two parts (λ_1 and λ_2) are independent of decision variables (p, T_C). Thus, there will be no change in making a decision regarding the concavity of these profit functions. However, in the numerical example, the concavity is presented numerically.

4.3. Case III: Without Advanced Payment

From Equations (13) and (19), one has:

$$\tau(p, T_C) = \frac{1}{T_C}\left(p(\psi - \gamma p + \eta R_C)T_C - \frac{1}{2}\hbar(\psi - \gamma p + \eta R_C)T_C^2 - \lambda_1\right)$$

$$\tau_N(p, T_C) = \frac{1}{T_C}\left(p(\psi - \gamma p + \eta R_C)T_C - \frac{1}{2}\hbar(\psi - \gamma p + \eta R_C)T_C^2 - \lambda_3\right)$$

where

$$\lambda_1 = \zeta + \Omega\omega + \frac{(n+1)}{2n}\phi\delta\omega\Omega\kappa + F_C N_t + 2\ell f_e \alpha N_t + \ell c_p w_p \Omega\alpha + 2\ell c_e N_t + \ell c_k \Omega + \chi R_C^2 - \frac{\Omega\omega v}{n}$$

$$\lambda_3 = \zeta + \Omega\omega + F_C N_t + 2\ell f_e \alpha N_t + \ell c_p w_p \Omega\alpha + 2\ell c_e N_t + \ell c_k \Omega + \chi R_C^2$$

The whole scenario of this case is similar to the previous Case II. Therefore, there will be no change in decision-making as in Case II. However, the concavity of the profit function is presented in the numerical example section.

5. Analysis and Discussion

5.1. Case Study

The choice of eco-friendly products is a growing trend that is being adopted by millions of people. A new addition in this category is an eco-friendly microwave oven (Figure 4), which draws the attention of business owners and customers. The higher the eco-friendliness, the higher the demand; although sometimes the price is slightly elevated, it satisfies all purposes of customers. A retailer who does not have enough capital can advance some purchase costs to the supplier to book the products. The supplier, in return, provides numerous discount amounts for him according to the retailer's payment. A case from a retailer shop is visited to fit in our model. The proposed problem is discussed with the shop manager, and he is asked to provide actual data accordingly. Those data are used in later numerical sections to validate the model and maintain a relationship with the data of the previously published article.

Figure 4. A retail shop of microwave oven. (Source: https://upload.wikimedia.org/wikipedia/commons/7/7e/Microwave_ovens%2C_Media_Markt%2C_Svagertorp%2C_Malmo.JPG, accessed on 26 December 2021).

5.2. Numerical Illustration

Here, we present three examples. We have collected secondary data from different published articles.

Example 1. (Case III) *In the first example, green carbon emission costs are considered with no payment in advance. For numerical illustration, the following parameters are considered: ordering costs per order placement ξ = \$1000/cycle, the demand combined with market potential Ψ = 220, price sensitivity coefficient γ = 0.65, low carbon preference coefficient η = 2, and manufacturer's carbon emission reduction level R_c = 0.5, carbon emission reduction investment χ = \$800. The purchase cost per unit ω = \$150, per unit holding cost h = \$2. Further, the fixed cost per trip F_c = \$200/trip, number of trips N_t = 3, fuel price α = \$0.3/liter, the empty vehicle fuel consumption f_c = 1 liter, travelled distance l = 100 km, product weight w_p = 0.5 kg, fuel consumption per ton of payload C_p = 1.5 liter/ton, carbon emission cost per unit distance c_e = \$0.03/km, and carbon emission cost per unit item per unit distance c_k = \$0.02/unit/km.*

We obtain optimal solutions per unit selling price p^* = \$260.36, replenishment time T_C^* = 6.21 months, order quantity Ω = 321.61 units, and total profit τ_N = \$3801.423 using Lingo 19 software with the aid of an exact optimization approach.

If we consider manufacturer's carbon emission reduction level (R_c) and selling price (p) as decision variables and cycle time (T_C = 6.21 months) as constant, then, we obtain the optimal solutions, per unit selling price p^* = \$260.54, manufacturer's carbon emission reduction level R_c^* = 0.62, order quantity Ω = 322.23 units, and total profit τ_N = \$3803.25.

Again, if we consider that manufacturer's carbon emission reduction level (R_c) and cycle time (T_C) are decision variables and selling price (p = \$260.35) is constant, then we obtain the optimal solutions, manufacturer's carbon emission reduction level R_c^* = 0.63, cycle time T_C^* = 6.38 months, order quantity Ω = 332.11 units, and total profit τ_N = \$3803.438.

From Figure 5, one can easily observe that the total profit function confirms the concavity nature in terms of the two decision variables, and the optimum profit is located at the blue dot point.

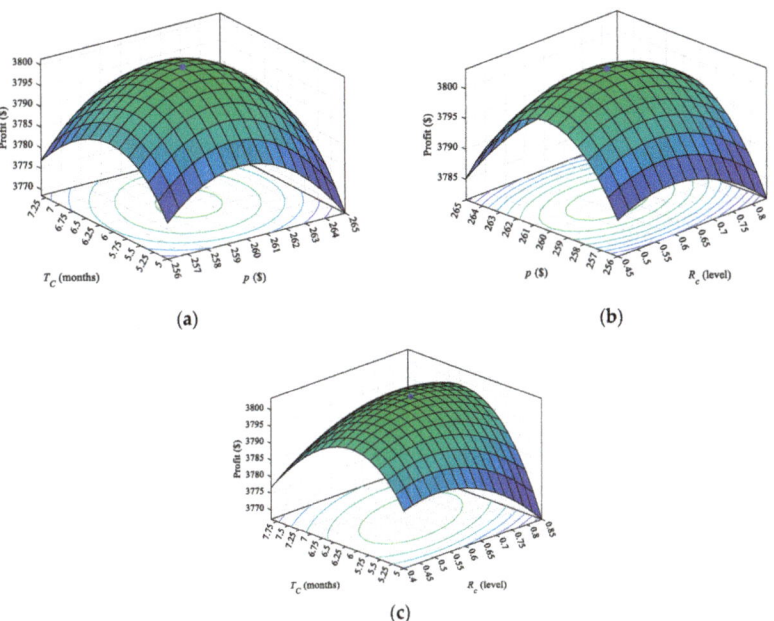

Figure 5. Profit function (τ) with regard to: (**a**) the selling price (p) and cycle time (T_C); (**b**) the manufacturer's carbon emission reduction level (R_c) and selling price (p); (**c**) the manufacturer's carbon emission reduction level (R_c) and cycle time (T_C).

When selling price (p), cycle time (T_C), and manufacturer's carbon emission reduction level (R_c) are decision variables, then optimal solutions are per-unit selling price $p^* = \$260.66$, cycle time $T_C^* = 6.40$ months, and manufacturer's carbon emission reduction level $R_c^* = 0.638$, order quantity $\Omega = 331.82$ units, and total profit $\tau_N = \$3803.499$.

Figure 6 shows that the profit function increases with respect to the increased selling price (p), cycle time (T_C), and manufacturer's carbon emission reduction level (R_c), and the profit function becomes optimum for optimum selling price $p^* = \$260.66$, optimum cycle time $T_c = 6.40$ months, and manufacturer's carbon emission reduction level $R_c^* = 0.638$. After the optimum point indicated by the green star marker, the profit function decreases, although the selling price and the cycle time increase. This behavior also confirms the concavity nature of the profit function.

Example 2. (Case II) *All the parameters are the same as in Example 1. For a single payment model, the supplier offers $v = 5\%$. Further, the length of time during prepayment $\kappa = 0.5$ years and retailer interest rate of loan from any financial institutes $\phi_L = 3\%$.*

We obtain the optimal solutions, per unit selling price $p^* = \$257.62$, replenishment time $T_C^* = 6.11$ months, order quantity $\Omega = 327.08$ units, and total profit $\tau_f = \$4083.795$.

If we consider manufacturer's carbon emission reduction level (R_c) and selling price (p) as decision variables and cycle time ($T_C = 6.12$ months) as constant, then we obtain the optimal solutions of per-unit selling price $p^* = \$257.83$, manufacturer's carbon emission reduction level $R_c^* = 0.63$, order quantity $\Omega = 328.48$ units, and total profit $\tau_N = \$4086.035$.

Again, if we consider manufacturer's carbon emission reduction level (R_c) and cycle time (T_C) as decision variables and selling price ($p = \$257.62$) as constant, then we obtain the optimal solutions as manufacturer's carbon emission reduction level $R_c^* = 0.65$, cycle time $T_C^* = 6.29$ months, order quantity $\Omega = 338.84$ units, and total profit $\tau_N = \$4086.235$.

When selling price (p), cycle time (T_C), and manufacturer's carbon emission reduction level (R_c) are decision variables, then optimal solutions are per-unit selling price $p^* = \$257.96$, cycle time $T_C^* = 6.31$ months, and manufacturer's carbon emission reduction level $R_c^* = 0.65$, order quantity $\Omega = 338.54$ units, and total profit $\tau_N = \$4086.305$.

For this example, Figure 7 confirms the concavity nature of the profit function with respect to the two decision variables.

Figure 8a shows that the profit declines due to growing lead time. Figure 8b confirms that the higher discount rate produces the higher profit gaining, and Figure 8c confirms that the total profit declines for the higher interest rate on the loan amount to collect capital.

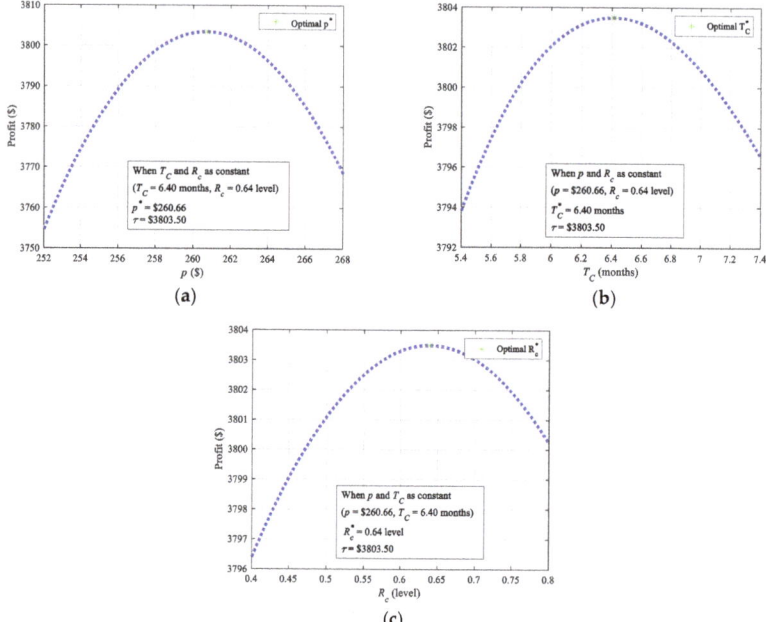

Figure 6. Profit function (τ) regarding (**a**) the selling price (p); (**b**) cycle time (T_C); (**c**) manufacturer's carbon emission reduction level (R_c).

Example 3. (Case I) *All the parameters are the same as Example 1. Some additional parameters are as follows: number of equal prepayments before receiving order quantity $n = 10$, the portion of total purchase cost $\delta = 0.8$, the interest rate of capital cost per year, length of time during which the prepayments are paid $\kappa = 0.5$ years, and discount rate for prepayment $v = 5\%$.*

We acquire the optimal solutions of per-unit selling price $p^* = \$276.86$, replenishment time $< \phi = 1\%$ months, order quantity $\Omega = 286.35$ units, and total profit $\tau = \$2304.672$.

If we consider manufacturer's carbon emission reduction level (R_c) and selling price (p) as decision variables and cycle time ($T_C = 6.21$ months) as constant, then we obtain the optimal solutions of per-unit selling price $p^* = \$276.45$, manufacturer's carbon emission reduction level $R_c^* = 0.49$, order quantity $\Omega = 256.35$ units, and total profit $\tau_N = \$2300.881$.

Again, if we consider manufacturer's carbon emission reduction level (R_c) and cycle time (T_C) as decision variables and selling price ($p = \$260.35$) as constant, then we obtain the optimal solutions of manufacturer's carbon emission reduction level $R_c^* = 0.36$, cycle time $T_C^* = 6.08$ months, order quantity $\Omega = 312.98$ units, and total profit $\tau_N = \$2134.147$.

When selling price (p), cycle time (T_C), and manufacturer's carbon emission reduction level (R_c) are decision variables, then optimal solutions are per-unit selling price $p^* = \$276.99$, cycle time $T_C^* = 7.06$ months, and manufacturer's carbon emission reduction level $R_c^* = 0.56$, order quantity $\Omega = 289.92$ units, and total profit $\tau_N = \$2305.004$.

For this example, Figure 9 confirms the concavity nature of the profit function with respect to the two decision variables.

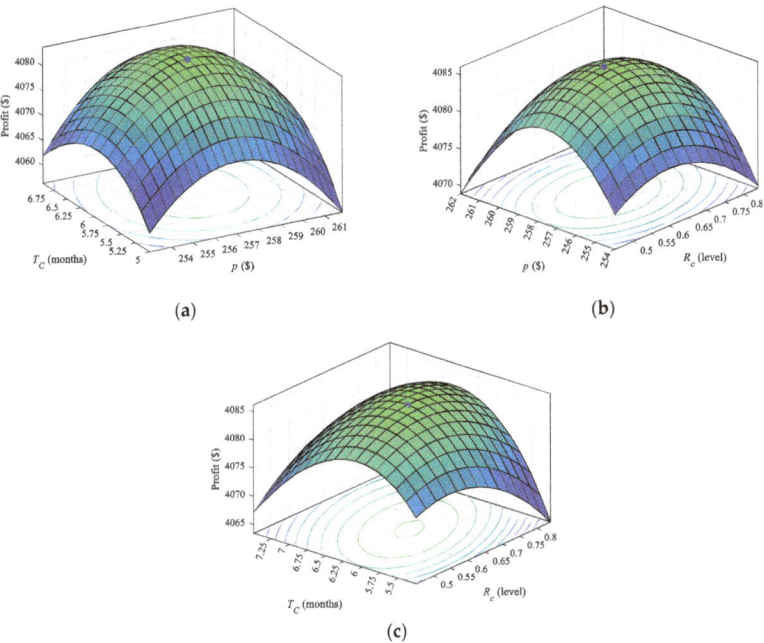

Figure 7. Profit function (τ) with regard to: (**a**) the selling price (p) and cycle time (T_C); (**b**) the manufacturer's carbon emission reduction level (R_c) and selling price (p); (**c**) the manufacturer's carbon emission reduction level (R_c) and cycle time (T_C).

Figure 10a shows that the profit is higher for a smaller number of instalments and is smaller for a larger number of instalments. As the retailer has to pay interest for instalment-based payment, the cost becomes higher, and the profit becomes lower. From Figure 10b, one can confirm that the higher lead time forces a lower profit to be gained. Figure 10c confirms that the total profit declines for higher portion payment in advance, whereas Figure 10d assures the fact that for a higher amount of discount rate for instalment payment in advance, the total profit increases.

5.3. Sensitivity Analysis

Table 2 shows the sensitivity analysis of the present work. One can easily observe the robustness among the parameters. For three cases, the test has been performed for the parameters within the range of -20% to $+20\%$. Some critical observations can be summarized based on the sensitivity table (Table 2):

a. The market potential (ψ) is positively correlated with the integrated profit. The selling price is correlated similarly, but the cycle length interacts negatively. One can detect the continuous rise in profit and selling price with growing market potential (ψ) for all these three cases, and the profit becomes maximum for Case II, whereas the selling price, as well as cycle length, become minimum.

b. The total profit and selling price decline for all three cases as the price elasticity parameter (γ) increases. The cycle length behaves in the opposite direction. One can observe the highest profit at the minimum value of the price elasticity parameter (γ) for a discount on a single-instalment payment (Case II).
c. The total profit increases for all three cases with higher values of carbon emission reduction level (R_C). The selling price and the cycle length show the same characteristics. The total profit is comparatively much lower in Case I as the instalment policy creates an extra cost. The profit is best in Case II, since the discount in purchasing cost influences higher profit gaining.
d. For all three cases, the ordering cost (ζ), as well as the holding cost ($\hbar$), have a direct impact on total profit. The higher values of those two costs create a lower profit and vice versa. The increasing ordering charge or holding charge means a decline in profit. It is easy to observe the significant consequence of this fact for all three cases. A similar type of effect has been noted for fluctuations of the per-unit purchase cost (ω).
e. The larger the number of trips (N_t) the lesser the profit becomes since an extra trip means it needs an additional fixed cost, variable cost, fuel, labor, etc. Therefore, the profit becomes lower for the intensifications of trips. The travel distance (ℓ), fuel cost (α), fuel consumption per ton of payload (c_p), and product weight (w_p) have similar impacts on profit as those can add additional expenses. Any longer distance brings additional cost in the expenses, so reduction of distance can optimize the profit, which is numerically true, as shown in the sensitivity table.
f. The implications of carbon emission cost on transportation cost have important roles in profit gaining. Increasing values of carbon emission cost per unit distance (c_e) and carbon emission cost per unit item per unit distance (c_k) force the total profit to be less in all three cases.

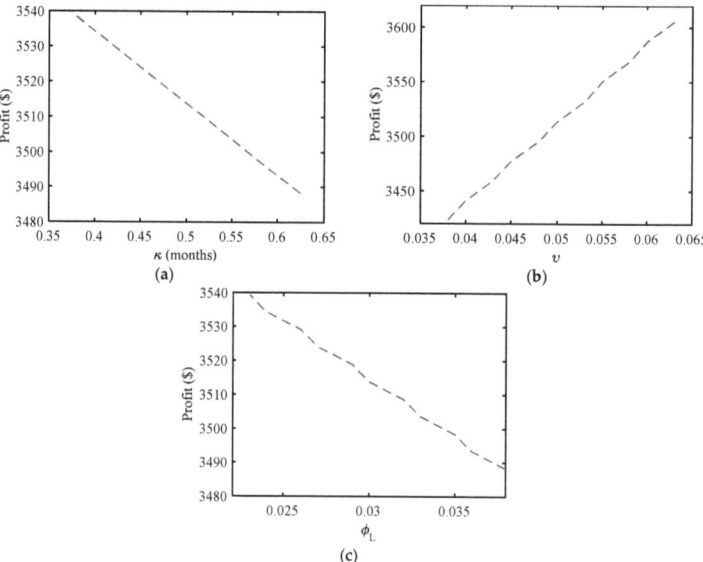

Figure 8. Total profit profile associated with single payment-based payment in advance parameters. (**a**) describes the total profit against different lead times, (**b**) describes the total profit for different discount rates due to instalment-based payment in advance, and (**c**) shows the profit vs. interest rate on the loan amount.

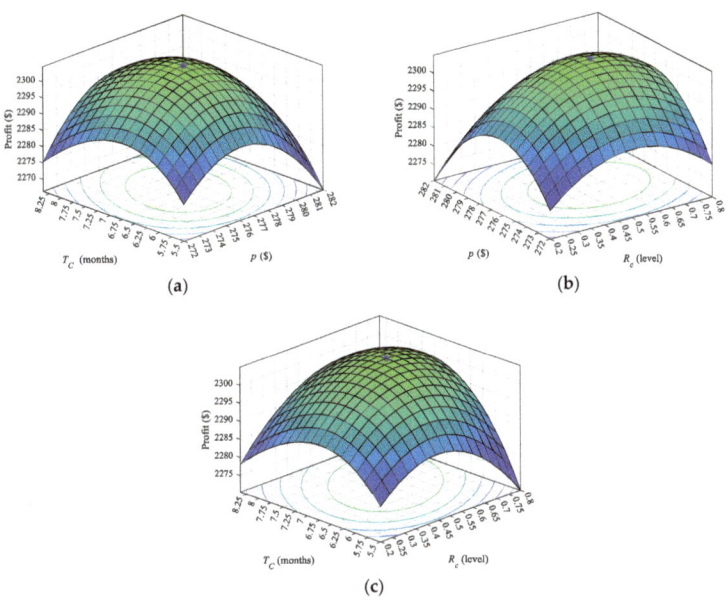

Figure 9. Profit function (τ) with regard to: (**a**) the selling price (p) and cycle time (T_C); (**b**) the manufacturer's carbon emission reduction level (R_c) and selling price (p); (**c**) the manufacturer's carbon emission reduction level (R_c) and cycle time (T_C).

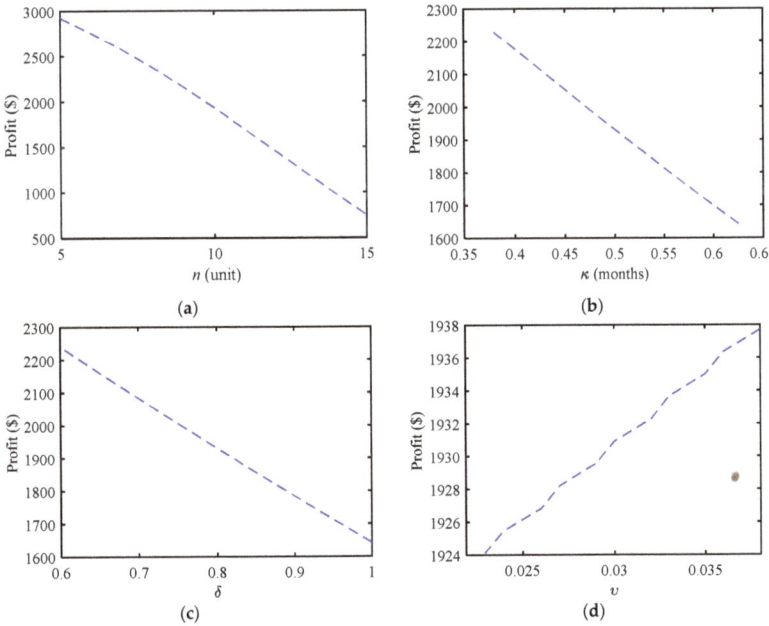

Figure 10. Total profit profile associated with instalment-based payment-in-advance parameters. (**a**) describes the total profit against the number of instalments, (**b**) shows the total profit for different lead times, (**c**) describes the profit vs. payment in advance portion, and (**d**) describes the total profit for different discount rates due to instalment based payment in advance.

Table 2. Sensitivity Analysis.

Parameter (Base)	Change in %	Changed Value	Case I			Case II			Case III		
			p	T_c	Profit (τ_N)	p	T_c	Profit (τ_f)	p	T_c	Profit (τ)
ψ (200)	20%	240.000	307.119	5.822	5033.220	288.058	5.311	7382.136	290.771	5.376	7020.964
	10%	220.000	290.755	6.428	3311.817	271.614	5.756	5278.759	274.336	5.839	4971.943
	−10%	180.000	258.518	8.618	893.981	238.977	7.150	2089.039	241.740	7.313	1891.779
	−20%	160.000	243.186	11.289	209.085	222.937	8.402	1007.654	225.755	8.676	866.120
γ (0.60)	20%	0.720	247.554	9.192	510.321	227.756	7.208	1636.208	230.539	7.411	1443.080
	10%	0.660	259.677	8.058	1090.069	240.206	6.729	2446.337	242.962	6.878	2223.500
	−10%	0.540	292.835	6.698	3116.149	273.686	6.012	4923.224	276.409	6.097	4641.829
	−20%	0.480	315.870	6.240	4772.288	296.810	5.732	6801.839	299.523	5.797	6491.457
η (2)	20%	2.400	274.676	7.269	1943.027	255.402	6.332	3529.771	258.138	6.443	3276.993
	10%	2.200	274.595	7.274	1936.966	255.320	6.336	3521.782	258.057	6.447	3269.278
	−10%	1.800	274.434	7.284	1924.869	255.157	6.342	3505.830	257.894	6.454	3253.873
	−20%	1.600	274.353	7.290	1918.833	255.075	6.346	3497.867	257.812	6.458	3246.183
R_C (0.50)	20%	0.600	274.729	7.375	1935.516	255.447	6.424	3521.148	258.185	6.537	3268.518
	10%	0.550	274.621	7.325	1933.370	255.342	6.380	3517.653	258.079	6.492	3265.220
	−10%	0.450	274.411	7.238	1928.139	255.137	6.302	3509.587	257.873	6.413	3257.564
	−20%	0.400	274.309	7.201	1925.044	255.037	6.270	3505.001	257.773	6.380	3253.193
χ (500)	20%	600.000	600.000	7.328	1927.490	255.259	6.381	3509.871	257.997	6.493	3257.708
	10%	550.000	550.000	7.304	1929.198	255.249	6.360	3511.833	257.986	6.472	3259.636
	−10%	450.000	450.000	7.255	1932.633	255.228	6.318	3515.777	257.965	6.429	3263.512
	−20%	400.000	400.000	7.230	1934.359	255.217	6.297	3517.759	257.954	6.408	3265.459
ζ (1000)	20%	1200.000	274.705	7.661	1904.138	255.402	6.667	3483.047	258.143	6.785	3231.349
	10%	1100.000	274.611	7.472	1917.355	255.321	6.505	3498.230	258.060	6.620	3246.269
	−10%	900.000	274.416	7.082	1944.839	255.153	6.169	3529.792	257.889	6.277	3277.285
	−20%	800.000	274.314	6.879	1959.165	255.066	5.994	3546.236	257.799	6.099	3293.445
$\hbar$ (2)	20%	2.400	274.874	6.665	1880.559	255.547	5.798	3455.949	258.290	5.901	3204.722
	10%	2.200	274.698	6.951	1905.144	255.396	6.050	3484.203	258.136	6.157	3232.485
	−10%	1.800	274.322	7.661	1958.047	255.072	6.675	3544.953	257.806	6.792	3292.184
	−20%	1.600	274.120	8.112	1986.783	254.898	7.072	3577.924	257.628	7.196	3324.587
ω (150)	20%	180.000	293.494	8.787	814.828	270.045	7.024	2257.755	273.355	7.211	2013.351
	10%	165.000	283.949	7.923	1321.012	262.628	6.655	2853.657	265.649	6.798	2602.859
	−10%	135.000	265.149	6.774	2643.576	247.869	6.065	4237.982	250.326	6.152	3989.230
	−20%	120.000	255.831	6.363	3458.375	240.517	5.824	5026.037	242.696	5.893	4785.642
F_C (200)	20%	240.000	274.630	7.510	1914.685	255.338	6.538	3495.164	258.077	6.653	3243.255
	10%	220.000	274.573	7.396	1922.735	255.288	6.439	3504.411	258.026	6.553	3252.342
	−10%	180.000	274.456	7.161	1939.223	255.187	6.237	3523.344	257.924	6.347	3270.948
	−20%	160.000	274.396	7.042	1947.672	255.136	6.134	3533.044	257.871	6.242	3280.480
N_t (6)	20%	10.000	276.047	10.344	1719.361	256.552	8.967	3270.326	259.315	9.130	3022.364
	10%	8.000	275.654	9.558	1772.825	256.217	8.296	3331.964	258.972	8.445	3082.910
	−10%	4.000	274.766	7.783	1895.591	255.455	6.772	3473.225	258.196	6.892	3221.698
	−20%	2.000	274.246	6.742	1968.856	255.007	5.876	3557.357	257.739	5.979	3304.374
ℓ (100)	20%	120.000	277.083	7.515	1751.504	257.773	6.508	3276.862	260.513	6.627	3032.791
	10%	110.000	275.798	7.396	1840.244	256.505	6.423	3394.380	259.244	6.538	3146.228
	−10%	90.000	273.233	7.166	2023.510	253.972	6.257	3635.126	256.708	6.365	3378.819
	−20%	80.000	271.953	7.056	2118.033	252.707	6.176	3758.353	255.441	6.281	3497.972
f_e (1)	20%	1.200	274.550	7.349	1925.991	255.268	6.399	3508.150	258.006	6.512	3256.016
	10%	1.100	274.532	7.314	1928.446	255.253	6.369	3510.969	257.991	6.481	3258.787
	−10%	0.900	274.497	7.244	1933.392	255.223	6.309	3516.648	257.960	6.420	3264.368
	−20%	0.800	274.479	7.209	1935.882	255.208	6.278	3519.508	257.944	6.389	3267.179
α (0.30)	20%	0.360	276.872	7.495	1765.910	257.565	6.493	3295.975	260.306	6.611	3051.237
	10%	0.330	275.693	7.386	1847.598	256.401	6.415	3404.086	259.140	6.530	3155.600
	−10%	0.270	273.338	7.176	2015.853	254.076	6.264	3625.122	256.811	6.372	3369.148
	−20%	0.240	272.162	7.075	2102.417	252.914	6.190	3738.045	255.648	6.296	3478.330

Table 2. Cont.

Parameter (Base)	Change in %	Changed Value	Case I			Case II			Case III		
			p	T_c	Profit (τ_N)	p	T_c	Profit (τ_f)	p	T_c	Profit (τ)
c_p (1.50)	20%	1.800	276.837	7.423	1770.736	257.535	6.432	3301.545	260.275	6.549	3056.708
	10%	1.650	275.675	7.350	1850.041	256.386	6.385	3406.898	259.125	6.499	3158.363
	−10%	1.350	273.355	7.211	2013.351	254.091	6.294	3622.255	256.827	6.403	3366.330
	−20%	1.200	272.197	7.144	2097.353	252.944	6.250	3732.257	255.678	6.356	3472.639
w_p (0.50)	20%	0.600	276.837	7.423	1770.736	257.535	6.432	3301.545	260.275	6.549	3056.708
	10%	0.550	275.675	7.350	1850.041	256.386	6.385	3406.898	259.125	6.499	3158.363
	−10%	0.450	273.355	7.211	2013.351	254.091	6.294	3622.255	256.827	6.403	3366.330
	−20%	0.400	272.197	7.144	2097.353	252.944	6.250	3732.257	255.678	6.356	3472.639
c_e (0.03)	20%	0.036	274.518	7.286	1930.418	255.241	6.345	3513.234	257.978	6.457	3261.013
	10%	0.033	274.516	7.283	1930.666	255.240	6.342	3513.518	257.977	6.454	3261.292
	−10%	0.027	274.513	7.276	1931.160	255.237	6.336	3514.086	257.974	6.447	3261.850
	−20%	0.024	274.511	7.272	1931.408	255.235	6.333	3514.370	257.972	6.444	3262.129
c_k (0.02)	20%	0.024	274.721	7.292	1916.421	255.442	6.347	3494.684	258.180	6.459	3243.109
	10%	0.022	274.618	7.286	1923.661	255.340	6.343	3504.237	258.077	6.455	3252.334
	−10%	0.018	274.412	7.273	1938.177	255.136	6.335	3523.379	257.873	6.446	3270.820
	−20%	0.016	274.308	7.267	1945.454	255.034	6.331	3532.969	257.771	6.442	3280.081

5.4. Managerial Implications

The managerial implications of this sustainable inventory management study in terms of pricing strategies, low carbon preferences, suitability of discount policy, and impact of payments in advance are vast:

(i) From the three observed cases, the lowest selling price is obtained when the payment in advance is performed in a single payment. Further study also confirms that profit is higher for a smaller number of instalments; hence, managers can optimize the number of installments in this direction considering their financial condition.
(ii) The case with a single payment also results in a lower selling price. It is beneficial for customers and increases the demand level.
(iii) One can take important pricing decisions from the study and maintain a healthy profit margin by incorporating these strategies and simultaneously observing the nature of the customers.
(iv) This study provides some insights into how preferences for low carbon can influence the sales of the retailer and in which way a manager can maintain an eco-friendly inventory. This study shows that the total profit increases with higher values of carbon emission reduction level and higher preferences for low carbon among customers.

6. Conclusions

This paper presents a low-carbon preference inventory model with selling price and carbon-emission-reduction-dependent demand. Some major issues solved through this model are:

(i) The optimal replenishment rate clinging to the commencement of payment in advance has been successfully integrated and offers some significant results.
(ii) Simultaneous integration of discount policy, payment in advance to the selling price, and reduction of carbon-emission-dependent demand work efficiently. It provides some techniques for the retailer to manage inventories profitably.
(iii) A smaller number of instalments of the payment in advance increase the profit. This study shows that the case with a single payment results in a higher total profit and a lower selling price.
(iv) With the increasing customers' preferences for environmentally friendly products, retailers should increase the effort for reducing emission levels.

Therefore, to maximize profit, this study recommends that retailers respond to the increasing customers' preferences for low carbon by promoting environmentally friendly products. Simultaneously, retailers should attract more customers by setting a lower price by minimizing the number of instalments to take advantage of the discounts offered.

However, this model has limitations in terms of exposition, choice of variables, incorporation of marketing strategies, etc. This model can easily be extended by incorporating trade-credit policy [40,44,45], including some carbon emission regulations [33,46] and taking more than one player, e.g., a vendor–buyer system [47,48]. This study also does not allow for shortages; hence, further research may consider shortages with a full or partial backlog. Moreover, the retailer can dynamically purchase the inventory from the outside supplier to reduce the financial risk and avail the full discount facilities.

Author Contributions: Conceptualization, S.S. and A.H.M.M.; methodology, S.S. and A.H.M.M.; software, S.S. and S.M.; validation, S.S., A.H.M.M., A.A., and I.M.H.; writing—original draft preparation, S.S., A.H.M.M., S.M., and Y.D.; writing—review and editing, S.S., A.H.M.M., Y.D., A.A., and I.M.H.; visualization, S.S., S.M., I.M.H. and A.H.M.M. All authors have read and agreed to the published version of the manuscript.

Funding: This work was supported by the Research Supporting Project No. (RSP-2021/389), King Saud University, Riyadh, Saudi Arabia.

Institutional Review Board Statement: Not applicable.

Informed Consent Statement: Not applicable.

Data Availability Statement: All data are given in the manuscript which is used to justify the proposed model.

Acknowledgments: We would like to thank the editors of the journal as well as the anonymous reviewers for their valuable suggestions that make the paper stronger and more consistent.

Conflicts of Interest: The authors declare no conflict of interest.

References

1. Borin, N.; Lindsey-Mullikin, J.; Krishnan, R. An analysis of consumer reactions to green strategies. *J. Prod. Brand Manag.* **2013**, *22*, 118–128. [CrossRef]
2. Lin, R.-J.; Tan, K.-H.; Geng, Y. Market demand, green product innovation, and firm performance: Evidence from Vietnam motorcycle industry. *J. Clean. Prod.* **2013**, *40*, 101–107. [CrossRef]
3. Zhang, Q.; Zhao, Q.; Zhao, X.; Tang, L. On the introduction of green product to a market with environmentally conscious consumers. *Comput. Ind. Eng.* **2020**, *139*, 106190. [CrossRef]
4. Fernando, Y.; Wah, W.X. The impact of eco-innovation drivers on environmental performance: Empirical results from the green technology sector in Malaysia. *Sustain. Prod. Consum.* **2017**, *12*, 27–43. [CrossRef]
5. Yalabik, B.; Fairchild, R.J. Customer, regulatory, and competitive pressure as drivers of environmental innovation. *Int. J. Prod. Econ.* **2011**, *131*, 519–527. [CrossRef]
6. Pang, Q.; Li, M.; Yang, T.; Shen, Y. Supply chain coordination with carbon trading price and consumers' environmental awareness dependent demand. *Math. Probl. Eng.* **2018**, *2018*, 8749251. [CrossRef]
7. Xia, L.; Hao, W.; Qin, J.; Ji, F.; Yue, X. Carbon emission reduction and promotion policies considering social preferences and consumers' low-carbon awareness in the cap-and-trade system. *J. Clean. Prod.* **2018**, *195*, 1105–1124. [CrossRef]
8. Xia, L.; Guo, T.; Qin, J.; Yue, X.; Zhu, N. Carbon emission reduction and pricing policies of a supply chain considering reciprocal preferences in cap-and-trade system. *Ann. Oper. Res.* **2017**, *268*, 149–175. [CrossRef]
9. Zanoni, S.; Mazzoldi, L.; Zavanella, L.E.; Jaber, M.Y. A joint economic lot size model with price and environmentally sensitive demand. *Prod. Manuf. Res.* **2014**, *2*, 341–354. [CrossRef]
10. Sana, S.S. A structural mathematical model on two echelon supply chain system. *Ann. Oper. Res.* **2021**. [CrossRef]
11. Hovelaque, V.; Bironneau, L. The carbon-constrained EOQ model with carbon emission dependent demand. *Int. J. Prod. Econ.* **2015**, *164*, 285–291. [CrossRef]
12. Tsai, P.-H.; Lin, G.-Y.; Zheng, Y.-L.; Chen, Y.-C.; Chen, P.-Z.; Su, Z.-C. Exploring the effect of Starbucks' green marketing on consumers' purchase decisions from consumers' perspective. *J. Retail. Consum. Serv.* **2020**, *56*, 102162. [CrossRef]
13. Nekmahmud, M.; Fekete-Farkas, M. Why not green marketing? Determinates of consumers' intention to green purchase decision in a new developing nation. *Sustainability* **2020**, *12*, 7880. [CrossRef]
14. Grant, J. Green marketing. *Strateg. Dir.* **2008**, *24*, 25–27. [CrossRef]
15. Ginsberg, J.; Bloom, P. Choosing the right green marketing strategy. *MIT Sloan Manag. Rev.* **2004**, 79–84.

16. Lampe, M.; Gazda, G.M. Green marketing in Europe and the United States: An evolving business and society interface. *Int. Bus. Rev.* **1995**, *4*, 295–312. [CrossRef]
17. Dong, G.; Liang, L.; Wei, L.; Xie, J.; Yang, G. Optimization model of trade credit and asset-based securitization financing in carbon emission reduction supply chain. *Ann. Oper. Res.* **2021**, *299*, 1–50. [CrossRef]
18. Zhang, Q.; Tsao, Y.-C.; Chen, T.-H. Economic order quantity under advance payment. *Appl. Math. Model.* **2014**, *38*, 5910–5921. [CrossRef]
19. Taleizadeh, A.A.; Tavakoli, S.; San-José, L.A. A lot sizing model with advance payment and planned backordering. *Ann. Oper. Res.* **2018**, *271*, 1001–1022. [CrossRef]
20. Li, R.; Skouri, K.; Teng, J.-T.; Yang, W.-G. Seller's optimal replenishment policy and payment term among advance, cash, and credit payments. *Int. J. Prod. Econ.* **2018**, *197*, 35–42. [CrossRef]
21. Maiti, A.K.; Maiti, M.K.; Maiti, M. Inventory model with stochastic lead-time and price dependent demand incorporating advance payment. *Appl. Math. Model.* **2009**, *33*, 2433–2443. [CrossRef]
22. Taleizadeh, A.A. An economic order quantity model for deteriorating item in a purchasing system with multiple prepayments. *Appl. Math. Model.* **2014**, *38*, 5357–5366. [CrossRef]
23. Mashud, A.H.M.; Roy, D.; Daryanto, Y.; Chakrabortty, R.K.; Tseng, M.-L. A sustainable inventory model with controllable carbon emissions, deterioration and advance payments. *J. Clean. Prod.* **2021**, *296*, 126608. [CrossRef]
24. Sepehri, A. Inventory management under carbon emission policies: A systematic literature review. *Decis. Mak. Invent. Manag.* **2021**, 187–218. [CrossRef]
25. Pattnaik, S.; Nayak, M.M.; Abbate, S.; Centobelli, P. Recent trends in sustainable inventory models: A literature review. *Sustainability* **2021**, *13*, 11756. [CrossRef]
26. Lin, T.-Y.; Sarker, B.R. A pull system inventory model with carbon tax policies and imperfect quality items. *Appl. Math. Model.* **2017**, *50*, 450–462. [CrossRef]
27. Kazemi, N.; Abdul-Rashid, S.H.; Ghazilla, R.A.R.; Shekarian, E.; Zanoni, S. Economic order quantity models for items with imperfect quality and emission considerations. *Int. J. Syst. Sci. Oper. Logist.* **2016**, *5*, 99–115. [CrossRef]
28. Shen, Y.J.; Shen, K.F.; Yang, C.T. A production inventory model for deteriorating items with collaborative preservation technology investment under carbon tax. *Sustainability* **2019**, *11*, 5027. [CrossRef]
29. Wee, H.-M.; Daryanto, Y. Imperfect quality item inventory models considering carbon emissions. *Optim. Invent. Manag. Asset* **2020**, 137–159. [CrossRef]
30. Daryanto, Y.; Wee, H.M.; Wu, K.H. Revisiting sustainable EOQ model considering carbon emission. *Int. J. Manuf. Technol. Manag.* **2021**, *35*. [CrossRef]
31. Paul, A.; Pervin, M.; Roy, S.K.; Maculan, N.; Weber, G.-W. A green inventory model with the effect of carbon taxation. *Ann. Oper. Res.* **2021**. [CrossRef]
32. As'ad, R.; Hariga, M.; Shamayleh, A. Sustainable dynamic lot sizing models for cold products under carbon cap policy. *Comput. Ind. Eng.* **2020**, *149*, 106800. [CrossRef] [PubMed]
33. Hasan, M.R.; Roy, T.C.; Daryanto, Y.; Wee, H.-M. Optimizing inventory level and technology investment under a carbon tax, cap-and-trade and strict carbon limit regulations. *Sustain. Prod. Consum.* **2021**, *25*, 604–621. [CrossRef]
34. Datta, T.K. Effect of green technology investment on a production-inventory system with carbon tax. *Adv. Oper. Res.* **2017**, *2017*, 4834839. [CrossRef]
35. Mashud, A.H.M.; Pervin, M.; Mishra, U.; Daryanto, Y.; Tseng, M.L.; Lim, M.K. A sustainable inventory model with controllable carbon emissions in green-warehouse farms. *J. Clean. Prod.* **2021**, *298*, 126777. [CrossRef]
36. Mishra, U.; Wu, J.-Z.; Tsao, Y.-C.; Tseng, M.-L. Sustainable inventory system with controllable non-instantaneous deterioration and environmental emission rates. *J. Clean. Prod.* **2020**, *244*, 118807. [CrossRef]
37. Lou, G.; Xia, H.; Zhang, J.; Fan, T. Investment strategy of emission-reduction technology in a supply chain. *Sustainability* **2015**, *7*, 10684–10708. [CrossRef]
38. Gao, X.; Zheng, H.; Zhang, Y.; Golsanami, N. Tax policy, environmental concern and level of emission reduction. *Sustainability* **2019**, *11*, 1047. [CrossRef]
39. Teng, J.-T.; Cárdenas-Barrón, L.E.; Chang, H.-J.; Wu, J.; Hu, Y. Inventory lot-size policies for deteriorating items with expiration dates and advance payments. *Appl. Math. Model.* **2016**, *40*, 8605–8616. [CrossRef]
40. Mashud, A.H.M.; Wee, H.-M.; Sarkar, B.; Chiang Li, Y.-H. A sustainable inventory system with the advanced payment policy and trade-credit strategy for a two-warehouse inventory system. *Kybernetes* **2020**, *50*, 1321–1348. [CrossRef]
41. Savaskan, R.C.; Van Wassenhove, L.N. Reverse channel design: The case of competing retailers. *Manag. Sci.* **2006**, *52*, 1–14. [CrossRef]
42. Swami, S.; Shah, J. Channel coordination in green supply chain management. *J. Oper. Res. Soc.* **2013**, *64*, 336–351. [CrossRef]
43. Taleizadeh, A.A. Lot-sizing model with advance payment pricing and disruption in supply under planned partial backordering. *Int. Trans. Oper. Res.* **2016**, *24*, 783–800. [CrossRef]
44. Wang, J.; Wang, K.; Li, X.; Zhao, R. Suppliers' trade credit strategies with transparent credit ratings: Null, exclusive, and nonchalant provision. *Eur. J. Oper. Res.* **2022**, *297*, 153–163. [CrossRef]
45. Tiwari, S.; Cárdenas-Barrón, L.E.; Malik, A.I.; Jaggi, C.K. Retailer's credit and inventory decisions for imperfect quality and deteriorating items under two-level trade credit. *Comput. Oper. Res.* **2022**, *138*, 105617. [CrossRef]

46. Rout, C.; Paul, A.; Kumar, R.S.; Chakraborty, D.; Goswami, A. Integrated optimization of inventory, replenishment and vehicle routing for a sustainable supply chain under carbon emission regulations. *J. Clean. Prod.* **2021**, *316*, 128256. [CrossRef]
47. Marchi, B.; Zanoni, S.; Jaber, M.Y. Credit-dependent demand in a vendor-buyer model with a two-level delay-in-payments contract under a consignment-stock policy agreement. *Appl. Math. Model.* **2021**, *99*, 585–605. [CrossRef]
48. Wangsa, I.D.; Tiwari, S.; Wee, H.M.; Reong, S. A sustainable vendor-buyer inventory system considering transportation, loading and unloading activities. *J. Clean. Prod.* **2020**, *271*, 122120. [CrossRef]

fractal and fractional

Article

Generalized *p*-Convex Fuzzy-Interval-Valued Functions and Inequalities Based upon the Fuzzy-Order Relation

Muhammad Bilal Khan [1,*], Savin Treanță [2,*] and Hüseyin Budak [3]

[1] Department of Mathematics, COMSATS University Islamabad, Islamabad 44000, Pakistan
[2] Department of Applied Mathematics, University Politehnica of Bucharest, 060042 Bucharest, Romania
[3] Department of Mathematics, Faculty of Science and Arts, Düzce University, Düzce 81620, Turkey; huseyinbudak@duzce.edu.tr
* Correspondence: bilal42742@gmail.com (M.B.K.); savin.treanta@upb.ro (S.T.)

Abstract: Convexity is crucial in obtaining many forms of inequalities. As a result, there is a significant link between convexity and integral inequality. Due to the significance of these concepts, the purpose of this study is to introduce a new class of generalized convex interval-valued functions called (p,s)-convex fuzzy interval-valued functions $((p,s)$-convex *F-I-V-F*s) in the second sense and to establish Hermite–Hadamard (H–H) type inequalities for (p,s)-convex *F-I-V-F*s using fuzzy order relation. In addition, we demonstrate that our results include a large class of new and known inequalities for (p,s)-convex *F-I-V-F*s and their variant forms as special instances. Furthermore, we give useful examples that demonstrate usefulness of the theory produced in this study. These findings and diverse approaches may pave the way for future research in fuzzy optimization, modeling, and interval-valued functions.

Keywords: (p,s)-convex fuzzy-interval-valued function; fuzzy Riemann integral; Jensen type inequality; Schur type inequality; Hermite–Hadamard type inequality; Hermite–Hadamard–Fejér type inequality

1. Introduction

A convex function has a convex set as its epigraph; therefore, the theory of inequality of convex functions falls under the umbrella of convexity. Nonetheless, it is a significant theory in and of itself, as it affects practically all fields of mathematics. The graphical analysis is most often the initial issue that necessitates the acquaintance with this theory. This is an opportunity to learn about the second derivative test of convexity, which is a useful tool for detecting convexity. The difficulty of identifying the extreme values of functions with many variables, as well as the application of Hessian as a higher dimensional generalization of the second derivative, follows. Holder, Jensen, and Minkowski all made early contributions to convex analysis. The next step is to go on to optimization issues in infinite dimensional spaces; however, despite the technological sophistication required to solve such problems, the fundamental concepts are quite similar to those underlying the one variable situation. Despite numerous applications, many contemporary difficulties in economics and engineering, the relevance of convex analysis is well recognized in optimization theory [1–3], and the idea of convexity no longer suffices.

Over the years, remarkable varieties of convexities, such as harmonic convexity [4], quasi convexity [5], Schur convexity [6], strong convexity [7,8], *p*-convexity [9], fuzzy convexity [10,11], fuzzy preinvexity [12] and generalized convexity [13], *p*-convexity [14] and so on, have been introduced to convex sets and convex functions. A fascinating field for research is the definition of convexity with an integral problem. Therefore, several authors have identified a great number of equalities or inequalities as applications of convex functions. The representative results include Gagliardo–Nirenberg-type inequality [15], Hardy-type inequality [16], Ostrowski-type inequality [17], Olsen-type inequality [18], and the most commonly known inequality of, namely, the H–H inequality [19]. Similarly, many authors have devoted themselves to study the fractional integral inequalities for single-valued and interval-valued functions, see [20–28].

In ref. [29], the enormous research work fuzzy set and system has been dedicated on development of different fields, and it plays an important role in the study of a wide class problems arising in pure mathematics and applied sciences including operation research, computer science, managements sciences, artificial intelligence, control engineering and decision sciences. Recently, fuzzy interval analysis and fuzzy interval-valued differential equations have been put forward to deal the ambiguity originate by insufficient data in some mathematical or computer models that determine real-world phenomena [30–40]. There are some integrals to deal with fuzzy-interval-valued functions (in short, F-I-V-Fs), where the integrands are F-I-V-Fs. For instance, Osuna-Gomez et al. [41], and Costa et al. [42] constructed Jensen's integral inequality for F-I-V-Fs through a Kulisch–Miranker order relation, see [43]. By using the same approach, Costa and Roman-Flores also presented Minkowski and Beckenbach's inequalities, where the integrands are F-I-V-Fs. This paper is motivated by [42–44] and especially by Costa et al. [45] because they established a relation between elements of fuzzy-interval space and interval space, and introduced level-wise fuzzy order relation on fuzzy-interval space through a Kulisch–Miranker order relation defined on interval space. For more information related to fuzzy interval calculus and generalized convex F-I-V-Fs, see [46–61].

Inspired by the ongoing research work, the new class of generalized convex F-I-V-Fs is introduced, which is known as (p,s)-convex F-I-V-Fs. With the help of this class and fuzzy Riemann integral operator, we introduce Jensen, Schur, and fuzzy interval H–H type inequalities via fuzzy order relation. Moreover, we show that our results include a wide class of new and known inequalities for (p,s)-convex F-I-V-Fs and their variant forms as special cases. Some useful examples are also presented to verify the validity of our main results.

2. Definitions and Basic Results

Let $\mathcal{K}_C$ and $\mathbb{F}_C(\mathbb{R})$ be the collection of all closed and bounded intervals, and fuzzy intervals of $\mathbb{R}$. We use $\mathcal{K}_C^+$ to represent the set of all positive intervals. The collection of all Riemann integrable real-valued functions, Riemann integrable I-V-Fs and fuzzy Riemann integrable F-I-V-Fs over $[t, s]$ is denoted by $\mathcal{R}_{[t, s]}$, $\mathcal{IR}_{[t, s]}$, and $\mathcal{FR}_{([t, s])}$, respectively. For more conceptions on interval-valued functions and fuzzy interval-valued functions, see [36,42–44]. Moreover, we have:

The inclusion "$\subseteq$" means that

$$\xi \subseteq \eta \text{ if and only if, } [\xi_*, \xi^*] \subseteq [\eta_*, \eta^*], \text{ if and only if } \eta_* \leq \xi_*, \xi^* \leq \eta^*, \quad (1)$$

for all $[\boldsymbol{r}_*, \boldsymbol{r}^*]$, $[\eta_*, \eta^*] \in \mathcal{K}_C$.

Remark 1 ([43]). The relation "$\leq_I$" defined on $\mathcal{K}_C$ by

$$[\boldsymbol{r}_*, \boldsymbol{r}^*] \leq_I [\eta_*, \eta^*] \text{ if and only if } \boldsymbol{r}_* \leq \eta_*, \boldsymbol{r}^* \leq \eta^*, \quad (2)$$

for all $[\boldsymbol{r}_*, \boldsymbol{r}^*]$, $[\eta_*, \eta^*] \in \mathcal{K}_C$; it is an order relation.

Proposition 1 ([7]). Let $\mathbb{F}_C(\mathbb{R})$ be a set of fuzzy numbers. If $\xi, \omega \in \mathbb{F}_C(\mathbb{R})$, then relation "$\preccurlyeq$" defined on $\mathbb{F}_C(\mathbb{R})$ by

$$\xi \preccurlyeq \omega \text{ if and only if, } [\xi]^\varphi \leq_I [\omega]^\varphi, \text{ for all } \varphi \in [0, 1]; \quad (3)$$

this relation is known as partial order relation.

Theorem 1 ([50]). Let $\mathfrak{U} : [t, s] \subset \mathbb{R} \to \mathbb{F}_C(\mathbb{R})$ be a F-I-V-F, whose φ-levels define the family of I-V-Fs $\mathfrak{U}_\varphi : [t, s] \subset \mathbb{R} \to \mathcal{K}_C$ are given by $\mathfrak{U}_\varphi(\varkappa) = [\mathfrak{U}_*(\varkappa, \varphi), \mathfrak{U}^*(\varkappa, \varphi)]$ for all $\in [t, s]$

and for all $\varphi \in (0, 1]$. Then, $\mathfrak{U}$ is fuzzy Riemann integrable over $[t, s]$ if, and only if, $\mathfrak{U}_*(\varkappa, \varphi)$ and $\mathfrak{U}^*(\varkappa, \varphi)$ both are Riemann integrable over $[t, s]$. Moreover, if $\mathfrak{U}$ is fuzzy Riemann integrable over $[t, s]$, then

$$((FR)\int_t^s \mathfrak{U}(\varkappa)d\varkappa)^\varphi = ((R)\int_t^s \mathfrak{U}_*(\varkappa,\varphi)d\varkappa, \ (R)\int_t^s \mathfrak{U}^*(\varkappa,\varphi)d\varkappa) = (IR)\int_t^s \mathfrak{U}_\varphi(\varkappa)d\varkappa, \quad (4)$$

for all $\varphi \in (0, 1]$.

Definition 1 ([10]). Let K be a convex set. Then, F-I-V-F $\mathfrak{U}: K \to \mathbb{F}_C(\mathbb{R})$ is named as a convex F-I-V-F on K if the coming inequality

$$\mathfrak{U}(\zeta\varkappa + (1-\zeta)y) \preccurlyeq \zeta\mathfrak{U}(\varkappa)\widetilde{+}(1-\zeta)\mathfrak{U}(y) \quad (5)$$

is valid for all $\varkappa, y \in K$, $\zeta \in [0, 1]$, where $\mathfrak{U}(\varkappa) \succcurlyeq \widetilde{0}$. If (5) is reversed, then $\mathfrak{U}$ is named as a concave on $[t, s]$. $\mathfrak{U}$ is affine if and only if it is both a convex and concave function.

Definition 2. Let K_p be a p-convex set and $s \in [0, 1]$. Then, F-I-V-F $\mathfrak{U}: K_p \to \mathbb{F}_C(\mathbb{R})$ is named as a (p,s)-convex F-I-V-F in the second sense on K_p such that

$$\mathfrak{U}\left([\zeta\varkappa^p + (1-\zeta)y^p]^{\frac{1}{p}}\right) \preccurlyeq \zeta^s\mathfrak{U}(\varkappa)\widetilde{+}(1-\zeta)^s\mathfrak{U}(y), \quad (6)$$

for all $\varkappa, y \in K_p$, $\zeta \in [0, 1]$, where $\mathfrak{U}(\varkappa) \succcurlyeq \widetilde{0}$. If (6) is reversed, then $\mathfrak{U}$ is named as a (p,s)-concave F-I-V-F in the second sense on $[t, s]$. $\mathfrak{U}$ is (p,s)-affine if and only if it is both (p,s)-convex and (p,s)-concave F-I-V-F in the second sense.

Remark 2. The (p,s)-convex F-I-V-Fs in the second sense have some very nice properties similar to convex F-I-V-F:

- If we attempt to take $\mathfrak{U}$ as (p,s)-convex F-I-V-F, then we can obtain that $Y\mathfrak{U}$ is also (p,s)-convex F-I-V-F, for $Y \geq 0$;
- If we attempt to take both $\mathcal{F}$ and $\mathfrak{U}$ both as (p,s)-convex F-I-V-Fs, then we can obtain that $\max(\mathcal{F}(\varkappa), \mathfrak{U}(\varkappa))$ is also a (p,s)-convex F-I-V-F.

We now discuss some new and known special cases of (p,s)-convex F-I-V-Fs in the second sense:

- If we attempt to take $s \equiv 1$, then from (p,s)-convex F-I-V-F, we achieve p-convex F-I-V-F, that is

$$\mathfrak{U}\left([\zeta\varkappa^p + (1-\zeta)y^p]^{\frac{1}{p}}\right) \preccurlyeq \zeta\mathfrak{U}(\varkappa)\widetilde{+}(1-\zeta)\mathfrak{U}(y), \ \forall \ \varkappa, y \in K, \zeta \in [0, 1]. \quad (7)$$

- If we attempt to take $p \equiv 1$, then from (p,s)-convex F-I-V-F, we achieve s-convex F-I-V-F, see [13]; that is,

$$\mathfrak{U}(\zeta\varkappa + (1-\zeta)y) \preccurlyeq \zeta^s\mathfrak{U}(\varkappa)\widetilde{+}(1-\zeta)^s\mathfrak{U}(y), \ \forall \varkappa, y \in K, \zeta \in [0, 1], s \in [0, 1]. \quad (8)$$

- If we attempt to take $p \equiv 1$ and $s \equiv 1$, then from (p,s)-convex F-I-V-F, we achieve convex F-I-V-F, see [13,36], that is

$$\mathfrak{U}(\zeta\varkappa + (1-\zeta)y) \preccurlyeq \zeta\mathfrak{U}(\varkappa)\widetilde{+}(1-\zeta)\mathfrak{U}(y), \ \forall \ \varkappa, y \in K, \zeta \in [0, 1]. \quad (9)$$

Theorem 2. Let K_p be p-convex set and $\mathfrak{U}: K_p \to \mathbb{F}_C(\mathbb{R})$ be a F-I-V-F, whose φ-levels define the family of IVFs $\mathfrak{U}_\varphi: K_p \subset \mathbb{R} \to \mathcal{K}_C^+ \subset \mathcal{K}_C$ are given by

$$\mathfrak{U}_\varphi(\varkappa) = [\mathfrak{U}_*(\varkappa, \varphi), \mathfrak{U}^*(\varkappa, \varphi)], \quad (10)$$

for all $\in K_p$ and for all $\varphi \in [0, 1]$. Then, $\mathfrak{U}$ is (p,s)-convex F-I-V-F in the second sense on K_p, if and only if, for all $\varphi \in [0, 1]$, $\mathfrak{U}_*(\varkappa, \varphi)$ and $\mathfrak{U}^*(\varkappa, \varphi)$ both are (p,s)-convex functions in the second sense.

Proof. Assume that, for each $\varphi \in [0, 1]$, $\mathfrak{U}_*(\varkappa, \varphi)$ and $\mathfrak{U}^*(\varkappa, \varphi)$ are (p,s)-convex function in the second sense on K_p. Then, from Equation (6), we have

$$\mathfrak{U}_*\left([\zeta\varkappa^p + (1-\zeta)y^p]^{\frac{1}{p}}, \varphi\right) \leq \zeta^s \mathfrak{U}_*(\varkappa, \varphi) + (1-\zeta)^s \mathfrak{U}_*(y, \varphi), \forall \varkappa, y \in K_p, \zeta \in [0, 1],$$

and

$$\mathfrak{U}^*\left([\zeta\varkappa^p + (1-\zeta)y^p]^{\frac{1}{p}}, \varphi\right) \leq \zeta^s \mathfrak{U}^*(\varkappa, \varphi) + (1-\zeta)^s \mathfrak{U}^*(y, \varphi), \forall \varkappa, y \in K_p, \zeta \in [0, 1].$$

Then, by Equation (10), we obtain

$$\mathfrak{U}_\varphi\left([\zeta\varkappa^p + (1-\zeta)y^p]^{\frac{1}{p}}\right) = \left[\mathfrak{U}_*\left([\zeta\varkappa^p + (1-\zeta)y^p]^{\frac{1}{p}}, \varphi\right), \mathfrak{U}^*\left([\zeta\varkappa^p + (1-\zeta)y^p]^{\frac{1}{p}}, \varphi\right)\right],$$

$$\leq_I [\zeta^s \mathfrak{U}_*(\varkappa, \varphi), \zeta^s \mathfrak{U}^*(\varkappa, \varphi)] + [(1-\zeta)^s \mathfrak{U}_*(y, \varphi), (1-\zeta)^s \mathfrak{U}^*(y, \varphi)],$$

that is

$$\mathfrak{U}\left([\zeta\varkappa^p + (1-\zeta)y^p]^{\frac{1}{p}}\right) \preccurlyeq \zeta^s \mathfrak{U}(\varkappa) \widetilde{+} (1-\zeta)^s \mathfrak{U}(y), \forall \varkappa, y \in K_p, \zeta \in [0, 1].$$

Hence, $\mathfrak{U}$ is (p,s)-convex F-I-V-F in the second sense on K_p.

Conversely, let $\mathfrak{U}$ be (p,s)-convex F-I-V-F in the second sense on K_p. Then, for all $\varkappa, y \in K_p$ and $\zeta \in [0, 1]$, we have

$$\mathfrak{U}\left([\zeta\varkappa^p + (1-\zeta)y^p]^{\frac{1}{p}}\right) \preccurlyeq \zeta^s \mathfrak{U}(\varkappa) \widetilde{+} (1-\zeta)^s \mathfrak{U}(y).$$

Therefore, from Equation (10), we have

$$\mathfrak{U}_\varphi\left([\zeta\varkappa^p + (1-\zeta)y^p]^{\frac{1}{p}}\right) = \left[\mathfrak{U}_*\left([\zeta\varkappa^p + (1-\zeta)y^p]^{\frac{1}{p}}, \varphi\right), \mathfrak{U}^*\left([\zeta\varkappa^p + (1-\zeta)y^p]^{\frac{1}{p}}, \varphi\right)\right].$$

Again, from Equation (10), we obtain

$$\zeta^s \mathfrak{U}_\varphi(\varkappa) \widetilde{+} (1-\zeta)^s \mathfrak{U}_\varphi(\varkappa) = [\zeta^s \mathfrak{U}_*(\varkappa, \varphi), \zeta^s \mathfrak{U}^*(\varkappa, \varphi)] + [(1-\zeta)^s \mathfrak{U}_*(y, \varphi), (1-\zeta)^s \mathfrak{U}^*(y, \varphi)],$$

Then, by (p,s)-convexity in the second sense of $\mathfrak{U}$, we have

$$\mathfrak{U}_*\left([\zeta\varkappa^p + (1-\zeta)y^p]^{\frac{1}{p}}, \varphi\right) \leq \zeta^s \mathfrak{U}_*(\varkappa, \varphi) + (1-\zeta)^s \mathfrak{U}_*(y, \varphi),$$

and

$$\mathfrak{U}^*\left([\zeta\varkappa^p + (1-\zeta)y^p]^{\frac{1}{p}}, \varphi\right) \leq \zeta^s \mathfrak{U}^*(\varkappa, \varphi) + (1-\zeta)^s \mathfrak{U}^*(y, \varphi),$$

for each $\varphi \in [0, 1]$. Hence, the result follows. □

Remark 3. On the basis of Theorem 2, we consider the special situation as below:

- If we attempt to take $\mathfrak{U}_*(\varkappa, \varphi) = \mathfrak{U}^*(\varkappa, \varphi)$ with $\varphi = 1$, then from Definition 2, we obtain the (p,s)-convex function, see [46];
- If we attempt to take $\mathfrak{U}_*(\varkappa, \varphi) = \mathfrak{U}^*(\varkappa, \varphi)$ with $\varphi = 1$ and $s = 1$, then from Definition 2, we obtain the p-convex function, see [9];

- If we attempt to take $\mathfrak{U}_*(\varkappa, \varphi) = \mathfrak{U}^*(\varkappa, \varphi)$ with $\varphi = 1$, $p = 1$ and $s = 0$, then from Definition 2, we obtain the P-function, see [47].

Example 1. We consider the F-I-V-F $\mathfrak{U} : [0, 1] \to \mathbb{F}_C(\mathbb{R})$ defined by

$$\mathfrak{U}(\varkappa)(\sigma) = \begin{cases} \frac{\sigma}{2\varkappa^p} & \sigma \in [0, 2\varkappa^p] \\ \frac{4\varkappa^p - \sigma}{2\varkappa^2} & \sigma \in (2\varkappa^p, 4\varkappa^p] \\ 0 & otherwise, \end{cases} \quad (11)$$

Then, for each $\varphi \in [0, 1]$, we have $\mathfrak{U}_\varphi(\varkappa) = [2\varphi\varkappa^p, (4 - 2\varphi)\varkappa^p]$. Since end point functions $\mathfrak{U}_*(\varkappa, \varphi)$ and $\mathfrak{U}^*(\varkappa, \varphi)$, both are (p, s)-convex functions in the second sense for each $\varphi \in [0, 1]$ and $s \in [0, 1]$. Hence, $\mathfrak{U}(\varkappa)$ is (p, s)-convex F-I-V-F in the second sense.

3. Discrete Inequalities for (p, s)-Convex F-I-V-F in the Second Sense

In the following, we establish the following result:

Theorem 3. (Discrete Jensen type inequality for (p, s)-convex F-I-V-F) Let $\omega_j \in \mathbb{R}^+$, $t_j \in [t, s]$, $(j = 1, 2, 3, \ldots, k, k \geq 2)$ and $\mathfrak{U} : [t, s] \to \mathbb{F}_C(\mathbb{R})$ be a (p, s)-convex F-I-V-F, whose φ-levels define the family of I-V-Fs $\mathfrak{U}_\varphi : [t, s] \subset \mathbb{R} \to \mathcal{K}_C^+$ are given by $\mathfrak{U}_\varphi(\varkappa) = [\mathfrak{U}_*(\varkappa, \varphi), \mathfrak{U}^*(\varkappa, \varphi)]$ for all $\in [t, s]$ and for all $\varphi \in [0, 1]$, then

$$\mathfrak{U}\left(\left[\frac{1}{W_k}\sum_{j=1}^{k}\omega_j t_j^p\right]^{\frac{1}{p}}\right) \preccurlyeq \sum_{j}^{k}\left(\frac{\omega_j}{W_k}\right)^s \mathfrak{U}(t_j), \quad (12)$$

where $W_k = \sum_{j=1}^{k}\omega_j$. If $\mathfrak{U}$ is (p, s)-concave F-I-V-F, then inequality Equation (29) is reversed.

Proof. When $k = 2$, then inequality Equation (12) is true. Considering that inequality Equation (29) is true for $k = n - 1$, then

$$\mathfrak{U}\left(\left[\frac{1}{W_{n-1}}\sum_{j=1}^{n-1}\omega_j t_j^p\right]^{\frac{1}{p}}\right) \preccurlyeq \sum_{j=1}^{n-1}\left(\frac{\omega_j}{W_{n-1}}\right)^s \mathfrak{U}(t_j)$$

Now, let us prove that inequality (12) holds for $k = n$.

$$\mathfrak{U}\left(\left[\frac{1}{W_n}\sum_{j=1}^{n}\omega_j t_j^p\right]^{\frac{1}{p}}\right)$$

$$= \mathfrak{U}\left(\left[\frac{W_{n-2}}{W_n}\frac{1}{W_{n-2}}\sum_{j=1}^{n-2}\omega_j t_j^p + \frac{\omega_{n-1} + \omega_n}{W_n}\left(\frac{\omega_{n-1}}{\omega_{n-1} + \omega_n}t_{n-1}^p + \frac{\omega_n}{\omega_{n-1} + \omega_n}t_n^p\right)\right]^{\frac{1}{p}}\right).$$

Therefore, for each $\varphi \in [0, 1]$, we have

$$\mathfrak{U}_*\left(\left[\frac{1}{W_n}\sum_{j=1}^{n}\omega_j t_j^p\right]^{\frac{1}{p}}, \varphi\right)$$

$$\mathfrak{U}^*\left(\left[\frac{1}{W_n}\sum_{j=1}^{n}\omega_j t_j^p\right]^{\frac{1}{p}}, \varphi\right)$$

$$= \mathfrak{U}_*\left(\left[\frac{W_{n-2}}{W_n}\frac{1}{W_{n-2}}\sum_{j=1}^{n-2}\omega_j t_j^p + \frac{\omega_{n-1}+\omega_n}{W_n}\left(\frac{\omega_{n-1}}{\omega_{n-1}+\omega_n}t_{n-1}^p + \frac{\omega_n}{\omega_{n-1}+\omega_n}t_n^p\right)\right]^{\frac{1}{p}},\varphi\right)$$

$$= \mathfrak{U}^*\left(\left[\frac{W_{n-2}}{W_n}\frac{1}{W_{n-2}}\sum_{j=1}^{n-2}\omega_j t_j^p + \frac{\omega_{n-1}+\omega_n}{W_n}\left(\frac{\omega_{n-1}}{\omega_{n-1}+\omega_n}t_{n-1}^p + \frac{\omega_n}{\omega_{n-1}+\omega_n}t_n^p\right)\right]^{\frac{1}{p}},\varphi\right)$$

$$\leq \sum_{j=1}^{n-2}\left(\frac{\omega_j}{W_n}\right)^s \mathfrak{U}_*(t_j,\varphi) + \left(\frac{\omega_{n-1}+\omega_n}{W_n}\right)^s \mathfrak{U}_*\left(\left[\frac{\omega_{n-1}}{\omega_{n-1}+\omega_n}t_{n-1}^p + \frac{\omega_n}{\omega_{n-1}+\omega_n}t_n^p\right]^{\frac{1}{p}},\varphi\right)$$

$$\leq \sum_{j=1}^{n-2}\left(\frac{\omega_j}{W_n}\right)^s \mathfrak{U}^*(t_j,\varphi) + \left(\frac{\omega_{n-1}+\omega_n}{W_n}\right)^s \mathfrak{U}^*\left(\left[\frac{\omega_{n-1}}{\omega_{n-1}+\omega_n}t_{n-1}^p + \frac{\omega_n}{\omega_{n-1}+\omega_n}t_n^p\right]^{\frac{1}{p}},\varphi\right)$$

$$\leq \sum_{j=1}^{n-2}\left(\frac{\omega_j}{W_n}\right)^s \mathfrak{U}_*(t_j,\varphi) + \left(\frac{\omega_{n-1}+\omega_n}{W_n}\right)^s \left[\left(\frac{\omega_{n-1}}{\omega_{n-1}+\omega_n}\right)^s \mathfrak{U}_*(t_{n-1},\varphi) + \left(\frac{\omega_n}{\omega_{n-1}+\omega_n}\right)^s \mathfrak{U}_*(t_n,\varphi)\right]$$

$$\leq \sum_{j=1}^{n-2}\left(\frac{\omega_j}{W_n}\right)^s \mathfrak{U}^*(t_j,\varphi) + \left(\frac{\omega_{n-1}+\omega_n}{W_n}\right)^s \left[\left(\frac{\omega_{n-1}}{\omega_{n-1}+\omega_n}\right)^s \mathfrak{U}^*(t_{n-1},\varphi) + \left(\frac{\omega_n}{\omega_{n-1}+\omega_n}\right)^s \mathfrak{U}^*(t_n,\varphi)\right]$$

$$\leq \sum_{j=1}^{n-2}\left(\frac{\omega_j}{W_n}\right)^s \mathfrak{U}_*(t_j,\varphi) + \left[\left(\frac{\omega_{n-1}}{W_n}\right)^s \mathfrak{U}_*(t_{n-1},\varphi) + \left(\frac{\omega_n}{W_n}\right)^s \mathfrak{U}_*(t_n,\varphi)\right]$$

$$\leq \sum_{j=1}^{n-2}\left(\frac{\omega_j}{W_n}\right)^s \mathfrak{U}^*(t_j,\varphi) + \left[\left(\frac{\omega_{n-1}}{W_n}\right)^s \mathfrak{U}^*(t_{n-1},\varphi) + \left(\frac{\omega_n}{W_n}\right)^s \mathfrak{U}^*(t_n,\varphi)\right]$$

$$= \sum_{j=1}^{n}\left(\frac{\omega_j}{W_n}\right)^s \mathfrak{U}_*(t_j,\varphi)$$

$$= \sum_{j=1}^{n}\left(\frac{\omega_j}{W_n}\right)^s \mathfrak{U}^*(t_j,\varphi).$$

From which, we have

$$\left[\mathfrak{U}_*\left(\left[\frac{1}{W_n}\sum_{j=1}^{n}\omega_j t_j^p\right]^{\frac{1}{p}},\varphi\right), \mathfrak{U}^*\left(\left[\frac{1}{W_n}\sum_{j=1}^{n}\omega_j t_j^p\right]^{\frac{1}{p}},\varphi\right)\right]$$

$$\leq_I \left[\sum_{j=1}^{n}\left(\frac{\omega_j}{W_n}\right)^s \mathfrak{U}_*(t_j,\varphi), \sum_{j=1}^{n}\left(\frac{\omega_j}{W_n}\right)^s \mathfrak{U}^*(t_j,\varphi)\right],$$

that is,

$$\mathfrak{U}\left(\left[\frac{1}{W_n}\sum_{j=1}^{n}\omega_j t_j^p\right]^{\frac{1}{p}}\right) \preccurlyeq \sum_{j=1}^{n}\left(\frac{\omega_j}{W_n}\right)^s \mathfrak{U}(t_j),$$

and the result follows. □

If $\omega_1 = \omega_2 = \omega_3 = \cdots = \omega_k = 1$, then Theorem 3 reduces to the following result:

Corollary 1. *Let $s \in [0, 1]$ $t_j \in [t, s]$, $(j = 1, 2, 3, \ldots, k, k \geq 2)$ and $\mathfrak{U} : [t, s] \to \mathbb{F}_C(\mathbb{R})$ be a (p,s)-convex F-I-V-F, whose φ-levels define the family of I-V-Fs $\mathfrak{U}_\varphi : [t, s] \subset \mathbb{R} \to \mathcal{K}_C^+$ that are given by $\mathfrak{U}_\varphi(\varkappa) = [\mathfrak{U}_*(\varkappa,\varphi), \mathfrak{U}^*(\varkappa,\varphi)]$ for all $\in [t, s]$ and for all $\varphi \in [0, 1]$; then,*

$$\mathfrak{U}\left(\left[\frac{1}{k}\sum_{j=1}^{k}t_j^p\right]^{\frac{1}{p}}\right) \preccurlyeq \sum_{j=1}^{k}\left(\frac{1}{k}\right)^s \mathfrak{U}(t_j). \tag{13}$$

If $\mathfrak{U}$ is a (p,s)-concave F-I-V-F, then inequality Equation (13) is reversed.

The next Theorem 4 gives the Schur-type inequality for (p,s)-convex F-I-V-Fs.

Theorem 4. *(Discrete Schur-type inequality for (p,s)-convex F-I-V-F) Let $s \in [0, 1]$ and $\mathfrak{U} : [t, s] \to \mathbb{F}_C(\mathbb{R})$ be a (p,s)-convex F-I-V-F, whose φ-levels define the family of IVFs $\mathfrak{U}_\varphi : [t, s] \subset \mathbb{R} \to \mathcal{K}_C^+$ are given by $\mathfrak{U}_\varphi(\varkappa) = [\mathfrak{U}_*(\varkappa,\varphi), \mathfrak{U}^*(\varkappa,\varphi)]$ for all $\in [t, s]$ and for all $\varphi \in [0, 1]$. If $t_1, t_2, t_3 \in [t, s]$, such that $t_1 < t_2 < t_3$ and $t_3^p - t_1^p, t_3^p - t_2^p, t_2^p - t_1^p \in [0, 1]$, we have*

$$(t_3^p - t_1^p)^s \mathfrak{U}(t_2) \preccurlyeq (t_3^p - t_2^p)^s \mathfrak{U}(t_1) + (t_2^p - t_1^p)^s \mathfrak{U}(t_3). \tag{14}$$

If $\mathfrak{U}$ is a (p,s)-concave F-I-V-F, then inequality Equation (14) is reversed.

Proof. Let t_j such that $L < t_j < U$ $(j = 1, 2, 3, \ldots, k)$, $(t_3{}^p - t_1{}^p)^s \rangle 0$. Then, by hypothesis, we have

$$\left(\frac{t_3{}^p - t_2{}^p}{t_3{}^p - t_1{}^p}\right)^s = \frac{(t_3{}^p - t_2{}^p)^s}{(t_3{}^p - t_1{}^p)^s} \text{ and } \left(\frac{t_2{}^p - t_1{}^p}{t_3{}^p - t_1{}^p}\right)^s = \frac{(t_2{}^p - t_1{}^p)^s}{(t_3{}^p - t_1{}^p)^s}.$$

Consider $\zeta = \frac{t_3{}^p - t_2{}^p}{t_3{}^p - t_1{}^p}$, then $t_2{}^p = \zeta t_1{}^p + (1 - \zeta) t_3{}^p$. Since $\mathfrak{U}$ is a (p,s)-convex F-I-V-F then, by hypothesis, we have

$$\mathfrak{U}(t_2) \preccurlyeq \left(\frac{t_3{}^p - t_2{}^p}{t_3{}^p - t_1{}^p}\right)^s \mathfrak{U}(t_1) + \left(\frac{t_2{}^p - t_1{}^p}{t_3{}^p - t_1{}^p}\right)^s \mathfrak{U}(t_3).$$

Therefore, for each $\varphi \in [0, 1]$, we have

$$\mathfrak{U}_*(t_2, \varphi) \leq \left(\frac{t_3{}^p - t_2{}^p}{t_3{}^p - t_1{}^p}\right)^s \mathfrak{U}_*(t_1, \varphi) + \left(\frac{t_2{}^p - t_1{}^p}{t_3{}^p - t_1{}^p}\right)^s \mathfrak{U}_*(t_3, \varphi),$$
$$\mathfrak{U}^*(t_2, \varphi) \leq \left(\frac{t_3{}^p - t_2{}^p}{t_3{}^p - t_1{}^p}\right)^s \mathfrak{U}^*(t_1, \varphi) + \left(\frac{t_2{}^p - t_1{}^p}{t_3{}^p - t_1{}^p}\right)^s \mathfrak{U}^*(t_3, \varphi)$$

(15)

$$= \frac{(t_3{}^p - t_2{}^p)^s}{(t_3{}^p - t_1{}^p)^s} \mathfrak{U}_*(t_1, \varphi) + \frac{(t_2{}^p - t_1{}^p)^s}{(t_3{}^p - t_1{}^p)^s} \mathfrak{U}_*(t_3, \varphi)$$
$$= \frac{(t_3{}^p - t_2{}^p)^s}{(t_3{}^p - t_1{}^p)^s} \mathfrak{U}^*(t_1, \varphi) + \frac{(t_2{}^p - t_1{}^p)^s}{(t_3{}^p - t_1{}^p)^s} \mathfrak{U}^*(t_3, \varphi).$$

(16)

From Equation (16), we have

$$(t_3{}^p - t_1{}^p)^s \mathfrak{U}_*(t_2, \varphi) \leq (t_3{}^p - t_2{}^p)^s \mathfrak{U}_*(t_1, \varphi) + (t_2{}^p - t_1{}^p)^s \mathfrak{U}_*(t_3, \varphi),$$
$$(t_3{}^p - t_1{}^p)^s \mathfrak{U}^*(t_2, \varphi) \leq (t_3{}^p - t_2{}^p)^s \mathfrak{U}^*(t_1, \varphi) + (t_2{}^p - t_1{}^p)^s \mathfrak{U}^*(t_3, \varphi),$$

that is

$$[(t_3{}^p - t_1{}^p)^s \mathfrak{U}_*(t_2, \varphi), (t_3{}^p - t_1{}^p)^s \mathfrak{U}^*(t_2, \varphi)]$$
$$\leq_I [(t_3{}^p - t_2{}^p)^s \mathfrak{U}_*(t_1, \varphi) + (t_2{}^p - t_1{}^p)^s \mathfrak{U}_*(t_3, \varphi), (t_3{}^p - t_2{}^p)^s \mathfrak{U}^*(t_1, \varphi) + (t_2{}^p - t_1{}^p)^s \mathfrak{U}^*(t_3, \varphi)].$$

Hence,

$$(t_3{}^p - t_1{}^p)^s \mathfrak{U}(t_2) \preccurlyeq (t_3{}^p - t_2{}^p)^s \mathfrak{U}(t_1) + (t_2{}^p - t_1{}^p)^s \mathfrak{U}(t_3).$$

□

A refinement of Jensen type inequality for (p,s)-convex F-I-V-F is given in the following theorem.

Theorem 5. Let $s \in [0, 1]$, $\omega_j \in \mathbb{R}^+$, $t_j \in [t, s]$, $(j = 1, 2, 3, \ldots, k, k \geq 2)$ and $\mathfrak{U} : [t, s] \to \mathbb{F}_C(\mathbb{R})$ be a (p,s)-convex F-I-V-F, whose φ-levels define the family of I-V-Fs $\mathfrak{U}_\varphi : [t, s] \subset \mathbb{R} \to \mathcal{K}_C^+$ are given by $\mathfrak{U}_\varphi(\varkappa) = [\mathfrak{U}_*(\varkappa, \varphi), \mathfrak{U}^*(\varkappa, \varphi)]$ for all $\in [t, s]$ and for all $\varphi \in [0, 1]$. If $(L, U) \subseteq [t, s]$, then

$$\sum_{j=1}^k \left(\frac{\omega_j}{W_k}\right)^s \mathfrak{U}(t_j) \preccurlyeq \sum_{j=1}^k \left(\left(\frac{U^p - t_j{}^p}{U^p - L^p}\right)^s \left(\frac{\omega_j}{W_k}\right)^s \mathfrak{U}(L, \varphi) + \left(\frac{t_j{}^p - L^p}{U^p - L^p}\right)^s \left(\frac{\omega_j}{W_k}\right)^s \mathfrak{U}(U, \varphi)\right), \quad (17)$$

where $W_k = \sum_{j=1}^k \omega_j$. If $\mathfrak{U}$ is (p,s)-concave F-I-V-F, then inequality Equation (17) is reversed.

Proof. Consider t_j such that $L < t_j < U$ $(j = 1, 2, 3, \ldots, k)$. Then, by hypothesis and inequality Equation (15), we have

$$\mathfrak{U}(t_j) \leq \left(\frac{U^p - t_j{}^p}{U^p - L^p}\right)^s \mathfrak{U}(L, \varphi) + \left(\frac{t_j{}^p - L^p}{U^p - L^p}\right)^s \mathfrak{U}(U, \varphi).$$

Therefore, for each $\varphi \in [0, 1]$, we have

$$\mathfrak{U}_*(t_j, \varphi) \leq \left(\frac{U^p - t_j^p}{U^p - L^p}\right)^s \mathfrak{U}_*(L, \varphi) + \left(\frac{t_j^p - L^p}{U^p - L^p}\right)^s \mathfrak{U}_*(U, \varphi),$$
$$\mathfrak{U}^*(t_j, \varphi) \leq \left(\frac{U^p - t_j^p}{U^p - L^p}\right)^s \mathfrak{U}^*(L, \varphi) + \left(\frac{t_j^p - L^p}{U^p - L^p}\right)^s \mathfrak{U}^*(U, \varphi).$$

The above inequality can be written as

$$\left(\frac{\omega_j}{W_k}\right)^s \mathfrak{U}_*(t_j, \varphi) \leq \left(\frac{U^p - t_j^p}{U^p - L^p}\right)^s \left(\frac{\omega_j}{W_k}\right)^s \mathfrak{U}_*(L, \varphi) + \left(\frac{t_j^p - L^p}{U^p - L^p}\right)^s \left(\frac{\omega_j}{W_k}\right)^s \mathfrak{U}_*(U, \varphi),$$
$$\left(\frac{\omega_j}{W_k}\right)^s \mathfrak{U}^*(t_j, \varphi) \leq \left(\frac{U^p - t_j^p}{U^p - L^p}\right)^s \left(\frac{\omega_j}{W_k}\right)^s \mathfrak{U}^*(L, \varphi) + \left(\frac{t_j^p - L^p}{U^p - L^p}\right)^s \left(\frac{\omega_j}{W_k}\right)^s \mathfrak{U}^*(U, \varphi) \quad (18)$$

Taking the sum of all inequalities (18) for $j = 1, 2, 3, \ldots, k$, we have

$$\sum_{j=1}^{k} \left(\frac{\omega_j}{W_k}\right)^s \mathfrak{U}_*(t_j, \varphi) \leq \sum_{j=1}^{k} \left(\left(\frac{U^p - t_j^p}{U^p - L^p}\right)^s \left(\frac{\omega_j}{W_k}\right)^s \mathfrak{U}_*(L, \varphi) + \left(\frac{t_j^p - L^p}{U^p - L^p}\right)^s \left(\frac{\omega_j}{W_k}\right)^s \mathfrak{U}_*(U, \varphi)\right),$$
$$\sum_{j=1}^{k} \left(\frac{\omega_j}{W_k}\right)^s \mathfrak{U}^*(t_j, \varphi) \leq \sum_{j=1}^{k} \left(\left(\frac{U^p - t_j^p}{U^p - L^p}\right)^s \left(\frac{\omega_j}{W_k}\right)^s \mathfrak{U}^*(L, \varphi) + \left(\frac{t_j^p - L^p}{U^p - L^p}\right)^s \left(\frac{\omega_j}{W_k}\right)^s \mathfrak{U}^*(U, \varphi)\right),$$

that is

$$\sum_{j=1}^{k} \left(\frac{\omega_j}{W_k}\right)^s \mathfrak{U}_\varphi(t_j) = \left[\sum_{j=1}^{k} \left(\frac{\omega_j}{W_k}\right)^s \mathfrak{U}_*(t_j, \varphi), \sum_{j=1}^{k} \left(\frac{\omega_j}{W_k}\right)^s \mathfrak{U}^*(t_j, \varphi)\right]$$

$$\preceq_I \left[\sum_{j=1}^{k} \left(\begin{array}{c}\left(\frac{U^p - t_j^p}{U^p - L^p}\right)^s \left(\frac{\omega_j}{W_k}\right)^s \mathfrak{U}_*(L, \varphi) \\ + \left(\frac{t_j^p - L^p}{U^p - L^p}\right)^s \left(\frac{\omega_j}{W_k}\right)^s \mathfrak{U}_*(U, \varphi)\end{array}\right), \sum_{j=1}^{k} \left(\begin{array}{c}\left(\frac{U^p - t_j^p}{U^p - L^p}\right)^s \left(\frac{\omega_j}{W_k}\right)^s \mathfrak{U}^*(L, \varphi) \\ + \left(\frac{t_j^p - L^p}{U^p - L^p}\right)^s \left(\frac{\omega_j}{W_k}\right)^s \mathfrak{U}^*(U, \varphi)\end{array}\right)\right]$$

$$= \sum_{j=1}^{k} \left(\frac{U^p - t_j^p}{U^p - L^p}\right)^s \left(\frac{\omega_j}{W_k}\right)^s [\mathfrak{U}_*(L, \varphi), \mathfrak{U}^*(L, \varphi)] + \sum_{j=1}^{k} \left(\frac{t_j^p - L^p}{U^p - L^p}\right)^s \left(\frac{\omega_j}{W_k}\right)^s [\mathfrak{U}_*(U, \varphi), \mathfrak{U}^*(U, \varphi)]$$

$$= \sum_{j=1}^{k} \left(\frac{U^p - t_j^p}{U^p - L^p}\right)^s \left(\frac{\omega_j}{W_k}\right)^s \mathfrak{U}_\varphi(L) + \sum_{j=1}^{k} \left(\frac{t_j^p - L^p}{U^p - L^p}\right)^s \left(\frac{\omega_j}{W_k}\right)^s \mathfrak{U}_\varphi(U).$$

Thus,

$$\sum_{j=1}^{k} \left(\frac{\omega_j}{W_k}\right)^s \mathfrak{U}(t_j) \preceq \sum_{j=1}^{k} \left(\left(\frac{U^p - t_j^p}{U^p - L^p}\right)^s \left(\frac{\omega_j}{W_k}\right)^s \mathfrak{U}(L) + \left(\frac{t_j^p - L^p}{U^p - L^p}\right)^s \left(\frac{\omega_j}{W_k}\right)^s \mathfrak{U}(U)\right),$$

and this completes the proof. □

We now consider some special cases of Theorems 3 and 5.

If $\mathfrak{U}_*(\varkappa, \varphi) = \mathfrak{U}_*(\varkappa, \varphi)$, then Theorems 3 and 5 reduce to the following results:

Corollary 2 ([21]). *(Jensen inequality for (p, s)-convex function) Let $s \in [0, 1]$, $\omega_j \in \mathbb{R}^+$, $t_j \in [t, s]$, $(j = 1, 2, 3, \ldots, k, k \geq 2)$ and let $\mathfrak{U} : [t, s] \to \mathbb{R}^+$ be a non-negative real-valued function. If $\mathfrak{U}$ is a (p, s)-convex function, then*

$$\mathfrak{U}\left(\left[\frac{1}{W_k} \sum_{j=1}^{k} \omega_j t_j^p\right]^{\frac{1}{p}}\right) \leq \sum_{j=1}^{k} \left(\frac{\omega_j}{W_k}\right)^s \mathfrak{U}(t_j), \quad (19)$$

where $W_k = \sum_{j=1}^{k} \omega_j$. If $\mathfrak{U}$ is (p, s)-concave function, then inequality (19) is reversed.

Corollary 3. Let $s \in [0, 1], \omega_j \in \mathbb{R}^+, t_j \in [t, s], (j = 1, 2, 3, \ldots, k, k \geq 2)$, and $\mathfrak{U} : [t, s] \to \mathbb{R}^+$ be a non-negative real-valued function. If $\mathfrak{U}$ is a (p, s)-convex function and $t_1, t_2, \ldots, t_j \in (L, U) \subseteq [t, s]$, then

$$\sum_{j=1}^{k} \left(\frac{\omega_j}{W_k}\right)^s \mathfrak{U}(t_j) \leq \sum_{j=1}^{k} \left(\left(\frac{U^p - t_j^p}{U^p - L^p}\right)^s \left(\frac{\omega_j}{W_k}\right)^s \mathfrak{U}(L) + \left(\frac{t_j^p - L^p}{U^p - L^p}\right)^s \left(\frac{\omega_j}{W_k}\right)^s \mathfrak{U}(U)\right), \quad (20)$$

where $W_k = \sum_{j=1}^{k} \omega_j$. If $\mathfrak{U}$ is a (p, s)-concave function, then inequality (20) is reversed.

4. Hermite–Hadamard Type Inequalities for (p, s)-Convex F-I-V-F in the Second Sense

In this section, we will continue with the H–H inequality for (p, s)-convex fuzzy-I-V-Fs as well as the fuzzy-interval H–H Fejér inequality for (p, s)-convex fuzzy-I-V-Fs using the fuzzy order relation. Firstly, we start with the following H–H inequality for (p, s)-convex fuzzy-I-V-Fs:

Theorem 6. Let $\mathfrak{U} : [t, s] \to \mathbb{F}_C(\mathbb{R})$ be a (p, s)-convex F-I-V-F, whose φ-levels define the family of I-V-Fs. $\mathfrak{U}_\varphi : [t, s] \subset \mathbb{R} \to \mathcal{K}_C^+$ are given by $\mathfrak{U}_\varphi(\varkappa) = [\mathfrak{U}_*(\varkappa, \varphi), \mathfrak{U}^*(\varkappa, \varphi)]$ for all $\in [t, s]$ and for all $\varphi \in [0, 1]$. If $\mathfrak{U} \in \mathcal{FR}_{([t, s])}$, then

$$2^{s-1} \mathfrak{U}\left(\left[\frac{t^p + s^p}{2}\right]^{\frac{1}{p}}\right) \preccurlyeq \frac{p}{s^p - t^p} (FR) \int_t^s \varkappa^{p-1} \mathfrak{U}(\varkappa) d\varkappa \preccurlyeq_p \frac{\mathfrak{U}(t) \widetilde{+} \mathfrak{U}(s)}{s + 1}. \quad (21)$$

If $\mathfrak{U}$ is a (p, s)-concave F-I-V-F, then

$$2^{s-1} \mathfrak{U}\left(\left[\frac{t^p + s^p}{2}\right]^{\frac{1}{p}}\right) \succcurlyeq \frac{p}{s^p - t^p} (FR) \int_t^s \varkappa^{p-1} \mathfrak{U}(\varkappa) d\varkappa \succcurlyeq \frac{\mathfrak{U}(t) \widetilde{+} \mathfrak{U}(s)}{s + 1}. \quad (22)$$

Proof. Let $\mathfrak{U}$ be a (p, s)-convex F-I-V-F. Then, by hypothesis, we have

$$2^s \mathfrak{U}\left(\left[\frac{t^p + s^p}{2}\right]^{\frac{1}{p}}\right) \preccurlyeq \mathfrak{U}\left([\zeta t^p + (1-\zeta)s^p]^{\frac{1}{p}}\right) \widetilde{+} \mathfrak{U}\left([(1-\zeta)t^p + \zeta s^p]^{\frac{1}{p}}\right).$$

Therefore, for each $\varphi \in [0, 1]$, we have

$$2^s \mathfrak{U}_*\left(\left[\frac{t^p+s^p}{2}\right]^{\frac{1}{p}}, \varphi\right) \leq \mathfrak{U}_*\left([\zeta t^p + (1-\zeta)s^p]^{\frac{1}{p}}, \varphi\right) + \mathfrak{U}_*((1-\zeta)t^p + \zeta s^p, \varphi),$$

$$2^s \mathfrak{U}^*\left(\left[\frac{t^p+s^p}{2}\right]^{\frac{1}{p}}, \varphi\right) \leq \mathfrak{U}^*\left([\zeta t^p + (1-\zeta)s^p]^{\frac{1}{p}}, \varphi\right) + \mathfrak{U}^*((1-\zeta)t^p + \zeta s^p, \varphi).$$

Then,

$$2^s \int_0^1 \mathfrak{U}_*\left(\left[\frac{t^p+s^p}{2}\right]^{\frac{1}{p}}, \varphi\right) d\zeta \leq \int_0^1 \mathfrak{U}_*\left([\zeta t^p + (1-\zeta)s^p]^{\frac{1}{p}}, \varphi\right) d\zeta + \int_0^1 \mathfrak{U}_*((1-\zeta)t^p + \zeta s^p, \varphi) d\zeta,$$

$$2^s \int_0^1 \mathfrak{U}^*\left(\left[\frac{t^p+s^p}{2}\right]^{\frac{1}{p}}, \varphi\right) d\zeta \leq \int_0^1 \mathfrak{U}^*\left([\zeta t^p + (1-\zeta)s^p]^{\frac{1}{p}}, \varphi\right) d\zeta + \int_0^1 \mathfrak{U}^*((1-\zeta)t^p + \zeta s^p, \varphi) d\zeta.$$

It follows that

$$2^{s-1} \mathfrak{U}_*\left(\left[\frac{t^p+s^p}{2}\right]^{\frac{1}{p}}, \varphi\right) \leq \frac{p}{s^p - t^p} \int_t^s \varkappa^{p-1} \mathfrak{U}_*(\varkappa, \varphi) d\varkappa,$$

$$2^{s-1} \mathfrak{U}^*\left(\left[\frac{t^p+s^p}{2}\right]^{\frac{1}{p}}, \varphi\right) \leq \frac{p}{s^p - t^p} \int_t^s \varkappa^{p-1} \mathfrak{U}^*(\varkappa, \varphi) d\varkappa.$$

That is,

$$2^{s-1}\left[\mathfrak{U}_*\left(\left[\frac{t^p+s^p}{2}\right]^{\frac{1}{p}},\varphi\right),\mathfrak{U}^*\left(\left[\frac{t^p+s^p}{2}\right]^{\frac{1}{p}},\varphi\right)\right]\leq_I \frac{p}{s^p-t^p}\left[\int_t^s \varkappa^{p-1}\mathfrak{U}_*(\varkappa,\varphi)d\varkappa, \int_t^s \varkappa^{p-1}\mathfrak{U}^*(\varkappa,\varphi)d\varkappa\right].$$

Thus,

$$2^{s-1}\mathfrak{U}\left(\left[\frac{t^p+s^p}{2}\right]^{\frac{1}{p}}\right) \preccurlyeq \frac{p}{s^p-t^p}(FR)\int_t^s \varkappa^{p-1}\mathfrak{U}(\varkappa)d\varkappa. \tag{23}$$

In a similar way as above, we have

$$\frac{p}{s^p-t^p}(FR)\int_t^s \varkappa^{p-1}\mathfrak{U}(\varkappa)d\varkappa \preccurlyeq \frac{1}{s+1}\left[\mathfrak{U}(t)\widetilde{+}\mathfrak{U}(s)\right]. \tag{24}$$

Combining Equations (23) and (24), we have

$$2^{s-1}\mathfrak{U}\left(\left[\frac{t^p+s^p}{2}\right]^{\frac{1}{p}}\right) \preccurlyeq \frac{p}{s^p-t^p}(FR)\int_t^s \varkappa^{p-1}\mathfrak{U}(\varkappa)d\varkappa \preccurlyeq \frac{1}{s+1}\left[\mathfrak{U}(t)\widetilde{+}\mathfrak{U}(s)\right].$$

Hence, we obtain the required result. □

Remark 4. On the basis of Theorem 6, we consider the certain the special situation as below:

- If we attempt to take $\mathfrak{U}_*(\varkappa,\varphi)=\mathfrak{U}^*(\varkappa,\varphi)$ with $\varphi=1$, then we achieve the (p,s)-convex function, see [9];
- If we attempt to take $s=1$, then we achieve the result for p-convex F-I-V-F-:

$$\mathfrak{U}\left(\left[\frac{t^p+s^p}{2}\right]^{\frac{1}{p}}\right) \preccurlyeq \frac{p}{s^p-t^p}(FR)\int_t^s \varkappa^{p-1}\mathfrak{U}(\varkappa)d\varkappa \preccurlyeq \frac{\mathfrak{U}(t)\widetilde{+}\mathfrak{U}(s)}{2}; \tag{25}$$

- If we attempt to take $p=1$, then we achieve the result for s-convex F-I-V-F, see [13]:

$$\mathfrak{U}\left(\frac{t+s}{2}\right) \preccurlyeq \frac{1}{s-t}(FR)\int_t^s \mathfrak{U}(\varkappa)d\varkappa \preccurlyeq \frac{\mathfrak{U}(t)\widetilde{+}\mathfrak{U}(s)}{s+1}; \tag{26}$$

- If we attempt to take $s=1$ and $p=1$, then we achieve the result for p-convex F-I-V-F, see [13]:

$$\mathfrak{U}\left(\frac{t+s}{2}\right) \preccurlyeq \frac{1}{s-t}(FR)\int_t^s \mathfrak{U}(\varkappa)d\varkappa \preccurlyeq \frac{\mathfrak{U}(t)\widetilde{+}\mathfrak{U}(s)}{2}; \tag{27}$$

- If we attempt to take $\mathfrak{U}_*(\varkappa,\varphi)=\mathfrak{U}^*(\varkappa,\varphi)$ with $\varphi=1$, then we acquire the result for classical (p,s)-convex function, see [21]:

$$2^{s-1}\mathfrak{U}\left(\left[\frac{t^p+s^p}{2}\right]^{\frac{1}{p}}\right) \leq \frac{p}{s^p-t^p}(R)\int_t^s \varkappa^{p-1}\mathfrak{U}(\varkappa)d\varkappa \leq \frac{1}{s+1}\left[\mathfrak{U}(t)\widetilde{+}\mathfrak{U}(s)\right]; \tag{28}$$

- If we attempt to take $\mathfrak{U}_*(\varkappa,\varphi)=\mathfrak{U}^*(\varkappa,\varphi)$ with $\varphi=1$ and $s=1$, then we acquire the result for classical p-convex function:

$$\mathfrak{U}\left(\left[\frac{t^p+s^p}{2}\right]^{\frac{1}{p}}\right) \leq \frac{p}{s^p-t^p}(R)\int_t^s \varkappa^{p-1}\mathfrak{U}(\varkappa)d\varkappa \leq \frac{\mathfrak{U}(t)+\mathfrak{U}(s)}{2}; \tag{29}$$

- If we attempt to take $\mathfrak{U}_*(\varkappa, \varphi) = \mathfrak{U}^*(\varkappa, \varphi)$ with, $\varphi = 1$, $p = 1$ and $s = 1$, then we acquire the result for classical convex function:

$$\mathfrak{U}\left(\frac{t+s}{2}\right) \leq \frac{1}{s-t} (R) \int_t^s \mathfrak{U}(\varkappa)d\varkappa \leq \frac{\mathfrak{U}(t) + \mathfrak{U}(s)}{2}. \quad (30)$$

Example 2. Let p be an odd number and $s \in [0, 1]$, and the F-I-V-F $\mathfrak{U} : [t, s] = [2, 3] \to \mathbb{F}_C(\mathbb{R})$ defined by

$$\mathfrak{U}(\varkappa)(\sigma) = \begin{cases} \dfrac{\sigma}{\left(2-\varkappa^{\frac{p}{2}}\right)}, & \sigma \in \left[0, 2-\varkappa^{\frac{p}{2}}\right] \\ \dfrac{2\left(2-\varkappa^{\frac{p}{2}}\right)-\sigma}{\left(2-\varkappa^{\frac{p}{2}}\right)}, & \sigma \in \left(2-\varkappa^{\frac{p}{2}}, 2\left(2-\varkappa^{\frac{p}{2}}\right)\right] \\ 0, & \text{otherwise}. \end{cases} \quad (31)$$

Then, for each $\varphi \in [0, 1]$, we have $\mathfrak{U}_\varphi(\varkappa) = \left[\varphi\left(2-\varkappa^{\frac{p}{2}}\right), (2-\varphi)\left(2-\varkappa^{\frac{p}{2}}\right)\right]$. Since end point functions $\mathfrak{U}_*(\varkappa, \varphi) = \varphi\left(2-\varkappa^{\frac{p}{2}}\right)$, $\mathfrak{U}^*(\varkappa, \varphi) = (2-\varphi)\left(2-\varkappa^{\frac{p}{2}}\right)$ are (p,s)-convex functions for each $\varphi \in [0, 1]$. Then, $\mathfrak{U}(\varkappa)$ is (p,s)-convex F-I-V-F. We now compute the following:

$$2^{s-1}\mathfrak{U}_*\left(\left[\tfrac{t^p+s^p}{2}\right]^{\frac{1}{p}}, \varphi\right) = \tfrac{4-\sqrt{10}}{2}\varphi,$$

$$2^{s-1}\mathfrak{U}^*\left(\left[\tfrac{t^p+s^p}{2}\right]^{\frac{1}{p}}, \varphi\right) = \tfrac{4-\sqrt{10}}{2}(2-\varphi),$$

$$\tfrac{p}{s^p-t^p} \int_t^s \varkappa^{p-1}\mathfrak{U}_*(\varkappa, \varphi)d\varkappa = \varphi \int_2^3 \left(2-\varkappa^{\frac{p}{2}}\right)d\varkappa = \tfrac{21}{50}\varphi,$$

$$\tfrac{p}{s^p-t^p} \int_t^s \varkappa^{p-1}\mathfrak{U}^*(\varkappa, \varphi)d\varkappa = (2-\varphi) \int_2^3 \left(2-\varkappa^{\frac{p}{2}}\right)d\varkappa = \tfrac{21}{50}(2-\varphi),$$

$$\tfrac{\mathfrak{U}_*(t, \varphi) + \mathfrak{U}_*(s, \varphi)}{s+1} = \tfrac{4-\sqrt{2}-\sqrt{3}}{2}\varphi,$$

$$\tfrac{\mathfrak{U}^*(t, \varphi) + \mathfrak{U}^*(s, \varphi)}{s+1} = \tfrac{4-\sqrt{2}-\sqrt{3}}{2}(2-\varphi),$$

for all $\varphi \in [0, 1]$. That means

$$\left[\tfrac{4-\sqrt{10}}{2}\varphi, \tfrac{4-\sqrt{10}}{2}(2-\varphi)\right] \leq_I \left[\tfrac{21}{50}\varphi, \tfrac{21}{50}(2-\varphi)\right] \leq_I \left[\tfrac{4-\sqrt{2}-\sqrt{3}}{2}\varphi, \tfrac{4-\sqrt{2}-\sqrt{3}}{2}(2-\varphi)\right], \text{ for all } \varphi \in [0, 1],$$

and the Theorem 6 has been verified.

Theorem 7. Let $\mathfrak{U} : [t, s] \to \mathbb{F}_C(\mathbb{R})$ be a (p,s)-convex F-I-V-F, whose φ-levels define the family of I-V-Fs $\mathfrak{U}_\varphi : [t, s] \subset \mathbb{R} \to \mathcal{K}_C^+$ are given by $\mathfrak{U}_\varphi(\varkappa) = [\mathfrak{U}_*(\varkappa, \varphi), \mathfrak{U}^*(\varkappa, \varphi)]$ for all $\in [t, s]$ and for all $\varphi \in [0, 1]$. If $\mathfrak{U} \in \mathcal{FR}_{([t, s])}$, then

$$4^{s-1} \mathfrak{U}\left(\left[\tfrac{t^p+s^p}{2}\right]^{\frac{1}{p}}\right) \preccurlyeq \rhd_2 \preccurlyeq \tfrac{p}{s^p-t^p} (FR) \int_t^s \varkappa^{p-1}\mathfrak{U}(\varkappa)d\varkappa \preccurlyeq \rhd_1 \preccurlyeq \tfrac{\mathfrak{U}(t)\widetilde{\mp}\mathfrak{U}(s)}{s+1}\left[\tfrac{1}{2} + \tfrac{1}{2^s}\right], \quad (32)$$

where

$$\rhd_1 = \dfrac{\tfrac{\mathfrak{U}(t)\widetilde{\mp}\mathfrak{U}(s)}{2} \widetilde{\mp} \mathfrak{U}\left(\left[\tfrac{t^p+s^p}{2}\right]^{\frac{1}{p}}\right)}{s+1}, \rhd_2 = 2^{s-2}\left[\mathfrak{U}\left(\left[\tfrac{3t^p+s^p}{4}\right]^{\frac{1}{p}}\right) \widetilde{\mp} \mathfrak{U}\left(\left[\tfrac{t^p+3s^p}{4}\right]^{\frac{1}{p}}\right)\right],$$

and $\rhd_1 = [\rhd_{1_*}, \rhd_1^*]$, $\rhd_2 = [\rhd_{2_*}, \rhd_2^*]$.

Proof. Take $\left[t^p, \frac{t^p+s^p}{2}\right]$, and we have

$$2^s \mathfrak{U}\left(\left[\frac{\zeta t^p + (1-\zeta)\frac{t^p+s^p}{2}}{2} + \frac{(1-\zeta)t^p + \zeta\frac{t^p+s^p}{2}}{2}\right]^{\frac{1}{p}}\right)$$
$$\preccurlyeq \mathfrak{U}\left(\left[\zeta t^p + (1-\zeta)\frac{t^p+s^p}{2}\right]^{\frac{1}{p}}\right) \widetilde{+} \mathfrak{U}\left(\left[(1-\zeta)t^p + \zeta\frac{t^p+s^p}{2}\right]^{\frac{1}{p}}\right).$$

Therefore, for each $\varphi \in [0,1]$, we have

$$2^s \mathfrak{U}_*\left(\left[\frac{\zeta t^p + (1-\zeta)\frac{t^p+s^p}{2}}{2} + \frac{(1-\zeta)t^p + \zeta\frac{t^p+s^p}{2}}{2}\right]^{\frac{1}{p}}, \varphi\right)$$
$$\leq \mathfrak{U}_*\left(\left[\zeta t^p + (1-\zeta)\frac{t^p+s^p}{2}\right]^{\frac{1}{p}}, \varphi\right) + \mathfrak{U}_*\left(\left[(1-\zeta)t^p + \zeta\frac{t^p+s^p}{2}\right]^{\frac{1}{p}}, \varphi\right),$$
$$2^s \mathfrak{U}^*\left(\left[\frac{\zeta t^p + (1-\zeta)\frac{t^p+s^p}{2}}{2} + \frac{(1-\zeta)t^p + \zeta\frac{t^p+s^p}{2}}{2}\right]^{\frac{1}{p}}, \varphi\right)$$
$$\leq \mathfrak{U}^*\left(\left[\zeta t^p + (1-\zeta)\frac{t^p+s^p}{2}\right]^{\frac{1}{p}}, \varphi\right) + \mathfrak{U}^*\left(\left[(1-\zeta)t^p + \zeta\frac{t^p+s^p}{2}\right]^{\frac{1}{p}}, \varphi\right).$$

Consequently, we obtain

$$2^{s-2}\mathfrak{U}_*\left(\left[\frac{3t^p+s^p}{4}\right]^{\frac{1}{p}}, \varphi\right) \leq \frac{p}{s^p-t^p} \int_t^{\frac{t^p+s^p}{2}} \varkappa^{p-1}\mathfrak{U}_*(\varkappa, \varphi)d\varkappa,$$
$$2^{s-2}\mathfrak{U}^*\left(\left[\frac{3t^p+s^p}{4}\right]^{\frac{1}{p}}, \varphi\right) \leq \frac{p}{s^p-t^p} \int_t^{\frac{t^p+s^p}{2}} \varkappa^{p-1}\mathfrak{U}^*(\varkappa, \varphi)d\varkappa.$$

That is,

$$2^{s-2}\left[\mathfrak{U}_*\left(\left[\frac{3t^p+s^p}{4}\right]^{\frac{1}{p}}, \varphi\right), \mathfrak{U}^*\left(\left[\frac{3t^p+s^p}{4}\right]^{\frac{1}{p}}, \varphi\right)\right]$$
$$\leq_I \frac{p}{s^p-t^p}\left[\int_t^{\frac{t^p+s^p}{2}} \varkappa^{p-1}\mathfrak{U}_*(\varkappa, \varphi)d\varkappa, \int_t^{\frac{t^p+s^p}{2}} \varkappa^{p-1}\mathfrak{U}^*(\varkappa, \varphi)d\varkappa\right].$$

It follows that

$$2^{s-2}\mathfrak{U}\left(\left[\frac{3t^p+s^p}{4}\right]^{\frac{1}{p}}\right) \preccurlyeq \frac{p}{s^p-t^p} \int_t^{\frac{t^p+s^p}{2}} \varkappa^{p-1}\mathfrak{U}(\varkappa)d\varkappa. \tag{33}$$

In a similar way as above, we have

$$2^{s-2}\mathfrak{U}\left(\left[\frac{t^p+3s^p}{4}\right]^{\frac{1}{p}}\right) \preccurlyeq \frac{p}{s^p-t^p} \int_{\frac{t^p+s^p}{2}}^{s} \varkappa^{p-1}\mathfrak{U}(\varkappa)d\varkappa. \tag{34}$$

Combining Equations (33) and (34), we have

$$2^{s-2}\left[\mathfrak{U}\left(\left[\frac{3t^p+s^p}{4}\right]^{\frac{1}{p}}\right) \widetilde{+} \mathfrak{U}\left(\left[\frac{t^p+3s^p}{4}\right]^{\frac{1}{p}}\right)\right] \preccurlyeq \frac{p}{s^p-t^p} \int_t^s \varkappa^{p-1}\mathfrak{U}(\varkappa)d\varkappa.$$

By using Theorem 6, we have

$$4^{s-1}\mathfrak{U}\left(\left[\frac{t^p+s^p}{2}\right]^{\frac{1}{p}}\right) = 4^{s-1}\mathfrak{U}\left(\left[\frac{1}{2}\cdot\frac{3t^p+s^p}{4} + \frac{1}{2}\cdot\frac{t^p+3s^p}{4}\right]^{\frac{1}{p}}\right).$$

Therefore, for each $\varphi \in [0, 1]$, we have

$$4^{s-1}\mathfrak{U}_*\left(\left[\frac{t^p+s^p}{2}\right]^{\frac{1}{p}}, \varphi\right) = 4^{s-1}\mathfrak{U}_*\left(\left[\frac{1}{2}\cdot\frac{3t^p+s^p}{4} + \frac{1}{2}\cdot\frac{t^p+3s^p}{4}\right]^{\frac{1}{p}}, \varphi\right),$$

$$4^{s-1}\mathfrak{U}^*\left(\left[\frac{t^p+s^p}{2}\right]^{\frac{1}{p}}, \varphi\right) = 4^{s-1}\mathfrak{U}^*\left(\left[\frac{1}{2}\cdot\frac{3t^p+s^p}{4} + \frac{1}{2}\cdot\frac{t^p+3s^p}{4}\right]^{\frac{1}{p}}, \varphi\right)$$

$$\leq 2^{s-2}\left[\mathfrak{U}_*\left(\left[\frac{3t^p+s^p}{4}\right]^{\frac{1}{p}}, \varphi\right) + \mathfrak{U}_*\left(\left[\frac{t^p+3s^p}{4}\right]^{\frac{1}{p}}, \varphi\right)\right]$$

$$\leq 2^{s-2}\left[\mathfrak{U}^*\left(\left[\frac{3t^p+s^p}{4}\right]^{\frac{1}{p}}, \varphi\right) + \mathfrak{U}^*\left(\left[\frac{t^p+3s^p}{4}\right]^{\frac{1}{p}}, \varphi\right)\right]$$

$$= \vartriangleleft_{2*}$$
$$= \vartriangleleft_2^*$$
$$\leq \frac{p}{s^p-t^p}\int_t^s \varkappa^{p-1}\mathfrak{U}_*(\varkappa, \varphi)d\varkappa$$
$$\leq \frac{p}{s^p-t^p}\int_t^s \varkappa^{p-1}\mathfrak{U}^*(\varkappa, \varphi)d\varkappa$$
$$\leq \frac{1}{s+1}\left[\frac{\mathfrak{U}_*(t,\varphi)+\mathfrak{U}_*(s,\varphi)}{2} + \mathfrak{U}_*\left(\left[\frac{t^p+s^p}{2}\right]^{\frac{1}{p}}, \varphi\right)\right]$$
$$\leq \frac{1}{s+1}\left[\frac{\mathfrak{U}^*(t,\varphi)+\mathfrak{U}^*(s,\varphi)}{2} + \mathfrak{U}^*\left(\left[\frac{t^p+s^p}{2}\right]^{\frac{1}{p}}, \varphi\right)\right]$$
$$= \vartriangleleft_{1*}$$
$$= \vartriangleleft_1^*$$
$$\leq \frac{1}{s+1}\left[\frac{\mathfrak{U}_*(t,\varphi)+\mathfrak{U}_*(s,\varphi)}{2} + \frac{1}{2^s}(\mathfrak{U}_*(t, \varphi) + \mathfrak{U}_*(s, \varphi))\right]$$
$$\leq \frac{1}{s+1}\left[\frac{\mathfrak{U}^*(t,\varphi)+\mathfrak{U}^*(s,\varphi)}{2} + \frac{1}{2^s}(\mathfrak{U}^*(t, \varphi) + \mathfrak{U}^*(s, \varphi))\right]$$
$$= \frac{1}{s+1}[\mathfrak{U}_*(t, \varphi) + \mathfrak{U}_*(s, \varphi)]\left[\frac{1}{2} + \frac{1}{2^s}\right]$$
$$= \frac{1}{s+1}[\mathfrak{U}^*(t, \varphi) + \mathfrak{U}^*(s, \varphi)]\left[\frac{1}{2} + \frac{1}{2^s}\right],$$

that is

$$4^{s-1}\mathfrak{U}\left(\left[\frac{t^p+s^p}{2}\right]^{\frac{1}{p}}\right) \preccurlyeq \vartriangleleft_2 \preccurlyeq \frac{p}{s^p-t^p}(FR)\int_t^s \varkappa^{p-1}\mathfrak{U}(\varkappa)d\varkappa \preccurlyeq \vartriangleleft_1 \preccurlyeq \frac{\mathfrak{U}(t)\widetilde{+}\mathfrak{U}(s)}{s+1}\left[\frac{1}{2} + \frac{1}{2^s}\right],$$

hence, the result follows. □

Example 3. Let p be an odd number and the F-I-V-F $\mathfrak{U}: [t, s] = [2, 3] \to \mathbb{F}_C(\mathbb{R})$ defined by,

$$\mathfrak{U}_\varphi(\varkappa) = \left[\varphi\left(2 - \varkappa^{\frac{p}{2}}\right), (2 - \varphi)\left(2 - \varkappa^{\frac{p}{2}}\right)\right],$$

as in Example 2, then $\mathfrak{U}(\varkappa)$ is (p, s)-convex F-I-V-F and satisfies Equation (21). We have

$$\mathfrak{U}_*(\varkappa, \varphi) = \varphi\left(2 - \varkappa^{\frac{p}{2}}\right) \text{ and } \mathfrak{U}^*(\varkappa, \varphi) = (2 - \varphi)\left(2 - \varkappa^{\frac{p}{2}}\right)$$

We now compute the following:

$$\frac{\mathfrak{U}_*(t,\varphi)+\mathfrak{U}_*(s,\varphi)}{s+1}\left[\frac{1}{2}+\frac{1}{2^s}\right] = \frac{4-\sqrt{2}-\sqrt{3}}{2}\varphi,$$

$$\frac{\mathfrak{U}^*(t,\varphi)+\mathfrak{U}^*(s,\varphi)}{s+1}\left[\frac{1}{2}+\frac{1}{2^s}\right] = \frac{4-\sqrt{2}-\sqrt{3}}{2}(2-\varphi),$$

$$\triangleright_{1*} = \frac{\frac{\mathfrak{U}_*(t,\varphi)+\mathfrak{U}_*(s,\varphi)}{2}+\mathfrak{U}_*\left(\left[\frac{t^p+s^p}{2}\right]^{\frac{1}{p}},\varphi\right)}{s+1} = \frac{8-\sqrt{2}-\sqrt{3}-\sqrt{10}}{4}\varphi,$$

$$\triangleright_1{}^* = \frac{\frac{\mathfrak{U}^*(t,\varphi)+\mathfrak{U}^*(s,\varphi)}{2}+\mathfrak{U}^*\left(\left[\frac{t^p+s^p}{2}\right]^{\frac{1}{p}},\varphi\right)}{s+1} = \frac{8-\sqrt{2}-\sqrt{3}-\sqrt{10}}{4}(2-\varphi),$$

$$\triangleright_{2*} = 2^{s-2}\left[\mathfrak{U}_*\left(\left[\frac{3t^p+s^p}{4}\right]^{\frac{1}{p}},\varphi\right)+\mathfrak{U}_*\left(\left[\frac{t^p+3s^p}{4}\right]^{\frac{1}{p}},\varphi\right)\right] = \frac{5-\sqrt{11}}{4}\varphi,$$

$$\triangleright_2{}^* = 2^{s-2}\left[\mathfrak{U}^*\left(\left[\frac{3t^p+s^p}{4}\right]^{\frac{1}{p}},\varphi\right)+\mathfrak{U}^*\left(\left[\frac{t^p+3s^p}{4}\right]^{\frac{1}{p}},\varphi\right)\right] = \frac{5-\sqrt{11}}{4}(2-\varphi),$$

$$4^{s-1}\mathfrak{U}_*\left(\left[\frac{t^p+s^p}{2}\right]^{\frac{1}{p}},\varphi\right) = \frac{4-\sqrt{10}}{2}\varphi,$$

$$4^{s-1}\mathfrak{U}^*\left(\left[\frac{t^p+s^p}{2}\right]^{\frac{1}{p}},\varphi\right) = \frac{4-\sqrt{10}}{2}(2-\varphi).$$

Then, we obtain that

$$\frac{4-\sqrt{10}}{2}\varphi \leq \frac{5-\sqrt{11}}{4}\varphi \leq \frac{21}{50}\varphi \leq \frac{8-\sqrt{2}-\sqrt{3}-\sqrt{10}}{4}\varphi \leq \frac{4-\sqrt{2}-\sqrt{3}}{2}\varphi,$$

$$\frac{4-\sqrt{10}}{2}(2-\varphi) \leq \frac{5-\sqrt{11}}{4}(2-\varphi) \leq \frac{21}{50}(2-\varphi) \leq \frac{8-\sqrt{2}-\sqrt{3}-\sqrt{10}}{4}(2-\varphi) \leq \frac{4-\sqrt{2}-\sqrt{3}}{2}(2-\varphi).$$

Hence, Theorem 7 is verified.

The next Theorems 8 and 9 give the second H–H Fejér inequality and the first H–H Fejér inequality for (p,s)-convex F-I-V-F, respectively.

Theorem 8. (Second H–H Fejér inequality for (p,s)-convex F-I-V-F) Let $\mathfrak{U}:[t,s] \to \mathbb{F}_C(\mathbb{R})$ be a (p,s)-convex F-I-V-F with $t < s$, whose φ-levels define the family of I-V-Fs $\mathfrak{U}_\varphi:[t,s] \subset \mathbb{R} \to \mathcal{K}_C^+$ are given by $\mathfrak{U}_\varphi(\varkappa) = [\mathfrak{U}_*(\varkappa,\varphi), \mathfrak{U}^*(\varkappa,\varphi)]$ for all $\varkappa \in [t,s]$ and for all $\varphi \in [0,1]$. If $\mathfrak{U} \in \mathcal{FR}_{([t,s])}$ and $\Psi:[t,s] \to \mathbb{R}$, $\Psi(\varkappa) \geq 0$, p-symmetric with respect to $\left[\frac{t^p+s^p}{2}\right]^{\frac{1}{p}}$, then

$$\frac{p}{s^p-t^p}(FR)\int_t^s \varkappa^{p-1}\mathfrak{U}(\varkappa)\Psi(\varkappa)d\varkappa \preccurlyeq [\mathfrak{U}(t)\widetilde{+}\mathfrak{U}(s)]\int_0^1 \zeta^s\Psi\left([(1-\zeta)t^p+\zeta s^p]^{\frac{1}{p}}\right)d\zeta. \quad (35)$$

If $\mathfrak{U}$ is (p,s)-concave F-I-V-F, then Equation (35) is reversed.

Proof. Let $\mathfrak{U}$ be a (p,s)-convex F-I-V-F. Then, for each $\varphi \in [0,1]$, we have

$$\mathfrak{U}_*\left([\zeta t^p+(1-\zeta)s^p]^{\frac{1}{p}},\varphi\right)\Psi\left([\zeta t^p+(1-\zeta)s^p]^{\frac{1}{p}}\right)$$
$$\leq (\zeta^s\mathfrak{U}_*(t,\varphi)+(1-\zeta)^s\mathfrak{U}_*(s,\varphi))\Psi\left([\zeta t^p+(1-\zeta)s^p]^{\frac{1}{p}}\right),$$
$$\mathfrak{U}^*\left([\zeta t^p+(1-\zeta)s^p]^{\frac{1}{p}},\varphi\right)\Psi\left([\zeta t^p+(1-\zeta)s^p]^{\frac{1}{p}}\right) \quad (36)$$
$$\leq (\zeta^s\mathfrak{U}^*(t,\varphi)+(1-\zeta)^s\mathfrak{U}^*(s,\varphi))\Psi\left([\zeta t^p+(1-\zeta)s^p]^{\frac{1}{p}}\right).$$

and

$$\mathfrak{U}_*\left([(1-\zeta)t^p+\zeta s^p]^{\frac{1}{p}},\varphi\right)\Psi\left([(1-\zeta)t^p+\zeta s^p]^{\frac{1}{p}}\right)$$
$$\leq ((1-\zeta)^s\mathfrak{U}_*(t,\varphi)+\zeta^s\mathfrak{U}_*(s,\varphi))\Psi\left([(1-\zeta)t^p+\zeta s^p]^{\frac{1}{p}}\right),$$
$$\mathfrak{U}^*\left([(1-\zeta)t^p+\zeta s^p]^{\frac{1}{p}},\varphi\right)\Psi\left([(1-\zeta)t^p+\zeta s^p]^{\frac{1}{p}}\right)$$
$$\leq ((1-\zeta)^s\mathfrak{U}^*(t,\varphi)+\zeta^s\mathfrak{U}^*(s,\varphi))\Psi\left([(1-\zeta)t^p+\zeta s^p]^{\frac{1}{p}}\right). \tag{37}$$

After adding Equations (36) and (37), and integrating over $[0,1]$, we get

$$\int_0^1 \mathfrak{U}_*\left([\zeta t^p+(1-\zeta)s^p]^{\frac{1}{p}},\varphi\right)\Psi\left([\zeta t^p+(1-\zeta)s^p]^{\frac{1}{p}}\right)d\zeta$$
$$+\int_0^1 \mathfrak{U}_*\left([(1-\zeta)t^p+\zeta s^p]^{\frac{1}{p}},\varphi\right)\Psi\left([(1-\zeta)t^p+\zeta s^p]^{\frac{1}{p}}\right)d\zeta$$
$$\leq \int_0^1 \begin{bmatrix}\mathfrak{U}_*(t,\varphi)\left\{\zeta^s\Psi\left([\zeta t^p+(1-\zeta)s^p]^{\frac{1}{p}}\right)+(1-\zeta)^s\Psi\left([(1-\zeta)t^p+\zeta s^p]^{\frac{1}{p}}\right)\right\}\\+\mathfrak{U}_*(s,\varphi)\left\{(1-\zeta)^s\Psi\left([\zeta t^p+(1-\zeta)s^p]^{\frac{1}{p}}\right)+\zeta^s\Psi\left([(1-\zeta)t^p+\zeta s^p]^{\frac{1}{p}}\right)\right\}\end{bmatrix}d\zeta,$$
$$\int_0^1 \mathfrak{U}^*\left([\zeta t^p+(1-\zeta)s^p]^{\frac{1}{p}},\varphi\right)\Psi\left([\zeta t^p+(1-\zeta)s^p]^{\frac{1}{p}}\right)d\zeta$$
$$+\int_0^1 \mathfrak{U}^*\left([(1-\zeta)t^p+\zeta s^p]^{\frac{1}{p}},\varphi\right)\Psi\left([(1-\zeta)t^p+\zeta s^p]^{\frac{1}{p}}\right)d\zeta$$
$$\leq \int_0^1 \begin{bmatrix}\mathfrak{U}^*(t,\varphi)\left\{\zeta^s\Psi\left([\zeta t^p+(1-\zeta)s^p]^{\frac{1}{p}}\right)+(1-\zeta)^s\Psi\left([(1-\zeta)t^p+\zeta s^p]^{\frac{1}{p}}\right)\right\}\\+\mathfrak{U}^*(s,\varphi)\left\{(1-\zeta)^s\Psi\left([\zeta t^p+(1-\zeta)s^p]^{\frac{1}{p}}\right)+\zeta^s\Psi\left([(1-\zeta)t^p+\zeta s^p]^{\frac{1}{p}}\right)\right\}\end{bmatrix}d\zeta.$$
$$=2\mathfrak{U}_*(t,\varphi)\int_0^1 \zeta^s\Psi\left([\zeta t^p+(1-\zeta)s^p]^{\frac{1}{p}}\right)d\zeta+2\mathfrak{U}_*(s,\varphi)\int_0^1 \zeta^s\Psi\left([(1-\zeta)t^p+\zeta s^p]^{\frac{1}{p}}\right)d\zeta$$
$$=2\mathfrak{U}^*(t,\varphi)\int_0^1 \zeta^s\Psi\left([\zeta t^p+(1-\zeta)s^p]^{\frac{1}{p}}\right)d\zeta+2\mathfrak{U}^*(s,\varphi)\int_0^1 \zeta^s\Psi\left([(1-\zeta)t^p+\zeta s^p]^{\frac{1}{p}}\right)d\zeta.$$

Since Ψ is symmetric, then

$$=2[\mathfrak{U}_*(t,\varphi)+\mathfrak{U}_*(s,\varphi)]\int_0^1 \zeta^s\Psi\left([(1-\zeta)t^p+\zeta s^p]^{\frac{1}{p}}\right)d\zeta$$
$$=2[\mathfrak{U}^*(t,\varphi)+\mathfrak{U}^*(s,\varphi)]\int_0^1 \zeta^s\Psi\left([(1-\zeta)t^p+\zeta s^p]^{\frac{1}{p}}\right)d\zeta. \tag{38}$$

Since

$$\int_0^1 \mathfrak{U}_*\left([\zeta t^p+(1-\zeta)s^p]^{\frac{1}{p}},\varphi\right)\Psi\left([\zeta t^p+(1-\zeta)s^p]^{\frac{1}{p}}\right)d\zeta$$
$$=\int_0^1 \mathfrak{U}_*\left([(1-\zeta)t^p+\zeta s^p]^{\frac{1}{p}},\varphi\right)\Psi\left([(1-\zeta)t^p+\zeta s^p]^{\frac{1}{p}}\right)d\zeta$$
$$=\frac{p}{s^p-t^p}\int_t^s \varkappa^{p-1}\mathfrak{U}_*(\varkappa,\varphi)\Psi(\varkappa)d\varkappa,$$
$$\int_0^1 \mathfrak{U}^*\left([\zeta t^p+(1-\zeta)s^p]^{\frac{1}{p}},\varphi\right)\Psi\left([\zeta t^p+(1-\zeta)s^p]^{\frac{1}{p}}\right)d\zeta$$
$$=\int_0^1 \mathfrak{U}^*\left([(1-\zeta)t^p+\zeta s^p]^{\frac{1}{p}},\varphi\right)\Psi\left([(1-\zeta)t^p+\zeta s^p]^{\frac{1}{p}}\right)d\zeta$$
$$=\frac{p}{s^p-t^p}\int_t^s \varkappa^{p-1}\mathfrak{U}^*(\varkappa,\varphi)\Psi(\varkappa)d\varkappa, \tag{39}$$

From Equation (39) and integrating with respect to ζ over $[0,1]$, we have

$$\frac{p}{s^p-t^p}\int_t^s \varkappa^{p-1}\mathfrak{U}_*(\varkappa,\varphi)\Psi(\varkappa)d\varkappa \leq [\mathfrak{U}_*(t,\varphi)+\mathfrak{U}_*(s,\varphi)]\int_0^1 \zeta^s\Psi\left([(1-\zeta)t^p+\zeta s^p]^{\frac{1}{p}}\right)d\zeta,$$
$$\frac{p}{s^p-t^p}\int_t^s \varkappa^{p-1}\mathfrak{U}^*(\varkappa,\varphi)\Psi(\varkappa)d\varkappa \leq [\mathfrak{U}^*(t,\varphi)+\mathfrak{U}^*(s,\varphi)]\int_0^1 \zeta^s\Psi\left([(1-\zeta)t^p+\zeta s^p]^{\frac{1}{p}}\right)d\zeta,$$

that is,

$$\frac{p}{s^p-t^p}\left[\int_t^s \varkappa^{p-1}\mathfrak{U}_*(\varkappa,\varphi)\Psi(\varkappa)d\varkappa, \int_t^s \varkappa^{p-1}\mathfrak{U}^*(\varkappa,\varphi)\Psi(\varkappa)d\varkappa\right]$$
$$\leq_I [\mathfrak{U}_*(t,\varphi)+\mathfrak{U}_*(s,\varphi), \mathfrak{U}^*(t,\varphi)+\mathfrak{U}^*(s,\varphi)]\int_0^1 \zeta^s\Psi\left([(1-\zeta)t^p+\zeta s^p]^{\frac{1}{p}}\right)d\zeta,$$

hence

$$\frac{p}{s^p - t^p} (FR) \int_t^s \varkappa^{p-1} \mathfrak{U}(\varkappa) \Psi(\varkappa) d\varkappa \preccurlyeq [\mathfrak{U}(t) \tilde{+} \mathfrak{U}(s)] \int_0^1 \zeta^s \Psi\left([(1-\zeta)t^p + \zeta s^p]^{\frac{1}{p}}\right) d\zeta.$$

□

Theorem 9. (First H–H Fejér inequality for (p, s)-convex F-I-V-F) Let $\mathfrak{U} : [t, s] \to \mathbb{F}_C(\mathbb{R})$ be a (p, s)-convex F-I-V-F with $t < s$, whose φ-levels define the family of I-V-Fs $\mathfrak{U}_\varphi : [t, s] \subset \mathbb{R} \to \mathcal{K}_C^+$ are given by $\mathfrak{U}_\varphi(\varkappa) = [\mathfrak{U}_*(\varkappa, \varphi), \mathfrak{U}^*(\varkappa, \varphi)]$ for all $\in [t, s]$ and for all $\varphi \in [0, 1]$. If $\mathfrak{U} \in \mathcal{FR}_{([t, s])}$ and $\Psi : [t, s] \to \mathbb{R}, \Psi(\varkappa) \geq 0$, p-symmetric with respect to $\left[\frac{t^p + s^p}{2}\right]^{\frac{1}{p}}$, and $\int_t^s \Psi(\varkappa) d\varkappa > 0$, then

$$2^{s-1} \mathfrak{U}\left(\left[\frac{t^p + s^p}{2}\right]^{\frac{1}{p}}\right) \preccurlyeq \frac{p}{\int_t^s \varkappa^{p-1} \Psi(\varkappa) d\varkappa} (FR) \int_t^s \varkappa^{p-1} \mathfrak{U}(\varkappa) \Psi(\varkappa) d\varkappa. \quad (40)$$

If $\mathfrak{U}$ is (p, s)-concave F-I-V-F, then inequality (40) is reversed.

Proof. Since $\mathfrak{U}$ is a (p, s)-convex F-I-V-F, then, for each $\varphi \in [0, 1]$, we have

$$\begin{aligned}
2^s \mathfrak{U}_*\left(\left[\frac{t^p + s^p}{2}\right]^{\frac{1}{p}}, \varphi\right) &\leq \mathfrak{U}_*\left([\zeta t^p + (1-\zeta)s^p]^{\frac{1}{p}}, \varphi\right) + \mathfrak{U}_*\left([(1-\zeta)t^p + \zeta s^p]^{\frac{1}{p}}, \varphi\right), \\
2^s \mathfrak{U}^*\left(\left[\frac{t^p + s^p}{2}\right]^{\frac{1}{p}}, \varphi\right) &\leq \mathfrak{U}^*\left([\zeta t^p + (1-\zeta)s^p]^{\frac{1}{p}}, \varphi\right) + \mathfrak{U}^*\left([(1-\zeta)t^p + \zeta s^p]^{\frac{1}{p}}, \varphi\right).
\end{aligned} \quad (41)$$

By multiplying Equation (41) by $\Psi\left([\zeta t^p + (1-\zeta)s^p]^{\frac{1}{p}}\right) = \Psi\left([(1-\zeta)t^p + \zeta s^p]^{\frac{1}{p}}\right)$ and integrating it by ζ over $[0, 1]$, we obtain

$$\begin{aligned}
2^s \mathfrak{U}_*\left(\left[\frac{t^p + s^p}{2}\right]^{\frac{1}{p}}, \varphi\right) \int_0^1 \Psi\left([(1-\zeta)t^p + \zeta s^p]^{\frac{1}{p}}, \varphi\right) d\zeta \\
\leq \left(\begin{array}{l} \int_0^1 \mathfrak{U}_*\left([\zeta t^p + (1-\zeta)s^p]^{\frac{1}{p}}, \varphi\right) \Psi\left([\zeta t^p + (1-\zeta)s^p]^{\frac{1}{p}}\right) d\zeta \\ + \int_0^1 \mathfrak{U}_*\left([(1-\zeta)t^p + \zeta s^p]^{\frac{1}{p}}, \varphi\right) \Psi\left([(1-\zeta)t^p + \zeta s^p]^{\frac{1}{p}}\right) d\zeta \end{array}\right), \\
2^s \mathfrak{U}^*\left(\left[\frac{t^p + s^p}{2}\right]^{\frac{1}{p}}, \varphi\right) \int_0^1 \Psi\left([(1-\zeta)t^p + \zeta s^p]^{\frac{1}{p}}\right) d\zeta \\
\leq \left(\begin{array}{l} \int_0^1 \mathfrak{U}^*\left([\zeta t^p + (1-\zeta)s^p]^{\frac{1}{p}}, \varphi\right) \Psi\left([\zeta t^p + (1-\zeta)s^p]^{\frac{1}{p}}\right) d\zeta \\ + \int_0^1 \mathfrak{U}^*\left([(1-\zeta)t^p + \zeta s^p]^{\frac{1}{p}}, \varphi\right) \Psi\left([(1-\zeta)t^p + \zeta s^p]^{\frac{1}{p}}\right) d\zeta \end{array}\right).
\end{aligned} \quad (42)$$

Since

$$\begin{aligned}
\int_0^1 \mathfrak{U}_*\left([\zeta t^p + (1-\zeta)s^p]^{\frac{1}{p}}, \varphi\right) \Psi\left([\zeta t^p + (1-\zeta)s^p]^{\frac{1}{p}}\right) d\zeta \\
= \int_0^1 \mathfrak{U}_*\left([(1-\zeta)t^p + \zeta s^p]^{\frac{1}{p}}, \varphi\right) \Psi\left([(1-\zeta)t^p + \zeta s^p]^{\frac{1}{p}}\right) d\zeta \\
= \frac{p}{s^p - t^p} \int_t^s \varkappa^{p-1} \mathfrak{U}_*(\varkappa, \varphi) \Psi(\varkappa) d\varkappa, \\
\int_0^1 \mathfrak{U}^*\left([\zeta t^p + (1-\zeta)s^p]^{\frac{1}{p}}, \varphi\right) \Psi\left([\zeta t^p + (1-\zeta)s^p]^{\frac{1}{p}}\right) d\zeta \\
= \int_0^1 \mathfrak{U}^*\left([(1-\zeta)t^p + \zeta s^p]^{\frac{1}{p}}, \varphi\right) \Psi\left([(1-\zeta)t^p + \zeta s^p]^{\frac{1}{p}}\right) d\zeta \\
= \frac{p}{s^p - t^p} \int_t^s \varkappa^{p-1} \mathfrak{U}^*(\varkappa, \varphi) \Psi(\varkappa) d\varkappa,
\end{aligned} \quad (43)$$

From Equation (43), we have

$$2^{s-1}\mathfrak{U}_*\left(\left[\frac{t^p+s^p}{2}\right]^{\frac{1}{p}},\varphi\right) \leq \frac{p}{\int_t^s \Psi(\varkappa)d\varkappa}\int_t^s \varkappa^{p-1}\mathfrak{U}_*(\varkappa,\varphi)\Psi(\varkappa)d\varkappa,$$

$$2^{s-1}\mathfrak{U}^*\left(\left[\frac{t^p+s^p}{2}\right]^{\frac{1}{p}},\varphi\right) \leq \frac{p}{\int_t^s \Psi(\varkappa)d\varkappa}\int_t^s \varkappa^{p-1}\mathfrak{U}^*(\varkappa,\varphi)\Psi(\varkappa)d\varkappa.$$

From this, we have

$$2^{s-1}\left[\mathfrak{U}_*\left(\left[\frac{t^p+s^p}{2}\right]^{\frac{1}{p}},\varphi\right), \mathfrak{U}^*\left(\left[\frac{t^p+s^p}{2}\right]^{\frac{1}{p}},\varphi\right)\right]$$
$$\leq_I \frac{p}{\int_t^s \Psi(\varkappa)d\varkappa}\left[\int_t^s \varkappa^{p-1}\mathfrak{U}_*(\varkappa,\varphi)\Psi(\varkappa)d\varkappa, \int_t^s \varkappa^{p-1}\mathfrak{U}^*(\varkappa,\varphi)\Psi(\varkappa)d\varkappa\right],$$

that is

$$2^{s-1}\mathfrak{U}\left(\left[\frac{t^p+s^p}{2}\right]^{\frac{1}{p}}\right) \preccurlyeq \frac{p}{\int_t^s \varkappa^{p-1}\Psi(\varkappa)d\varkappa}\ (FR)\int_t^s \varkappa^{p-1}\mathfrak{U}(\varkappa,\varphi)\Psi(\varkappa)d\varkappa,$$

and this completes the proof. □

Remark 5. If we attempt to take $s = 1$ in Theorems 8 and 9, then we achieve the appropriate theorems for p-convex F-I-V-Fs, see [13]:

- If we attempt to take $\mathfrak{U}_*(\varkappa,\varphi) = \mathfrak{U}^*(\varkappa,\varphi)$ with $\varphi = 1$, then, from Theorems 8 and 9, we achieve classical first and second H–H Fejér inequality for (p,s)-convex function, [21];
- If in Theorems 8 and 9, we attempt to take $\mathfrak{U}_*(\varkappa,\varphi) = \mathfrak{U}^*(\varkappa,\varphi)$ with $\varphi = 1$ and $s = 1$, then we acquire the classical appropriate theorems for p-convex function, see [49];
- If, in Theorems 8 and 9, we attempt to take $\mathfrak{U}_*(\varkappa,\varphi) = \mathfrak{U}^*(\varkappa,\varphi)$ with $\varphi = 1, s = 1$ and $p = 1$, then we acquire the appropriate theorems for a convex function [48];
- If we attempt to take $\Psi(\varkappa) = 1$, then combining Theorem 8 and Theorem 9, we acquire Theorem 4.1.

Example 4. We consider the F-I-V-F $\mathfrak{U}:[1,4] \to \mathbb{F}_C(\mathbb{R})$ defined by

$$\mathfrak{U}(\varkappa)(\sigma) = \begin{cases} \frac{\sigma - e^{\varkappa p}}{e^{\varkappa p}}, & \sigma \in [e^{\varkappa p}, 2e^{\varkappa p}], \\ \frac{4e^{\varkappa p}-\sigma}{2e^{\varkappa p}}, & \sigma \in (2e^{\varkappa p}, 4e^{\varkappa p}], \\ 0, & otherwise, \end{cases} \quad (44)$$

Then, for each $\varphi \in [0,1]$, we have $\mathfrak{U}_\varphi(\varkappa) = [(1+\varphi)e^{\varkappa p}, 2(2-\varphi)e^{\varkappa p}]$. Since end point functions $\mathfrak{U}_*(\varkappa,\varphi), \mathfrak{U}^*(\varkappa,\varphi)$ are (p,s)-convex functions, for each $s, \varphi \in [0,1]$, then $\mathfrak{U}(\varkappa)$ is (p,s)-convex F-I-V-F. If

$$\Psi(\varkappa) = \begin{cases} \varkappa^p - 1, & \sigma \in [1, \frac{5}{2}], \\ 4 - \varkappa^p, & \sigma \in (\frac{5}{2}, 4], \end{cases} \quad (45)$$

where $p = 1$. Then, we have

$$\frac{p}{s^p-t^p}\int_1^4 \varkappa^{p-1}\mathfrak{U}_*(\varkappa,\varphi)\Psi(\varkappa)d\varkappa = \frac{1}{3}\int_1^4 \varkappa^{p-1}\mathfrak{U}_*(\varkappa,\varphi)\Psi(\varkappa)d\varkappa$$
$$= \frac{1}{3}\int_1^{\frac{5}{2}} \varkappa^{p-1}\mathfrak{U}_*(\varkappa,\varphi)\Psi(\varkappa)d\varkappa + \frac{1}{3}\int_{\frac{5}{2}}^4 \varkappa^{p-1}\mathfrak{U}_*(\varkappa,\varphi)\Psi(\varkappa)d\varkappa$$
$$= \frac{1}{3}(1+\varphi)\int_1^{\frac{5}{2}} e(-1)d\varkappa + \frac{1}{3}(1+\varphi)\int_{\frac{5}{2}}^4 e(4-)d\varkappa \approx 11(1+\varphi),$$
$$\frac{p}{s^p-t^p}\int_1^4 \varkappa^{p-1}\mathfrak{U}^*(\varkappa,\varphi)\Psi(\varkappa)d\varkappa = \frac{1}{3}\int_1^4 \varkappa^{p-1}\mathfrak{U}^*(\varkappa,\varphi)\Psi(\varkappa)d\varkappa$$
$$= \frac{1}{3}\int_1^{\frac{5}{2}} \varkappa^{p-1}\mathfrak{U}^*(\varkappa,\varphi)\Psi(\varkappa)d\varkappa + \frac{1}{3}\int_{\frac{5}{2}}^4 \varkappa^{p-1}\mathfrak{U}^*(\varkappa,\varphi)\Psi(\varkappa)d\varkappa$$
$$= \frac{2}{3}(2-\varphi)\int_1^{\frac{5}{2}} e(-1)d\varkappa + \frac{2}{3}(2-\varphi)\int_{\frac{5}{2}}^4 e(4-)d\varkappa \approx 22(2-\varphi), \quad (46)$$

and

$$\begin{aligned}
[\mathfrak{U}^*(t,\varphi) &+ \mathfrak{U}^*(s,\varphi)] \int_0^1 \zeta^s \Psi\left(\left[(1-\tau)t^p + \tau s^p\right]^{\frac{1}{p}}\right) d\tau \\
[\mathfrak{U}_*(t,\varphi) &+ \mathfrak{U}_*(s,\varphi)] \int_0^1 \zeta^s \Psi\left(\left[(1-\tau)t^p + \tau s^p\right]^{\frac{1}{p}}\right) d\tau \\
&= (1+\varphi)[e+e^4] \left[\int_0^{\frac{1}{2}} 3\tau^2 d + \int_{\frac{1}{2}}^1 \tau(3-3\tau)d\tau\right] \approx \tfrac{43}{2}(1+\varphi) \\
&= 2(2-\varphi)[e+e^4] \left[\int_0^{\frac{1}{2}} 3\tau^2 d + \int_{\frac{1}{2}}^1 \tau(3-3\tau)d\tau\right] \approx 43(2-\varphi).
\end{aligned} \qquad (47)$$

From Equations (46) and (47), we have

$$[11(1+\varphi), 22(2-\varphi)] \leq_I \left[\tfrac{43}{2}(1+\varphi), 43(2-\varphi)\right], \text{ for each } \varphi \in [0,1].$$

Hence, Theorem 8 is verified.

For Theorem 9, we have

$$\begin{aligned}
2^{s-1}\mathfrak{U}_*\left(\left[\tfrac{t^p+s^p}{2}\right]^{\frac{1}{p}}, \varphi\right) &\approx \tfrac{61}{5}(1+\varphi), \\
2^{s-1}\mathfrak{U}^*\left(\left[\tfrac{t^p+s^p}{2}\right]^{\frac{1}{p}}, \varphi\right) &\approx \tfrac{122}{5}(2-\varphi),
\end{aligned} \qquad (48)$$

$$\int_t^s \varkappa^{p-1}\Psi(\varkappa)d\varkappa = \int_1^{\frac{5}{2}}(\varkappa-1)d\varkappa \int_{\frac{5}{2}}^4 (4-)d\varkappa = \tfrac{9}{4},$$

$$\begin{aligned}
\tfrac{p}{\int_t^s \varkappa^{p-1}\Psi(\varkappa)d\varkappa} \int_1^4 \varkappa^{p-1}\mathfrak{U}_*(\varkappa,\varphi)\Psi(\varkappa)d\varkappa &\approx \tfrac{73}{5}(1+\varphi), \\
\tfrac{p}{\int_t^s \varkappa^{p-1}\Psi(\varkappa)d\varkappa} \int_1^4 \varkappa^{p-1}\mathfrak{U}^*(\varkappa,\varphi)\Psi(\varkappa)d\varkappa &\approx \tfrac{293}{10}(2-\varphi).
\end{aligned} \qquad (49)$$

From Equations (48) and (49), we have

$$\left[\tfrac{61}{5}(1+\varphi), \tfrac{122}{5}(2-\varphi)\right] \leq_I \left[\tfrac{73}{5}(1+\varphi), \tfrac{293}{10}(2-\varphi)\right].$$

Hence, Theorem 9 has been demonstrated.

5. Conclusions and Future Developments

Through this study, we have provided a reformative version of the different inequalities in the frame of fuzzy interval space, which offers a better approximation than the interval integral inequalities.

Then, for mappings satisfying the property "the product of two (p,s)-convex F-I-V-Fs is a (p,s)-convex F-I-V-F", we created certain fuzzy interval integral inequalities in terms of the fuzzy interval H–H type inequalities. It is a fascinating topic to apply these fuzzy interval inequalities to φ-type special means, numerical integration, and probability density functions. With the methods and ideas provided in this article, the interested readers are encouraged to further excavation on fuzzy interval inequalities. In the future, we will try to explore this concept and its generalizations with the help of fuzzy fractional integral operators.

Author Contributions: Conceptualization, M.B.K.; validation, H.B. and S.T.; formal analysis, H.B. and S.T.; investigation, M.B.K. and S.T.; resources, M.B.K. and H.B.; writing—original draft, M.B.K. and H.B.; writing—review and editing, M.B.K. and S.T.; visualization, M.B.K., H.B. and S.T.; supervision, M.B.K. and S.T.; project administration, H.B., M.B.K. and S.T. All authors have read and agreed to the published version of the manuscript.

Funding: This research received no external funding.

Institutional Review Board Statement: Not applicable.

Informed Consent Statement: Not applicable.

Data Availability Statement: Not Applicable.

Acknowledgments: The authors would like to thank the Rector, COMSATS University Islamabad, Islamabad, Pakistan, for providing excellent research.

Conflicts of Interest: The authors declare no conflict of interest.

Abbreviations

$\mathcal{K}_C$	Collection of all closed and bounded intervals
$\mathcal{K}_C^+$	Collection of all closed and bounded positive intervals
$\mathbb{F}_C(\mathbb{R})$	Collection of all closed and bounded fuzzy intervals
F-I-V-Fs	Fuzzy-interval-valued functions
I-V-Fs	Interval-valued functions
$\leq_I$	order relation
$\preccurlyeq$	fuzzy order relation
(p,s)-convex F-I-V-Fs	(p,s)-Convex fuzzy-interval-valued functions
H–H inequality	Hermite–Hadamard inequality
H–H Fejér inequality	Hermite–Hadamard–Fejér inequality
(FR)-integrable	Fuzzy Riemann integrable
$\mathcal{R}_{[t,s]}$	Riemann integrable real-valued functions
$\mathcal{IR}_{[t,s]}$	Riemann integrable I-V-Fs
$\mathcal{FR}_{([t,s])}$	Riemann integrable F-I-V-Fs

References

1. Dragomir, S.S.; Pearce, V. *Selected Topics on Hermite-Hadamard Inequalities and Applications*; RGMIA Monographs: Victoria, Australia, 2000.
2. Mehrez, K.; Agarwal, P. New Hermite–Hadamard type integral inequalities for convex functions and their applications. *J. Comput. Appl. Math.* **2019**, *50*, 274–285. [CrossRef]
3. Mitrinović, D.S.; Pečarić, J.E.; Fink, A.M. *Classical and New Inequalities in Analysis. Mathematics and Its Applications*; (East European Series); Kluwer Academic Publishers Group: Dordrecht, The Netherlands, 1993; Volume 61.
4. Awan, M.U.; Akhtar, N.; Iftikhar, S.; Noor, M.A.; Chu, Y.-M. New Hermite–Hadamard type inequalities for n-polynomial harmonically convex functions. *J. Inequal. Appl.* **2020**, *2020*, 125. [CrossRef]
5. Latif, M.A.; Rashid, S.; Dragomir, S.S.; Chu, Y.-M. Hermite–Hadamard type inequalities for co-ordinated convex and quasi-convex functions and their applications. *J. Inequal. Appl.* **2019**, *2019*, 317. [CrossRef]
6. Chu, Y.-M.; Xia, W.-F.; Zhang, X.-H. The Schur concavity, Schur multiplicative and harmonic convexities of the second dual form of the Hamy symmetric function with applications. *J. Multivar. Anal.* **2012**, *105*, 412–442. [CrossRef]
7. Ullah, S.Z.; Khan, M.A.; Khan, Z.A.; Chu, Y.-M. Integral majorization type inequalities for the functions in the sense of strong convexity. *J. Funct. Spaces* **2019**, *2019*, 9487823.
8. Ullah, S.Z.; Khan, M.A.; Chu, Y.-M. Majorization theorems for strongly convex functions. *J. Inequal. Appl.* **2019**, *2019*, 58.
9. Zhang, K.-S.; Wan, J.-P. P-convex functions and their properties. *Pure Appl. Math.* **2007**, *23*, 130–133.
10. Chang, S.S.; Zhu, Y.G. On variational inequalities for fuzzy mappings. *Fuzzy Sets Syst.* **1989**, *32*, 359–367. [CrossRef]
11. Nanda, S.; Kar, K. Convex fuzzy mappings. *Fuzzy Sets Syst.* **1992**, *48*, 129–132. [CrossRef]
12. Noor, M.A. Fuzzy preinvex functions. *Fuzzy Sets Syst.* **1994**, *64*, 95–104. [CrossRef]
13. Khan, M.B.; Noor, M.A.; Noor, K.I.; Chu, Y.-M. New Hermite-Hadamard Type Inequalities for-Convex Fuzzy-Interval-Valued Functions. *Adv. Diff. Equ.* **2021**, *2021*, 6–20. [CrossRef]
14. Liu, W. New integral inequalities involving beta function via P-convexity. *Miskolc Math. Notes* **2014**, *15*, 585–591. [CrossRef]
15. Sawano, Y.; Wadade, H. On the Gagliardo-Nirenberg type inequality in the critical Sobolev-Orrey space. *J. Fourier Anal. Appl.* **2013**, *19*, 20–47. [CrossRef]
16. Cowling, P.C.M.G.; Ricci, F. Hardy and uncertainty inequalities on stratified Lie groups. *Adv. Math.* **2015**, *277*, 365–387.
17. Gavrea, B.; Gavrea, I. On some Ostrowski type inequalities. *Gen. Math.* **2010**, *18*, 33–44.
18. Gunawan, H. Fractional integrals and generalized Olsen inequalities. *Kyungpook Math. J.* **2009**, *49*, 31–39. [CrossRef]
19. Hadamard, J. Étude sur les propriétés des fonctions entières en particulier d'une fonction considérée par Riemann. *J. Math. Pure Appl.* **1893**, *58*, 171–215.
20. Khan, M.B.; Treanță, S.; Soliman, M.S.; Nonlaopon, K.; Zaini, H.G. Some Hadamard–Fejér Type Inequalities for LR-Convex Interval-Valued Functions. *Fractal Fract.* **2022**, *6*, 6. [CrossRef]
21. Fang, Z.-B.; Shi, R.-J. On the (p,h)-convex function and some integral inequalities. *J. Inequal. Appl.* **2014**, *2014*, 45. [CrossRef]

22. Khan, M.B.; Zaini, H.G.; Treanță, S.; Soliman, M.S.; Nonlaopon, K. Riemann–Liouville Fractional Integral Inequalities for Generalized Pre-Invex Functions of Interval-Valued Settings Based upon Pseudo Order Relation. *Mathematics* **2022**, *10*, 204. [CrossRef]
23. Kara, H.; Ali, M.A.; Budak, H. Hermite–Hadamard-type inequalities for interval-valued coordinated convex functions involving generalized fractional integrals. *Math. Methods Appl. Sci.* **2021**, *44*, 104–123. [CrossRef]
24. Liu, P.; Khan, M.B.; Noor, M.A.; Noor, K.I. New Hermite–Hadamard and Jensen inequalities for log-s-convex fuzzy-interval-valued functions in the second sense. *Complex. Intell. Syst.* **2021**, *2021*, 155. [CrossRef]
25. Khan, M.B.; Noor, M.A.; Abdullah, L.; Chu, Y.-M. Some New Classes of Preinvex Fuzzy-Interval-Valued Functions and Inequalities. *Int. J. Comput. Intell. Syst.* **2021**, *14*, 1403–1418. [CrossRef]
26. Sarikaya, M.Z.; Set, E.; Yaldiz, H.; BaSak, N. Hermite–Hadamard's inequalities for fractional integrals and related fractional inequalities. *Math. Comput. Model.* **2013**, *57*, 2403–2407. [CrossRef]
27. Toplu, T.; Set, E.; Scan, I.; Maden, S. Hermite–Hadamard type inequalities for p-convex functions via Katugampola fractional integrals. *Facta Univ. Ser. Math. Inform.* **2019**, *34*, 149–164. [CrossRef]
28. Bede, B. Mathematics of fuzzy sets and fuzzy logic. *Stud. Fuzziness Soft Comput.* **2013**, *295*, 1–288.
29. Thaiprayoon, C.; Ntouyas, S.K.; Tariboon, J. On the nonlocal Katugampola fractional integral conditions for fractional Langevin equation. *Adv. Diff. Equ.* **2015**, *2015*, 374. [CrossRef]
30. RAgarwal, P.; Baleanu, D.; Nieto, J.J.; Torres, D.F.M.; Zhou, Y. A survey on fuzzy fractional differential and optimal control nonlocal evolution equations. *J. Comput. Appl. Math.* **2018**, *339*, 3–29. [CrossRef]
31. Ahmad, M.Z.; Hasan, M.K.; de Baets, B. Analytical and numerical solutions of fuzzy differential equations. *Inf. Sci.* **2013**, *236*, 156–167. [CrossRef]
32. Alijani, Z.; Baleanu, D.; Shiri, B.; Wu, G.C. Spline collocation methods for systems of fuzzy fractional differential equations. *Chaos Solitons Fract.* **2020**, *131*, 109510. [CrossRef]
33. Bede, B.; Gal, S.G. Generalizations of the differentiability of fuzzy-number-valued functions with applications to fuzzy differential equations. *Fuzzy Sets Syst.* **2005**, *151*, 581–599. [CrossRef]
34. Bede, B.; Stefanini, L. Generalized differentiability of fuzzy-valued functions. *Fuzzy Sets Syst.* **2013**, *230*, 119–141. [CrossRef]
35. Diamond, P.; Kloeden, P.E. *Metric Spaces of Fuzzy Sets: Theory and Applications*; World Scientific: Singapore, 1994.
36. Goetschel, R., Jr.; Voxman, W. Elementary fuzzy calculus. *Fuzzy Sets Syst.* **1986**, *18*, 31–43. [CrossRef]
37. Hoa, N.V. On the initial value problem for fuzzy differential equations of non-integer order $\alpha \in (1, 2)$. *Soft Comput.* **2020**, *24*, 935–954. [CrossRef]
38. Mazandarani, M.; Najariyan, M. Type-2 fuzzy fractional derivatives. *Commun. Nonlinear Sci. Numer. Simul.* **2014**, *19*, 2354–2372. [CrossRef]
39. Stefanini, L. A generalization of Hukuhara difference and division for interval and fuzzy arithmetic. *Fuzzy Sets Syst.* **2010**, *161*, 1564–1584. [CrossRef]
40. Zadeh, L.A. Fuzzy sets. *Inf. Control* **1965**, *8*, 338–353. [CrossRef]
41. Osuna-Gomez, R.; Jimenez-Gamero, M.D.; Chalco-Cano, Y.; Rojas-Medar, M.A. Hadamard and Jensen Inequalities for s-Convex Fuzzy Processes. In *Soft Methodology and Random Information Systems*; Advances in Soft Computing; Springer: Berlin/Heidelberg, Germany, 2004; Volume 26, pp. 1–15.
42. Costa, T.M. Jensen's inequality type integral for fuzzy-interval-valued functions. *Fuzzy Sets Syst.* **2017**, *327*, 31–47. [CrossRef]
43. Kulish, U.; Miranker, W. *Computer Arithmetic in Theory and Practice*; Academic Press: New York, NY, USA, 2014.
44. Zhao, D.F.; An, T.Q.; Ye, G.J.; Liu, W. New Jensen and Hermite-Hadamard type inequalities for h-convex interval-valued functions. *J. Inequal. Appl.* **2018**, *2018*, 302. [CrossRef]
45. Costa, T.M.; Roman-Flores, H. Some integral inequalities for fuzzy-interval-valued functions. *Inf. Sci.* **2017**, *420*, 110–125. [CrossRef]
46. Breckner, W.W. Stetigkeitsaussagen für eine Klasse verallgemeinerter konvexer funktionen in topologischen linearen Räumen. *Pupl. Inst. Math.* **1978**, *23*, 13–20.
47. SDragomir, S.; Pecaric, J.; Persson, L.E. Some inequalities of Hadamard type. *Soochow J. Math.* **1995**, *21*, 335–341.
48. Fejer, L. Uberdie Fourierreihen, II. *Math. Naturwiss Anz. Ungar. Akad. Wiss.* **1906**, *24*, 369–390.
49. Kunt, M.; İşcan, İ. Hermite–Hadamard–Fejer type inequalities for p-convex functions. *Arab. J. Math. Sci.* **2017**, *23*, 215–230. [CrossRef]
50. Kaleva, O. Fuzzy differential equations. *Fuzzy Sets Syst.* **1987**, *24*, 301–317. [CrossRef]
51. Khan, M.B.; Noor, M.A.; Noor, K.I.; Nisar, K.S.; Ismail, K.A.; Elfasakhany, A. Some Inequalities for LR-$$\left ({h}_{1},{h}_{2}\right) $$ h 1, h 2-Convex Interval-Valued Functions by Means of Pseudo Order Relation. *Int. J. Comput. Intell. Syst.* **2021**, *14*, 1–15. [CrossRef]
52. Sana, G.; Khan, M.B.; Noor, M.A.; Mohammed, P.O.; Chu, Y.M. Harmonically convex fuzzy-interval-valued functions and fuzzy-interval Riemann–Liouville fractional integral inequalities. *Int. J. Comput. Intell. Syst.* **2021**, *14*, 1809–1822. [CrossRef]
53. Khan, M.B.; Noor, M.A.; Al-Bayatti, H.M.; Noor, K.I. Some new inequalities for LR-log-h-convex interval-valued functions by means of pseudo order relation. *Appl. Math. Inf. Sci.* **2021**, *15*, 459–470.
54. Khan, M.B.; Mohammed, P.O.; Noor, M.A.; Hamed, Y.S. New Hermite-Hadamard inequalities in fuzzy-interval fractional calculus and related inequalities. *Symmetry* **2021**, *13*, 673. [CrossRef]

55. Guo, Y.; Ye, G.; Liu, W.; Zhao, D.; Treanţă, S. Optimality conditions and duality for a class of generalized convex interval-valued optimization problems. *Mathematics* **2021**, *9*, 2979. [CrossRef]
56. Treanţă, S. Characterization results of solutions in interval-valued optimization problems with mixed constraints. *J. Glob. Optim.* **2020**, *2020*, 1–14. [CrossRef]
57. JMacías-Díaz, E.; Khan, M.B.; Noor, M.A.; Allah, A.M.A.; Alghamdi, S.M. Hermite-Hadamard inequalities for generalized convex functions in interval-valued calculus. *AIMS Math.* **2022**, *7*, 4266–4292. [CrossRef]
58. Khan, M.B.; Noor, M.A.; Al-Shomrani, M.M.; Abdullah, L. Some Novel Inequalities for LR-h-Convex Interval-Valued Functions by Means of Pseudo Order Relation. *Math. Methods Appl. Sci.* **2022**, *45*, 1310–1410. [CrossRef]
59. Khan, M.B.; Noor, M.A.; Abdeljawad, T.; Abdalla, B.; Althobaiti, A. Some fuzzy-interval integral inequalities for harmonically convex fuzzy-interval-valued functions. *AIMS Math.* **2022**, *7*, 349–370. [CrossRef]
60. Mesiar, R.; Li, J.; Pap, E. The Choquet integral as Lebesgue integral and related inequalities. *Kybernetika* **2010**, *46*, 1098–1107.
61. Agahi, H.; Mesiar, R.; Ouyang, Y.; Pap, E.; Štrboja, M. Berwald type inequality for Sugeno integral. *Appl. Math. Comput.* **2010**, *217*, 4100–4108. [CrossRef]

 *fractal and fractional*

Article

The Method of Fundamental Solutions for the 3D Laplace Inverse Geometric Problem on an Annular Domain

Mojtaba Sajjadmanesh [1], Hassen Aydi [2,3,4,*], Eskandar Ameer [5] and Choonkil Park [6,*]

[1] Faculty of Basic Science, University of Bonab, Bonab P.O. Box 55517-61167, Iran; m.sajjadmanesh@ubonab.ac.ir
[2] Institut Supérieur d'Informatique et des Techniques de Communication, Université de Sousse, Sousse 4000, Tunisia
[3] China Medical University Hospital, China Medical University, Taichung 40402, Taiwan
[4] Department of Mathematics and Applied Mathematics, Sefako Makgatho Health Sciences University, Ga-Rankuwa 0204, South Africa
[5] Department of Mathematics, Taiz University, Taiz 6803, Yemen; eskanderameer@taiz.edu.ye
[6] Research Institute for Natural Sciences, Hanyang University, Seoul 04763, Korea
* Correspondence: hassen.aydi@isima.rnu.tn (H.A.); baak@hanyang.ac.kr (C.P.)

Abstract: In this paper, we are interested in an inverse geometric problem for the three-dimensional Laplace equation to recover an inner boundary of an annular domain. This work is based on the method of fundamental solutions (MFS) by imposing the boundary Cauchy data in a least-square sense and minimisation of the objective function. This approach can also be considered with noisy boundary Cauchy data. The simplicity and efficiency of this method is illustrated in several numerical examples.

Keywords: inverse geometric problem; Laplace equation; method of fundamental solution; least-square problem

1. Introduction

The inverse geometry problems, as an important subclass of inverse problems, can be subdivided into two subclasses, depending on the location of the unknown boundary. In the first kind, the portion of the outer boundary of the solution domain is unknown, whilst in the second kind, the inner boundary is unknown.

There are many methods for solving the inverse geometry problems, such as the boundary element regularisation method by Lesnic et al. [1], the method of fundamental solutions and moving pseudo-boundary method by Karageorghis et al. [2–4], the boundary function method by Wang et al. [5], the conjugate gradient method (CGM) and the boundary element technique by Huang et al. [6,7].

Bin-Mohsin and Lesnic in 2012 utilised the method of fundamental solutions (MFS) to the modified Helmholtz inverse geometry problem on an annular domain [8].

The purpose of this paper is to extend the aforementioned current approach to the three-dimensional Laplace equation based on the method of fundamental solutions. Finally, two examples are presented to show the simplicity and efficiency of this method.

2. Formulation of the Inverse Geometric Problem

Let $D \subset \mathbb{R}^3$ be a simply connected domain with an unknown boundary ∂D which is compactly contained in a simply connected domain $\Omega \subset \mathbb{R}^3$ with the boundary $\partial \Omega$. Let us consider the following inverse problem:

$$\Delta u = 0 \quad \text{in } \Omega \setminus \overline{D}, \qquad (1)$$

subject to the boundary conditions,

$$u = f \quad \text{on} \quad \partial\Omega, \tag{2}$$

$$\frac{\partial u}{\partial n} = g \quad \text{on} \quad \partial\Omega, \tag{3}$$

$$u = h \quad \text{on} \quad \partial D, \tag{4}$$

where $f \in H^{1/2}(\partial\Omega)$ and $g \in H^{-1/2}(\partial\Omega)$ are given functions and n is an outward unit normal vector on $\partial\Omega$. Moreover, the function $h \in H^{1/2}(\partial D)$ is given on the unknown boundary ∂D. Without loss of generality, we can suppose that Ω is the unit disk $B(\underline{0}; 1)$; otherwise we can conformally map the exterior of the simply connected domain Ω onto the exterior of the unit disk.

The unknown boundary ∂D can be expressed in spherical coordinates as

$$\partial D = \{\, r(\theta, \varphi)(\cos\theta \sin\varphi, \sin\theta \sin\varphi, \cos\varphi);\ \theta \in [0, 2\pi),\ \varphi \in [0, \pi]\,\}$$

where $r(\theta, \varphi)$ is a 2π-periodic and π-periodic smooth function with respect to θ and φ, respectively, with values in the interval $(0, 1)$.

The inverse problem we are concerned with is to determine geometrically the domain boundary ∂D by utilising the method of fundamental solutions.

3. The Least-Square Problem Based on the MFS

In the classic MFS, the solution of a homogeneous linear partial differential equation (PDE) is approximated by a linear combination of the fundamental solutions with the set of sources located outside the problem domain and a set of points on the domain boundary. The linear combination coefficients are determined by collocation or, alternatively, with a least-squares fit of the boundary conditions.

Based on the MFS, one can approximate the solution of (1) by a linear combination of its fundamental solutions, which is given by [9]

$$U(\mathbf{z}, \mathbf{s}) = \frac{1}{4\pi r};\ r = \|\mathbf{z} - \mathbf{s}\|. \tag{5}$$

i.e.,

$$u(\mathbf{z}) = \sum_{j=1}^{n_s} c_j U(\mathbf{z}, \mathbf{s}_j), \tag{6}$$

where the collocation points $\mathbf{z}_i$ and $\mathbf{z}_{i+M}$ are uniformly located on $\partial\Omega$ and ∂D, respectively, i.e.,

$$\mathbf{z}_i = (\cos\hat{\theta}_i \sin\hat{\varphi}_i,\ \sin\hat{\theta}_i \sin\hat{\varphi}_i,\ \cos\hat{\varphi}_i),\ i = \overline{1, M} \tag{7}$$

$$\mathbf{z}_{i+M} = r_i(\cos\theta_i \sin\varphi_i,\ \sin\theta_i \sin\varphi_i,\ \cos\varphi_i),\ i = \overline{1, N} \tag{8}$$

Further, the $n_s := M + N$ source points $\mathbf{s}_j$ and $\mathbf{s}_{j+M}$ are uniformly located on the outside of $\overline{\Omega}$ and the inside of D, respectively, i.e.,

$$\mathbf{s}_j = R_1(\cos\hat{\theta}_j \sin\hat{\varphi}_j, \sin\hat{\theta}_j \sin\hat{\varphi}_j, \cos\hat{\varphi}_j),\ j = \overline{1, M} \tag{9}$$

$$\mathbf{s}_{j+M} = \frac{r_j}{R_2}(\cos\theta_j \sin\varphi_j, \sin\theta_j \sin\varphi_j, \cos\varphi_j),\ j = \overline{1, N} \tag{10}$$

where $R_1, R_2 > 1$.

The coefficients vector $\mathbf{c} = (c_j)_{j=\overline{1,M+N}}$ in linear combination (6) and also, the radial vector $\mathbf{r} = (r_j)_{j=\overline{1,N}}$ can be determined by imposing the boundary conditions (2)–(4) in a least-square sense, which recasts into minimising the objective function

$$T(\mathbf{c},\mathbf{r}) = \|u - f\|^2_{L^2(\partial\Omega)} + \left\|\frac{\partial u}{\partial n} - g\right\|^2_{L^2(\partial\Omega)} + \|u - h\|^2_{L^2(\partial D)}. \tag{11}$$

Upon discretisation, Equation (11) yields

$$T(\mathbf{c},\mathbf{r}) = \sum_{i=1}^{M}\left[\sum_{j=1}^{M+N} c_j U(\mathbf{z}_i, \mathbf{s}_j) - f(\mathbf{z}_i)\right]^2 + \sum_{i=M+1}^{2M}\left[\sum_{j=1}^{M+N} c_j \frac{\partial U}{\partial n}(\mathbf{z}_{i-M}, \mathbf{s}_j) - g(\mathbf{z}_{i-M})\right]^2$$
$$+ \sum_{i=2M+1}^{2M+N}\left[\sum_{j=1}^{M+N} c_j U(\mathbf{z}_{i-M}, \mathbf{s}_j) - h(\mathbf{z}_{i-M})\right]^2. \tag{12}$$

In general, the boundary data $F \in \{f, g, h\}$ are measured noisy data satisfying

$$F_i^\delta = F_i + \delta\, rand(i)\, F_i, \tag{13}$$

where δ is the percentage noise and the number $rand(i)$ is a random number drawn from the standard uniform distribution on the interval $[-1, 1]$ generated by the MATLAB code $-1 + 2rand(i)$.

Imposing noise on all measured data implies

$$T^\delta(\mathbf{c},\mathbf{r}) = \sum_{i=1}^{M}\left[\sum_{j=1}^{M+N} c_j U(\mathbf{z}_i, \mathbf{s}_j) - f^\delta(\mathbf{z}_i)\right]^2 + \sum_{i=M+1}^{2M}\left[\sum_{j=1}^{M+N} c_j \frac{\partial U}{\partial n}(\mathbf{z}_{i-M}, \mathbf{s}_j) - g^\delta(\mathbf{z}_{i-M})\right]^2$$
$$+ \sum_{i=2M+1}^{2M+N}\left[\sum_{j=1}^{M+N} c_j U(\mathbf{z}_{i-M}, \mathbf{s}_j) - h^\delta(\mathbf{z}_{i-M})\right]^2. \tag{14}$$

The minimisation of (12) or (14) imposes $2M + N$ nonlinear equations for the $2N + M$ unknowns $(\mathbf{c}, \mathbf{r})$, and for a unique solution, it is necessary that $M \geq N$.

4. Error Analysis and the Regularisation

The accuracy of the presented method is evaluated by the normalised relative root mean square error (RMSE) and L_∞-error:

$$\text{RMSE} = \frac{\left\{\frac{1}{N}\sum_{i=1}^{N}|r_i^{(an)} - r_i^{(num)}|^2\right\}^{\frac{1}{2}}}{\max_{1\leq i\leq N}|r_i^{(an)}|}, \quad L_\infty\text{-error} = \max_{1\leq i\leq N}|r_i^{(an)} - r_i^{(num)}|$$

where $r_i^{(an)}$ and $r_i^{(num)}$ denote the analytical and numerical radial vectors, respectively, at the ith collocation point on the boundary ∂D.

The obtained numerical radial vectors from the presented method are unstable, especially when noise is added to the boundary data, and so the regularisation is needed. For this, we can add the following regularisation terms via standard zeroth- and first-order Tikhonov's regularisation with parameters $\lambda_1, \lambda_2 \geq 0$ to the functional (14), i.e.,

$$Reg(\underline{a},\underline{r}) = \sum_{j=2M+N+1}^{3M+2N}\left(\sqrt{\lambda_1}a_{j-2M-N}\right)^2 + \sum_{j=3M+2N+1}^{3M+3N-1}\left(\sqrt{\lambda_2}(r_{j-3M-2N+1} - r_{j-3M-2N})\right)^2, \tag{15}$$

5. Numerical Examples

In this section, we give some examples to check the effectiveness of the presented method. We consider a three-dimensional annular domain with an outer boundary as the unit sphere $\partial\Omega = B(\underline{0};1)$, $R_1 = R_2 = 2$ and $M = N \in \{25, 50\}$ in (7)–(10). Moreover, the percentage noise $\delta = 5\%$ is added to every measured boundary data.

The minimisation of functional (12) or (14) is carried out using the MATLAB optimisation toolbox routine lsqnonlin, which solves nonlinear least-squares problems.

Example 1. *Consider a three-dimensional annular domain with an unknown inner boundary $\partial D = B(\underline{0}; r^{(an)})$ of radius $r^{(an)} = 0.7$. The boundary data are given as follows:*

$$u|_{\partial\Omega} = f(\theta, \varphi) = \frac{1}{2}\left\{\sin 2\varphi \left(\cos\theta + \sin\theta\right) + \sin 2\theta \sin^2\varphi\right\},$$

$$\frac{\partial u}{\partial n}|_{\partial\Omega} = g(\theta, \varphi) = \sin 2\varphi \left(\cos\theta + \sin\theta\right) + \sin 2\theta \sin^2\varphi,$$

$$u|_{\partial D} = h(\theta, \varphi) = \frac{49}{200}\left\{\sin 2\varphi \left(\cos\theta + \sin\theta\right) + \sin 2\theta \sin^2\varphi\right\}.$$

The exact solution for these input boundary data is $u(x) = x_1 x_2 + x_1 x_3 + x_2 x_3$.

Table 1 gives the values of the objective functions and the corresponding errors obtained using the optimal initial guess r_0 and $M = N \in \{25, 50\}$ without using regularisation parameters. It can be seen that the values of the corresponding errors increase with the number of collocation points and so regularisation is needed.

Table 1. The values of the optimal initial guess, r_0, objective functions and the corresponding errors with $M = N \in \{25, 50\}$ and no regularisation parameters for Example 1.

	$M = N$	r_0	Objective Functions	L_∞-Error	RMSE
Without noise	25	0.5	4.9013×10^{-3}	4.3884	1.4418
	50	0.6	6.1682	5.0114	1.7985
With noise 5%	25	0.6	2.1799	2.1210	7.1903×10^{-1}
	50	0.6	1.8326×10^{1}	1.0113	9.1243×10^{-1}

In Tables 2 and 3, we present the values of the objective functions and the corresponding errors with initial guess, r_0, obtained using the regularisation parameters $\lambda_1, \lambda_2 \in \{0, 10^{-6}, 10^{-3}, 10^{-1}\}$ with $M = N \in \{25, 50\}$ and so, in Table 4, we give the minimal objective functions and the corresponding errors with initial guess r_0.

Table 2. The values of the optimal initial guess, r_0, objective functions and the corresponding errors using the regularisation parameters λ_1, λ_2 with $M = N = 25$ for Example 1.

			r_0	Objective Functions	L_∞-Error	RMSE
Without noise	$\lambda_1 = 0$	$\lambda_2 = 10^{-6}$	0.7	1.8102×10^{-2}	2.6100	1.0249
		$\lambda_2 = 10^{-3}$	0.5	2.5783×10^{-3}	1.2932×10^{-1}	1.3835×10^{-1}
		$\lambda_2 = 10^{-1}$	0.3	2.2118×10^{-3}	1.0442×10^{-1}	1.3354×10^{-1}
	$\lambda_1 = 10^{-6}$	$\lambda_2 = 0$	0.6	5.8822×10^{-3}	2.9436×10^{-1}	1.5451×10^{-1}
	$\lambda_1 = 10^{-3}$		0.5	1.2470	2.8593	8.8216×10^{-1}
	$\lambda_1 = 10^{-1}$		0.7	5.8688	5.1408×10^{3}	1.4688×10^{-3}
With noise 5%	$\lambda_1 = 0$	$\lambda_2 = 10^{-6}$	0.9	4.1621×10^{-2}	3.5650×10^{-1}	1.1951×10^{-1}
		$\lambda_2 = 10^{-3}$	0.6	1.3166×10^{-2}	1.6224×10^{-1}	1.3880×10^{-1}
		$\lambda_2 = 10^{-1}$	0.5	1.2386×10^{-2}	1.0866×10^{-1}	1.1611×10^{-1}
	$\lambda_1 = 10^{-6}$	$\lambda_2 = 0$	0.4	3.1242×10^{-2}	6.5845×10^{-1}	2.8637×10^{-1}
	$\lambda_1 = 10^{-3}$		0.2	1.2775	8.0410×10^{-1}	3.8164×10^{-1}
	$\lambda_1 = 10^{-1}$		0.2	5.9897	3.3646	2.0861

Table 3. The values of the optimal initial guess, r_0, objective functions and the corresponding errors using the regularisation parameters λ_1, λ_2 with $M = N = 50$ for Example 1.

			r_0	Objective Functions	L_∞-Error	RMSE
Without noise	$\lambda_1 = 0$	$\lambda_2 = 10^{-6}$ $\lambda_2 = 10^{-3}$ $\lambda_2 = 10^{-1}$	0.6 0.7 0.7	3.4092 2.4650 1.6998×10^{-5}	4.6998 8.8865×10^{-1} 9.8441×10^{-3}	1.726 4.5447×10^{-1} 1.1249×10^{-2}
	$\lambda_1 = 10^{-6}$ $\lambda_1 = 10^{-3}$ $\lambda_1 = 10^{-1}$	$\lambda_2 = 0$	0.6 0.5 0.2	4.3927×10^{-2} 9.7102×10^{-1} 1.4145×10^{1}	4.7363 4.6546 1.2983	1.8483 1.9311 1.5821
With noise 5%	$\lambda_1 = 0$	$\lambda_2 = 10^{-6}$ $\lambda_2 = 10^{-3}$ $\lambda_2 = 10^{-1}$	0.6 0.8 0.6	1.2600 1.2761×10^{-2} 6.2499×10^{-3}	1.5259 1.0640 6.2367×10^{-2}	4.6191×10^{-1} 4.4252×10^{-1} 3.1760×10^{-2}
	$\lambda_1 = 10^{-6}$ $\lambda_1 = 10^{-3}$ $\lambda_1 = 10^{-1}$	$\lambda_2 = 0$	0.6 0.7 0.6	1.1694 9.7398×10^{-1} 1.0836×10^{1}	4.8164 3.2081 4.4080	1.6659 1.0049 1.9817

Example 2. *Consider a three-dimensional annular domain with an unknown inner boundary of radius $r^{(an)} = \frac{1}{4}(1 + \cos\theta \sin 2\varphi)$. The boundary data are given as follows:*

$$u|_{\partial\Omega} = f(\theta, \varphi) = 3\sin^2\varphi - 2,$$

$$\frac{\partial u}{\partial n}\Big|_{\partial\Omega} = g(\theta, \varphi) = 6\sin^2\varphi - 4,$$

$$u|_{\partial D} = h(\theta, \varphi) = \frac{1}{16}(3\sin^2\varphi - 2)(\sin 2\varphi \cos\theta + 1)^2.$$

The exact solution for these input boundary data is $u(x) = x_1^2 + x_2^2 - 2x_3^2$.

Table 4. The values of the minimal objective functions and the corresponding errors with initial guess r_0, obtained (with/no) selecting the optimal regularisation parameters with $M = N \in \{25, 50\}$ for Example 1.

$M = N$	With/No Noise	With/No Regularisation	λ_1	λ_2	r_0	Objective Function	L_∞-Error	RMSE
25	no	no with	0	10^{-1}	0.5 0.3	4.9013×10^{-3} 2.2118×10^{-3}	4.3884 1.0442×10^{-1}	1.4418 1.3354×10^{-1}
	with	no with	0	10^{-1}	0.6 0.5	2.1799 1.2386×10^{-2}	2.1210 1.0866×10^{-1}	7.1903×10^{-1} 1.1611×10^{-1}
50	no	no with	0	10^{-1}	0.6 0.7	6.1682 1.6998×10^{-5}	5.0114 9.8441×10^{-3}	1.7985 1.1249×10^{-2}
	with	no with	0	10^{-1}	0.6 0.6	1.8326×10^{1} 6.2499×10^{-3}	1.0113 6.2367×10^{-2}	9.1243×10^{-1} 3.1760×10^{-2}

Table 5 gives the values of the objective functions and the corresponding errors obtained using the optimal initial guess r_0, $M = N \in \{25, 50\}$ without using regularisation parameters, whilst Tables 6 and 7 are obtained using the regularisation parameters λ_1, λ_2 and so, in Table 8 we give the minimal objective functions and the corresponding errors with initial guess r_0.

Table 5. The values of the optimal initial guess, r_0, objective functions and the corresponding errors for Example 2 with $M = N \in \{25, 50\}$ and no regularisation parameters.

	$M = N$	r_0	Objective Functions	L_∞-Error	RMSE
Without noise	25 50	0.5 0.6	5.0884×10^{1} 1.0837×10^{2}	4.1355×10^{-1} 8.1024×10^{-1}	6.0305×10^{-1} 1.2263
With noise 5%	25 50	0.8 0.7	3.2489×10^{1} 5.4166×10^{1}	1.9251 4.5130	1.9305 1.8355

Table 6. The values of the optimal initial guess, r_0, objective functions and the corresponding errors using the regularisation parameters $\lambda_1, \lambda_2 \in \{0, 10^{-6}, 10^{-3}, 10^{-1}\}$ with $M = N = 25$ for Example 2.

			r_0	Objective Functions	L_∞-Error	RMSE
Without noise	$\lambda_1 = 0$	$\lambda_2 = 10^{-6}$ $\lambda_2 = 10^{-3}$ $\lambda_2 = 10^{-1}$	0.5 0.4 0.8	5.3972×10^1 4.1825×10^1 2.0433×10^{-1}	5.0754×10^{-1} 4.0134×10^{-1} 4.0078×10^{-1}	6.3467×10^{-1} 4.7861×10^{-1} 4.5618×10^{-1}
	$\lambda_1 = 10^{-6}$ $\lambda_1 = 10^{-3}$ $\lambda_1 = 10^{-1}$	$\lambda_2 = 0$	0.6 0.1 0.4	4.0430×10^1 5.9128 2.5867×10^1	1.5634 4.9809×10^{-1} 1.9664	1.3413 5.6395×10^{-1} 2.9192
With noise 5%	$\lambda_1 = 0$	$\lambda_2 = 10^{-6}$ $\lambda_2 = 10^{-3}$ $\lambda_2 = 10^{-1}$	0.8 0.2 0.1	3.3672×10^1 5.4494 1.9182×10^{-1}	1.9251 5.3183×10^{-1} 4.3568×10^{-1}	1.9396 4.8568×10^{-1} 4.7239×10^{-1}
	$\lambda_1 = 10^{-6}$ $\lambda_1 = 10^{-3}$ $\lambda_1 = 10^{-1}$	$\lambda_2 = 0$	0.8 0.2 0.4	3.4257 5.9130 2.5843×10^1	4.6055 6.4875×10^{-1} 1.9654	2.1547 6.2476×10^{-1} 2.9119

Table 7. The values of the optimal initial guess, r_0, objective functions and the corresponding errors using the regularisation parameters $\lambda_1, \lambda_2 \in \{0, 10^{-6}, 10^{-3}, 10^{-1}\}$ with $M = N = 50$ for Example 2.

			r_0	Objective Functions	L_∞-Error	RMSE
Without noise	$\lambda_1 = 0$	$\lambda_2 = 10^{-6}$ $\lambda_2 = 10^{-3}$ $\lambda_2 = 10^{-1}$	0.4 0.2 0.8	8.3494×10^1 1.4682×10^{-4} 3.8174	8.2159×10^{-1} 8.0200×10^{-1} 9.5145×10^{-1}	1.2749 1.0192 1.3288
	$\lambda_1 = 10^{-6}$ $\lambda_1 = 10^{-3}$ $\lambda_1 = 10^{-1}$	$\lambda_2 = 0$	0.5 0.7 0.4	2.0732 3.6352 4.7401×10^1	6.4408 5.2101 1.1726×10^1	2.9474 3.879 7.6963
With noise 5%	$\lambda_1 = 0$	$\lambda_2 = 10^{-6}$ $\lambda_2 = 10^{-3}$ $\lambda_2 = 10^{-1}$	0.8 0.9 0.7	3.6080×10^1 2.1617×10^1 3.2234×10^{-2}	8.5674×10^{-1} 5.2427×10^{-1} 4.7546×10^{-1}	1.1744 6.5482×10^{-1} 4.9726×10^{-1}
	$\lambda_1 = 10^{-6}$ $\lambda_1 = 10^{-3}$ $\lambda_1 = 10^{-1}$	$\lambda_2 = 0$	0.6 0.6 0.7	1.0459×10^1 7.0388 7.5548×10^1	6.7322×10^{-1} 9.6421×10^{-1} 3.4477×10^3	6.447×10^{-1} 1.1551 1.2448×10^3

Table 8. The values of the minimal objective functions and the corresponding errors with initial guess, r_0, obtained (with/no) selecting the optimal regularisation parameters λ_1, λ_2 with $M = N \in \{25, 50\}$ for Example 2.

$M = N$	With/No Noise	With/No Regularisation	λ_1	λ_2	r_0	Objective Function	L_∞-Error	RMSE
25	no	no with	0	10^{-1}	0.5 0.8	5.0884×10^1 2.0433×10^{-1}	4.1355×10^{-1} 4.0078×10^{-1}	6.0305×10^{-1} 4.5618×10^{-1}
	with	no with	0	10^{-1}	0.8 0.1	3.2489×10^1 1.9182×10^{-1}	1.9251 4.3568×10^{-1}	1.9305 4.7239×10^{-1}
50	no	no with	0	10^{-3}	0.6 0.2	1.0837×10^2 1.4682×10^{-4}	8.1024×10^{-1} 8.0200×10^{-1}	1.2263 1.0192
	with	no with	0	10^{-1}	0.7 0.7	5.4166×10^1 3.2234×10^{-2}	4.5130 4.7546×10^{-1}	1.8355 4.9726×10^{-1}

6. Conclusions

In this paper, we extended the aforementioned method presented in [8], based on the method of fundamental solutions to solve numerically the three-dimensional inverse geometry problem on an annular domain. To obtain the stable and accuracy results, Tikhonov's regularisation parameters were used combined with the problem of the minimising an objective function. From the examples, we can see that our proposed method is effective and stable, even for the boundary data added with noise.

Author Contributions: Validation, H.A.; formal analysis, E.A., C.P.; investigation, M.S., H.A., E.A., C.P.; writing—original draft preparation, M.S.; writing—review and editing, H.A.; supervision, M.S.; funding acquisition, C.P. All authors have read and agreed to the published version of the manuscript.

Funding: This work does not receive any external funding.

Institutional Review Board Statement: Not applicable.

Informed Consent Statement: Not applicable.

Data Availability Statement: Data sharing is not applicable to this article, as no data set was generated or analysed during the current study.

Conflicts of Interest: The authors declare no confilct of interests.

References

1. Lesnic, D.; Berger, J.R.; Martin, P.A. A boundary element regularization method for the boundary determination in potential corrosion damage. *Inverse Probl. Eng.* **2002**, *10*, 163–182. [CrossRef]
2. Karageorghis, A.; Lesnic, D.; Marin, L. The method of fundamental solutions for the detection of rigid inclusions and cavities in plane linear elastic bodies. *Comput. Struct.* **2012**, *106*, 176–188. [CrossRef]
3. Karageorghis, A.; Lesnic, D.; Marin, L. A moving pseudo-boundary method of fundamental solutions for void detection. *Numer. Methods Partial Diff. Equat.* **2013**, *29*, 935–960. [CrossRef]
4. Karageorghis, A.; Lesnic, D.; Marin, L. The method of fundamental solutions for three-dimensional inverse geometric elasticity problems. *Comput. Struct.* **2016**, *166*, 51–59. [CrossRef]
5. Wanga, F.; Huaa, Q.; Liub, C.S. Boundary function method for inverse geometry problem in two-dimensional anisotropic heat conduction equation. *Appl. Math. Lett.* **2018**, *84*, 130–136. [CrossRef]
6. Huang, C.H.; Yan, J.Y. An inverse problem in simultaneously measuring temperature dependent thermal conductivity and heat capacity. *Int. J. Heat Mass Transf.* **1995**, *38*, 3433–3441. [CrossRef]
7. Huang, C.H.; Yeh, C.Y.; Orlande, H.R.B. A non-linear inverse problem in simultaneously estimating the heat and mass production rates for a chemically reacting fluid. *Chem. Eng. Sci.* **2003**, *58*, 3741–3752. [CrossRef]
8. Bin-Mohsin, B.; Lesnic, D. Determination of inner boundaries in modified Helmholtz inverse geometric problems using the method of fundamental solutions. *Math. Comput. Simul.* **2012**, *82*, 1445–1458. [CrossRef]
9. Vladimirov, V.S. *Equations of Mathematical Physics*; Mir Publishers: Moscow, Russia, 1981.

fractal and fractional

Article

Multistability of the Vibrating System of a Micro Resonator

Yijun Zhu and Huilin Shang *

School of Mechanical Engineering, Shanghai Institute of Technology, Shanghai 201418, China; zyjmain@163.com
* Correspondence: suliner60@hotmail.com

Abstract: Multiple attractors and their fractal basins of attraction can lead to the loss of global stability and integrity of Micro Electro Mechanical Systems (MEMS). In this paper, multistability of a class of electrostatic bilateral capacitive micro-resonator is researched in detail. First, the dynamical model is established and made dimensionless. Second, via the perturbating method and the numerical description of basins of attraction, the multiple periodic motions under primary resonance are discussed. It is found that the variation of AC voltage can induce safe jump of the micro resonator. In addition, with the increase of the amplitude of AC voltage, hidden attractors and chaos appear. The results may have some potential value in the design of MEMS devices.

Keywords: micro resonator; fractal; multistability; safe jump; hidden attractor; chaos; basin of attraction

1. Introduction

Multistability, i.e., the coexistence of multiple attractors, is a common dynamical phenomenon in MEMS/NEMS [1,2]. Based on it, there are many applications such as MEMS-based memory [3] and switches [4]. In addition, considering the loss of global stability that multistability may trigger, there are some devices that should avoid the appearance of multiple attractors in their vibrating systems, such as filters [5], microvalves [6], and micro-relays [7]. As one of the fastest developing MEMS products [8], electrostatic micro-resonators should assure that the resonators undergo periodic vibration whose amplitudes vary continuously with the driven voltages. However, in practical applications of electrostatic micro-resonators [9], there are many complex dynamic behaviors such as multistability [10,11], quasi-periodic motion [12] and periodic-n motion [13], chaos [14,15], and pull-in instability.

It is of great significance to study the multistability and necessary conditions for inducing it either for avoiding this phenomenon or making use of it. Thus, multistability of vibrating systems of micro resonators has been studied experimentally and numerically during these decades [16]. Via experiments, Mohammadreza investigated the dynamic response of an electrostatic micro-actuator in the vicinity of the primary resonance and the parametric one [17]. Siewe et al. [18] studied the vibration of a double-side MEMS resonator numerically and found the variation of the driven voltage could induce the coexistence of chaos and quasi-periodic motions. Shang et al. [19] found the coexisting chaos and dynamical pull-in in the vibrating system of a single-side electrostatic micro sensor. Haghighi et al. [20] found the coexisting periodic-n motion and the chaotic motion of micromechanical resonators with electrostatic forces on both sides, and then discussed the global bifurcation of its vibrating system by approximately expressing its homoclinic orbits as the ones of a typical duffing equation. When amplifying signals of a nanomechanical duffing resonator, Almog et al. [21] found that multistability was an interesting dynamical phenomenon of nonlinear systems and could be explored for many applications. Gusso et al. [22] studied chaos of a typical micro/nanoelectromechanical beam resonator with two-sided electrodes experimentally and observed multiple attractors in a significant region of the relevant parameter space, involving periodic and chaotic attractors.

By applying cell-mapping method to depict the basins of attraction for all the attractors, they also found that the basin boundaries were fractal under certain conditions of the excitations, indicating that the attractors are strongly intermingled. Liu et al. [23] applied the method of multiple scales (MMS) to analyze the multiple periodic motions induced by the local bifurcation, and used the Melnikov method to predict necessary conditions for chaos and its control. The corresponding numerical results were also presented by the basins of attraction and spectrum diagrams. Angelo et al. [24] investigated the effect of the linear and nonlinear stiffness terms and damping coefficients on dynamical behaviors of a microelectromechanical resonator and controlled the chaotic motion by forcing it into an orbit obtained analytically via the harmonic balance method. However, most study concentrated on describing or observing the phenomenon itself rather than studying its mechanism, which is still not that clear yet.

To this end, we consider a typical electrostatic driven bilateral capacitive microresonator and study the possible multistability and its mechanism in its vibrating system. The paper is organized as follows. In Section 2, the dynamical model is constructed and made dimensionless. In Sections 3 and 4, two different cases for coexisting multiple periodic attractors, fractal basins of attraction, and other complex attractors of the systems are discussed both theoretically and numerically. In Section 5, the conclusions are presented.

2. Dynamical Model

We choose to study a class of bilateral micro resonator whose simplified diagram is shown in Figure 1. The driven forces on the resonator are electrostatic ones between the moving electrode and the fixed electrode [25]. The driven voltage in Figure 1 is the combination of alternate current (AC) and direct current (DC) actuation. In the figure, x is the vertical displacement of the moving electrode at moment t, d the initial gap width between the moving electrode and each fixed one, V_b the DC bias voltage, $V_{AC}sin\Omega t$ the AC voltage where V_{AC} is the amplitude and Ω the frequency. Suppose that the amplitude of the AC voltage V_{AC} is much lower than the bias DC voltage V_b, i.e., $V_{AC} \ll V_b$. According to the Second Law of Newton, the vibrating system of the moving electrode can be expressed as a nonlinear system as follows:

$$m\frac{d^2x}{dt^2} + c\frac{dx}{dt} + k_1 x + k_2 x^3 = \frac{C_0}{2(d-x)^2}(V_b + V_{AC}sin\Omega t)^2 - \frac{C_0 V_b^2}{2(d+x)^2} \quad (1)$$

where m represents the effective lumped mass of the moving electrode, k_1 its linear mechanical stiffness, k_2 its cubic nonlinear stiffness, c the damping coefficient, C_0 the initial capacitance of the parallel-plate structure.

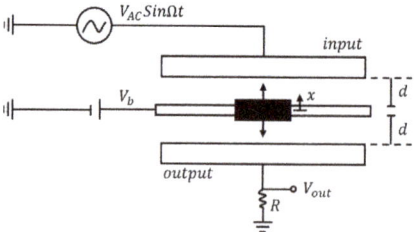

Figure 1. Simplified diagram of a bilateral MEMS resonator.

Introducing the following dimensionless variables

$$\omega_0 = \sqrt{\frac{k_1}{m}}, \omega = \frac{\Omega}{\omega_0}, \mu = \frac{c}{m\omega_0}, \alpha = \frac{k_2 d^2}{m\omega_0^2}, \beta = \frac{C_0 V_b^2}{2k_1 d^3}, \gamma = \frac{V_{AC}}{V_b}, T = \omega_0 t, u = \frac{x}{d}, \dot{u} = \frac{du}{dT} \quad (2)$$

and substituting Equation (2) into Equation (1), one can obtain that

$$\ddot{u} + \mu\dot{u} + u + \alpha u^3 = \frac{\beta}{(1-u)^2}(1+\gamma sin\omega T)^2 - \frac{\beta}{(1+u)^2} \quad (3)$$

which is a dimensionless system. Since in the original system (1), the viscous damping coefficient of air c is very tiny, and $V_{AC} \ll V_b$, the parameters μ and γ in (3) will be both small and can be considered as perturbed parameters. Thus, considering $\mu = 0$ and $\gamma = 0$ in Equation (3), one has the unperturbed system that can be expressed as below:

$$\dot{u} = v, \ \dot{v} = -u - \alpha u^3 + \frac{\beta}{(1-u)^2} - \frac{\beta}{(1+u)^2}. \quad (4)$$

Letting the right side of Equation (4) be zero, one can determine equilibria of the dimensionless system (3). Equation (5) is a Hamilton system with the Hamiltonian

$$H(u,v) = \frac{1}{2}v^2 + \frac{1}{2}u^2 + \frac{\alpha}{4}u^4 - \frac{\beta}{1-u} - \frac{\beta}{1+u} + 2\beta \quad (5)$$

and the function of potential energy (P.E.)

$$V(u) = \frac{1}{2}u^2 + \frac{\alpha}{4}u^4 - \frac{\beta}{1-u} - \frac{\beta}{1+u} + 2\beta. \quad (6)$$

Concerning Equation (4), the number of the equilibria, and the shapes and positions of the possible potential wells of the unperturbed system (4) depend on the parameters α and β. The same as in [20], the values of the parameters in the system (1) are given by:

$$m = 5 \times 12^{-12} \text{ kg}, c = 5 \times 12^{-8} \text{ kg/s}, k_1 = 5 \text{ μN/μm}, k_2 = 15 \text{ μN/μm}^3, d = 2\text{μm}, C_0 = 1.875 \times 10^{-18} \text{ mF}. \quad (7)$$

Accordingly, in system (4), $\alpha = 12$.

Different equilibria and potential energy diagrams of the unperturbed system under different values of the parameter β can be seen in Figure 2. It shows that there are three P.E. poles when $\beta = 0.211$, five P.E. poles when β increases to 0.338, and only one P.E. pole when β increases to 0.6. Under different values of β, the potential wells and unperturbed orbits are shown in Figure 3. When $\beta = 0.211$, there are three equilibria (two non-trivial equilibria are saddles and the origin is a center) as well as one well surrounded by heteroclinic orbits (see Figure 3a). As β increases to 0.338, there will be five equilibria among which two non-trivial equilibria S_1 (−0.196339,0) and S_2 (0.196339,0) are centers of the two wells surrounded by homoclinic orbits; the other three equilibria are unstable. When $\beta = 0.6$, no wells or non-trivial equilibria of the unperturbed system (4) exist. The P.E. poles in Figure 2 correspond to the fixed points shown in Figure 3. Therefore, according to Equations (2) and (7), when the structural parameters are fixed, the number of centers will depend on the value of DC bias voltage V_b: when the DC bias voltage is very low, there will be a center of the system (4) as well as a stable point attractor of the system (3) without AC voltage. Under a higher DC bias voltage, there may be two centers of the system (4). As is well known, periodic vibration can often be attributed to the perturbation of the centers. Since the number and the location of the centers in Figure 3a,b are totally different, the mechanism for the possible multiple periodic attractors of the vibrating system of the micro-resonator can be different as well. Therefore, in Sections 3 and 4, we discuss the different mechanism of multi-stability for these two different cases, i.e., the only center (the origin) and the two non-trivial centers, respectively.

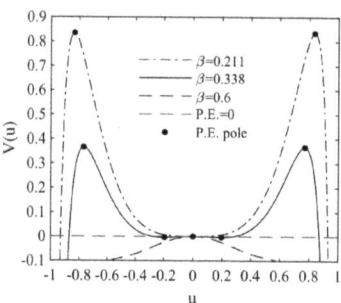

Figure 2. Potential energy of the unperturbed system (4) under different values of parameter β.

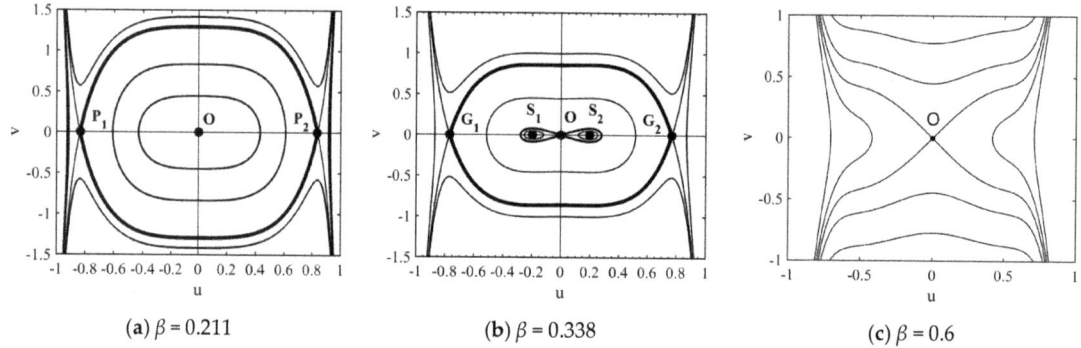

Figure 3. Orbits of the unperturbed system (4) under different values of parameter β.

3. Multiple Periodic Attractors in the Neighborhood of the Origin

Considering the case where the DC bias voltage is low, and the periodic vibration of the microstructure is induced by the perturbation of the only center (see Figure 3a, where $V_b = 3$ V), one may use the Method of Multiple Scales (MMS) to analyze the periodic solutions in the neighborhood of the origin. Expanding the fractional terms of the dimensionless system (3) as Taylor series in the neighborhood of $u = 0$, and neglecting the higher-order-than-three terms of u, one has:

$$\ddot{u} + \mu\dot{u} + u + \alpha u^3 = 2\beta\gamma \sin\omega T + 4\beta u + 4u\beta\gamma \sin\omega T + 6u^2\beta\gamma \sin\omega T + 8\beta u^3 + 8u^3\beta\gamma \sin\omega T. \tag{8}$$

As mentioned in Section 2, the values of the parameters μ and γ in the above system are small; one can introduce a small parameter ε satisfying $0 < \varepsilon \ll 1$, and can re-scale the two parameters in the system (8) as:

$$\mu = \varepsilon^2 \tilde{\mu}, \quad \gamma = \varepsilon^2 \tilde{\gamma}. \tag{9}$$

Then Equation (8) becomes

$$\ddot{u} + \tilde{\omega}^2 u = -\varepsilon^2 \tilde{\mu}\dot{u} + 2\varepsilon^2 \beta\tilde{\gamma} \sin\omega T + 4u\varepsilon^2 \beta\tilde{\gamma} \sin\omega T + 6u^2\varepsilon^2 \beta\tilde{\gamma} \sin\omega T - P_1 u^3 + 8u^3\varepsilon^2 \beta\tilde{\gamma} \sin\omega T. \tag{10}$$

where

$$\tilde{\omega}^2 = 1 - 4\beta, \quad P_1 = \alpha - 8\beta. \tag{11}$$

To apply MMS, one may rescale some terms in the system (10) that

$$\omega = \tilde{\omega} + \varepsilon\sigma, \quad u = \varepsilon u_1 + \varepsilon^2 u_2 + \cdots, \quad \sigma = O(1). \tag{12}$$

and
$$T_i = \varepsilon^i T, D_i = \frac{\partial}{\partial T_i}, \frac{d}{dT} = \sum_{i=0}^{n} \varepsilon^i D_i \ (i = 0, 1, 2, \cdots) \tag{13}$$

Comparing the coefficients of $\varepsilon^1, \varepsilon^2$, and ε^3 in the system (10), respectively, one obtains that
$$\varepsilon^1: \ D_0^2 u_1 + \omega^2 u_1 = 0, \tag{14}$$
$$\varepsilon^2: \ D_0^2 u_2 + \omega^2 u_2 = -2D_1 D_0 u_1 + 2\omega\sigma u_1 + 2\beta\widetilde{\gamma}\sin\omega T, \tag{15}$$

and
$$\varepsilon^3: D_0^2 u_3 + \omega^2 u_3 = -2D_1 D_0 u_2 - \widetilde{\mu} D_0 u_1 - D_1^2 u_1 + 2u_2\omega\sigma - \sigma^2 u_1 - 2D_2 D_0 u_1 - P_1 u_1^3 + 4\beta\widetilde{\gamma} u_1 \sin\omega T. \tag{16}$$

To solve Equation (14), one can assume that
$$u_1 = A_1(T_1, T_2)e^{i\omega T_0} + \overline{A}_1(T_1, T_2)e^{-i\omega T_0}, \tag{17}$$

where
$$A_1 = \frac{a(T_1, T_2)}{2} e^{i\theta(T_1, T_2)}. \tag{18}$$

Substituting Equations (17) and (18) into Equation (15), and eliminating the secular terms of Equation (15), one will have:
$$D_1 A_1 = -\frac{\beta\widetilde{\gamma}}{2\omega} - i\sigma A_1. \tag{19}$$

Solving Equation (15), one may assume:
$$u_2 = A_2(T_2)e^{i\omega T_0} + \overline{A}_2(T_2)e^{-i\omega T_0}. \tag{20}$$

Substituting Equation (20) into Equation (16), and eliminating secular terms of Equation (16), one will obtain:
$$D_2 A_1 = -\frac{\widetilde{\mu}}{2} A_1 + \frac{\beta\widetilde{\gamma}}{2\omega} - \frac{\sigma\beta\widetilde{\gamma}}{4\omega^2} + \frac{3iP_1 A_1^2 \overline{A}_1}{2\omega}. \tag{21}$$

Since
$$\dot{A}_1 \approx D_0 A_1 + \varepsilon D_1 A_1 + \varepsilon^2 D_2 A_1, \tag{22}$$

Substituting Equations (19) and (21) into Equation (22), and expressing it by the original dimensionless parameters of Equation (3), one has:
$$\begin{array}{l}\varepsilon\dot{a} = -\frac{\mu}{2}(\varepsilon a) - P_2 \cos\theta, \\ (\varepsilon a)\dot{\theta} = -(\omega - \widetilde{\omega})(\varepsilon a) + \frac{3P_1(\varepsilon a)^3}{8\omega} + P_2 \sin\theta.\end{array} \tag{23}$$

where
$$P_2 = \frac{(3\omega - \widetilde{\omega})\beta\gamma}{2\omega^2}. \tag{24}$$

According to Equation (18), it is obvious that the amplitude of the periodic solution a is the function of the time scale T_1 where T_1 is a one-order term of ε; thus, one can assume the amplitude of the solution u of Equation (10) $\widetilde{a}$ as:
$$\varepsilon a = \widetilde{a}. \tag{25}$$

Letting $\dot{a} = 0$, and $\dot{\theta} = 0$, one can obtain:
$$-\frac{\mu}{2}\widetilde{a} = P_2 \cos\theta, \ (\omega - \widetilde{\omega})\widetilde{a} - \frac{3P_1 \widetilde{a}^3}{8\omega} = P_2 \sin\theta. \tag{26}$$

Eliminating the triangulation function of Equation (26), one can get:

$$\frac{\mu^2}{4}a^2 + \left(\omega - \widetilde{\omega} - \frac{3P_1 a^2}{8\omega}\right)^2 a^2 = \frac{(3\omega - \widetilde{\omega})^2}{4\omega^4}\beta^2 \gamma^2. \qquad (27)$$

According to Equations (17), (18) and (25), the periodic solution can be expressed as:

$$u \approx \widetilde{a}\cos(\omega T + \theta). \qquad (28)$$

To determine the stability of the periodic solutions, one can get the corresponding characteristic equation of the periodic solution based on Equation (23). It shows that the periodic solution will lose its stability when its amplitude $\widetilde{a}$ satisfies:

$$\left(8(\omega - \widetilde{\omega}) - \frac{9\widetilde{a}^2 P_1}{\omega}\right)\left(8(\omega - \widetilde{\omega}) - \frac{3\widetilde{a}^2 P_1}{\omega}\right) \geq 16\mu^2. \qquad (29)$$

Based on Equations (26)–(29), the variation of the amplitude of the periodic solutions of the system (3) and their stability with AC voltage is shown in Figure 4 where the frequency and the amplitude of AC voltage are considered as the control parameters in Figure 4a and Figure 4b, respectively. In Figure 4a, where $V_{AC} = 0.01$ V, when ω is lower than 0.45, there is only one periodic attractor in the system (3) whose amplitude changes continuously with the increase in ω. Comparatively, when ω ranges from 0.46 to 0.69, the global dynamical behaviors of the system (3) will change to bistable periodic attractors, which can be attributed to Hopf bifurcation. As ω continues to increase from 0.7, the periodic attractor with the higher amplitude will disappear, only the periodic attractor with the lower amplitude will exist, and its amplitude will decrease continuously with the increase of ω. Similarly, the change in global dynamical behaviors in Figure 4b also shows that in a certain range of V_{AC}, there will be two periodic attractors coexisting, which can be due to Hopf bifurcation of the system (3). The accuracy of the theoretical prediction in Figure 4 are verified by the numerical results.

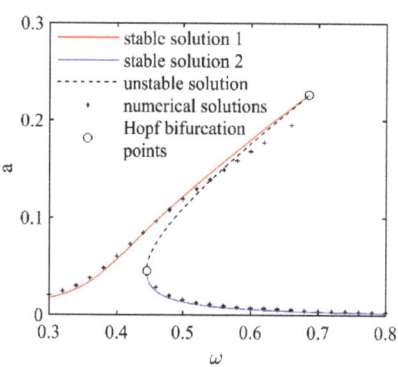

(a) Amplitude of the periodic solution vs. ω when $V_{AC} = 0.01$ V

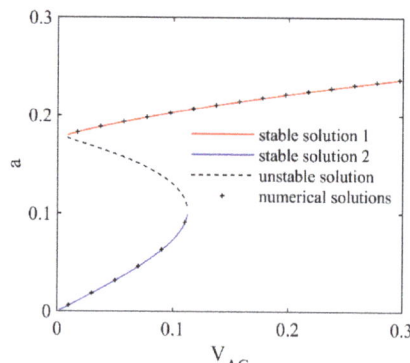

(b) Amplitude of the periodic solution vs. V_{AC} when $\omega = 0.6$

Figure 4. Variation of the amplitude of the periodic solution with the change in AC voltage.

Figure 4 demonstrates that the parameters ω and VAC can induce the coexistence of bistability, meaning that under fixed values of parameters of system (3), different initial conditions may lead to different periodic attractors. Accordingly, it is necessary to classify the basins of attraction for the two different periodic attractors. Here, the 4th order Runge-Kutta approach and the cell-mapping method are applied to depict the basins of attraction of the system (3). The time step is taken as $1/10^2$ of the period of excitation. To investigate the long-term dynamical behaviors, it is supposed that an initial condition will be safe if the vibration in this initial condition keeps satisfying $|u(T)| < 1$ within 10^5 excited circles; otherwise, the micro resonator will undergo pull in [20]. The union of all initial conditions leading to the same periodic motion will be the basin of attraction for that attractor which will surely be marked in the same color in the initial plane. The basins of attraction of system (3) are drawn in sufficiently large ranges for the initial position and velocity of the proof mass defined as $|u(0)| < 1$ and $|\dot{u}(0)| < 1.5$ by generating an 200×100 array of initial points. The change of attractors and the area and nature of their basins of attraction with frequency ω is shown in Figure 5 where the amplitude of AC voltage V_{AC} is fixed as 0.01 V.

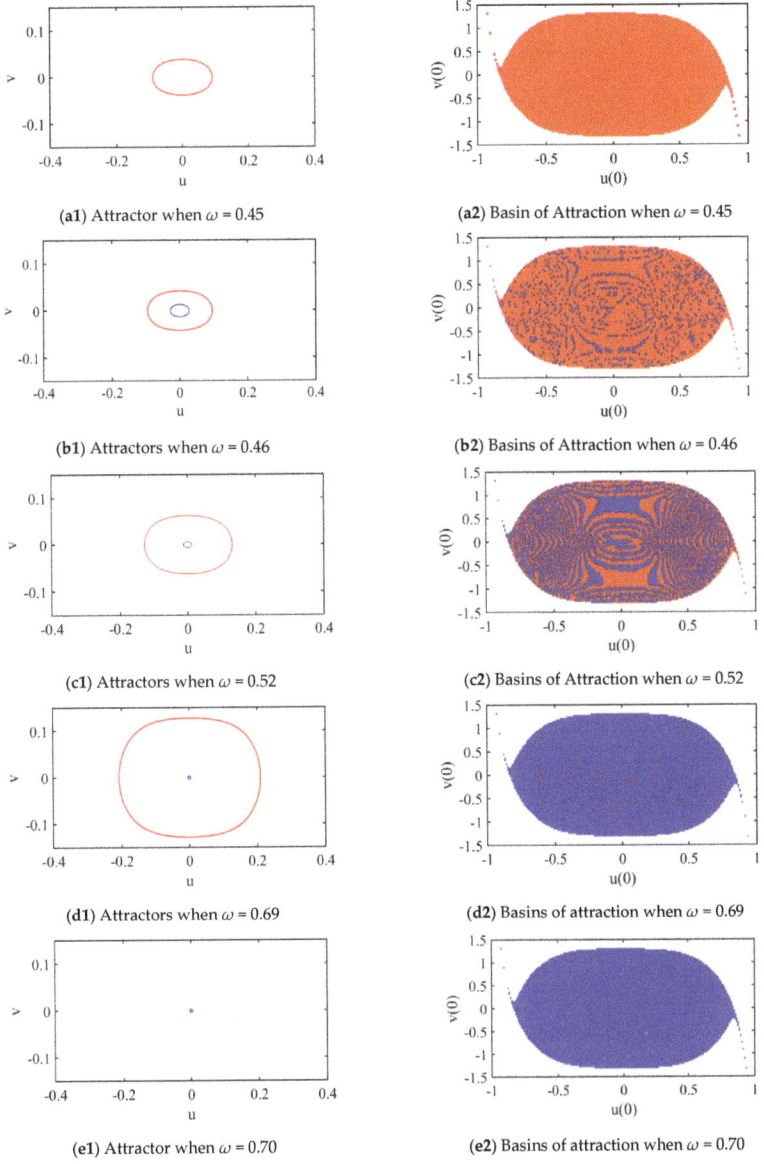

Figure 5. Evolution of multiple attractors and their basins of attraction under different values of ω.

According to Figure 5, with the increase in parameter ω, the number of attractors and the boundary of basins of attraction will both change. When $\omega = 0.45$ in system (3), there will only be one periodic attractor whose basin of attraction is comparatively bigger with a smooth boundary (see Figure 5a1,a2). However, with a small increase of ω, i.e., $\omega = 0.45$, the global dynamics are totally different, as shown in Figure 5b1,b2 where two periodic attractors coexist, whose basins of attraction mix each other and are both fractal. It means that the dynamical behavior of system (3) is highly sensitive to initial conditions. In other words, system (3) may undergo a safe jump. A similar phenomenon can be seen in Figure 5c1,c2 under a higher ω. As ω increases, the basin of attraction of the periodic attractor with the higher amplitude becomes small (see the red regions in Figure 5b2,c2,d2). Specifically, in Figure 5d2, the regions of attraction are almost blue, and there is very little area of basin of attraction for that periodic attractor. As $\omega = 0.70$ (see Figure 5e1,e2), the periodic attractor with the higher amplitude disappears, and there is only the other attractor whose basin of attraction is almost the same as that in Figure 4b, showing that when the frequency ω increases enough, the periodic attractor with the lower amplitude replaces the initial one.

4. Multistability in the Neighborhood of Non-Trivial Equilibria

In this section, the case that the DC voltage is higher, and the periodic vibration of the microstructure is induced by the perturbation of the two non-trivial centers (see Figure 3b) is considered; thus, we set $\beta = 0.338$, i.e., $V_b = 3.8V$. In addition, we consider the effect of AC voltage on the global dynamics of the system (3). To begin with, setting

$$\varepsilon \hat{u} = u \mp u_c \tag{30}$$

where u_c is the abscissa of the right center (see S_1 in Figure 3), rescaling he two parameters μ and γ in the system (3) by Equation (9), expanding the fractional terms of the dimensionless system (3) as a Taylor series in the neighborhood of the non-trivial equilibria and ignoring the higher-order-than-cubic terms of $\hat{u}$, the system (3) becomes

$$\ddot{\hat{u}} = -\varepsilon^2 \tilde{\mu} \dot{\hat{u}} - \hat{\omega}^2 \hat{u} + \varepsilon Q_1 \hat{u}^2 + \varepsilon^2 Q_2 \hat{u}^3 + \frac{2\varepsilon \beta \tilde{\gamma} \sin \omega T}{(1 \mp u_c)^2} + \frac{4\varepsilon^2 \beta \tilde{\gamma} \sin \omega T}{(1 \mp u_c)^3} \hat{u} + \frac{6\varepsilon^3 \beta \tilde{\gamma} \sin \omega T}{(1 \mp u_c)^4} \hat{u}^2, \tag{31}$$

where

$$\hat{\omega}^2 = 1 + 3\alpha u_c^2 - \frac{4\beta(1 + 3u_c^2)}{(1 - u_c^2)^3}, \quad Q_1 = \pm 3u_c\left(-\alpha + \frac{8\beta(1 + u_c^2)}{(1 - u_c^2)^4}\right), \quad Q_2 = -\alpha + \frac{8\beta(1 + 10u_c^2 + 5u_c^4)}{(1 - u_c^2)^5}. \tag{32}$$

To apply the Method of Multiple Scale in Equation (32), one can assume in this equation that:

$$\hat{\omega} = \omega + \varepsilon \hat{\sigma}, \quad \hat{u} = \hat{u}_0 + \varepsilon \hat{u}_1 + \varepsilon^2 \hat{u}_2 + \cdots, \quad \hat{\sigma} = O(1). \tag{33}$$

Comparing the coefficients of $\varepsilon^1, \varepsilon^2$ and ε^3, one has:

$$\varepsilon^0: D_0^2 \hat{u}_0 + \omega^2 \hat{u}_0 = 0, \tag{34}$$

$$\varepsilon^1: D_0^2 \hat{u}_1 + \omega^2 \hat{u}_1 = -2D_1 D_0 \hat{u}_0 + 2\omega \hat{\sigma} \hat{u}_0 + \frac{2\beta \tilde{\gamma} \sin \omega T}{(1 \mp u_c)^2} + Q_1 \hat{u}_0^2, \tag{35}$$

and

$$\varepsilon^2: D_0^2 \hat{u}_2 + \omega^2 \hat{u}_2 = -2D_1 D_0 \hat{u}_1 - \tilde{\mu} D_0 \hat{u}_0 - D_1^2 \hat{u}_0 + 2\hat{u}_1 \omega \hat{\sigma} - \hat{\sigma}^2 \hat{u}_0 - 2D_2 D_0 \hat{u}_0 + 2Q_1 \hat{u}_0 \hat{u}_1 + Q_2 \hat{u}_0^3 + \frac{4\beta \tilde{\gamma} \sin \omega T}{(1 \mp u_c)^3} \hat{u}_0. \tag{36}$$

One can set the solution of Equation (34) as:

$$\hat{u}_0 = B_1(T_1) e^{i\omega T_0} + \overline{B}_1(T_1) e^{-i\omega T_0}. \tag{37}$$

Substituting Equation (37) into Equation (35), and eliminating the secular terms of Equation (35), one can obtain that:

$$D_1 B_1 = \frac{-\beta\tilde{\gamma}}{2\omega(1\mp u_c)^2} - i\sigma B_1,$$
$$\hat{u}_1 = -\frac{B_1^2 P}{3\omega^2}e^{i2\omega T_0} - \frac{\overline{B}_1^2 P}{3\omega^2}e^{-i2\omega T_0} + \frac{2B_1\overline{B}_1 P}{\omega^2}. \quad (38)$$

Substituting the equation above into Equation (36) and eliminating its secular terms, one can have:

$$D_2 B_1 = -\frac{\tilde{\mu}B_1}{2} - \frac{\hat{\sigma}\beta\tilde{\gamma}}{4\omega^2(1\mp u_c)^2} - i\left(\frac{5Q_1^2}{3\omega^3} + \frac{3Q_2}{2\omega}\right)B_1^2\overline{B}_1. \quad (39)$$

Now setting

$$B_1 = \frac{1}{2}\varepsilon b(T_1, T_2)e^{i\varphi(T_1, T_2)}, \quad (40)$$

considering

$$\dot{B}_1 \approx D_0 B_1 + \varepsilon D_1 B_1 + \varepsilon^2 D_2 B_1, \quad (41)$$

and substituting Equations (38) and (39) into Equation (41), and expressing Equation (41) by the original dimensionless parameters of Equation (3), one has:

$$\dot{b} = \frac{-(3\omega-\hat{\omega})\beta\gamma\cos\varphi}{2\omega^2(1\mp u_c)^2} - \frac{\mu b}{2},$$
$$b\dot{\varphi} = \frac{(3\omega-\hat{\omega})\beta\gamma\sin\varphi}{2\omega^2(1\mp u_c)^2} - (\omega-\hat{\omega})b - \frac{5b^3 Q_1^2}{12\omega^3} - \frac{3b^3 Q_2}{8\omega}. \quad (42)$$

The periodic solution of the system (3) satisfies $\dot{b} = 0$, and $\dot{\varphi} = 0$, i.e.,

$$-\frac{\mu b}{2} = \frac{(3\omega-\hat{\omega})\beta\gamma\cos\varphi}{2\omega^2(1\mp u_c)^2}, \quad (\omega-\hat{\omega})b + \left(\frac{5Q_1^2}{12\omega^3} + \frac{3Q_2}{8\omega}\right)b^3 = \frac{(3\omega-\hat{\omega})\beta\gamma\sin\varphi}{2\omega^2(1\mp u_c)^2}. \quad (43)$$

The periodic solution can be expressed analytically as:

$$u = \pm u_c + \frac{2b^2 Q_1}{3\omega^2} + b\cos(\omega T + \varphi) - \frac{b^2 Q_1}{3\omega^2}\cos^2(\omega T + \varphi). \quad (44)$$

According to the characteristic solutions of Equation (42), it shows that the theoretical periodic solution expressed by Equation (44) will become unstable if:

$$\left(\omega - \hat{\omega} - \left(\frac{5Q_1^2}{12\omega^3} + \frac{9Q_2}{8\omega}\right)b^2\right)\left(\omega - \hat{\omega} + \left(\frac{5Q_1^2}{12\omega^3} + \frac{3Q_2}{8\omega}\right)b^2\right) \geq \frac{\mu^2}{4}. \quad (45)$$

Based on Equations (43)–(45), the evolution of the periodic solutions of system (3) with the amplitude of AC voltage when $\omega = 0.6$ is shown in Figure 6. Obviously, when V_{AC} increases from 0, the two non-trivial equilibria lose their stability; instead, there are two periodic attractors coexisting. The amplitudes of the two periodic attractors increases with the amplitude of AC voltage. The coexistence of multiple periodic attractors can be attributed to the disturbance of the bistable non-trivial equilibria of the system (3) when $V_{AC} = 0$ V.

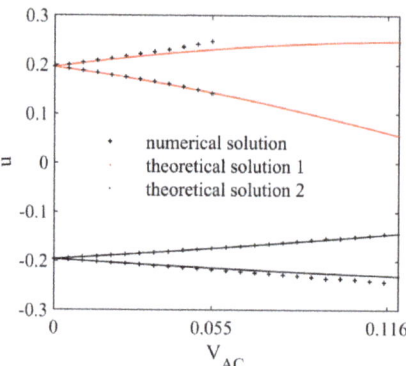

Figure 6. Variation of the periodic solutions with the amplitude of AC voltage when $\omega = 0.6$.

In Figure 6, when V_{AC} varies from 0 to 0.055 V, the numerical simulation is in great agreement with the theoretical solution. However, when V_{AC} exceeds 0.056 V, the theoretical prediction of the periodic attractor in the neighborhood of the right non-trivial equilibria is not that accurate, which may be due to the limitation of the Method of Multiple Scale. It will then be essential for us to apply numerical simulation to investigate the evolution of the attractors with the change in AC voltage. The basic settings for the simulation, such as the time step and initial plane, are the same as that in Section 3. The change of the attractors and the area and nature of their basins of attraction with V_{AC} are shown in Figure 7, where $\omega = 0.6$. The evolution of global dynamics of system (3) with the increase in V_{AC} can be separated into the following five stages.

Firstly, when $V_{AC} = 0$ V, there are two point attractors coexisting whose basins of attraction are fractal and trigger each other (see Figure 7a1,a2). According to Figure 7a2, in a small neighborhood of each point attractor, the attractor of system (3) is locally stable. Otherwise, a small disturbance of initial conditions will lead to a different point attractor, meaning that it is easy to induce a safe jump.

Secondly, when V_{AC} increases from 0 to 0.01 V (see Figure 7b1–d2), the number of the periodic attractors increases with V_{AC}. At $V_{AC} = 0.005$ V, the two point attractors become two periodic attractors; apart from these two periodic attractors predicted theoretically, a new periodic attractor appears suddenly, marked by the yellow curve in Figure 7b1, and its basin of attraction is discrete (see the yellow regions in Figure 7b2). It shows that the new periodic attractor is a hidden attractor [26]. When V_{AC} increases to 0.006 V, another hidden attractor appears, which is almost symmetric to the former one (see the blue curve of Figure 7c1 and the blue regions of Figure 7c2). When $V_{AC} = 0.01$ V, there are five periodic attractors coexisting, as shown in Figure 7d1. A new periodic attractor appears (see the green curve in Figure 7d1), whose amplitude is much bigger than the other ones.

Thirdly, as V_{AC} increases from 0.01 V to 0.116 V, the number of attractors will decrease. Comparing Figure 7e1 with Figure 7d1, it is obvious that when V_{AC} increases to 0.02 V, the yellow periodic attractor disappears whose basin of attraction is eroded by that of the green attractor; thus, the basin of attraction of the green attractor can be much bigger in Figure 7e2 than in Figure 7d2. When V_{AC} continues to increase, the other three periodic attractors, i.e., the blue attractor, the red one, and the black one, disappear successively (see Figure 7f1,h1,j1) whose basins of attraction are aggressed by the basin of attraction of the green attractor, as shown in Figure 7e1–j2. Till V_{AC} becomes 0.116 V, there will be a single periodic attractor left whose basin of attraction is not fractal but with a smooth boundary (see Figure 7j1,j2).

Besides, when V_{AC} increases to 0.128 V, there will be a new complex attractor coexisting with the former green periodic attractor. It is a period-3 attractor (see the purple curve in Figure 7k1) whose basin of attraction is fractal and eroded to the basin of attraction of the periodic attractor (see Figure 7k2). It follows that a small change of initial conditions possibly shifts the dynamical behavior of the system (3) from a periodic motion to a period-3 motion, which is another type of safe jump.

Finally, as V_{AC} continues to increase, another type of complex dynamical behavior is induced. According to the phase map, Poincare map, and frequency spectrum in Figure 8a–c, there is only a chaotic attractor when V_{AC} = 0.28 V, and the boundary of its basin of attraction is not fractal (see Figure 8d).

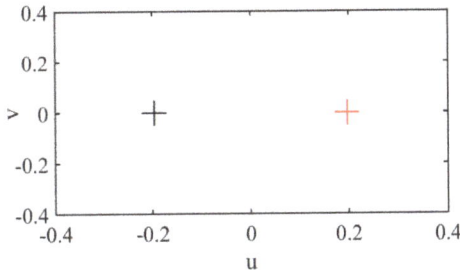
(**a1**) Attractors when V_{AC} = 0 V

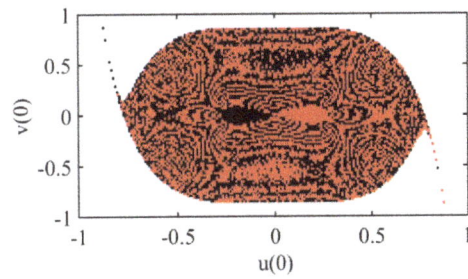
(**a2**) Basins of attraction when V_{AC} = 0 V

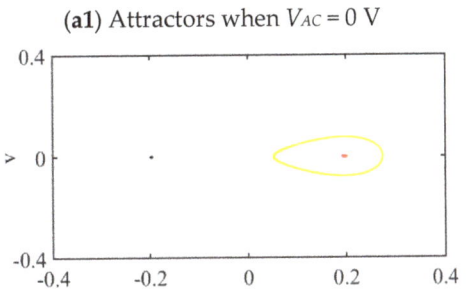
(**b1**) Attractors when V_{AC} = 0.005 V

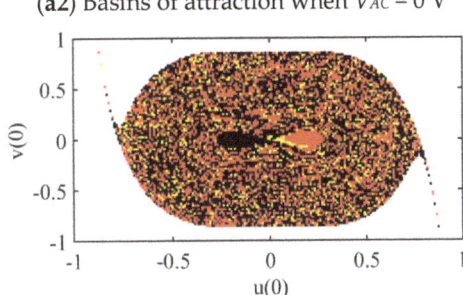
(**b2**) Basins of attraction when V_{AC} = 0.005 V

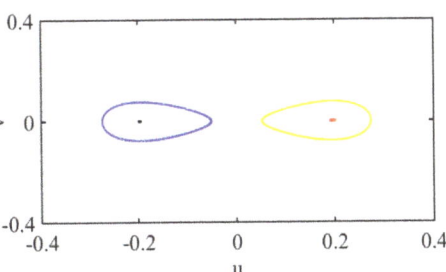
(**c1**) Attractors when V_{AC} = 0.006 V

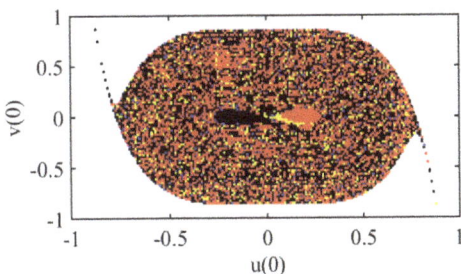
(**c2**) Basins of attraction when V_{AC} = 0.006 V

Figure 7. *Cont.*

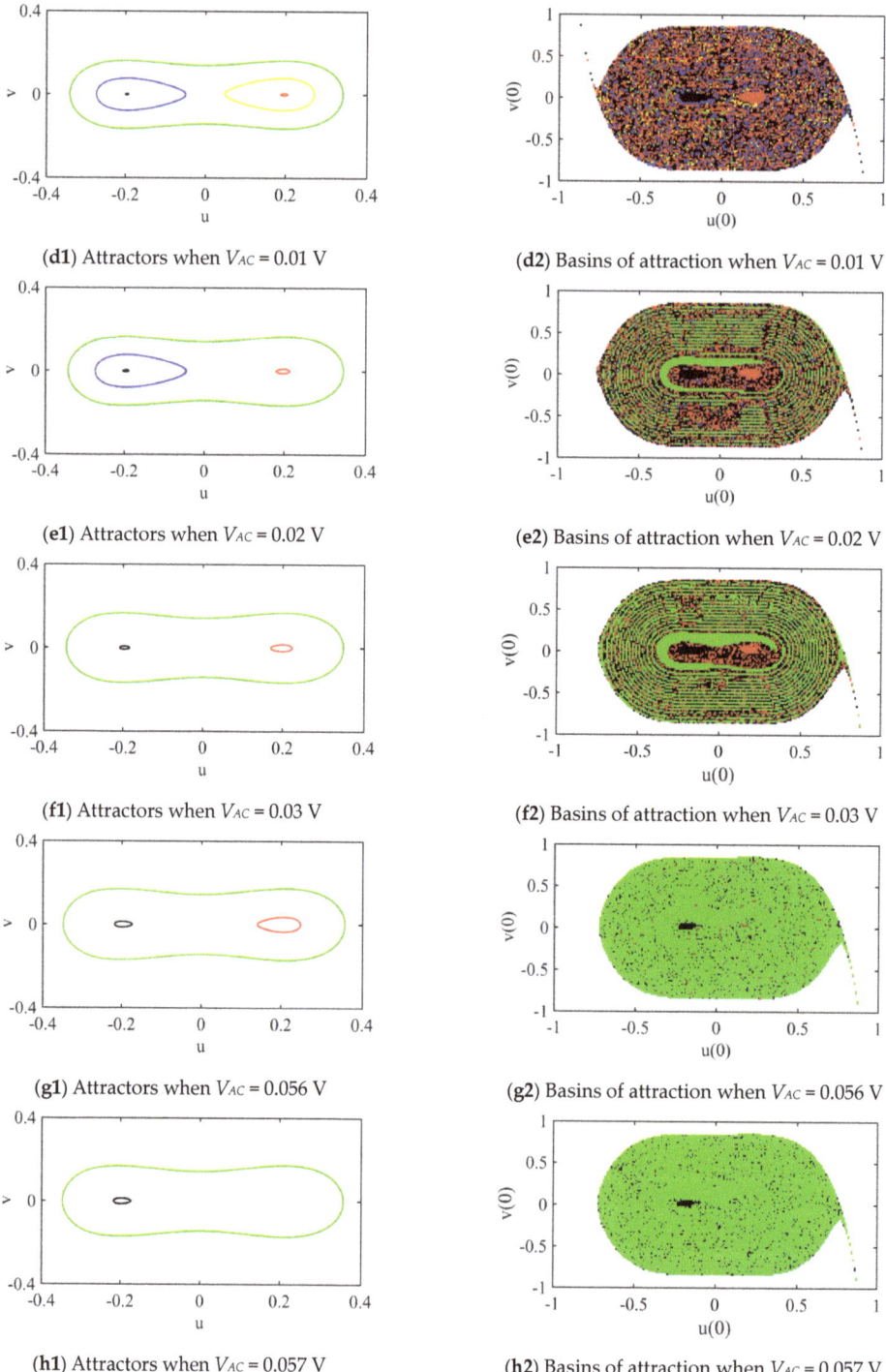

(d1) Attractors when V_{AC} = 0.01 V

(d2) Basins of attraction when V_{AC} = 0.01 V

(e1) Attractors when V_{AC} = 0.02 V

(e2) Basins of attraction when V_{AC} = 0.02 V

(f1) Attractors when V_{AC} = 0.03 V

(f2) Basins of attraction when V_{AC} = 0.03 V

(g1) Attractors when V_{AC} = 0.056 V

(g2) Basins of attraction when V_{AC} = 0.056 V

(h1) Attractors when V_{AC} = 0.057 V

(h2) Basins of attraction when V_{AC} = 0.057 V

Figure 7. *Cont.*

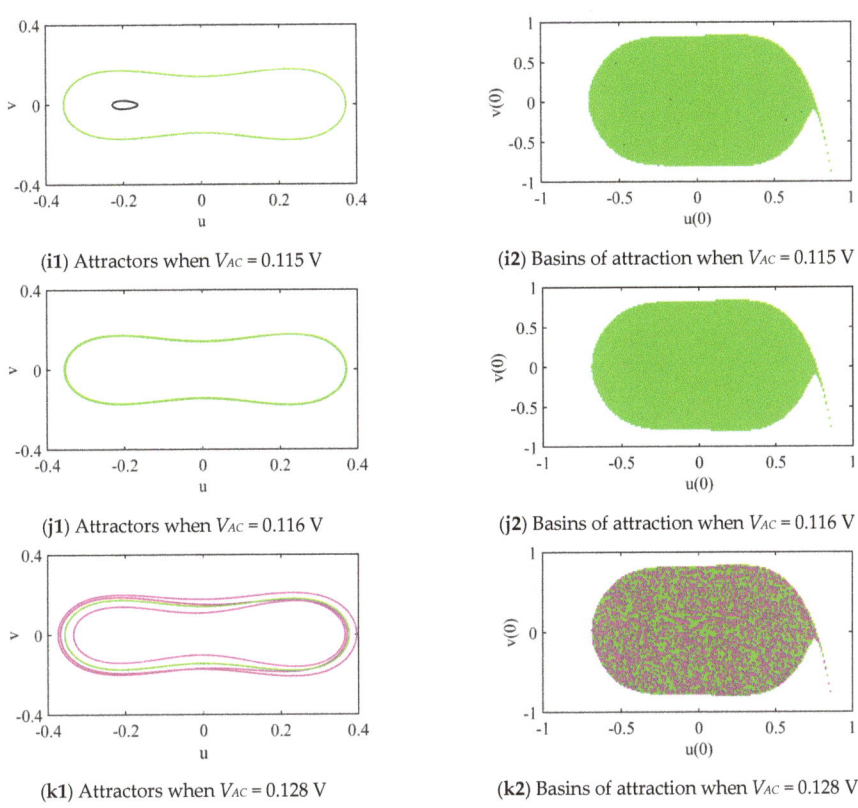

Figure 7. Evolution of multiple attractors and their basins of attraction under different values of V_{AC}.

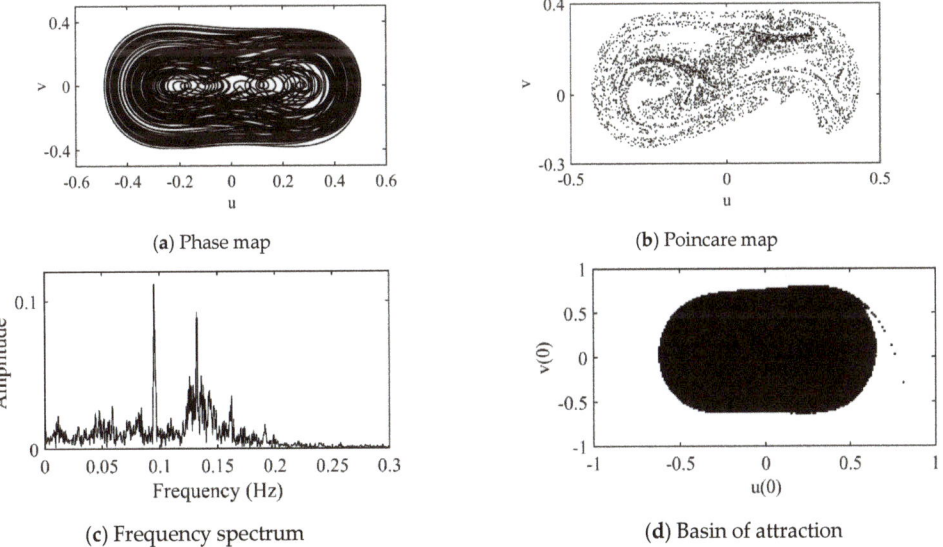

Figure 8. Attractor and its basin of attraction when $V_{AC} = 0.28$.

5. Conclusions

In this paper, a typical electrostatic bilateral micro-resonator is considered. The theory of local bifurcation and numerical approaches are applied to analyze the global dynamics of the vibrating system of the micro resonator. The main conclusions are presented as follows:

(1) DC bias voltage has some effect on the dynamics of the micro resonator. Without AC voltage, when the DC bias voltage is low, there will be only one stable point attractor in its vibrating system; when the DC bias voltage increases, there may be bistable point attractors.

(2) In the case of a low bias DC voltage, multiple periodic attractors and the corresponding safe jump occur due to Hopf bifurcation when varying the frequency or amplitude of AC voltage in certain ranges.

(3) Under a higher bias DC voltage that can induce bistable point attractors, when increasing the value of the amplitude of AC voltage, there will be multiple periodic attractors attributed to the loss of stability of the two non-trivial point attractors; apart from this, there will be some other complex dynamical behaviors of the micro-resonator vibrating system, such as safe jump, hidden attractors, period-n attractor, and chaos.

Our results provide some theoretical reference in avoiding complex dynamics of micro resonators, thus having some potential values in the design of micro sensors. The hidden attractors are depicted numerically, but their mechanism is still not that clear, which will be discussed in our future study.

Author Contributions: Conceptualization, H.S.; methodology, H.S.; software, Y.Z.; validation, H.S.; formal analysis, H.S.; investigation, Y.Z.; writing—original draft preparation, Y.Z. and H.S.; writing—review and editing, H.S.; visualization, Y.Z.; supervision, H.S.; project administration, H.S.; funding acquisition, H.S. All authors have read and agreed to the published version of the manuscript.

Funding: This research was funded by the National Natural Science Foundation of China, grant number 11472176.

Institutional Review Board Statement: Not applicable.

Informed Consent Statement: Not applicable.

Data Availability Statement: Not applicable.

Acknowledgments: Huilin Shang acknowledges the support of the National Natural Science Foundation of China under grant number 11472176. The authors are grateful for the valuable comments of the reviewers.

Conflicts of Interest: The authors declare no conflict of interest.

References

1. Luo, S.H.; Li, S.B.; Tajaddodianfar, F. Adaptive chaos control of the fractional-order arch MEMS resonator. *Nonlinear Dyn.* **2018**, *91*, 539–547. [CrossRef]
2. Michael, G.; Derek, E.M.; Dominic, V. Delayed pull-in transitions in overdamped MEMS devices. *J. Micromech. Microeng.* **2018**, *28*, 015006.
3. Uranga, A.; Verd, J.; Marigó, E.; Giner, J.; Muñóz-Gamarra, J.; Barniol, N. Exploitation of non-linearities in CMOS NEMs electrostatic resonators for mechanical memories. *Sens. Actuators A Phys.* **2013**, *197*, 88–95. [CrossRef]
4. Wu, Y.B.; Ding, G.F.; Zhang, C.C.; Wang, J.; Mao, S.P.; Wang, H. Magneto static bistable mems switch with electro thermal actuators. *Electron. Lett.* **2010**, *46*, 1074–1075. [CrossRef]
5. Ouakad, H.M. An electrostatically actuated mems arch band-pass filter. *Shock Vib.* **2013**, *20*, 809–819. [CrossRef]
6. Pernod, P.; Preobrazhensky, V.; Merlen, A.; Ducloux, O.; Talbi, A.; Gimeno, L.; Viard, R.; Tiercelin, N. MEMS magneto-mechanical microvalves (MMMS) for aerodynamic active flow control. *J. Magn. Magn. Mater.* **2010**, *322*, 1642–1646. [CrossRef]
7. Miao, X.; Dai, X.; Wang, P.; Ding, G.; Zhao, X. Design, fabrication and characterization of a bistable electromagnetic micro relay with large displacement. *Microelectron. J.* **2011**, *42*, 992–998. [CrossRef]
8. Younis, M.I. *MEMS Linear and Nonlinear Statics and Dynamics*; Springer: New York, NY, USA, 2011.
9. Alsaleem, F.M.; Younis, M.I. Integrity analysis of electrically actuated resonators with delayed feedback controller. *J. Dyn. Syst. Meas. Control.* **2011**, *133*, 031011. [CrossRef]

10. Ruzziconi, L.; Younis, M.I.; Lenci, S. An electrically actuated imperfect microbeam: Dynamical integrity for interpreting and predicting the device response. *Meccanica* **2013**, *48*, 1761–1775. [CrossRef]
11. Ruzziconi, L.; Bataineh, A.M.; Younis, M.I.; Cui, W.; Lenci, S. Nonlinear dynamics of an electrically actuated imperfect microbeam resonator: Experimental investigation and reduced-order modeling. *J. Micromech. Microeng.* **2013**, *23*, 075012. [CrossRef]
12. Zhang, W.M.; Meng, G.; Wei, K.X. Dynamics of nonlinear coupled electrostatic micromechanical resonators under two frequency parametric and external excitations. *Shock Vib.* **2010**, *17*, 759–770. [CrossRef]
13. Luo, A.C.J.; Wang, F.Y. Chaotic motion in a micro-electro-mechanical system with non-linearity from capacitors. *Commun. Nonlinear Sci. Numer. Simul.* **2002**, *7*, 31–49. [CrossRef]
14. Chavarette, F.R.; Balthazar, J.M.; Felix, J.L.P.; Rafikov, M. A reducing of a chaotic movement to a periodic orbit, of a micro-electro-mechanical system, by using an optimal linear control design. *Commun. Nonlinear Sci. Numer. Simul.* **2009**, *14*, 1844–1853. [CrossRef]
15. Caruntu, D.I.; Martin, A.B.; Christian, A.R.; Beatriz, J. Frequency-amplitude response of superharmonic resonance of second order of electrostatically actuated MEMS cantilever resonators. *Int. J. Non-Linear Mech.* **2021**, *133*, 103719. [CrossRef]
16. Han, J.X.; Zhang, Q.C.; Wang, W.; Jin, G.; Li, B. Stability and perturbation analysis of a one-degree-of-freedom doubly clamped micro-resonator with delayed velocity feedback control. *J. Vib. Control.* **2018**, *24*, 2454–2470. [CrossRef]
17. Mohammadreza, Z.; Hassen, M.O.; Saber, A. Theoretical and experimental investigations of the primary and parametric resonances in repulsive force based MEMS actuators. *Sens. Actuators A Phys.* **2020**, *303*, 111635.
18. Siewe, M.S.; Hegazy, U.H. Homoclinic bifurcation and chaos control in MEMS resonators. *Appl. Math. Model.* **2011**, *35*, 5533–5552. [CrossRef]
19. Shang, H. Pull-in instability of a typical electrostatic MEMS resonator and its control by delayed feedback. *Nonlinear Dyn.* **2017**, *90*, 171–183. [CrossRef]
20. Haghighi, H.S.; Markazi, A.H.D. Chaos prediction and control in MEMS resonators. *Commun. Nonlinear Sci. Numer. Simul.* **2010**, *15*, 3091–3099. [CrossRef]
21. Almog, R.; Zaitsev, S.; Shtempluck, O.; Buks, E. Signal amplification in a nanomechanical duffing resonator via stochastic resonance. *Appl. Phys. Lett.* **2007**, *90*, 013508. [CrossRef]
22. Gusso, A.; Viana, R.L.; Mathias, A.C.; Caldas, L.I. Nonlinear dynamics and chaos in micro/nanoelectromechanical beam resonators actuated by two-sided electrodes. *Chaos Solitons Fractals* **2019**, *122*, 6–16. [CrossRef]
23. Liu, C.X.; Yan, Y.; Wang, W.Q. Resonance and chaos of micro and nano electro mechanical resonators with time delay feedback. *Chaos Solitons Fractals* **2020**, *131*, 109512. [CrossRef]
24. Angelo, M.T.; Jose, M.B.; Rodrigo, T.R.; Ribeiro, M.A.; Lenz, W.B. On suppression of chaotic motion of a nonlinear MEMS oscillator. *Nonlinear Dyn.* **2020**, *99*, 537–557.
25. Mestrom, R.M.C.; Fey, R.H.B.; van Beek, J.T.M.; Phan, K.; Nijmeijer, H. Modeling the dynamics of a MEMS resonator: Simulations and experiments. *Sens. Actuators A Phys.* **2008**, *142*, 306–315. [CrossRef]
26. Leonov, G.A.; Kuznetsov, N.V.; Vagaitsev, V.I. Localization of hidden Chua's attractors. *Phys. Lett. A* **2011**, *375*, 2230–2233. [CrossRef]

fractal and fractional

Article

Hermite-Hadamard Inequalities in Fractional Calculus for Left and Right Harmonically Convex Functions via Interval-Valued Settings

Muhammad Bilal Khan [1], Jorge E. Macías-Díaz [2,3,*], Savin Treanţă [4,*], Mohammed S. Soliman [5] and Hatim Ghazi Zaini [6]

1. Department of Mathematics, COMSATS University Islamabad, Islamabad 44000, Pakistan; bilal42742@gmail.com
2. Departamento de Matemáticas y Física, Universidad Autónoma de Aguascalientes, Avenida Universidad 940, Ciudad Universitaria, Aguascalientes 20131, Mexico
3. Department of Mathematics, School of Digital Technologies, Tallinn University, Narva Rd. 25, 10120 Tallinn, Estonia
4. Department of Applied Mathematics, University Politehnica of Bucharest, 060042 Bucharest, Romania
5. Department of Electrical Engineering, College of Engineering, Taif University, P.O. Box 11099, Taif 21944, Saudi Arabia; soliman@tu.edu.sa
6. Department of Computer Science, College of Computers and Information Technology, Taif University, P.O. Box 11099, Taif 21944, Saudi Arabia; h.zaini@tu.edu.sa
* Correspondence: jemacias@correo.uaa.mx (J.E.M.-D.); savin.treanta@upb.ro (S.T.)

Abstract: The purpose of this study is to define a new class of harmonically convex functions, which is known as left and right harmonically convex interval-valued function (LR-$\mathcal{H}$-convex IV-F), and to establish novel inclusions for a newly defined class of interval-valued functions (IV-Fs) linked to Hermite–Hadamard (H-H) and Hermite-Hadamard-Fejér (H-H-Fejér) type inequalities via interval-valued Riemann–Liouville fractional integrals (IV-RL-fractional integrals). We also attain some related inequalities for the product of two LR-$\mathcal{H}$-convex IV-Fs. These findings enable us to identify a new class of inclusions that may be seen as significant generalizations of results proved by Iscan and Chen. Some examples are included in our findings that may be used to determine the validity of the results. The findings in this work can be seen as a considerable advance over previously published findings.

Keywords: interval-valued function; LR-Harmonically convexity; fractional integral operator; Hermite–Hadamard type inequalities

1. Introduction

The concept of convexity of functions is a useful instrument that is used to solve a wide range of pure and applied scientific issues. Many researchers have recently committed themselves to investigate the attributes and inequalities of convexity in various directions, as evidenced by [1–6] and the references therein. The Hermite–Hadamard inequality (H-H inequality), which is also used frequently in many other parts of practical mathematics, notably in optimization and probability, is one of the most important mathematical inequalities relevant to convex maps. Let us elicit it as follows:

Suppose that the mapping: $[t, v] \to \mathbb{R}$. For every for all $\varkappa, \mu \in [t, v]$ and $s \in [0, 1]$, if the successive inequality

$$\mathfrak{A}((1-s)\varkappa + s\mu) \leq (1-s)\mathfrak{A}(\varkappa) + s\mathfrak{A}(\mu) \qquad (1)$$

Then, $\mathfrak{A}$ is named as convex function on the convex interval $[t, v]$. If (1) is reversed, then, $\mathfrak{A}$ is named as a concave function on $[t, v]$.

This famous inequality gives error bounds for the mean value of a continuous convex mapping: $[t, v] \to \mathbb{R}$, which has gotten a lot of attention from a lot of authors. Many investigations have been conducted on the H-H type inequalities for additional forms of convex mappings. For example, s-convex mappings may be found in Kórus [7], N-quasi-convex mappings in Abramovich and Persson [8], h-convex mappings in Delavar and De La Sen [9], etc. Kadakal and Bekar [10], Işcan [11], Marinescu and Monea [12], Kadakal et al. [13], and the references therein provide new developments on this important issue.

Fractional calculus has shown to be an important cornerstone in mathematics and applied sciences as a very valuable tool. As a result of this fruitful interaction of various approaches to fractional calculus, many authors have studied some prominent integral inequalities, including [14] in the study of the H-H inequality for Riemann–Liouville fractional integrals, [15] in the H-H Fejér type inequality for Katugampola fractional integrals, and [16] in the extensions of trapezium inequalities for k-fractional integrals. We recommend interested readers to [17,18] and the references therein for other significant conclusions relating to fractional integral operators.

Set-valued analysis is a subset of interval analysis. There is no denying that interval analysis is important in both pure and practical research. The error limits of numerical solutions of finite state machines were one of the first applications of interval analysis. However, interval analysis, as one of the strategies for resolving interval uncertainty, has been a key component of mathematical and computer models for the past fifty years. Several applications in automated error analysis [19], computer graphics [20], and neural network output optimization [21] have been described. Furthermore, Refs. [22,23] has several optimization theory applications involving *IV-Fs*. The interested reader is recommended to Zhao et al. [24] and Román-Flores et al. [25] and their references for current developments in the area of *IV-Fs*. We recommend interested readers to [26–34] and the references therein for other significant conclusions relating to inequalities and fractional integral inequalities.

We structured the article in the following manner in response to the aforementioned tendency and invigorated by ongoing research activity in this fascinating topic. To prove fractional integral inclusions, firstly, we have generalized the class of $\mathcal{H}$-convex functions in terms of LR-$\mathcal{H}$-convex *IV-Fs*. Then, a class of *IV-RL*-fractional integrals inequalities is presented to achieve this aim. Some inclusion relations for convex *IV-Fs* in connection with the renowned H-H, H-H-Fejér type inequalities are found in this paper utilizing the newly presented class of $\mathcal{H}$-convex functions.

2. Preliminaries

Let us begin the rest of this part by outlining the theory of interval analysis, which is mostly due to [28]. The sets of all closed intervals of $\mathbb{R}$, the sets of all negative closed intervals of $\mathbb{R}$, and the sets of all positive closed intervals of $\mathbb{R}$ are denoted by $\mathcal{K}_C$, $\mathcal{K}_C^-$, and $\mathcal{K}_C^+$, respectively. For more conceptions on *IV-Fs*, see [24]. Moreover, we have:

Remark 1 ([29]). *(i) The relation " $\leq_p$ " defined on $\mathcal{K}_C$ by*

$$[\mathcal{Q}_*, \mathcal{Q}^*] \leq_p [\mathcal{Z}_*, \mathcal{Z}^*] \text{ if and only if } \mathcal{Q}_* \leq \mathcal{Z}_*, \mathcal{Q}^* \leq \mathcal{Z}^*, \tag{2}$$

for all $[\mathcal{Q}_*, \mathcal{Q}^*], [\mathcal{Z}_*, \mathcal{Z}^*] \in \mathcal{K}_C$, it is a pseudo order relation. For given $[\mathcal{Q}_*, \mathcal{Q}^*], [\mathcal{Z}_*, \mathcal{Z}^*] \in \mathcal{K}_C$, we say that $[\mathcal{Q}_*, \mathcal{Q}^*] \leq_p [\mathcal{Z}_*, \mathcal{Z}^*]$ if and only if $\mathcal{Q}_* \leq \mathcal{Z}_*, \mathcal{Q}^* \leq \mathcal{Z}^*$ or $\mathcal{Q}_* \leq \mathcal{Z}_*, \mathcal{Q}^* < \mathcal{Z}^*$. The relation $[\mathcal{Q}_*, \mathcal{Q}^*] \leq_p [\mathcal{Z}_*, \mathcal{Z}^*]$ coincident to $[\mathcal{Q}_*, \mathcal{Q}^*] \leq [\mathcal{Z}_*, \mathcal{Z}^*]$ on $\mathcal{K}_C$.

(ii) It can be easily seen that " $\leq_p$ " looks like "left and right" on the real line $\mathbb{R}$, so we call " $\leq_p$ " is "left and right" (or "LR" order, in short).

Theorem 1 ([28]). *If $\mathfrak{A} : [t, v] \subset \mathbb{R} \to \mathcal{K}_C$ is an I-V·F on such that $\mathfrak{A}(\varkappa) = [\mathfrak{A}_*(\varkappa), \mathfrak{A}^*(\varkappa)]$, then, $\mathfrak{A}$ is Riemann integrable over $[t, v]$ if and only if, $\mathfrak{A}_*$ and $\mathfrak{A}^*$ both are Riemann integrable over $[t, v]$ such that*

$$(IR)\int_t^v \mathfrak{A}(\varkappa)d\varkappa = \left[(R)\int_t^v \mathfrak{A}_*(\varkappa)d\varkappa, \ (R)\int_t^v \mathfrak{A}^*(\varkappa)d\varkappa\right].$$

The following interval-valued Riemann–Liouville fractional integral (IV-RL-fractional integral) operators were presented by Buduk et al. [1]:

Let $\beta > 0$ and $L([t, v], \mathcal{K}_C^+)$ be the collection of all Lebesgue measurable I-V-Fs on $[t, v]$. Then, the IV-RL-fractional integrals of $\mathfrak{A} \in L([t, v], \mathcal{K}_C^+)$ with order $\beta > 0$ are defined by

$$\mathfrak{T}_{t^+}^\beta \mathfrak{A}(\varkappa) = \frac{1}{\Gamma(\beta)} \int_t^\varkappa (\varkappa - s)^{\beta-1} \mathfrak{A}(s) ds, \ (\varkappa > t), \tag{3}$$

and

$$\mathfrak{T}_{v^-}^\beta \mathfrak{A}(\varkappa) = \frac{1}{\Gamma(\beta)} \int_\varkappa^v (s - \varkappa)^{\beta-1} \mathfrak{A}(s) ds, \ (\varkappa < v), \tag{4}$$

respectively, where $\Gamma(\beta) = \int_0^\infty s^{\varkappa-1} e^{-s} ds$ is the Euler gamma function.

Definition 1 ([27]). *A set $K = [t, v] \subset \mathbb{R}^+ = (0, \infty)$ is said to be harmonically convex set, if, for all $\varkappa, \mu \in K, s \in [0, 1]$, we have:*

$$\frac{\varkappa \mu}{s\varkappa + (1-s)\mu} \in K. \tag{5}$$

Definition 2 ([27]). *Suppose that the mapping: $[t, v] \to \mathbb{R}$. For every $\varkappa, \mu \in [t, v]$ and $s \in [0, 1]$, if the successive inequality*

$$\mathfrak{A}\left(\frac{\varkappa \mu}{s\varkappa + (1-s)\mu}\right) \leq (1-s)\mathfrak{A}(\varkappa) + s\mathfrak{A}(\mu), \tag{6}$$

Then, $\mathfrak{A}$ is named as harmonically convex function ($\mathcal{H}$-convex function) on interval $[t, v]$. If (6) is reversed, then, $\mathfrak{A}$ is named as a $\mathcal{H}$-concave function on $[t, v]$.

Definition 3 ([29]). *Suppose that the mapping: $[t, v] \to \mathcal{K}_C$. For every $\varkappa, \mu \in [t, v]$ and $s \in [0, 1]$, if the successive inequality*

$$\mathfrak{A}((1-s)\varkappa + s\mu) \leq_p (1-s)\mathfrak{A}(\varkappa) + s\mathfrak{A}(\mu), \tag{7}$$

Then, $\mathfrak{A}$ is named as LR-convex IV-F on the convex interval $[t, v]$. If (7) is reversed, then, $\mathfrak{A}$ is named as a concave function on $[t, v]$.

Definition 4. *Suppose that the mapping $\mathfrak{A} : [t, v] \to \mathcal{K}_C$. For all $\varkappa, \mu \in [t, v]$ and $s \in [0, 1]$, if the successive inequality*

$$\mathfrak{A}\left(\frac{\varkappa \mu}{s\varkappa + (1-s)\mu}\right) \leq_p (1-s)\mathfrak{A}(\varkappa) + s\mathfrak{A}(\mu), \tag{8}$$

is valid, then, $\mathfrak{A}$ is named as LR-harmonically convex IV-F (LR-$\mathcal{H}$-convex IV-F) defined on interval $[t, v]$. If (8) is reversed, then, $\mathfrak{A}$ is called LR-$\mathcal{H}$-concave IV-F on $[t, v]$. The set of all LR-$\mathcal{H}$-convex (LR-$\mathcal{H}$-concave IV-F) is denoted

$$LRHSX([t, v], \mathcal{K}_C)(LRHSV([t, v], \mathcal{K}_C)).$$

Theorem 2. *Let K be harmonically convex set, and let $\mathfrak{A} : K \to \mathcal{K}_C$ be an IV-F is given by*

$$\mathfrak{A}(\varkappa) = [\mathfrak{A}_*(\varkappa), \mathfrak{A}^*(\varkappa)], \ \forall \ \varkappa, \tag{9}$$

for all $\varkappa \in K$. Then, $\mathfrak{A}$ is LR-$\mathcal{H}$-convex function on K, if and only if, $\mathfrak{A}_*(\varkappa)$ and $\mathfrak{A}^*(\varkappa)$ are $\mathcal{H}$-convex functions.

Proof. Assume that $\mathfrak{A}_*(\varkappa)$ and $\mathfrak{A}^*(\varkappa)$ are $\mathcal{H}$-convex on K. Then, from (6), we have

$$\mathfrak{A}_*\left(\frac{\varkappa\mu}{s\varkappa+(1-s)\mu}\right) \leq (1-s)\mathfrak{A}_*(\varkappa) + s\mathfrak{A}_*(\mu),$$

and

$$\mathfrak{A}^*\left(\frac{\varkappa\mu}{s\varkappa+(1-s)\mu}\right) \leq (1-s)\mathfrak{A}^*(\varkappa) + s\mathfrak{A}^*(\mu).$$

Then, by (9), we obtain

$$\mathfrak{A}\left(\frac{\varkappa\mu}{s\varkappa+(1-s)\mu}\right) = [\mathfrak{A}_*(s\varkappa+(1-s)\mu), \mathfrak{A}^*(s\varkappa+(1-s)\mu)] \leq_p (1-s)[\mathfrak{A}_*(\varkappa), \mathfrak{A}^*(\varkappa)] + s[\mathfrak{A}_*(\mu), \mathfrak{A}^*(\mu)],$$

that is

$$\mathfrak{A}\left(\frac{\varkappa\mu}{s\varkappa+(1-s)\mu}\right) \leq_p (1-s)\mathfrak{A}(\varkappa) + s\mathfrak{A}(\mu), \forall\, \varkappa,\mu \in K,\, s \in [0,1].$$

Hence, $\mathfrak{A}$ is LR-$\mathcal{H}$-convex IV-F on K.

Conversely, let $\mathfrak{A}$ be LR-$\mathcal{H}$-convex IV-F on K. Then, for all $\varkappa,\mu \in K, s \in [0,1]$, we have

$$\mathfrak{A}\left(\frac{\varkappa\mu}{s\varkappa+(1-s)\mu}\right) \leq_p (1-s)\mathfrak{A}(\varkappa) + s\mathfrak{A}(\mu).$$

Therefore, from (9), left side of above inequality, we have

$$\mathfrak{A}\left(\frac{\varkappa\mu}{s\varkappa+(1-s)\mu}\right) = \left[\mathfrak{A}_*\left(\frac{\varkappa\mu}{s\varkappa+(1-s)\mu}\right), \mathfrak{A}^*\left(\frac{\varkappa\mu}{s\varkappa+(1-s)\mu}\right)\right].$$

Again, from (9), we obtain

$$(1-s)\mathfrak{A}(\varkappa) + s\mathfrak{A}(\varkappa) = (1-s)[\mathfrak{A}_*(\varkappa), \mathfrak{A}^*(\varkappa)] + s[\mathfrak{A}_*(\mu), \mathfrak{A}^*(\mu)],$$

for all $\varkappa,\mu \in K, s \in [0,1]$. Then, by $\mathcal{H}$-convexity of $\mathfrak{A}$, we have for all $\varkappa,\mu \in K, s \in [0,1]$ such that

$$\mathfrak{A}_*\left(\frac{\varkappa\mu}{s\varkappa+(1-s)\mu}\right) \leq (1-s)\mathfrak{A}_*(\varkappa) + s\mathfrak{A}_*(\mu),$$

and

$$\mathfrak{A}^*\left(\frac{\varkappa\mu}{s\varkappa+(1-s)\mu}\right) \leq (1-s)\mathfrak{A}^*(\varkappa) + s\mathfrak{A}^*(\mu),$$

this concludes the proof. □

Remark 2. *If one attempts to take $\mathfrak{A}_*(\varkappa) = \mathfrak{A}^*(\varkappa)$, then, from Definition 3, we achieve Definition 2.*

Example 1. *We consider the IV-Fs $\mathfrak{A}:[1,2] \to \mathcal{K}_C$ defined by $\mathfrak{A}(\varkappa) = [\ln(\varkappa), 2\sqrt{\varkappa}]$. Since end point functions $\mathfrak{A}_*(\varkappa), \mathfrak{A}^*(\varkappa)$ are $\mathcal{H}$-convex functions. Hence, $\mathfrak{A}(\varkappa)$ is LR-$\mathcal{H}$-convex IV-F.*

In next result, we will establish a relation between LR-convex IV-F and LR-$\mathcal{H}$-convex IV-F.

Theorem 3. *Let $\mathfrak{A}:K \to \mathcal{K}_C$ be an IV-F such that $\mathfrak{A}(\varkappa) = [\mathfrak{A}_*(\varkappa), \mathfrak{A}^*(\varkappa)]$, for all $\varkappa \in K$. Then, $\mathfrak{A}(\varkappa)$ is LR-$\mathcal{H}$-convex IV-F on K, if and only if, $\mathfrak{A}\left(\frac{1}{\varkappa}\right)$ is LR-convex IV-F on K.*

Proof. Since $\mathfrak{A}(\varkappa)$ is a LR-$\mathcal{H}$-convex IV-F, then, for $\varkappa,\mu \in [t,v],\, s \in [0,1]$, we have

$$\mathfrak{A}\left(\frac{\varkappa\mu}{s\varkappa+(1-s)\mu}\right) \leq_p (1-s)\mathfrak{A}(\varkappa) + s\mathfrak{A}(\mu).$$

Therefore, we have

$$\mathfrak{A}_*\left(\frac{\varkappa\mu}{s\varkappa+(1-s)\mu}\right) \leq (1-s)\mathfrak{A}_*(\varkappa) + s\mathfrak{A}_*(\mu),$$
$$\mathfrak{A}^*\left(\frac{\varkappa\mu}{s\varkappa+(1-s)\mu}\right) \leq (1-s)\mathfrak{A}^*(\varkappa) + s\mathfrak{A}^*(\mu). \quad (10)$$

Consider $\theta(\varkappa) = \mathfrak{A}\left(\frac{1}{\varkappa}\right)$. Taking $m = \frac{1}{\varkappa}$ and $n = \frac{1}{\mu}$ to replace $\varkappa$ and μ, respectively. Then, applying (10)

$$\mathfrak{A}_*\left(\frac{\frac{1}{\varkappa\mu}}{s\frac{1}{\varkappa}+(1-s)\frac{1}{\mu}}\right) = \mathfrak{A}_*\left(\frac{1}{(1-s)\varkappa+s\mu}\right)$$
$$= \theta_*((1-s)\varkappa+s\mu)$$
$$\leq s\mathfrak{A}_*\left(\frac{1}{\mu}\right) + (1-s)\mathfrak{A}_*\left(\frac{1}{\varkappa}\right)$$
$$= s\theta_*(\mu) + (1-s)\theta_*(\varkappa),$$
$$\mathfrak{A}^*\left(\frac{\frac{1}{\varkappa\mu}}{s\frac{1}{\varkappa}+(1-s)\frac{1}{\mu}}\right) = \mathfrak{A}^*\left(\frac{1}{(1-s)\varkappa+s\mu}\right)$$
$$= \theta^*((1-s)\varkappa+s\mu)$$
$$\leq s\mathfrak{A}^*\left(\frac{1}{\mu}\right) + (1-s)\mathfrak{A}^*\left(\frac{1}{\varkappa}\right)$$
$$= s\theta^*(\mu) + (1-s)\theta^*(\varkappa).$$

It follows that

$$\left[\mathfrak{A}_*\left(\frac{\frac{1}{\varkappa\mu}}{s\frac{1}{\varkappa}+(1-s)\frac{1}{\mu}}\right), \mathfrak{A}^*\left(\frac{\frac{1}{\varkappa\mu}}{s\frac{1}{\varkappa}+(1-s)\frac{1}{\mu}}\right)\right] =$$
$$[\theta_*((1-s)\varkappa+s\mu), \theta^*((1-s)\varkappa+s\mu)] \leq_p s[\theta_*(\mu), \theta^*(\mu)] + (1-s)[\theta_*(\varkappa), \theta^*(\varkappa)].$$

which implies that

$$\theta((1-s)\varkappa+s\mu) \leq_p s\theta(\mu) + (1-s)\theta(\varkappa).$$

This concludes that $\theta(\varkappa)$ is a LR-convex IV-F.

Conversely, let θ is LR-convex IV-F on K. Then, for all $\varkappa, \mu \in K, s \in [0, 1]$, we have

$$\theta(s\varkappa+(1-s)\mu) \leq_p s\theta(\varkappa) + (1-s)\theta(\mu).$$

By using the same steps as above, we have

$$\theta_*\left(s\frac{1}{\varkappa}+(1-s)\frac{1}{\mu}\right) = \mathfrak{A}_*\left(\frac{1}{s\frac{1}{\varkappa}+(1-s)\frac{1}{\mu}}\right)$$
$$= \mathfrak{A}_*\left(\frac{\varkappa\mu}{(1-s)\varkappa+s\mu}\right)$$
$$\leq s\theta_*\left(\frac{1}{\varkappa}\right) + (1-s)\theta_*\left(\frac{1}{\mu}\right)$$
$$= s\mathfrak{A}_*(\varkappa) + (1-s)\mathfrak{A}_*(\mu)$$
$$\theta^*\left(s\frac{1}{\varkappa}+(1-s)\frac{1}{\mu}\right) = \mathfrak{A}_*\left(\frac{1}{s\frac{1}{\varkappa}+(1-s)\frac{1}{\mu}}\right)$$
$$= \mathfrak{A}_*\left(\frac{\varkappa\mu}{(1-s)\varkappa+s\mu}\right)$$
$$\leq s\theta^*\left(\frac{1}{\varkappa}\right) + (1-s)\theta^*\left(\frac{1}{\mu}\right)$$
$$= s\mathfrak{A}_*(\varkappa) + (1-s)\mathfrak{A}_*(\mu)$$

It follows that

$$\mathfrak{A}\left(\frac{\varkappa\mu}{s\varkappa+(1-s)\mu}\right) \leq_p (1-s)\mathfrak{A}(\varkappa) + s\mathfrak{A}(\mu).$$

This completes the proof. □

Remark 3. If one attempts to take $\mathfrak{A}_*(\varkappa) = \mathfrak{A}^*(\varkappa)$, then, from Theorem 3, we acquire the Lemma 2.1 of [30].

3. Main Results

Budak et al. [1] introduced the notion of *IV-RL*-fractional integrals. As may be seen, fractional integral definitions and *IV-RL*-fractional integral definitions have comparable configurations. As a result of this observation, we may state the *H-H* inequality for *LR*-harmonically *IV-Fs* using *IV-RL*-fractional integrals.

Theorem 4. *Let* $\mathfrak{A} \in LRHSX([t, v], \mathcal{K}_C^+)$, *and defined on the interval* $[t, v]$ *such that* $\mathfrak{A}(\varkappa) = [\mathfrak{A}_*(\varkappa), \mathfrak{A}^*(\varkappa)]$ *for all* $\varkappa \in [t, v]$. *If* $\mathfrak{A} \in L([t, v], \mathcal{K}_C^+)$ *and fractional integral over* $[t, v]$, *then*

$$\mathfrak{A}\left(\frac{2tv}{t+v}\right) \leq_p \frac{\Gamma(\beta+1)}{2(v-t)^\beta}\left[\mathfrak{I}^\beta_{\frac{1}{t}-}(\mathfrak{A} \circ \Psi)\left(\frac{1}{v}\right) + \mathfrak{I}^\beta_{\frac{1}{v}+}(\mathfrak{A} \circ \Psi)\left(\frac{1}{t}\right)\right] \leq_p \frac{\mathfrak{A}(t) + \mathfrak{A}(v)}{2}. \quad (11)$$

If $\mathfrak{A}(\varkappa)$ *is LR-$\mathcal{H}$-concave IV-F, then*

$$\mathfrak{A}\left(\frac{2tv}{t+v}\right) \geq_p \frac{\Gamma(\beta+1)}{2(v-t)^\beta}\left[\mathfrak{I}^\beta_{\frac{1}{t}-}(\mathfrak{A} \circ \Psi)\left(\frac{1}{v}\right) + \mathfrak{I}^\beta_{\frac{1}{v}+}(\mathfrak{A} \circ \Psi)\left(\frac{1}{t}\right)\right] \geq_p \frac{\mathfrak{A}(t) + \mathfrak{A}(v)}{2}. \quad (12)$$

where $\Psi(\varkappa) = \frac{1}{\varkappa}$.

Proof. Let $\mathfrak{A} \in LRHSX([t, v], \mathcal{K}_C^+)$. Then, by hypothesis, we have

$$2\mathfrak{A}\left(\frac{2tv}{t+v}\right) \leq_p \mathfrak{A}\left(\frac{tv}{st+(1-s)v}\right) + \mathfrak{A}\left(\frac{tv}{(1-s)t+sv}\right).$$

Therefore, we have

$$2\mathfrak{A}_*\left(\frac{2tv}{t+v}\right) \leq \mathfrak{A}_*\left(\frac{tv}{st+(1-s)v}\right) + \mathfrak{A}_*\left(\frac{tv}{(1-s)t+sv}\right),$$
$$2\mathfrak{A}^*\left(\frac{2tv}{t+v}\right) \leq \mathfrak{A}^*\left(\frac{tv}{st+(1-s)v}\right) + \mathfrak{A}^*\left(\frac{tv}{(1-s)t+sv}\right).$$

Consider $\theta(\varkappa) = \mathfrak{A}\left(\frac{1}{\varkappa}\right)$. By Theorem 3, we have $\theta(\varkappa)$ is *LR*-convex *IV-F*. Then, above inequality, we have

$$2\theta_*\left(\frac{t+v}{2tv}\right) \leq \theta_*\left(\frac{st+(1-s)v}{tv}\right) + \theta_*\left(\frac{(1-s)t+sv}{tv}\right).$$

Multiplying both sides by $s^{\beta-1}$ and integrating the obtained result with respect to s over $(0, 1)$, we have

$$2\int_0^1 s^{\beta-1}\theta_*\left(\frac{t+v}{2tv}\right)ds \leq \int_0^1 s^{\beta-1}\theta_*\left(\frac{st+(1-s)v}{tv}\right)ds + \int_0^1 s^{\beta-1}\theta_*\left(\frac{(1-s)t+sv}{tv}\right)ds.$$

Let $\varkappa = \frac{(1-s)t+sv}{tv}$ and $\mu = \frac{st+(1-s)v}{tv}$. Then, we have

$$\frac{2}{\beta}\theta_*\left(\frac{t+v}{2tv}\right) \leq \left(\frac{tv}{v-t}\right)^\beta \int_{\frac{1}{v}}^{\frac{1}{t}} \left(\frac{1}{t} - \mu\right)^{\beta-1} \theta_*(\mu)d\mu + \left(\frac{tv}{v-t}\right)^\beta \int_{\frac{1}{v}}^{\frac{1}{t}} \left(\varkappa - \frac{1}{v}\right)^{\beta-1} \theta_*(\varkappa)d\varkappa = \Gamma(\beta)\left(\frac{tv}{v-t}\right)^\beta \left[\mathfrak{I}^\beta_{\left(\frac{1}{t}\right)-}\theta_*\left(\frac{1}{v}\right) + \mathfrak{I}^\beta_{\left(\frac{1}{v}\right)+}\theta_*\left(\frac{1}{t}\right)\right].$$

Similarly, for $\theta^*(\varkappa)$, we have

$$\frac{2}{\beta}\theta^*\left(\frac{t+v}{2tv}\right) \leq \Gamma(\beta)\left(\frac{tv}{v-t}\right)^\beta \left[\mathfrak{I}^\beta_{\left(\frac{1}{t}\right)-}\theta^*\left(\frac{1}{v}\right) + \mathfrak{I}^\beta_{\left(\frac{1}{v}\right)+}\theta^*\left(\frac{1}{t}\right)\right]$$

It follows that

$$2\left[\theta_*\left(\frac{t+v}{2tv}\right), \theta^*\left(\frac{t+v}{2tv}\right)\right] \leq_P \Gamma(\beta+1)\left(\frac{tv}{v-t}\right)^\beta \left[\mathfrak{T}^\beta_{(\frac{1}{t})^-}\theta_*\left(\frac{1}{v}\right) + \mathfrak{T}^\beta_{(\frac{1}{v})^+}\theta_*\left(\frac{1}{t}\right), \mathfrak{T}^\beta_{(\frac{1}{t})^-}\theta^*\left(\frac{1}{v}\right) + \mathfrak{T}^\beta_{(\frac{1}{v})^+}\theta^*\left(\frac{1}{t}\right)\right].$$

That is,

$$2\,\theta\left(\frac{t+v}{2tv}\right) \leq_P \Gamma(\beta+1)\left(\frac{tv}{v-t}\right)^\beta \left[\mathfrak{T}^\beta_{(\frac{1}{t})^-}\theta\left(\frac{1}{v}\right) + \mathfrak{T}^\beta_{(\frac{1}{v})^+}\theta\left(\frac{1}{t}\right)\right]. \tag{13}$$

In a similar way as above, we have

$$\Gamma(\beta)\left(\frac{tv}{v-t}\right)^\beta \left[\mathfrak{T}^\beta_{(\frac{1}{t})^-}\theta\left(\frac{1}{v}\right) + \mathfrak{T}^\beta_{(\frac{1}{v})^+}\theta\left(\frac{1}{t}\right)\right] \leq_P \frac{\theta\left(\frac{1}{t}\right)+\theta\left(\frac{1}{v}\right)}{\beta}. \tag{14}$$

Combining (31) and (32), we have

$$\theta\left(\frac{t+v}{2tv}\right) \leq_P \frac{\Gamma(\beta+1)\left(\frac{tv}{v-t}\right)^\beta}{2}\left[\mathfrak{T}^\beta_{(\frac{1}{t})^-}\theta\left(\frac{1}{v}\right) + \mathfrak{T}^\beta_{(\frac{1}{v})^+}\theta\left(\frac{1}{t}\right)\right] \leq_P \frac{\theta\left(\frac{1}{t}\right)+\theta\left(\frac{1}{v}\right)}{2},$$

that is

$$\mathfrak{A}\left(\frac{2tv}{t+v}\right) \leq_P \frac{\Gamma(\beta+1)}{2(v-t)^\beta}\left[\mathfrak{T}^\beta_{\frac{1}{t}^-}(\mathfrak{A}\circ\Psi)\left(\frac{1}{v}\right) + \mathfrak{T}^\beta_{\frac{1}{v}^+}(\mathfrak{A}\circ\Psi)\left(\frac{1}{t}\right)\right] \leq_P \frac{\mathfrak{A}(t)+\mathfrak{A}(v)}{2}.$$

Hence, the required result. □

Remark 4. *On the basic of the inequality (29), we consider certain special cases as below. If we attempt to take $\beta = 1$, then, we achieve the coming inequality which is also new one:*

$$\mathfrak{A}\left(\frac{2tv}{t+v}\right) \leq_P \frac{tv}{v-t}\int_t^v \frac{\mathfrak{A}(\varkappa)}{\varkappa^2}d\varkappa \leq_P \frac{\mathfrak{A}(t)+\mathfrak{A}(v)}{2}. \tag{15}$$

If we attempt to take $\mathfrak{A}_*(\varkappa) = \mathfrak{A}^*(\varkappa)$, then, we achieve the coming inequality, see [30]:

$$\mathfrak{A}\left(\frac{2tv}{t+v}\right) \leq \frac{\Gamma(\beta+1)}{2(v-t)^\beta}\left[\mathfrak{T}^\beta_{\frac{1}{t}^-}(\mathfrak{A}\circ\Psi)\left(\frac{1}{v}\right) + \mathfrak{T}^\beta_{\frac{1}{v}^+}(\mathfrak{A}\circ\Psi)\left(\frac{1}{t}\right)\right] \leq \frac{\mathfrak{A}(t)+\mathfrak{A}(v)}{2}. \tag{16}$$

If we attempt to take $\mathfrak{A}_*(\varkappa) = \mathfrak{A}^*(\varkappa)$ with $\beta = 1$, then, we acquire the coming inequality, see [27].

$$\mathfrak{A}\left(\frac{2tv}{t+v}\right) \leq \frac{tv}{v-t}\int_t^v \frac{\mathfrak{A}(\varkappa)}{\varkappa^2}d\varkappa \leq \frac{\mathfrak{A}(t)+\mathfrak{A}(v)}{2}. \tag{17}$$

Example 2. *If we consider taking the IV-Fs $\mathfrak{A} : [0, 2] \to \mathbb{F}_C(\mathbb{R})$ such that $[1, 2]\sqrt{\varkappa}$, then, all assumptions mentioned in Theorem 4 are met. Since $\mathfrak{A}_*(\varkappa) = \sqrt{\varkappa}$, $\mathfrak{A}^*(\varkappa, \theta) = 2\sqrt{\varkappa}$. If $\beta = 1$, then, we compute the following:*

$$\mathfrak{A}_*\left(\tfrac{2tv}{t+v}\right) \leq \tfrac{\Gamma(\beta+1)}{2(v-t)^\beta}\left[\mathfrak{T}^\beta_{\frac{1}{t}^-}(\mathfrak{A}_*\circ\Psi)\left(\tfrac{1}{v}\right) + \mathfrak{T}^\beta_{\frac{1}{v}^+}(\mathfrak{A}_*\circ\Psi)\left(\tfrac{1}{t}\right)\right] \leq \tfrac{\mathfrak{A}_*(t)+\mathfrak{A}_*(v)}{2}.$$

$$\mathfrak{A}_*\left(\tfrac{2tv}{t+v}\right) = \mathfrak{A}_*(0) = 0,$$

$$\tfrac{\Gamma(\beta+1)}{2(v-t)^\beta}\left[\mathfrak{T}^\beta_{\frac{1}{t}^-}(\mathfrak{A}_*\circ\Psi)\left(\tfrac{1}{v}\right) + \mathfrak{T}^\beta_{\frac{1}{v}^+}(\mathfrak{A}_*\circ\Psi)\left(\tfrac{1}{t}\right)\right] = 0,$$

$$\tfrac{tv}{v-t}\int_t^v \tfrac{\mathfrak{A}_*(\varkappa)}{\varkappa^2}d\varkappa = \tfrac{0}{2}\int_0^2 \tfrac{\sqrt{\varkappa}}{\varkappa^2}d\varkappa = 0,$$

$$\tfrac{\mathfrak{A}_*(t)+\mathfrak{A}_*(v)}{2} = \tfrac{1}{\sqrt{2}}.$$

That means

$$0 \leq 0 \leq \frac{1}{\sqrt{2}}.$$

Similarly, it can be easily shown that

$$\mathfrak{A}^*\left(\frac{2tv}{t+v}\right) \leq \frac{\Gamma(\beta+1)}{2(v-t)^\beta}\left[\mathfrak{T}^\beta_{\frac{1}{t}-}(\mathfrak{A}^* \circ \Psi)\left(\frac{1}{v}\right) + \mathfrak{T}^\beta_{\frac{1}{v}+}(\mathfrak{A}^* \circ \Psi)\left(\frac{1}{t}\right)\right] \leq \frac{\mathfrak{A}^*(t) + \mathfrak{A}^*(v)}{2}.$$

Now

$$\mathfrak{A}^*\left(\frac{2tv}{t+v}\right) = \mathfrak{A}_*(0) = 0,$$

$$\frac{\Gamma(\beta+1)}{2(v-t)^\beta}\left[\mathfrak{T}^\beta_{\frac{1}{t}-}(\mathfrak{A}^* \circ \Psi)\left(\frac{1}{v}\right) + \mathfrak{T}^\beta_{\frac{1}{v}+}(\mathfrak{A}^* \circ \Psi)\left(\frac{1}{t}\right)\right] = 0,$$

$$\frac{\mathfrak{A}^*(t) + \mathfrak{A}^*(v)}{2} = \sqrt{2}.$$

From which, we have

$$0 \leq 0 \leq \sqrt{2},$$

that is

$$[0, 0] \leq_p [0, 0] \leq_p \left[\frac{1}{\sqrt{2}}, \sqrt{2}\right].$$

Hence,

$$\mathfrak{A}\left(\frac{2tv}{t+v}\right) \leq_p \frac{\Gamma(\beta+1)}{2(v-t)^\beta}\left[\mathfrak{T}^\beta_{\frac{1}{t}-}(\mathfrak{A} \circ \Psi)\left(\frac{1}{v}\right) + \mathfrak{T}^\beta_{\frac{1}{v}+}(\mathfrak{A} \circ \Psi)\left(\frac{1}{t}\right)\right] \leq_p \frac{\mathfrak{A}(t) + \mathfrak{A}(v)}{2}.$$

Based on the IV-RL-fractional integrals, our next main results in association with the H-H type inequalities for product of two LR-harmonically IV-Fs are presented as follows.

Theorem 5. *Let* $\mathfrak{A}, \Psi \in LRHSX([t, v], \mathcal{K}_C^+)$, *and defined on the interval* $[t, v]$ *such that* $\mathfrak{A}(\varkappa) = [\mathfrak{A}_*(\varkappa), \mathfrak{A}^*(\varkappa)]$ *and* $\Psi(\varkappa) = [\Psi_*(\varkappa), \Psi^*(\varkappa)]$ *for all* $\varkappa \in [t, v]$. *If* $\mathfrak{A} \times \Psi \in L([t, v], \mathcal{K}_C^+)$, *and fractional integral over* $[t, v]$, *then*

$$\frac{\Gamma(\beta+1)}{2}\left(\frac{tv}{v-t}\right)^\beta\left[\mathfrak{T}^\beta_{(\frac{1}{v})^+}\mathfrak{A} \circ \Psi\left(\frac{1}{t}\right) \times \Psi \circ \Psi\left(\frac{1}{t}\right) + \mathfrak{T}^\beta_{(\frac{1}{t})^-}\mathfrak{A} \circ \Psi\left(\frac{1}{v}\right) \times \Psi \circ \Psi\left(\frac{1}{v}\right)\right] \leq_p \left(\frac{1}{2} - \frac{\beta}{(\beta+1)(\beta+2)}\right)\mathfrak{D}(t,v) + \left(\frac{\beta}{(\beta+1)(\beta+2)}\right)\mathcal{Q}(t,v),$$

where $\mathfrak{D}(t,v) = \mathfrak{A}(t) \times \Psi(t) + \mathfrak{A}(v) \times \Psi(v)$, $\mathcal{Q}(t,v) = \mathfrak{A}(t) \times \Psi(v) + \mathfrak{A}(v) \times \Psi(t)$, *and* $\mathfrak{D}(t,v) = [\mathfrak{D}_*(t,v), \mathfrak{D}^*(t,v)]$ *and* $\mathcal{Q}(t,v) = [\mathcal{Q}_*(t,v), \mathcal{Q}^*(t,v)]$.

Proof. Since $\mathfrak{A}, \Psi \in LRHSX([t, v], \mathcal{K}_C^+)$, then, we have

$$\mathfrak{A}_*\left(\frac{tv}{st+(1-s)v}\right) \leq (1-s)\mathfrak{A}_*(t) + s\mathfrak{A}_*(v),$$

and

$$\Psi_*\left(\frac{tv}{st+(1-s)v}\right) \leq (1-s)\Psi_*(t) + s\Psi_*(v) .$$

From the definition of LR-$\mathcal{H}$-convex IV-Fs it follows that $0 \leq_p \mathfrak{A}(\varkappa)$ and $0 \leq_p \Psi(\varkappa)$, so

$$\begin{aligned}&\mathfrak{A}_*\left(\frac{tv}{st+(1-s)v}\right) \times \Psi_*\left(\frac{tv}{st+(1-s)v}\right)\\ &\leq \left((1-s)\mathfrak{A}_*(t) + s\mathfrak{A}_*(v)\right)\left((1-s)\Psi_*(t) + s\Psi_*(v)\right)\\ &= (1-s)^2\mathfrak{A}_*(t) \times \Psi_*(t) + s^2\mathfrak{A}_*(v) \times \Psi_*(v)\\ &\quad + s(1-s)\mathfrak{A}_*(t) \times \Psi_*(v) + s(1-s)\mathfrak{A}_*(v) \times \Psi_*(t)\end{aligned} . \quad (18)$$

Analogously, we have

$$\mathfrak{A}_*\left(\frac{tv}{(1-s)t+sv}\right)\Psi_*\left(\frac{tv}{(1-s)t+sv}\right)$$
$$\leq s^2\mathfrak{A}_*(t) \times \Psi_*(t) + (1-s)^2\mathfrak{A}_*(v) \times \Psi_*(v) \qquad (19)$$
$$+s(1-s)\mathfrak{A}_*(t) \times \Psi_*(v) + s(1-s)\mathfrak{A}_*(v) \times \Psi_*(t)$$

Adding (18) and (19), we have

$$\mathfrak{A}_*\left(\frac{tv}{st+(1-s)v}\right) \times \Psi_*\left(\frac{tv}{st+(1-s)v}\right) + \mathfrak{A}_*\left(\frac{tv}{(1-s)t+sv}\right) \times \Psi_*\left(\frac{tv}{(1-s)t+sv}\right)$$
$$\leq \left[s^2 + (1-s)^2\right]\left[\mathfrak{A}_*(t) \times \Psi_*(t) + \mathfrak{A}_*(v) \times \Psi_*(v)\right] \qquad (20)$$
$$+2s(1-s)\left[\mathfrak{A}_*(v) \times \Psi_*(t) + \mathfrak{A}_*(t) \times \Psi_*(v)\right]$$

Taking multiplication of (20) by $s^{\beta-1}$ and integrating the obtained result with respect to s over (0, 1), we have

$$\int_0^1 s^{\beta-1}\mathfrak{A}_*\left(\frac{tv}{st+(1-s)v}\right) \times \Psi_*\left(\frac{tv}{st+(1-s)v}\right)ds$$
$$+ \int_0^1 s^{\beta-1}\mathfrak{A}_*\left(\frac{tv}{(1-s)t+sv}\right) \times \Psi_*\left(\frac{tv}{(1-s)t+sv}\right)ds$$
$$\leq \mathfrak{D}_*(t,v) \int_0^1 s^{\beta-1}\left[s^2 + (1-s)^2\right]ds + 2\mathcal{Q}_*(t,v) \int_0^1 s^{\beta-1}s(1-s)ds.$$

It follows that,

$$\Gamma(\beta)\left(\frac{tv}{v-t}\right)^\beta\left[\mathfrak{T}^\beta_{(\frac{1}{v})^+}\mathfrak{A}_*\left(\frac{1}{t}\right) \times \Psi_*\left(\frac{1}{t}\right) + \mathfrak{T}^\beta_{(\frac{1}{t})^-}\mathfrak{A}_*\left(\frac{1}{v}\right) \times \Psi_*\left(\frac{1}{v}\right)\right]$$
$$\leq \frac{2}{\beta}\left(\frac{1}{2} - \frac{\beta}{(\beta+1)(\beta+2)}\right)\mathfrak{D}_*(t,v) + \frac{2}{\beta}\left(\frac{\beta}{(\beta+1)(\beta+2)}\right)\mathcal{Q}_*(t,v)$$

Similarly, for $\mathfrak{A}^*(\varkappa)$, we have

$$\Gamma(\beta)\left(\frac{tv}{v-t}\right)^\beta\left[\mathfrak{T}^\beta_{(\frac{1}{v})^+}\mathfrak{A}_*\left(\frac{1}{t}\right) \times \Psi_*\left(\frac{1}{t}\right) + \mathfrak{T}^\beta_{(\frac{1}{t})^-}\mathfrak{A}_*\left(\frac{1}{v}\right) \times \Psi_*\left(\frac{1}{v}\right)\right]$$
$$\leq \frac{2}{\beta}\left(\frac{1}{2} - \frac{\beta}{(\beta+1)(\beta+2)}\right)\mathfrak{D}_*(t,v) + \frac{2}{\beta}\left(\frac{\beta}{(\beta+1)(\beta+2)}\right)\mathcal{Q}_*(t,v)$$

that is

$$\Gamma(\beta)\left(\frac{tv}{v-t}\right)^\beta\left[\mathfrak{T}^\beta_{(\frac{1}{v})^+}\mathfrak{A}_*\left(\frac{1}{t}\right) \times \Psi_*\left(\frac{1}{t}\right) + \mathfrak{T}^\beta_{(\frac{1}{t})^-}\mathfrak{A}_*\left(\frac{1}{v}\right) \times \Psi_*\left(\frac{1}{v}\right), \mathfrak{T}^\beta_{(\frac{1}{v})^+}\mathfrak{A}^*\left(\frac{1}{t}\right) \times \right.$$
$$\left.\Psi^*\left(\frac{1}{t}\right) + \mathfrak{T}^\beta_{(\frac{1}{t})^-}\mathfrak{A}^*\left(\frac{1}{v}\right) \times \Psi^*\left(\frac{1}{v}\right)\right] \leq_p \frac{2}{\beta}\left(\frac{1}{2} - \frac{\beta}{(\beta+1)(\beta+2)}\right)[\mathcal{D}_*(t,v), \mathcal{D}^*(t,v)]$$
$$+ \frac{2}{\beta}\left(\frac{\beta}{(\beta+1)(\beta+2)}\right)[\mathcal{Q}_*(t,v), \mathcal{Q}^*(t,v)].$$

Thus,

$$\frac{\Gamma(\beta+1)}{2}\left(\frac{tv}{v-t}\right)^\beta\left[\mathfrak{T}^\beta_{(\frac{1}{v})^+}\mathfrak{A}\circ\Psi\left(\frac{1}{t}\right) \times \Psi\circ\Psi\left(\frac{1}{t}\right) + \mathfrak{T}^\beta_{(\frac{1}{t})^-}\mathfrak{A}\circ\Psi\left(\frac{1}{v}\right) \times \Psi\circ\right.$$
$$\left.\Psi\left(\frac{1}{v}\right)\right] \leq_p \left(\frac{1}{2} - \frac{\beta}{(\beta+1)(\beta+2)}\right)\mathfrak{D}(t,v) + \left(\frac{\beta}{(\beta+1)(\beta+2)}\right)\mathcal{Q}(t,v).$$

and the theorem has been established. □

Theorem 6. *Let* $\mathfrak{A}, \Psi \in LRHSX([t, v], \mathcal{K}_C^+)$, *and defined on the interval* $[t, v]$ *such that* $\mathfrak{A}(\varkappa) = [\mathfrak{A}_*(\varkappa), \mathfrak{A}^*(\varkappa)]$ *and* $\Psi(\varkappa) = [\Psi_*(\varkappa), \Psi^*(\varkappa)]$ *for all* $\varkappa \in [t, v]$. *If* $\mathfrak{A} \times \Psi \in L([t, v], \mathcal{K}_C^+)$ *and fractional integral over* $[t, v]$, *then*

$$\mathfrak{A}\left(\tfrac{2tv}{t+v}\right)\times\Psi\left(\tfrac{2tv}{t+v}\right)\leq_P \tfrac{\Gamma(\beta+1)}{4}\left(\tfrac{tv}{v-t}\right)^\beta\left[\mathfrak{T}^\beta_{\left(\tfrac{1}{v}\right)^+}\mathfrak{A}\left(\tfrac{1}{t}\right)\times\Psi\left(\tfrac{1}{t}\right)+\mathfrak{T}^\beta_{\left(\tfrac{1}{t}\right)^-}\mathfrak{A}\left(\tfrac{1}{v}\right)\times\Psi\left(\tfrac{1}{v}\right)\right]+\tfrac{1}{2}\left(\tfrac{1}{2}-\tfrac{\beta}{(\beta+1)(\beta+2)}\right)\mathcal{Q}(t,v)+$$
$$\tfrac{1}{2}\left(\tfrac{\beta}{(\beta+1)(\beta+2)}\right)\mathfrak{D}(t,v),$$

where $\mathfrak{D}(t,v)=\mathfrak{A}(t)\times\Psi(t)+\mathfrak{A}(v)\times\Psi(v)$, $\mathcal{Q}(t,v)=\mathfrak{A}(t)\times\Psi(v)+\mathfrak{A}(v)\times\Psi(t)$, and $\mathfrak{D}(t,v)=[\mathfrak{D}_*(t,v),\mathfrak{D}^*(t,v)]$ and $\mathcal{Q}(t,v)=[\mathcal{Q}_*(t,v),\mathcal{Q}^*(t,v)]$.

Proof. Consider $\mathfrak{A},\Psi\in LRHSX([t,v],\mathcal{K}_C^+)$. Then, by hypothesis, we have

$$\mathfrak{A}_*\left(\tfrac{2tv}{t+v}\right)\times\Psi_*\left(\tfrac{2tv}{t+v}\right)$$
$$\leq\tfrac{1}{4}\left[\begin{array}{c}\mathfrak{A}_*\left(\tfrac{tv}{st+(1-s)v}\right)\times\Psi_*\left(\tfrac{tv}{st+(1-s)v}\right)\\+\mathfrak{A}_*\left(\tfrac{tv}{st+(1-s)v}\right)\times\Psi_*\left(\tfrac{tv}{(1-s)t+sv}\right)\end{array}\right]+\tfrac{1}{4}\left[\begin{array}{c}\mathfrak{A}_*\left(\tfrac{tv}{(1-s)t+sv}\right)\times\Psi_*\left(\tfrac{tv}{st+(1-s)v}\right)\\+\mathfrak{A}_*\left(\tfrac{tv}{(1-s)t+sv}\right)\times\Psi_*\left(\tfrac{tv}{(1-s)t+sv}\right)\end{array}\right],$$

$$\leq\tfrac{1}{4}\left[\begin{array}{c}\mathfrak{A}_*\left(\tfrac{tv}{st+(1-s)v}\right)\times\Psi_*\left(\tfrac{tv}{st+(1-s)v}\right)\\+\mathfrak{A}_*\left(\tfrac{tv}{(1-s)t+sv}\right)\times\Psi_*\left(\tfrac{tv}{(1-s)t+sv}\right)\end{array}\right]+\tfrac{1}{4}\left[\begin{array}{c}(s\mathfrak{A}_*(t)+(1-s)\mathfrak{A}_*(v))\\\times((1-s)\Psi_*(t)+s\Psi_*(v))\\+((1-s)\mathfrak{A}_*(t)+s\mathfrak{A}_*(v))\\\times(s\Psi_*(t)+(1-s)\Psi_*(v))\end{array}\right],\quad(21)$$

$$=\tfrac{1}{4}\left[\begin{array}{c}\mathfrak{A}_*\left(\tfrac{tv}{st+(1-s)v}\right)\times\Psi_*\left(\tfrac{tv}{st+(1-s)v}\right)\\+\mathfrak{A}_*\left(\tfrac{tv}{(1-s)t+sv}\right)\times\Psi_*\left(\tfrac{tv}{(1-s)t+sv}\right)\end{array}\right]+\tfrac{1}{4}\left[\begin{array}{c}\{s^2+(1-s)^2\}\mathcal{Q}_*(t,v)\\+\{s(1-s)+(1-s)s\}\mathfrak{D}_*(t,v)\end{array}\right].$$

Multiplying inequality (21) by $s^{\beta-1}$ and integrating over $(0,1)$,

$$\mathfrak{A}_*\left(\tfrac{2tv}{t+v}\right)\times\Psi_*\left(\tfrac{2tv}{t+v}\right)$$
$$\leq\tfrac{1}{4}\left[\begin{array}{c}\int_0^1 s^{\beta-1}\mathfrak{A}_*\left(\tfrac{tv}{st+(1-s)v}\right)\times\Psi_*\left(\tfrac{tv}{st+(1-s)v}\right)ds\\+\int_0^1 s^{\beta-1}\mathfrak{A}_*\left(\tfrac{tv}{(1-s)t+sv}\right)\times\Psi_*\left(\tfrac{tv}{(1-s)t+sv}\right)ds\end{array}\right]+\left[\begin{array}{c}\tfrac{1}{4}\mathcal{Q}_*(t,v)\int_0^1 s^{\beta-1}[s^2+(1-s)^2]ds\\+2\mathfrak{D}_*(t,v)\int_0^1 s^{\beta-1}s(1-s)ds\end{array}\right].$$

Taking $\varkappa=\tfrac{tv}{st+(1-s)v}$ and $\mu=\tfrac{tv}{(1-s)t+sv}$

$$\tfrac{1}{\beta}\mathfrak{A}_*\left(\tfrac{2tv}{t+v}\right)\times\Psi_*\left(\tfrac{2tv}{t+v}\right)$$
$$\leq \tfrac{\Gamma(\beta)}{4}\left(\tfrac{tv}{v-t}\right)^\beta\left[\mathfrak{T}^\beta_{\left(\tfrac{1}{v}\right)^+}\mathfrak{A}_*\circ\Psi\left(\tfrac{1}{t}\right)\times\Psi_*\circ\Psi\left(\tfrac{1}{t}\right)+\mathfrak{T}^\beta_{\left(\tfrac{1}{t}\right)^-}\mathfrak{A}_*\circ\Psi\left(\tfrac{1}{v}\right)\times\Psi_*\circ\Psi\left(\tfrac{1}{v}\right)\right]$$
$$+\tfrac{1}{2\beta}\left(\tfrac{1}{2}-\tfrac{\beta}{(\beta+1)(\beta+2)}\right)\mathcal{Q}_*(t,v)+\tfrac{1}{2\beta}\left(\tfrac{\beta}{(\beta+1)(\beta+2)}\right)\mathfrak{D}_*(t,v),$$

$$\tfrac{1}{\beta}\mathfrak{A}^*\left(\tfrac{2tv}{t+v}\right)\times\Psi^*\left(\tfrac{2tv}{t+v}\right)$$
$$\leq \tfrac{\Gamma(\beta)}{4}\left(\tfrac{tv}{v-t}\right)^\beta\left[\mathfrak{T}^\beta_{\left(\tfrac{1}{v}\right)^+}\mathfrak{A}^*\circ\Psi\left(\tfrac{1}{t}\right)\times\Psi^*\circ\Psi\left(\tfrac{1}{t}\right)+\mathfrak{T}^\beta_{\left(\tfrac{1}{t}\right)^-}\mathfrak{A}^*\circ\Psi\left(\tfrac{1}{v}\right)\times\Psi^*\circ\Psi\left(\tfrac{1}{v}\right)\right]$$
$$+\tfrac{1}{2\beta}\left(\tfrac{1}{2}-\tfrac{\beta}{(\beta+1)(\beta+2)}\right)\mathcal{Q}^*(t,v)+\tfrac{1}{2\beta}\left(\tfrac{\beta}{(\beta+1)(\beta+2)}\right)\mathfrak{D}^*(t,v),$$

Similarly, for $\mathfrak{A}^*(\varkappa)$, we have

$$\tfrac{1}{\beta}\mathfrak{A}^*\left(\tfrac{2tv}{t+v}\right)\times\Psi^*\left(\tfrac{2tv}{t+v}\right)$$
$$\leq \tfrac{\Gamma(\beta)}{4}\left(\tfrac{tv}{v-t}\right)^\beta\left[\mathfrak{T}^\beta_{\left(\tfrac{1}{v}\right)^+}\mathfrak{A}^*\circ\Psi\left(\tfrac{1}{t}\right)\times\Psi^*\circ\Psi\left(\tfrac{1}{t}\right)+\mathfrak{T}^\beta_{\left(\tfrac{1}{t}\right)^-}\mathfrak{A}^*\circ\Psi\left(\tfrac{1}{v}\right)\times\Psi^*\circ\Psi\left(\tfrac{1}{v}\right)\right],$$
$$+\tfrac{1}{2\beta}\left(\tfrac{1}{2}-\tfrac{\beta}{(\beta+1)(\beta+2)}\right)\mathcal{Q}^*(t,v)+\tfrac{1}{2\beta}\left(\tfrac{\beta}{(\beta+1)(\beta+2)}\right)\mathfrak{D}^*(t,v),$$

that is

$$\mathfrak{A}\left(\tfrac{2tv}{t+v}\right)\widetilde{\times}\Psi\left(\tfrac{2tv}{t+v}\right)\leq_P\tfrac{\Gamma(\beta+1)}{4}\left(\tfrac{tv}{v-t}\right)^\beta\left[\mathfrak{T}^\beta_{\left(\tfrac{1}{v}\right)^+}\mathfrak{A}\left(\tfrac{1}{t}\right)\times\Psi\left(\tfrac{1}{t}\right)+\mathfrak{T}^\beta_{\left(\tfrac{1}{t}\right)^-}\mathfrak{A}\left(\tfrac{1}{v}\right)\times\Psi\left(\tfrac{1}{v}\right)\right]+\tfrac{1}{2}\left(\tfrac{1}{2}-\tfrac{\beta}{(\beta+1)(\beta+2)}\right)\mathcal{Q}(t,v)+$$
$$\tfrac{1}{2}\left(\tfrac{\beta}{(\beta+1)(\beta+2)}\right)\mathfrak{D}(t,v).$$

Hence, the required result. □

Theorem 7. Let $\mathfrak{A}, \Psi \in LRHSX([t, v], \mathcal{K}_C^+)$, and defined on the interval $[t, v]$ such that $\mathfrak{A}(\varkappa) = [\mathfrak{A}_*(\varkappa), \mathfrak{A}^*(\varkappa)]$ and $\Psi(\varkappa) = [\Psi_*(\varkappa), \Psi^*(\varkappa)]$ for all $\varkappa \in [t, v]$. If $\mathfrak{A} \times \Psi \in L([t, v], \mathcal{K}_C^+)$ and fractional integral over $[t, v]$, then

$$2\mathfrak{A}\left(\tfrac{2tv}{t+v}\right) \times \Psi\left(\tfrac{2tv}{t+v}\right) \leq_p \tfrac{\Gamma(\beta+1)}{2^{1-\beta}}\left(\tfrac{tv}{v-t}\right)^\beta \left[\mathcal{I}^\beta_{\left(\tfrac{t+v}{2tv}\right)^+} \mathfrak{A}\circ\Psi\left(\tfrac{1}{t}\right) \times \Psi\circ\Psi\left(\tfrac{1}{t}\right) + \mathcal{I}^\beta_{\left(\tfrac{t+v}{2tv}\right)^-} \mathfrak{A}\circ\Psi\left(\tfrac{1}{v}\right) \times \Psi\circ\Psi\left(\tfrac{1}{v}\right)\right] +$$
$$\left(\tfrac{1}{2} - \tfrac{\beta^2+3\beta}{4(\beta+1)(\beta+2)}\right)\mathcal{Q}(t,v) + \tfrac{\beta^2+3\beta}{4(\beta+1)(\beta+2)}\mathfrak{D}(t,v),$$

where $\mathfrak{D}(t,v) = \mathfrak{A}(t) \times \Psi(t) + \mathfrak{A}(v) \times \Psi(v)$, $\mathcal{Q}(t,v) = \mathfrak{A}(t) \times \Psi(v) + \mathfrak{A}(v) \times \Psi(t)$, and $\mathfrak{D}(t,v) = [\mathfrak{D}_*(t,v), \mathfrak{D}^*(t,v)]$ and $\mathcal{Q}(t,v) = [\mathcal{Q}_*(t,v), \mathcal{Q}^*(t,v)]$.

Proof. Consider $\mathfrak{A}, \Psi \in LRHSX([t, v], \mathcal{K}_C^+)$. Then, by hypothesis, we have

$$\mathfrak{A}_*\left(\tfrac{2tv}{t+v}\right) \times \Psi_*\left(\tfrac{2tv}{t+v}\right)$$
$$\leq \tfrac{1}{4}\begin{bmatrix} \mathfrak{A}_*\left(\tfrac{tv}{st+(1-s)v}\right) \times \Psi_*\left(\tfrac{tv}{st+(1-s)v}\right) \\ +\mathfrak{A}_*\left(\tfrac{tv}{st+(1-s)v}\right) \times \Psi_*\left(\tfrac{tv}{(1-s)t+sv}\right) \end{bmatrix} + \tfrac{1}{4}\begin{bmatrix} \mathfrak{A}_*\left(\tfrac{tv}{(1-s)t+sv}\right) \times \Psi_*\left(\tfrac{tv}{st+(1-s)v}\right) \\ +\mathfrak{A}_*\left(\tfrac{tv}{(1-s)t+sv}\right) \times \Psi_*\left(\tfrac{tv}{(1-s)t+sv}\right) \end{bmatrix},$$
$$\leq \tfrac{1}{4}\begin{bmatrix} \mathfrak{A}_*\left(\tfrac{tv}{st+(1-s)v}\right) \times \Psi_*\left(\tfrac{tv}{st+(1-s)v}\right) \\ +\mathfrak{A}_*\left(\tfrac{tv}{(1-s)t+sv}\right) \times \Psi_*\left(\tfrac{tv}{(1-s)t+sv}\right) \end{bmatrix} + \tfrac{1}{4}\begin{bmatrix} (s\mathfrak{A}_*(t)+(1-s)\mathfrak{A}_*(v)) \\ \times((1-s)\Psi_*(t)+s\Psi_*(v)) \\ +((1-s)\mathfrak{A}_*(t)+s\mathfrak{A}_*(v)) \\ \times(s\Psi_*(t)+(1-s)\Psi_*(v)) \end{bmatrix} \quad (22)$$
$$= \tfrac{1}{4}\begin{bmatrix} \mathfrak{A}_*\left(\tfrac{tv}{st+(1-s)v}\right) \times \Psi_*\left(\tfrac{tv}{st+(1-s)v}\right) \\ +\mathfrak{A}_*\left(\tfrac{tv}{(1-s)t+sv}\right) \times \Psi_*\left(\tfrac{tv}{(1-s)t+sv}\right) \end{bmatrix} + \tfrac{1}{4}\begin{bmatrix} \{s^2+(1-s)^2\}\mathcal{Q}_*(t,v) \\ +2s(1-s)\mathfrak{D}_*(t,v) \end{bmatrix}.$$

Multiplying inequality (22) by $2^{1+\beta}\beta s^{\beta-1}$ and then, integrating the obtain outcome over $\left[0, \tfrac{1}{2}\right]$,

$$\mathfrak{A}_*\left(\tfrac{2tv}{t+v}\right) \times \Psi_*\left(\tfrac{2tv}{t+v}\right)$$
$$\leq \tfrac{1}{4}\int_0^{\tfrac{1}{2}} 2^{1+\beta}\beta s^{\beta-1}\left[\mathfrak{A}_*\left(\tfrac{tv}{st+(1-s)v}\right) \times \Psi_*\left(\tfrac{tv}{st+(1-s)v}\right) + \mathfrak{A}_*\left(\tfrac{tv}{(1-s)t+sv}\right) \times \Psi_*\left(\tfrac{tv}{(1-s)t+sv}\right)\right]ds$$
$$+\tfrac{1}{4}\left[\mathcal{Q}_*(t,v)\int_0^{\tfrac{1}{2}} 2^{1+\beta}\beta s^{\beta-1}[s^2+(1-s)^2]ds + 2\mathfrak{D}_*(t,v)\int_0^{\tfrac{1}{2}} 2^{1+\beta}\beta s^{\beta-1}s(1-s)ds\right]$$

Taking $\varkappa = \tfrac{tv}{st+(1-s)v}$ and $\mu = \tfrac{tv}{(1-s)t+sv}$, then, we get

$$2\mathfrak{A}_*\left(\tfrac{2tv}{t+v}\right) \times \Psi_*\left(\tfrac{2tv}{t+v}\right)$$
$$\leq \tfrac{\Gamma(\beta+1)}{2^{1-\beta}}\left(\tfrac{tv}{v-t}\right)^\beta \left[\mathcal{I}^\beta_{\left(\tfrac{1}{v}\right)^+} \mathfrak{A}_*\circ\Psi\left(\tfrac{1}{t}\right) \times \Psi_*\circ\Psi\left(\tfrac{1}{t}\right) + \mathcal{I}^\beta_{\left(\tfrac{1}{t}\right)^-} \mathfrak{A}_*\circ\Psi\left(\tfrac{1}{v}\right) \times \Psi_*\circ\Psi\left(\tfrac{1}{v}\right)\right] \quad (23)$$
$$+\left(\tfrac{1}{2} - \tfrac{\beta}{(\beta+1)(\beta+2)}\right)\mathcal{Q}_*(t,v) + \left(\tfrac{\beta}{(\beta+1)(\beta+2)}\right)\mathfrak{D}_*(t,v).$$

Similarly, for $\mathfrak{A}^*(\varkappa)$, we have

$$2\mathfrak{A}^*\left(\tfrac{2tv}{t+v}\right) \times \Psi^*\left(\tfrac{2tv}{t+v}\right)$$
$$\leq \tfrac{\Gamma(\beta+1)}{2^{1-\beta}}\left(\tfrac{tv}{v-t}\right)^\beta \left[\mathcal{I}^\beta_{\left(\tfrac{1}{v}\right)^+} \mathfrak{A}^*\circ\Psi\left(\tfrac{1}{t}\right) \times \Psi^*\circ\Psi\left(\tfrac{1}{t}\right) + \mathcal{I}^\beta_{\left(\tfrac{1}{t}\right)^-} \mathfrak{A}^*\circ\Psi\left(\tfrac{1}{v}\right) \times \Psi^*\circ\Psi\left(\tfrac{1}{v}\right)\right] \quad (24)$$
$$+\left(\tfrac{1}{2} - \tfrac{\beta^2+3\beta}{4(\beta+1)(\beta+2)}\right)\mathcal{Q}^*(t,v) + \tfrac{\beta^2+3\beta}{4(\beta+1)(\beta+2)}\mathfrak{D}^*(t,v).$$

From (23) and (24), we have

$$2\mathfrak{A}\left(\tfrac{2tv}{t+v}\right) \times \Psi\left(\tfrac{2tv}{t+v}\right) \leq_p \tfrac{\Gamma(\beta+1)}{2^{1-\beta}}\left(\tfrac{tv}{v-t}\right)^\beta \left[\mathcal{I}^\beta_{\left(\tfrac{t+v}{2tv}\right)^+} \mathfrak{A}\circ\Psi\left(\tfrac{1}{t}\right) \times \Psi\circ\Psi\left(\tfrac{1}{t}\right) + \mathcal{I}^\beta_{\left(\tfrac{t+v}{2tv}\right)^-} \mathfrak{A}\circ\Psi\left(\tfrac{1}{v}\right) \times \Psi\circ\Psi\left(\tfrac{1}{v}\right)\right] +$$
$$\left(\tfrac{1}{2} - \tfrac{\beta^2+3\beta}{4(\beta+1)(\beta+2)}\right)\mathcal{Q}(t,v) + \tfrac{\beta^2+3\beta}{4(\beta+1)(\beta+2)}\mathfrak{D}(t,v).$$

Now, we present the reformative version of the generalized *IV-RL*-fractional integral H-H Fejér inequality on convex interval. □

Theorem 8. Let $\mathfrak{A} \in LRHSX([t, v], \mathcal{K}_C^+)$, and defined on the interval $[t, v]$ such that $\mathfrak{A}(\varkappa) = [\mathfrak{A}_*(\varkappa), \mathfrak{A}^*(\varkappa)]$ for all $\varkappa \in [t, v] \in [0, 1]$ and let $\mathfrak{A} \in L([t, v], \mathcal{K}_C^+)$ and fractional integral over $[t, v]$. If $\mathfrak{D} : [t, v] \to \mathbb{R}$, $\mathfrak{D}\left(\frac{1}{\frac{1}{t}+\frac{1}{v}-\frac{1}{\varkappa}}\right) = \mathfrak{D}(\varkappa) \geq 0$, then

$$\mathfrak{A}\left(\frac{2tv}{t+v}\right)\left[\mathfrak{T}^{\beta}_{\left(\frac{1}{v}\right)^+}(\mathfrak{D} \circ \Psi)\left(\frac{1}{t}\right) + \mathfrak{T}^{\beta}_{\left(\frac{1}{t}\right)^-}(\mathfrak{D} \circ \Psi)\left(\frac{1}{v}\right)\right] \leq_p \left[\mathfrak{T}^{\beta}_{\left(\frac{1}{v}\right)^+}(\mathfrak{A}\mathfrak{D} \circ \Psi)\left(\frac{1}{t}\right) + \mathfrak{T}^{\beta}_{\left(\frac{1}{t}\right)^-}(\mathfrak{A}\mathfrak{D} \circ \Psi)\left(\frac{1}{v}\right)\right] \leq_p \frac{\mathfrak{A}(t)+\mathfrak{A}(v)}{2}\left[\mathfrak{T}^{\beta}_{\frac{1}{v}^+}(\mathfrak{D} \circ \Psi)\left(\frac{1}{t}\right) + \mathfrak{T}^{\beta}_{\frac{1}{t}^-}(\mathfrak{D} \circ \Psi)\left(\frac{1}{v}\right)\right]. \quad (25)$$

If $\mathfrak{A}$ is LR-$\mathcal{H}$-concave IV-F, then, inequality (25) is reversed.

Proof. Since $\mathfrak{A} \in LRHSX([t, v], \mathcal{K}_C^+)$, then, we have

$$\mathfrak{A}_*\left(\frac{2tv}{t+v}\right) \leq \frac{1}{2}\left(\mathfrak{A}_*\left(\frac{tv}{st+(1-s)v}\right) + \mathfrak{A}_*\left(\frac{tv}{(1-s)t+sv}\right)\right). \quad (26)$$

Multiplying both sides by (26) by $s^{\beta-1}\mathfrak{D}\left(\frac{tv}{(1-s)t+sv}\right)$ and then, integrating the resultant with respect to s over $[0, 1]$, we obtain

$$\mathfrak{A}_*\left(\frac{2tv}{t+v}\right)\int_0^1 s^{\beta-1}\mathfrak{D}\left(\frac{tv}{(1-s)t+sv}\right)ds \leq \frac{1}{2}\left(\begin{array}{l}\int_0^1 s^{\beta-1}\mathfrak{A}_*\left(\frac{tv}{st+(1-s)v}\right)\mathfrak{D}\left(\frac{tv}{(1-s)t+sv}\right)ds \\ +\int_0^1 s^{\beta-1}\mathfrak{A}_*\left(\frac{tv}{(1-s)t+sv}\right)\mathfrak{D}\left(\frac{tv}{(1-s)t+sv}\right)ds\end{array}\right). \quad (27)$$

Let $\varkappa = \frac{tv}{st+(1-s)v}$. Then, we have

$$2\left(\frac{tv}{v-t}\right)^{\beta}\mathfrak{A}_*\left(\frac{2tv}{t+v}\right)\int_{\frac{1}{v}}^{\frac{1}{t}}\left(\varkappa - \frac{1}{v}\right)^{\beta-1}\mathfrak{D}\left(\frac{1}{\varkappa}\right)d\varkappa$$
$$\leq \left(\frac{tv}{v-t}\right)^{\beta}\int_{\frac{1}{v}}^{\frac{1}{t}}\left(\varkappa - \frac{1}{v}\right)^{\beta-1}\mathfrak{A}_*\left(\frac{1}{\frac{1}{t}+\frac{1}{v}-\varkappa}\right)\mathfrak{D}\left(\frac{1}{\varkappa}\right)d\varkappa + \left(\frac{tv}{v-t}\right)^{\beta}\int_{\frac{1}{v}}^{\frac{1}{t}}\left(\varkappa - \frac{1}{v}\right)^{\beta-1}\mathfrak{A}_*\left(\frac{1}{\varkappa}\right)\mathfrak{D}\left(\frac{1}{\varkappa}\right)d\varkappa$$
$$= \left(\frac{tv}{v-t}\right)^{\beta}\int_{\frac{1}{v}}^{\frac{1}{t}}\left(\frac{1}{t} - \varkappa\right)^{\beta-1}\mathfrak{A}_*(\varkappa)\mathfrak{D}\left(\frac{1}{\frac{1}{t}+\frac{1}{v}-\varkappa}\right)d\varkappa + \left(\frac{tv}{v-t}\right)^{\beta}\int_{\frac{1}{v}}^{\frac{1}{t}}\left(\varkappa - \frac{1}{v}\right)^{\beta-1}\mathfrak{A}_*\left(\frac{1}{\varkappa}\right)\mathfrak{D}\left(\frac{1}{\varkappa}\right)d\varkappa$$
$$= \Gamma(\beta)\left(\frac{tv}{v-t}\right)^{\beta}\left[\mathfrak{T}^{\beta}_{\left(\frac{1}{v}\right)^+}\mathfrak{A}_*\mathfrak{D}\left(\frac{1}{t}\right) + \mathfrak{T}^{\beta}_{\left(\frac{1}{t}\right)^-}\mathfrak{A}_*\mathfrak{D}\left(\frac{1}{v}\right)\right] \quad (28)$$

Similarly, for $\mathfrak{A}^*(\varkappa)$, we have

$$2\left(\frac{tv}{v-t}\right)^{\beta}\mathfrak{A}^*\left(\frac{2tv}{t+v}\right)\int_{\frac{1}{v}}^{\frac{1}{t}}\left(\varkappa - \frac{1}{v}\right)^{\beta-1}\mathfrak{D}\left(\frac{1}{\varkappa}\right)d\varkappa \leq \Gamma(\beta)\left(\frac{tv}{v-t}\right)^{\beta}\left[\mathfrak{T}^{\beta}_{\left(\frac{1}{v}\right)^+}\mathfrak{A}^*\mathfrak{D}\left(\frac{1}{t}\right) + \mathfrak{T}^{\beta}_{\left(\frac{1}{t}\right)^-}\mathfrak{A}^*\mathfrak{D}\left(\frac{1}{v}\right)\right]. \quad (29)$$

From (28) and (29), we have

$$\Gamma(\beta)\left(\frac{tv}{v-t}\right)^{\beta}\left[\mathfrak{A}_*\left(\frac{2tv}{t+v}\right), \mathfrak{A}^*\left(\frac{2tv}{t+v}\right)\right] \cdot \left[\mathfrak{T}^{\beta}_{\left(\frac{1}{v}\right)^+}\mathfrak{D}\left(\frac{1}{t}\right) + \mathfrak{T}^{\beta}_{\left(\frac{1}{t}\right)^-}\mathfrak{D}\left(\frac{1}{v}\right)\right]$$
$$\leq {}_p\Gamma(\beta)\left(\frac{tv}{v-t}\right)^{\beta}\left[\mathfrak{T}^{\beta}_{\left(\frac{1}{v}\right)^+}\mathfrak{A}_*D\left(\frac{1}{t}\right) + \mathfrak{T}^{\beta}_{\left(\frac{1}{t}\right)^-}\mathfrak{A}_*D\left(\frac{1}{v}\right), \mathfrak{T}^{\beta}_{\left(\frac{1}{v}\right)^+}\mathfrak{A}^*D\left(\frac{1}{t}\right) + \mathfrak{T}^{\beta}_{\left(\frac{1}{t}\right)^-}\mathfrak{A}^*D\left(\frac{1}{v}\right)\right],$$

that is

$$\mathfrak{A}\left(\frac{2tv}{t+v}\right)\left[\mathfrak{T}^{\beta}_{\left(\frac{1}{v}\right)^+}(\mathfrak{D} \circ \Psi)\left(\frac{1}{t}\right) + \mathfrak{T}^{\beta}_{\left(\frac{1}{t}\right)^-}(\mathfrak{D} \circ \Psi)\left(\frac{1}{v}\right)\right] \leq_p \left[\mathfrak{T}^{\beta}_{\left(\frac{1}{v}\right)^+}(\mathfrak{A}\mathfrak{D} \circ \Psi)\left(\frac{1}{t}\right) + \mathfrak{T}^{\beta}_{\left(\frac{1}{t}\right)^-}(\mathfrak{A}\mathfrak{D} \circ \Psi)\left(\frac{1}{v}\right)\right]. \quad (30)$$

Similarly, if $\mathfrak{A}$ be a LR-$\mathcal{H}$-convex IV-F and $s^{\beta-1}\mathfrak{D}\left(\frac{tv}{st+(1-s)v}\right) \geq 0$, then, we have

$$s^{\beta-1}\mathfrak{A}_*\left(\frac{tv}{st+(1-s)v}\right)\mathfrak{D}\left(\frac{tv}{st+(1-s)v}\right) \leq s^{\beta-1}((1-s)\mathfrak{A}_*(t) + s\mathfrak{A}_*(v))\mathfrak{D}\left(\frac{tv}{st+(1-s)v}\right). \quad (31)$$

And

$$s^{\beta-1}\mathfrak{A}_*\left(\frac{tv}{(1-s)t+sv}\right)\mathfrak{D}\left(\frac{tv}{st+(1-s)v}\right) \leq s^{\beta-1}(s\mathfrak{A}_*(t) + (1-s)\mathfrak{A}_*(v))\mathfrak{D}\left(\frac{tv}{st+(1-s)v}\right). \quad (32)$$

After adding (31) and (32), and integrating the resultant over $[0, 1]$, we get

$$\int_0^1 s^{\beta-1}\mathfrak{A}_*\left(\frac{tv}{st+(1-s)v}\right)\mathfrak{D}\left(\frac{tv}{st+(1-s)v}\right)ds + \int_0^1 s^{\beta-1}\mathfrak{A}_*\left(\frac{tv}{(1-s)t+sv}\right)\mathfrak{D}\left(\frac{tv}{st+(1-s)v}\right)ds$$
$$\leq \int_0^1\left[s^{\beta-1}\mathfrak{A}_*(t)\{s+(1-s)\}\left(\frac{tv}{st+(1-s)v}\right) + s^{\beta-1}\mathfrak{A}_*(v)\{(1-s)+s\}\left(\frac{tv}{st+(1-s)v}\right)\right]ds,$$
$$= \mathfrak{A}_*(t)\int_0^1 s^{\beta-1}D\left(\frac{tv}{st+(1-s)v}\right)ds + \mathfrak{A}_*(v)\int_0^1 s^{\beta-1}D\left(\frac{tv}{st+(1-s)v}\right)ds.$$

Similarly, for $\mathfrak{A}^*(\varkappa)$, we have

$$\int_0^1 s^{\beta-1}\mathfrak{A}^*\left(\frac{tv}{st+(1-s)v}\right)\mathfrak{D}\left(\frac{tv}{st+(1-s)v}\right)ds + \int_0^1 s^{\beta-1}\mathfrak{A}^*\left(\frac{tv}{(1-s)t+sv}\right)\mathfrak{D}\left(\frac{tv}{st+(1-s)v}\right)ds =$$
$$\mathfrak{A}^*(t)\int_0^1 s^{\beta-1}D\left(\frac{tv}{st+(1-s)v}\right)ds + \mathfrak{A}^*(v)\int_0^1 s^{\beta-1}D\left(\frac{tv}{st+(1-s)v}\right)ds.$$

From which, we have

$$\Gamma(\beta)\left(\frac{tv}{v-t}\right)^\beta\left[\mathfrak{T}^\beta_{\frac{1}{v}^+}\mathfrak{A}\mathfrak{D}\circ\Psi(v) + \mathfrak{T}^\beta_{(\frac{1}{t})^-}\mathfrak{A}\mathfrak{D}\circ\Psi\left(\frac{1}{v}\right)\right] \leq_p \Gamma(\beta)\left(\frac{tv}{v-t}\right)^\beta\frac{\mathfrak{A}(t)+\mathfrak{A}(v)}{2}\left[\mathfrak{T}^\beta_{\frac{1}{v}^+}(\mathfrak{D}\circ\Psi)\left(\frac{1}{t}\right) + \mathfrak{T}^\beta_{(\frac{1}{t})^-}(\mathfrak{D}\circ\Psi)\left(\frac{1}{v}\right)\right],$$

that is

$$\left[\mathfrak{T}^\beta_{(\frac{1}{v})^+}\mathfrak{A}\mathfrak{D}\circ\Psi\left(\frac{1}{t}\right) + \mathfrak{T}^\beta_{(\frac{1}{t})^-}\mathfrak{A}\mathfrak{D}\circ\Psi\left(\frac{1}{v}\right)\right] \leq_p \frac{\mathfrak{A}(t)+\mathfrak{A}(v)}{2}\left[\mathfrak{T}^\beta_{(\frac{1}{v})^+}(\mathfrak{D}\circ\Psi)\left(\frac{1}{t}\right) + \mathfrak{T}^\beta_{(\frac{1}{t})^-}(\mathfrak{D}\circ\Psi)\left(\frac{1}{v}\right)\right]. \quad (33)$$

By combining (30) and (33), we obtain the required inequality (25). □

Remark 5. *Let one attempt to take $\beta = 1$. Then, from (25), we acquire the coming inequality, which is also new one:*

$$\mathfrak{A}\left(\frac{2tv}{t+v}\right)\int_t^v \frac{\mathfrak{D}(\varkappa)}{\varkappa^2}d\varkappa \leq_p \int_t^v \frac{\mathfrak{A}(\varkappa)}{\varkappa^2}\mathfrak{D}(\varkappa)d\varkappa \leq_p \frac{\mathfrak{A}(t)+\mathfrak{A}(v)}{2}\int_t^v \frac{\mathfrak{D}(\varkappa)}{\varkappa^2}d\varkappa.$$

Let one attempt to take $\mathfrak{D}(\varkappa) = 1$. Then, from (25), we obtain inequality (11).

Let one attempt to take $\mathfrak{D}(\varkappa) = 1$ and $\beta = 1$, then, from (25), we get H-H inequality for LR-$\mathcal{H}$-convex IV-F.

$$\mathfrak{A}\left(\frac{2tv}{t+v}\right) \leq_p \frac{tv}{v-t}\int_t^v \frac{\mathfrak{A}(\varkappa)}{\varkappa^2}d\varkappa \leq_p \frac{\mathfrak{A}(t)+\mathfrak{A}(v)}{2}.$$

If one attempts to take $\mathfrak{A}_*(\varkappa) = \mathfrak{A}^*(\varkappa)$, then, from (40), we acquire the fractional H-H Fejér inequality, see [31].

Let one attempt to take $\mathfrak{A}_*(\varkappa) = \mathfrak{A}^*(\varkappa)$ with $\beta = 1$. Then, from (25), we achieve the coming inequality, see [3].

$$\mathfrak{A}\left(\frac{2tv}{t+v}\right)\int_t^v \frac{\mathfrak{D}(\varkappa)}{\varkappa^2}d\varkappa \leq \int_t^v \frac{\mathfrak{A}(\varkappa)}{\varkappa^2}\mathfrak{D}(\varkappa)d\varkappa \leq \frac{\mathfrak{A}(t)+\mathfrak{A}(v)}{2}\int_t^v \frac{\mathfrak{D}(\varkappa)}{\varkappa^2}d\varkappa.$$

If one attempts to take $\mathfrak{A}_*(\varkappa) = \mathfrak{A}^*(\varkappa)$ with $\mathfrak{D}(\varkappa) = 1$ then, from (25), we acquire the coming classical inequality for $\mathcal{H}$-convex function.

$$\mathfrak{A}\left(\frac{2tv}{t+v}\right) \leq \frac{\Gamma(\beta+1)}{2(v-t)^\beta}\left[\mathfrak{T}^\beta_{\frac{1}{v}^-}(\mathfrak{A}\circ\Psi)\left(\frac{1}{v}\right) + \mathfrak{T}^\beta_{\frac{1}{v}^+}(\mathfrak{A}\circ\Psi)\left(\frac{1}{t}\right)\right] \leq \frac{\mathfrak{A}(t)+\mathfrak{A}(v)}{2}.$$

If one attempts to take $\mathfrak{A}_*(\varkappa) = \mathfrak{A}^*(\varkappa)$ and $\mathfrak{D}(\varkappa) = \beta = 1$ then, from (25), we acquire the coming classical inequality for $\mathcal{H}$-convex function.

$$\mathfrak{A}\left(\frac{2tv}{t+v}\right) \leq \frac{tv}{v-t}\int_t^v \frac{\mathfrak{A}(\varkappa)}{\varkappa^2}d\varkappa \leq \frac{\mathfrak{A}(t)+\mathfrak{A}(v)}{2}.$$

4. Conclusions

We use *IV-RL*-fractional integral operators to infer various inclusions in the H-H, H-H-Fejér type inequalities, and some related inequalities in this paper. We show the relationships between the examined results and previously published ones to show their generic properties. In addition, some nontrivial examples are given to demonstrate the accuracy of the results derived in the study. The point we wish to make here is that interval-valued analyses are commonly used in practical mathematics, particularly in the field of optimality analysis (see [22,23]). This important subject in interval-valued analysis using fractional integral operators deserves to be explored further.

In our final view, we believe that our work can be generalized to other models of fractional calculus, such as Atangana–Baleanu and Prabhakar fractional operators with Mittag–Liffler functions in their kernels. We have left this consideration as an open problem for the researchers who are interested in this field. The interested researchers can proceed as done in references [15,16].

Author Contributions: Conceptualization, M.B.K.; methodology, M.B.K.; validation, J.E.M.-D., M.S.S., S.T. and H.G.Z.; formal analysis, M.B.K.; investigation, M.S.S.; resources, J.E.M.-D.; data curation, H.G.Z.; writing—original draft preparation, M.B.K. and H.G.Z.; writing—review and editing, M.B.K. and J.E.M.-D.; visualization, H.G.Z.; supervision, M.B.K. and M.S.S.; project administration, M.B.K.; funding acquisition, S.T.; M.S.S. and H.G.Z. All authors have read and agreed to the published version of the manuscript.

Funding: This work was funded by Taif University Researchers Supporting Project number (TURSP-2020/345), Taif University, Taif, Saudi Arabia and this work was also supported by the National Council of Science and Technology of Mexico (CONACYT) through grant A1-S-45928.

Data Availability Statement: Not Applicable.

Acknowledgments: The authors would like to thank the Rector, COMSATS University Islamabad, Islamabad, Pakistan, for providing excellent research.

Conflicts of Interest: The authors declare no conflict of interest.

References

1. Khan, M.B.; Mohammed, P.O.; Noor, M.A.; Hamed, Y.S. New Hermite-Hadamard inequalities in fuzzy-interval fractional calculus and related inequalities. *Symmetry* **2021**, *13*, 673. [CrossRef]
2. Sana, G.; Khan, M.B.; Noor, M.A.; Mohammed, P.O.; Chu, Y.M. Harmonically convex fuzzy-interval-valued functions and fuzzy-interval Riemann–Liouville fractional integral inequalities. *Int. J. Comput. Intell. Syst.* **2021**, *14*, 1809–1822. [CrossRef]
3. Hadamard, J. Étude sur les propriétés des fonctions entières et en particulier d'une fonction considérée par Riemann. *J. Mathématiques Pures Appliquées* **1893**, *7*, 171–215.
4. Hermite, C. Sur deux limites d'une intégrale définie. *Mathesis* **1883**, *3*, 82–97.
5. Khan, M.B.; Treanță, S.; Soliman, M.S.; Nonlaopon, K.; Zaini, H.G. Some Hadamard-Fejér Type Inequalities for LR-Convex Interval-Valued Functions. *Fractal Fract.* **2022**, *6*, 6. [CrossRef]
6. Macías-Díaz, J.E.; Khan, M.B.; Noor, M.A.; Abd Allah, A.M.; Alghamdi, S.M. Hermite-Hadamard inequalities for generalized convex functions in interval-valued calculus. *AIMS Math.* **2022**, *7*, 4266–4292. [CrossRef]
7. Kórus, P. An extension of the Hermite-Hadamard inequality for convex and *s*-convex functions. *Aequ. Math.* **2019**, *93*, 527–534. [CrossRef]
8. Abramovich, S.; Persson, L.E. Fejér and Hermite-Hadamard type inequalities for N-quasi-convex functions. *Math. Notes* **2017**, *102*, 599–609. [CrossRef]
9. Delavar, M.R.; De La Sen, M. A mapping associated to h-convex version of the Hermite–Hadamard inequality with applications. *J. Math. Inequal.* **2020**, *14*, 329–335. [CrossRef]
10. Kadakal, H.; Bekar, K. New inequalities for AH-convex functions using beta and hypergeometric functions. *Poincare J. Anal. Appl.* **2019**, *2*, 105–114. [CrossRef]
11. İşcan, İ. Weighted Hermite–Hadamard–Mercer type inequalities for convex functions. *Numer. Methods Part. Difer. Equ.* **2021**, *37*, 118–130. [CrossRef]
12. Marinescu, D.Ş.; Monea, M. A very short proof of the Hermite-Hadamard inequalities. *Am. Math. Month.* **2020**, *127*, 850–851. [CrossRef]
13. Kadakal, M.; Karaca, H.; İşcan, İ. Hermite-Hadamard type inequalities for multiplicatively geometrically P-functions. *Poincare J. Anal. Appl.* **2018**, *2*, 77–85. [CrossRef]

14. Khan, M.B.; Noor, M.A.; Noor, K.I.; Chu, Y.M. New Hermite-Hadamard type inequalities for -convex fuzzy-interval-valued functions. *Adv. Differ. Equ.* **2021**, *2021*, 6–20. [CrossRef]
15. Chen, H.; Katugampola, U.N. Hermite-Hadamard and Hermite-Hadamard-Fejér type inequalities for generalized fractional integrals. *J. Math. Anal. Appl.* **2017**, *446*, 1274–1291. [CrossRef]
16. Du, T.S.; Awan, M.U.; Kashuri, A.; Zhao, S.S. Some k-fractional extensions of the trapezium inequalities through generalized relative semi-(m,h)-preinvexity. *Appl. Anal.* **2021**, *100*, 642–662. [CrossRef]
17. Mehrez, K.; Agarwal, P. New Hermite-Hadamard type integral inequalities for convex functions and their applications. *J. Comput. Appl. Math.* **2019**, *350*, 274–285. [CrossRef]
18. Kunt, M.; İşcan, İ.; Turhan, S.; Karapinar, D. Improvement of fractional Hermite–Hadamard type inequality for convex functions. *Miskolc. Math. Notes* **2018**, *19*, 1007–1017. [CrossRef]
19. Rothwell, E.J.; Cloud, M.J. Automatic error analysis using intervals. *IEEE Trans. Ed.* **2012**, *55*, 9–15. [CrossRef]
20. Snyder, J.M. Interval analysis for computer graphics. *SIGGRAPH Comput. Graph.* **1992**, *26*, 121–130. [CrossRef]
21. de Weerdt, E.; Chu, Q.P.; Mulder, J.A. Neural network output optimization using interval analysis. *IEEE Trans. Neural Netw.* **2009**, *20*, 638–653. [CrossRef] [PubMed]
22. Ghosh, D.; Debnath, A.K.; Pedrycz, W. A variable and a fxed ordering of intervals and their application in optimization with interval-valued functions. *Int. J. Approx. Reason.* **2020**, *121*, 187–205. [CrossRef]
23. Singh, D.; Dar, B.A.; Kim, D.S. KKT optimality conditions in interval valued multiobjective programming with generalized differentiable functions. *Eur. J. Oper. Res.* **2016**, *254*, 29–39. [CrossRef]
24. Zhao, D.F.; An, T.Q.; Ye, G.J.; Torres, D.F.M. On Hermite–Hadamard type inequalities for harmonical h-convex interval-valued functions. *Math. Inequal. Appl.* **2020**, *23*, 95–105.
25. Román-Flores, H.; Chalco-Cano, Y.; Lodwick, W.A. Some integral inequalities for interval-valued functions. *Comput. Appl. Math.* **2018**, *37*, 1306–1318. [CrossRef]
26. Khan, M.B.; Noor, M.A.; Abdeljawad, T.; Mousa, A.A.A.; Abdalla, B.; Alghamdi, S.M. LR-Preinvex Interval-Valued Functions and Riemann-Liouville Fractional Integral Inequalities. *Fractal Fract.* **2021**, *5*, 243. [CrossRef]
27. Iscan, I. Hermite-Hadamard type inequalities for harmonically convex functions. *Hacet. J. Math. Stat.* **2014**, *43*, 935–942. [CrossRef]
28. Moore, R.E. *Interval Analysis*; Prentice Hall: Englewood Cliffs, NJ, USA, 1966.
29. Zhang, D.; Guo, C.; Chen, D.; Wang, G. Jensen's inequalities for set-valued and fuzzy set-valued functions. *Fuzzy Sets Syst.* **2020**, *2020*, 1–27. [CrossRef]
30. Chen, F. Extensions of the Hermite–Hadamard inequality for harmonically convex functions via fractional integrals. *Appl. Math. Comput.* **2015**, *268*, 121–128. [CrossRef]
31. Kunt, M.; İşcan, İ.; Gözütok, U. On new inequalities of Hermite–Hadamard-Fejér type for harmonically convex functions via fractional integrals. *SpringerPlus* **2016**, *5*, 1–19. [CrossRef]
32. Chen, F.; Wu, S. Fejér and Hermite-Hadamard type inequalities for harmonically convex functions. *J. Appl. Math.* **2014**, *2014*, 386806. [CrossRef]
33. Kunt, M.; Iscan, I.; Yazici, N. Hermite-Hadamard type inequalities for product of harmonically convex functions via Riemann-Liouville fractional integrals. *J. Math. Anal.* **2016**, *7*, 74–82.
34. Chen, F. A note on Hermite-Hadamard inequalities for Products of convex functions. *J. Appl. Math.* **2013**, *935020*, 1–5. [CrossRef]

fractal and fractional

Article

Non-Dominated Sorting Manta Ray Foraging Optimization for Multi-Objective Optimal Power Flow with Wind/Solar/Small-Hydro Energy Sources

Fatima Daqaq [1,2,*], Salah Kamel [3], Mohammed Ouassaid [2], Rachid Ellaia [1,2] and Ahmed M. Agwa [4,5,*]

1. Laboratory of Study and Research for Applied Mathematics, Mohammadia School of Engineers, Mohammed V University in Rabat, Rabat 10090, Morocco; ellaia@emi.ac.ma
2. Engineering for Smart and Sustainable Systems Research Center, Mohammadia School of Engineers, Mohammed V University in Rabat, Rabat 10090, Morocco; ouassaid@emi.ac.ma
3. Electrical Engineering Department, Faculty of Engineering, Aswan University, Aswan 81542, Egypt; skamel@aswu.edu.eg
4. Department of Electrical Engineering, College of Engineering, Northern Border University, Arar 1321, Saudi Arabia
5. Prince Faisal bin Khalid bin Sultan Research Chair in Renewable Energy Studies and Applications (PFCRE), Northern Border University, Arar 1321, Saudi Arabia
* Correspondence: fati.daqaq@gmail.com (F.D.); ahmad.agua@nbu.edu.sa (A.M.A.)

Abstract: This present study describes a novel manta ray foraging optimization approach based non-dominated sorting strategy, namely (NSMRFO), for solving the multi-objective optimization problems (MOPs). The proposed powerful optimizer can efficiently achieve good convergence and distribution in both the search and objective spaces. In the NSMRFO algorithm, the elitist non-dominated sorting mechanism is followed. Afterwards, a crowding distance with a non-dominated ranking method is integrated for the purpose of archiving the Pareto front and improving the optimal solutions coverage. To judge the NSMRFO performances, a bunch of test functions are carried out including classical unconstrained and constrained functions, a recent benchmark suite known as the completions on evolutionary computation 2020 (CEC2020) that contains twenty-four multimodal optimization problems (MMOPs), some engineering design problems, and also the modified real-world issue known as IEEE 30-bus optimal power flow involving the wind/solar/small-hydro power generations. Comparison findings with multimodal multi-objective evolutionary algorithms (MMMOEAs) and other existing multi-objective approaches with respect to performance indicators reveal the NSMRFO ability to balance between the coverage and convergence towards the true Pareto front (PF) and Pareto optimal sets (PSs). Thus, the competing algorithms fail in providing better solutions while the proposed NSMRFO optimizer is able to attain almost all the Pareto optimal solutions.

Keywords: multimodal multi-objective optimization; manta ray foraging optimizer; non-dominated solution; crowing distance; engineering design problem; optimal power flow; renewable energy sources

Citation: Daqaq, F.; Kamel, S.; Ouassaid, M.; Ellaia, R.; Agwa, A.M. Non-Dominated Sorting Manta Ray Foraging Optimization for Multi-Objective Optimal Power Flow with Wind/Solar/Small-Hydro Energy Sources. *Fractal Fract.* **2022**, *6*, 194. https://doi.org/10.3390/fractalfract6040194

Academic Editor: Savin Treanţă

Received: 17 February 2022
Accepted: 22 March 2022
Published: 31 March 2022

Publisher's Note: MDPI stays neutral with regard to jurisdictional claims in published maps and institutional affiliations.

Copyright: © 2022 by the authors. Licensee MDPI, Basel, Switzerland. This article is an open access article distributed under the terms and conditions of the Creative Commons Attribution (CC BY) license (https://creativecommons.org/licenses/by/4.0/).

1. Introduction

Nowadays, meta-heuristics become popular in different research areas for resolving challenging optimization issues. These stochastic approaches are among the best and effective strategies in finding optimal solutions, conflicting with the classical (deterministic) optimization approaches which are devalued due to their drawbacks as local optima stagnation [1], etc. In spite of the benefits of the intelligence algorithms, they require some improvement to satisfy the diverse characteristics of complex real-world applications. Features that mostly faced in real issues are uncertainty [2], dynamicity [3], combinatorial, multiple objectives [4,5], constraints, etc. Along these lines, it is obviously seen that no approach is qualified in resolving the diverse kind of optimization problems. In that regard, the No-Free Lunch (NFL) theorem [6] validates this and opens the way for developers to create the newest approaches and enhances the quality of the existing ones.

Some of well-regarded meta-heuristic algorithms are: a genetic algorithm (GA), which is the first stochastic algorithm inspired by John Holland in 1960 [7], followed by simulated annealing (SA) in 1983 [8], particle swarm optimization (PSO) in 1995 by Kennedy [9], and more approaches that were developed later such as ant bee colony (ABC) [10], arithmetic optimization algorithm (AOA) [11], Harris hawks optimization (HHO) [12], sine cosine algorithm (SCA) [13], black widow optimization (BWO) [14], dynamic differential annealed optimization (DDAO) [15], levy flight distribution (LFD) [16], Salp swarm algorithm (SSA) [17], Henry gas solubility optimization (HGSO) [18], manta ray foraging optimization (MRFO) [19], and so on. All these developed stochastic algorithms are frequently single-objective; therefore, the researchers improve them according to the nature and complexity of their problems. Hence, in line with the aforementioned nature of applications, we have improved the new bio-inspired approach called manta ray foraging optimization (MRFO) [19] with a view to cope with the multi-objective problems (MOPs), which are the main focus of this paper.

During the two last years, several studies have guaranteed the superiority and efficiency of the MRFO algorithm in solving global optimization problems, such as: Fahd et al. [20], who applied the standard MRFO to perform the dynamic operation for connecting PV into the grid system. The authors in [21] have examined the global maximum power point of a partially shaded MJSC photovoltaic (PV) array applying the MRFO algorithm. Regarding the work of Selem et al. [22], the MRFO was applied to define the unknown electrical parameters of proton exchange membrane fuel cells stacks, which is considered as a constrained optimization problem. In addition, El-Hameed et al. [23] used MRFO to solve the solar module parameters' identifications of a three diode equivalent model of PV. In addition, in an attempt to ameliorate the performance of this suggested approach, Dalia et al. [24] introduced a modified MRFO by using the fractional-order optimization algorithms, in order to enhance its exploitation ability. Referring to [25], a binary version of MRFO has been proposed using four S-Shaped and four V-Shaped transfer functions for the feature selection problem. In the bio-medical area, Karrupusamy utilized a hybrid MRFO to identify the issue in an existing brain tumor by using a convolutional neural network as a classifier that classifies the features and supplies optimal classification results [26], etc. The authors in [27] have used the multi-objective manta ray foraging optimization (MOMRFO) based weighted sum to handle the optimal power flow (OPF) problem for hybrid AC and multi-terminal direct current power grids. In Ref. [28], the authors applied the IMOMRFO to solve the IEEE-30 and IEEE-57 OPF issues.

In accordance with the literature review, the multi-objective algorithms (MOAs) are divided into two techniques: a priori versus a posteriori [29]. The first class converts the multi-objective problem to a single one, by aggregating all objectives in one function using a set of weights that are chosen by an expert in the problem domain (decision makers). The drawback of this method appears when we generate the Pareto optimal set, we should run the algorithm multiple times. Alternatively, the second class is the posteriori technique which does not require any addition weights. In this method, the multi-objective formulation is maintained and the Pareto optimal set is obtained in just one run; then, the decision-making occurs after the optimization. In addition, the Pareto front of all kinds of problems can be determined utilizing this posteriori technique, which is the focus of this work, in which a new multi-objective version of MRFO based non-dominated sorting approach named NSMRFO was developed.

Different shapes of fronts exist in multi-objective problems: linear, convex, concave, separated, and so on. Therefore, to obtain an accurate approximation Pareto optimal front for every multi-objective optimization issues, three fundamental challenges should be addressed: distribution of solutions (coverage), accuracy (convergence), and local fronts [30]. Thus, an efficient algorithm is the one that has the ability to balance between them: avoid a premature convergence and extract a uniform distribution front covering the entire true Pareto optimal front. Some of the most popular multi-objective algorithms are: Non-dominated sorting genetic algorithm (NSGA) [31,32], strength Pareto evolutionary algorithm (SPEA) [33,34], multi-objective particle swarm optimization (MOPSO) [35], and multi-objective evolutionary algorithm based on decomposition (MOEA/D) [36].

Since the multi-objective optimization problem appears, the non-dominated sorting strategy with crowding distance and non-dominated ranking has been known as the efficient and most significant mechanisms in handling the algorithms for solving the multi-objective problems. The significant advantages of the NSGA-II and its borrowed MOAs motivated us to suggest a novel multi-objective variant of the MRFO approach, which is based on the NSGA-II outstanding operators. The search mechanism in MRFO is kept similar in the NSMRFO optimizer. Furthermore, in order to assess the NSMRFO success, various MMMOEAs with other MOEAs were investigated for comparisons with respect to diverse indicator metrics in search and objective spaces. Thus, with accordance to the statistical outcomes, the proposed NSMRFO outperformed its competitors and even the existing MOMRFOs for different kinds of problems.

The main contributions of this paper are as follows:

- A new multi-objective version of manta ray foraging optimization based the crowding distance and non-dominated sorting operators has been introduced.
- Various performance metrics have been employed in order to affirm the NSMRFO effectiveness.
- The suggested NSMRFO was benchmarked on the standard unconstrained, constrained multi-objective test suites, CEC2020 multi-modal multi-objective optimization functions as well as engineering design problems to verify its validity.
- The IEEE 30-bus OPF as one of the most significant real-world issues in the power system is investigated with wind/solar/small-hydro energy sources for the first time with the multi-objective case.

The remainder of this paper is arranged in four sections as follows: Section 2 summarizes the basic definitions of the multi-objective problems, and then describes the proposed algorithm MRFO and the structure of its multi-objective version NSMRFO. Simulation results, analyses, and competing algorithms are discussed in Section 3. As a final point, Section 4 concludes this work and proposes some future research directions.

2. Multi-Objective Optimization

As mentioned before, the multi-objective problem is the subject of handling problems that need optimizing more than one objective simultaneously, which are mostly in conflict. The basic mathematical formulation of the multi-objective optimization for such minimization problem can be defined as:

$$\begin{aligned}
\text{Minimize:} \quad & F(\vec{x}) = \left\{ f_1(\vec{x}), f_2(\vec{x}), \ldots, f_{N_{obj}}(\vec{x}) \right\} \\
\text{Subject to:} \quad & g_i(\vec{x}) \geq 0, \quad i = 1, 2, \ldots, m \\
& h_i(\vec{x}) = 0, \quad i = 1, 2, \ldots, p \\
& L_i \leq x_i \leq U_i, \quad i = 1, 2, \ldots, n
\end{aligned} \quad (1)$$

where $F(\vec{x})$ is the objective function to be optimized, $h_i(\vec{x})$ is the equality constraints, $g_i(\vec{x})$ is the inequality constraints, N_{obj}, m, p, and n are the numbers of objective functions, inequality constraints, equality constraints, and variables. L_i and U_i are the boundaries of the ith variable.

The arithmetic relational operators cannot be effective in multi-objective optimization to compare the search space of different solutions. Alternatively, the Pareto optimal dominance concept is utilized to determine which solution is better than another. The essential definitions of dominance relation are defined as follows [37,38]:

Let us take two vectors $\vec{x} = (x_1, x_2, \ldots, x_n)$ and $\vec{y} = (y_1, y_2, \ldots, y_n)$

Definition 1 (Pareto Dominance). *$\vec{x}$ is said to dominate $\vec{y}$ if and only if $\vec{x}$ is partially less than $\vec{y}$ (i.e., $\vec{x} \preceq \vec{y}$):*

$$\forall i \in \{1, 2, \ldots, N_{obj}\} : f_i(\vec{x}) \leq f_i(\vec{y}) \land \exists i \in \{1, 2, \ldots, N_{obj}\} : f_i(\vec{x}) < f_i(\vec{y}) \tag{2}$$

Definition 2 (Pareto Optimality). *$\vec{x} \in X$ is called a Pareto-optimal solution iff:*

$$\nexists \vec{y} \in X \mid F(\vec{y}) < F(\vec{x}) \tag{3}$$

Definition 3 (Pareto Optimal Set). *The Pareto optimal set is a set that comprises all Pareto optimal solutions (neither $\vec{x}$ dominates $\vec{y}$ nor $\vec{y}$ dominates $\vec{x}$):*

$$P_s := \{x, y \in X \mid \exists F(\vec{y}) > F(\vec{x})\} \tag{4}$$

Definition 4 (Pareto Optimal Front). *The Pareto optimal front is defined as:*

$$P_f := \{F(\vec{x}) \mid \vec{x} \in P_s\} \tag{5}$$

In such multi-objective optimization problem, a solution is the set of best non-dominated solutions. Therefore, the Pareto optimal solutions projection in the objective space are kept in a set called Pareto optimal front as illustrated in Figure 1. The solutions of both spaces obviously reveal that the green shapes are better than the others, since they dominate all other colors.

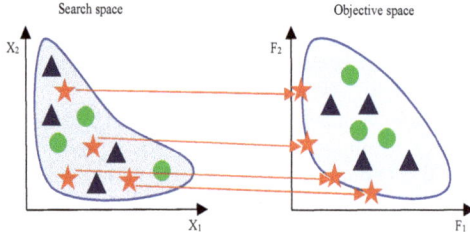

Figure 1. Parameter and objective spaces.

The concept of the MRFO standard version is explained briefly in the following section.

2.1. Manta Ray Foraging Optimizer (MRFO)

MRFO is among the recent algorithms proposed in 2020, inspired by giant known critters of the sea called Manta Rays [19]. Figure 2 depicts the shape of a manta ray. To establish this algorithm, the authors mimic three feeding behaviors of Manta Rays: chain, somersault, and cyclone feeding. Furthermore, the manta rays are assumed as search agents which explore the planktons' location and proceed towards them. Then, the planktons at significant concentration represent the best solution. The source code of MRFO is given in https://www.mathworks.com/matlabcentral/fileexchange/73130-manta-ray-foraging-optimization-mrfo (accessed on 24 May 2021).

Following the population-based optimization algorithms, the steps of MRFO are randomly initialized as illustrated below:

$$x_i = Lb_i + rand \times (Ub_i - Lb_i), \quad i = 1, \ldots, N \tag{6}$$

where Ub and Lb are the maximum and minimum bounds of variables in the search space, rand is a random number between 0 and 1, $rand \in [0, 1]$.

The three main operators are mathematically clarified in the next subsections.

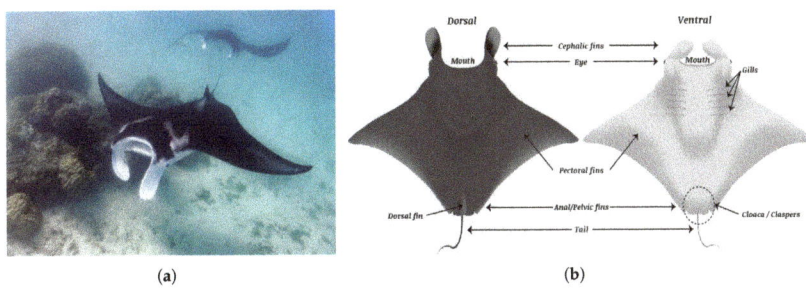

Figure 2. Manta ray body form. (**a**) Manta ray in the ocean; (**b**) parts of a manta ray, dorsal, and ventral.

2.1.1. Chain Foraging

In this foraging strategy, about 50 Mantas line up head to tail forming an orderly line. The chain swims towards the position of intense concentration of plankton with a fully open mouth. The missing plankton by the leader (manta at the top of the chain) will be devoured by the followers. In the course of the foraging process, the position of each follower is updated towards the best source of plankton and individuals in front of it. This foraging phase is depicted in Figure 3. The mathematical updating formulas are presented as follows:

$$x_{i,j}^{t+1} = \begin{cases} x_{i,j}^t + r_1 \cdot \left(x_{best,j}^t - x_{i,j}^t\right) + \alpha \cdot \left(x_{best,j}^t - x_{i,j}^t\right), & i = 1 \\ x_{i,j}^t + r_2 \cdot \left(x_{i-1,j}^t - x_{i,j}^t\right) + \alpha \cdot \left(x_{best,j}^t - x_{i,j}^t\right), & i = 2, \ldots, N \end{cases} \quad (7)$$

where $x_{i,j}$ is the position of ith manta ray in the jth dimension, r_1 and r_2 are the random vector in range [0, 1], $x_{best,j}^t$ is the best plankton concentration position, and α is a weight coefficient that is expressed as:

$$\alpha = 2 \cdot r_3 \cdot \sqrt{|\log(r_4)|} \quad (8)$$

where r_3 and r_4 introduce the random vector in range [0, 1].

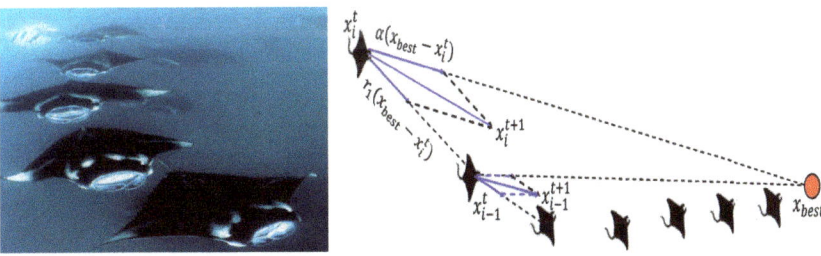

Figure 3. Simulation model of chain foraging behavior.

2.1.2. Cyclone Foraging

Cyclone foraging phase follows the feeding strategy in WOA [39] in terms of spiral movement. After discovering a significant amount of plankton in the profundity of the ocean, the mantas move one behind another towards plankton making a spiral shape. This foraging phase is illustrated in Figure 4. The manta updates its position based on its best previous position and the manta in front of it.

The spiral-shaped movement is mathematically modeled as:

$$x_{i,j}^{t+1} = \begin{cases} x_{best,j} + r_5 \cdot \left(x_{best,j}^t - x_{i,j}^t\right) + \beta \cdot \left(x_{best,j}^t - x_{i,j}^t\right), & i = 1 \\ x_{best,j} + r_6 \cdot \left(x_{i-1,j}^t - x_{i,j}^t\right) + \beta \cdot \left(x_{best,j}^t - x_{i,j}^t\right), & i = 2, \ldots, N \end{cases} \quad (9)$$

where r_5 and r_6 present the random value in [0, 1]; β is the weight coefficient that is formulated as:

$$\beta = 2e^{r_7 \frac{T-t+1}{T}} \cdot \sin(2\pi r_7) \tag{10}$$

where r_7 denotes the random vector in range [0, 1], and T and t are the maximum and current iteration, respectively.

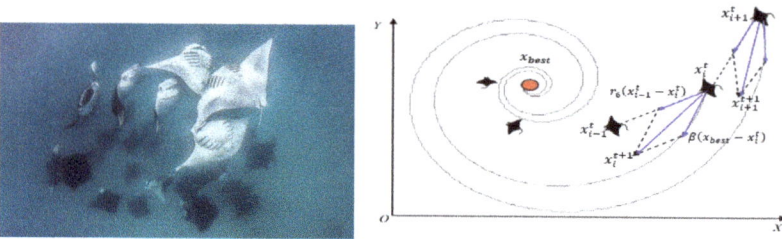

Figure 4. Simulation model of cyclone foraging behavior.

The cyclone foraging can be considered as the main phase in MRFO, in which it performs the intensification (exploitation) and diversification (exploration) mechanisms. The exploitation improvement is achieved based on considering the best plankton found so far as a reference point. On the other hand, the exploration phase incites MRFO to reach the overall optimal solution in accordance with the mathematical equations described below:

$$x_{rand,j} = \text{Lb}_j + r_8 \cdot (\text{Ub}_j - \text{Lb}_j) \tag{11}$$

$$x_{i,j}^{t+1} = \begin{cases} x_{rand,j} + r_9 \left(x_{rand,j} - x_{i,j}^t \right) + \beta \left(x_{rand,j} - x_{i,j}^t \right), & i = 1 \\ x_{rand,j} + r_{10} \left(x_{i-1,j}^t - x_{ij}^t \right) + \beta \left(x_{rand,j} - x_{i,j}^t \right), & i = 2, \ldots, N \end{cases} \tag{12}$$

where $x_{rand,j}$ is the random position generated inside the search space.

2.1.3. Somersault Foraging

The last phase in MRFO is somersault feeding, wherein the manta ray swims to and fro around a pivot point and somersaults around itself to a new position. Figure 5 illustrates this feeding behavior. The manta updates its position using the following mathematical model:

$$x_{i,j}^{t+1} = x_{i,j}^t + S \cdot \left(r_{11} \cdot x_{best,j} - r_{12} \cdot x_{i,j}^t \right), \quad i = 1, \ldots, N \tag{13}$$

where r_{11}, r_{12} depict the random values between 0 and 1. S is the somersault factor, $S = 2$.

Figure 5. Simulation model of somersault foraging behavior.

MRFO's diversification and intensification phases are balanced using the variations value t/T, which is gradually increased. The expression $t/T > rand$ denotes the exploration stage, reversibly, and the exploitation process is adopted. The main steps followed in MRFO are demonstrated in Figure 6.

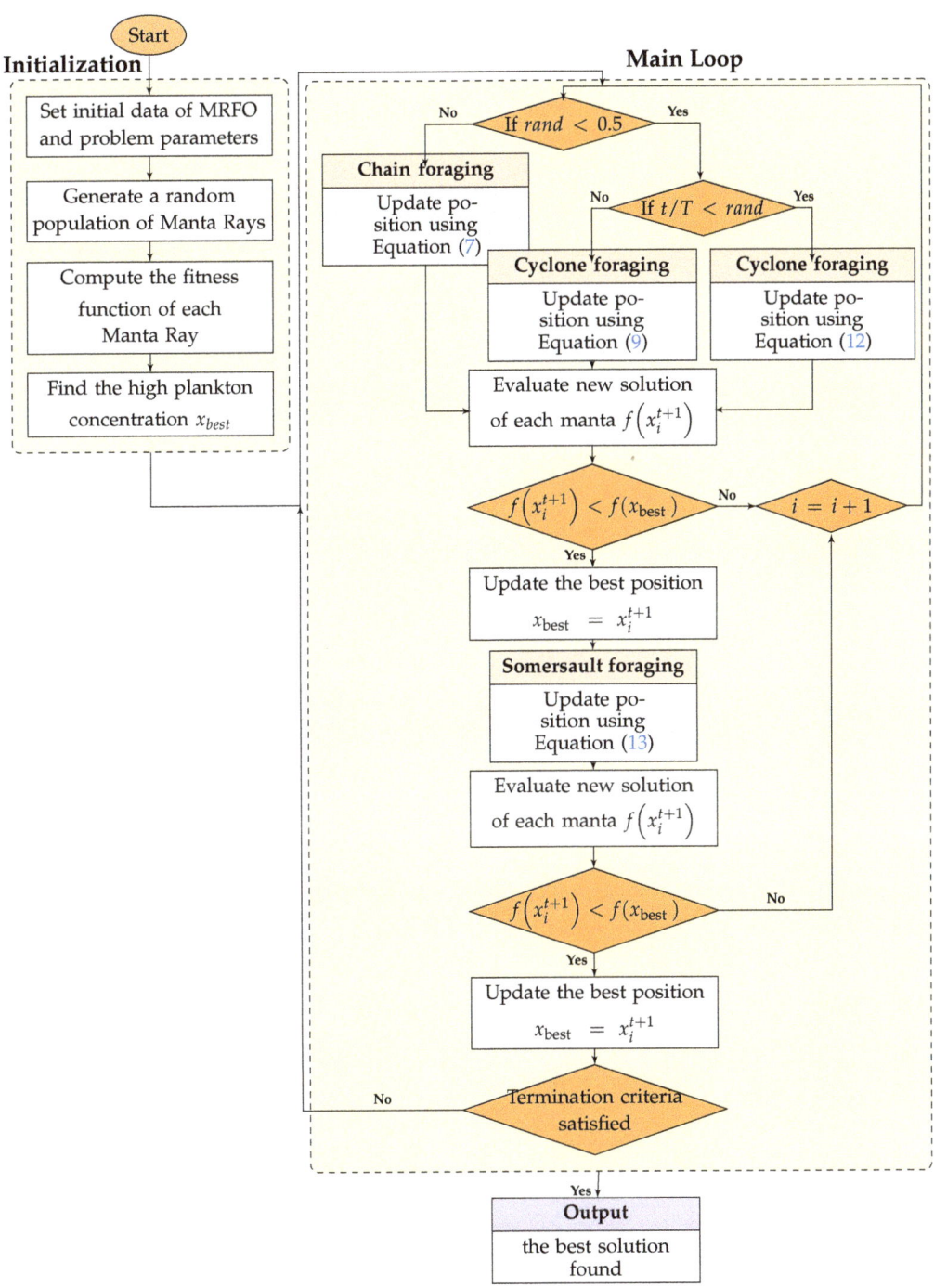

Figure 6. MRFO flowchart for the minimization problem.

2.2. Proposed (NSMRFO)

As MRFO is relevant for single objective issues, we have developed a multi-objective version of MRFO to handle problems with many fitness functions by applying the Pareto dominance strategy. This variant is inspired from the non-dominated sorting genetic algorithm (NSGA-II) approach, which is the most popular and efficient algorithm in the area of multi-objective optimization in the literature. The non-dominated sorting (NDS) technique employs the crowding distance to define an ordering among individuals and preserve the diversity and the elitist mechanism. To compute all non-dominated solutions, a ranking process is applied called non-dominated ranking (NDR), in which the front that is not dominated by any solutions is assigned to rank 1, while rank 2 is in accordance with the front that is dominated by at least one of the solutions, and so on; the ranking scheme is described in Figure 7a. The crowding distance value of a particular solution is the average distance of its two neighboring solutions as illustrated in Figure 7b. Therefore, the less value of crowding distance denotes comparatively the higher crowded space and conversely. The formulation of this crowding distance mechanism is defined as below:

$$CD_i(j) = \frac{f_i(j+1) - f_i(j-1)}{f_i^{max} - f_i^{min}} \quad (14)$$

where f_i^{min} and f_i^{max} are the minimum and maximum values of the ith objective function.

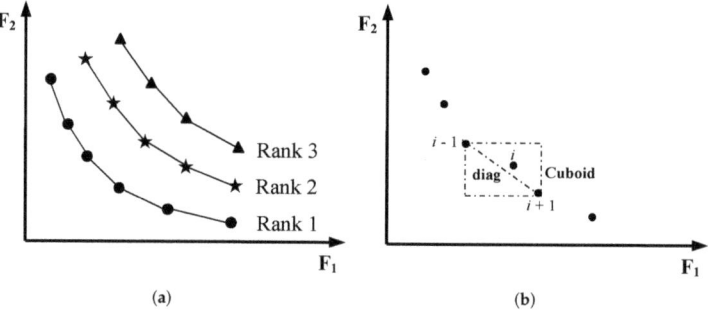

Figure 7. Non-dominated ranking fonts (**a**); crowding-distance calculation (**b**).

It is worth discussing here that the (NDS) provides a probability to the dominated solutions to be chosen as well, which enhances the diversification of the suggested algorithm. The pseudo-code of NSMRFO is depicted in Algorithm 1. The computational space complexity of NSMRFO is the same as NSGA-II of the order of $O(MN^2)$, where M is the number of objectives, and N is the number of manta rays, while the computational complexity was found to be much better than that of some of the approaches such as SPEA and NSGA, which are of $O(MN^3)$.

Algorithm 1 Non-Dominated Sorting Manta Ray Foraging Optimization

1: Initialize the NSMRFO parameters: design variables, bounds(Ub, Lb), population, and termination criteria.
2: Generate a uniform random initial population of mantas x with respect to Ub and Lb.
3: Compute the fitness function of each manta and sort all of them.
4: Determine the non-dominated solutions in the initial population and save them in Pareto archive.
5: Compute crowding distance for each Pareto archive member.
6: Select a position vector based on crowding distance value.
7: Compute the position vector and update the position of mantas following the MRFO procedure.
8: Compute the fitness values of all the updated positions of mantas.
9: Determine the new non-dominated solutions in the population, save them in a Pareto archive, then remove any dominated solutions in the Pareto archive.
10: Compute the crowding distance value for each Pareto archive member and eliminate as many as necessary according to archive size with the lowest crowding distance value.
11: Perform non-dominated sorting according to the crowding distance mechanism, then select the global best solution using the ranking scheme.
12: Display Pareto optimal set.

2.3. Evaluation Criteria

The employed performance metrics are described in this section. The performance indicators are one of the techniques employed for measuring the potential of a multi-objective algorithm in terms of the diversity and coverage. In this work, various metrics are used such as generational distance (GD) [40], inverted generational distance (IGD) [41] in search [42] and objective [42] spaces, spacing (SP) [43], Pareto sets proximity (rPSP) [44] and reciprocal of hypervolume (rHV) [45], which are formulated as follows:

- Generational Distance (GD) [40]:

$$GD = \frac{\sqrt{\sum_{i=1}^{n_{pf}} d_i^2}}{n_{pf}} \quad (15)$$

where n_{pf} is the number of obtained Pareto optimal solutions, and d_i indicates the Euclidean distance between the ith Pareto optimal solution attained and the closest true Pareto optimal solution in the reference set:

- Inverted Generational Distance (IGD) [41]:

$$IGD = \frac{\sqrt{\sum_{i=1}^{n_{tpf}} (d_i')^2}}{n_{tpf}} \quad (16)$$

where n_{tpf} is the number of true Pareto optimal solutions and d_i' indicates the Euclidean distance between the ith true Pareto optimal solution and the closest Pareto optimal solution obtained in the reference set.
$IGDX$ is the IGD in search space. $IGDF$ is the IGD in objective space:

- Spacing (SP) [43]:

$$SP = \sqrt{\frac{1}{n_{pf}-1} \sum_{i=1}^{n_{pf}} (\bar{d} - d_i)^2} \quad (17)$$

where n_{pf} is the number of Pareto optimal solutions obtained. d_i indicates the Euclidean distance between the ith Pareto optimal solution attained and the closest true Pareto optimal solution in the reference set. $\bar{d}$ is the average of all d_i.

- Reciprocal of Pareto sets proximity (rPSP) [44]:

$$\text{rPSP} = \frac{IGDX}{CR} \tag{18}$$

$$CR = \left(\prod_{i=1}^{m} \delta_i\right)^{1/2D} \tag{19}$$

$$\delta_i = \left(\frac{\min\left(PFe_i^{\max}, PFt_i^{\max}\right) - \max\left(PFe_i^{\min}, PFt_i^{\min}\right)}{PFt_i^{\max} - PFt_i^{\min}}\right)^2 \tag{20}$$

where CR is the cover rate. m is the number of objective functions. D is the number of decision variables. PFt, PFe are the true and obtained Pareto front, respectively:

- Reciprocal of hypervolume (rHV) [45]:

$$rHV(S, w) = \frac{1}{HV(S, w)} \tag{21}$$

$$HV(S, w) = \lambda_D\left(\bigcup_{z \in S}[z; w]\right) \tag{22}$$

where λ_D is the D-dimensional Lebesgue measure.

3. Experimental Results and Analysis

In this section, the effectiveness of the proposed multi-objective approach is carried out by using 18 different unconstrained benchmark functions, the CEC2020 benchmark test that contains 24 functions, four constrained problems, four engineering design problems, and the IEEE 30-bus optimal power flow issue incorporating wind/solar/small-hydro power. These test suites have different shapes of front like linear, convex, concave, connected, disconnected, etc., as indicated in Table 1. Five analyses are investigated to prove the robustness of the developed NSMRFO algorithm, the first one aims to assess the convergence by using generation distance (GD) metric, the second evaluates the diversity by computing the spacing (SP) metric, the inverse generation distance (IGD) in the search and objective spaces metric, which intends to affirm the NSMRFO's efficacy in balancing between convergence and diversity, and the reciprocal of Pareto sets proximity (rPSP) and hypervolume (rHV). Moreover, for evaluating the NSMRFO approach, 11 significant multi-objective optimization algorithms are re-implemented, which are named: multi-objective slime mould algorithm (MOSMA) [46], multi-objective bonobo optimizer based decomposition (MOBO/D) [47], multi-objective multi-verse optimization (MOMVO) [48], multi-objective water cycle algorithm (MOWCA) [49], non-dominated sorting grey wolf optimizer (NSGWO) [50], multi-objective manta ray foraging optimizer (MOMRFO) [51], improved multi-objective manta ray foraging optimizer (IMOMRFO) [28], non-sorting genetic algorithm II (NSGA-II) [32], double niched non sorting genetic algorithm II (DN-NSGA) [52], omni optimizer (OMNI) [53], a multi-objective particle swarm optimizer using ring topology (MO_ Ring_ PSO_ SCD) [44], and their characteristics are shown in Table 2. The MATLAB codes for these algorithms were downloaded from: https://aliasgharheidari.com/SMA.html (SMA), https://www.mathworks.com/matlabcentral/fileexchange/79843-multi-objective-bonobo-optimizer-with-decomposition-method (BO), https://seyedalimirjalili.com/mvo (MVO), https://ali-sadollah.com/water-cycle-algorithm-wca / (WCA), https://www.mathworks.com/matlabcentral/fileexchange/75259-multi-objective-non-sorted-grey-wolf-mogwo-nsgwo?s_tid=srchtitle_nsgwo_1 (GWO), https://www.mathworks.com/matlabcentral/fileexchange/103530-momrfo-multi-objective-manta-ray-foraging-optimizer?s_tid=srchtitle_MOMRFO_1 (MRFO), https://www.mathworks.com/matlabcentral/fileexchange/103895-improved-multi-objective-manta-ray-foraging-optimization?s_tid=srchtitle_MOMRFO_2 (IMRFO), and the codes of the other algorithms and the CEC2020 test suite can be found https://www.mathworks.com/matlabcentral/fileexchange/103895-improved-multi-objective-manta-ray-foraging-optimization?s_tid=srchtitle_MO

MRFO_2, respectively (all accessed on 7 August 2021). Each approach is executed on a personal computer, Windows 8.1 (64-bit), core i5 with 4GB-RAM Processor @1:8 GHz using MATLAB R2020a. All benchmark functions are executed 20 times for 1000 iterations and 100 populations, except the CEC2020 problem, which is executed 21 times for a population $pop = 200$, and a maximum number of function evaluations equal to $10,000 * pop$. In addition, the OPF problem was repeated 20 time for $pop = 100$ and 200 iterations. Note that the best performing algorithm is assessed based on the mean and standard deviation outcomes. The quantitative and qualitative performance outcomes are illustrated in Tables 3–15 and Figures 8–16, respectively. The outcomes of each set of benchmark functions are outlined and discussed in the following sections.

Table 1. Descriptions of the unconstrained, constrained, and CEC2020 benchmark functions.

Case	Name	No. Objs	Description
Classical test problems			
F1	Schaffer1	2	Convex
F2	Schaffer2	2	Disconnected
F3	Kursawe	2	Degenerate, disconnected
F4	Poloni	2	Disconnected
F5	Fonseca2	2	Concave
F6	Viennet2	3	Connected
F7	Viennet3	3	Connected and asymmetric
ZDT test suite			
F8	ZDT1	2	Convex
F9	ZDT2	2	Concave
F10	ZDT3	2	Disconnected
F11	ZDT4	2	Convex, many local optima
F12	ZDT6	2	Concave, nonuniform fitness landscape
YTD test suite			
F13	YTD1	2	Disconnected
F14	YTD2	2	Disconnected
F15	YTD3	2	Disconnected
F16	YTD4	2	Disconnected
F17	YTD5	2	Disconnected
F18	YTD6	2	Disconnected
Constrained test suite			
F19	OZY	2	Convex, mixed
F20	BNH	2	Convex
F21	SRN	2	Linear, degenerate
F22	CONSTR	2	Convex
CEC-2020 test suite			
F23	MMF1	2	Convex
F24	MMF2	2	Convex
F25	MMF4	2	Concave
F26	MMF5	2	Convex
F27	MMF7	2	Convex
F28	MMF8	2	Convex
F29	MMF10	2	Convex
F30	MMF11	2	Convex
F31	MMF12	2	Convex
F32	MMF13	2	Convex
F33	MMF14	3	Concave
F34	MMF15	3	Concave
F35	MMF1_e	2	Convex
F36	MMF14_a	3	Concave
F37	MMF15_a	3	Concave
F38	MMF10_l	2	Convex
F39	MMF11_l	2	Convex
F40	MMF12_l	2	Convex
F41	MMF13_l	2	Convex
F42	MMF15_l	3	Concave
F43	MMF15_a_l	3	Concave
F44	MMF16_l1	3	Concave
F45	MMF16_l2	3	Concave
F46	MMF16_l3	3	Concave

Table 2. Parameter settings of the tested algorithms.

Algorithms	Parameters	Values
NSMRFO	Somersault factor S	2
MOSMA [46]	-	-
MOBO/D [47]	alpha-bonobo (scab)	1.55
	selected bonobo (scsb)	1.4
	probability (rcpp)	0.004
	Initial probability (pxgm-initial)	0.08
	subgroup size factor (tsgsfactor-max)	0.07
	Number of grids per dimension	7
	Grid inflation rate	0.1
	Leader selection pressure	2
	Deletion selection pressure	2
MOMVO [48]	maximum worm hole existence probability	1
	minimum worm hole existence probability	0.2
MOWCA [49]	number of rivers + sea (Nsr)	4
NSGWO [50]	-	-
MOMRFO [51]	Somersault factor S	2
	nGrid	30
IMOMRFO [28]	Somersault factor S	2

3.1. Evaluation on Unconstrained Benchmark Functions

As mentioned above, the proposed approach is tested firstly on the classical unconstrained test problems with two and three objectives. The achieved mean and STD values of 20 runs of each parameter metrics from NSMRFO and different approaches are presented in Table 3 and indicated in bold. It is worth noting here that the better algorithm is the one with the lower metric value. The suggested approach NSMRFO managed to outperform the MOSMA [46], MOBO/D [47], MOMVO [48], MOWCA [49], MOMRFO [51], (IMOMRFO) [28], and (NSGA-II) [32] optimizer significantly on eight functions out of 18 cases for GD, 14 out of 10 cases for IGD, and 14 out of 18 cases for SP metrics. By comparison, MOSMA is better on SCH2 for all metrics, the MOWCA is best on FON2 for GD, on POL for IGD, on POL, and VNT3 for SP, the MOMVO is best on just SCH1 for SP metric, the IMOMRFO is best on SCH1 for GD, and VNT2 for IGD; however, the NSGA-II and MOMRFO offered a good solution on four and eight functions, respectively. By contrast, the MOBO/D optimizer provides the worst results. Therefore, it may be observed from this table that the NSMRFO approach is able to outperform all competitors on most cases. Furthermore, it is also evident from Figures 8–10 that NSMRFO converges better toward a true Pareto front with different features from diverse perspectives. In addition, the Pareto optimal solutions have been well distributed over the true PF on the classical functions.

Table 3. GD, IGD, and SP metrics comparison based on unconstrained test suites.

Case	Name	NSMRFO	MOSMA	MOBO/D	MOMVO	MOWCA	NSGA-II	MOMRFO	IMOMRFO
GD Metric									
F1	SCH1	1.5186×10^{-3} $\mathbf{6.21 \times 10^{-5}}$	1.5408×10^{-3} 7.73×10^{-5}	1.1205×10^{-3} 1.51×10^{-4}	1.6145×10^{-3} 6.31×10^{-4}	5.0232×10^{-3} 2.14×10^{-3}	1.4988×10^{-3} 6.44×10^{-5}	1.9728×10^{-3} 1.41×10^{-4}	$\mathbf{1.0480 \times 10^{-3}}$ 1.12×10^{-4}
F2	SCH2	3.7839×10^{-3} 1.83×10^{-4}	$\mathbf{3.1145 \times 10^{-3}}$ 9.21×10^{-4}	4.4024×10^{-3} 2.77×10^{-3}	3.5789×10^{-3} 1.23×10^{-4}	3.9098×10^{-3} 9.19×10^{-5}	$3.0103 \times 10^{+0}$ $1.0829 \times 10^{+0}$	3.5790×10^{-3} 3.62×10^{-4}	3.5958×10^{-3} 3.20×10^{-4}
F3	KUR	$\mathbf{1.5001 \times 10^{-3}}$ $\mathbf{1.52 \times 10^{-4}}$	4.8631×10^{-3} 6.02×10^{-3}	8.4540×10^{-3} 8.89×10^{-3}	3.3798×10^{-3} 1.99×10^{-3}	6.5932×10^{-3} 4.22×10^{-3}	6.6901×10^{-2} 5.30×10^{-2}	3.1843×10^{-3} 2.91×10^{-4}	3.4414×10^{-3} 3.27×10^{-4}
F4	POL	$\mathbf{2.2210 \times 10^{-2}}$ 1.56×10^{-2}	5.6752×10^{-2} 1.99×10^{-2}	6.9886×10^{-2} 4.43×10^{-2}	1.2298×10^{-1} 5.05×10^{-2}	7.6231×10^{-2} 1.75×10^{-2}	5.8359×10^{-2} 4.06×10^{-3}	4.9070×10^{-2} 1.37×10^{-2}	7.1044×10^{-2} 1.24×10^{-2}
F5	FON2	2.3155×10^{-4} 2.49×10^{-5}	1.4972×10^{-4} 1.82×10^{-5}	6.5563×10^{-4} 9.49×10^{-5}	1.1445×10^{-3} 5.71×10^{-4}	$\mathbf{1.4387 \times 10^{-4}}$ 3.65×10^{-5}	1.9174×10^{-4} 2.61×10^{-5}	2.8196×10^{-4} 4.1343×10^{-5}	2.9866×10^{-4} 3.1249×10^{-5}

Table 3. Cont.

Case	Name	NSMRFO	MOSMA	MOBO/D	MOMVO	MOWCA	NSGA-II	MOMRFO	IMOMRFO
F6	VNT2	**4.0305**×10^{-3} $9.00×10^{-4}$	$2.1878×10^{-2}$ $6.02×10^{-3}$	$1.8094×10^{-2}$ $1.02×10^{-3}$	$1.6617×10^{-2}$ $6.12×10^{-3}$	$1.9053×10^{-2}$ $7.44×10^{-4}$	$1.8746×10^{-2}$ $1.20×10^{-3}$	$1.1303×10^{-2}$ $1.34×10^{-3}$	$1.2150×10^{-2}$ $1.86×10^{-3}$
F7	VNT3	$2.3873×10^{-2}$ $1.27×10^{-3}$	$2.3977×10^{-2}$ $1.12×10^{-3}$	$2.8189×10^{-2}$ $3.31×10^{-2}$	$2.3966×10^{-2}$ $1.06×10^{-3}$	$2.0508×10^{-2}$ $6.24×10^{-3}$	$2.3930×10^{-2}$ $1.49×10^{-3}$	**1.6262**×10^{-2} $1.77×10^{-3}$	$2.2563×10^{-2}$ $4.08×10^{-3}$
F8	ZDT1	$2.3319×10^{-4}$ $3.23×10^{-5}$	$3.6551×10^{-4}$ $4.74×10^{-5}$	$5.0011×10^{-4}$ $2.51×10^{-5}$	$3.8157×10^{-3}$ $3.82×10^{-4}$	$2.6413×10^{-4}$ $4.24×10^{-5}$	$3.6790×10^{-2}$ $2.69×10^{-2}$	**1.9395**×10^{-4} $5.12×10^{-5}$	$3.7296×10^{-2}$ $1.80×10^{-2}$
F9	ZDT2	**9.3101**×10^{-5} $4.36×10^{-6}$	$3.0210×10^{-4}$ $5.52×10^{-5}$	$2.3881×10^{-4}$ $8.69×10^{-5}$	$5.7435×10^{-3}$ $3.93×10^{-4}$	$5.8786×10^{-3}$ $9.36×10^{-3}$	$8.8345×10^{-2}$ $7.92×10^{-2}$	$1.1687×10^{-4}$ $3.11×10^{-5}$	$4.2670×10^{-2}$ $3.74×10^{-2}$
F10	ZDT3	$1.6294×10^{-4}$ $1.07×10^{-5}$	$2.1877×10^{-4}$ $2.36×10^{-5}$	$8.7873×10^{-4}$ $5.25×10^{-4}$	$1.9516×10^{-2}$ $4.98×10^{-2}$	$4.6262×10^{-4}$ $6.01×10^{-4}$	$2.0248×10^{-2}$ $1.41×10^{-2}$	**1.5360**×10^{-4} $2.23×10^{-5}$	$3.6556×10^{-2}$ $3.63×10^{-2}$
F11	ZDT4	**2.2409**×10^{-4} $4.91×10^{-5}$	$2.7059×10^{-4}$ $5.52×10^{-5}$	$9.3797×10^{-2}$ $4.58×10^{-2}$	$1.3951×10^{+0}$ $7.42×10^{-1}$	$1.2212×10^{+0}$ $1.26×10^{+0}$	$2.9861×10^{+0}$ $1.04×10^{+0}$	$3.8600×10^{-4}$ $5.23×10^{-5}$	$3.4435×10^{-4}$ $2.54×10^{-4}$
F12	ZDT6	$4.4226×10^{-2}$ $4.93×10^{-2}$	$1.9098×10^{-2}$ $2.54×10^{-2}$	$1.0946×10^{-4}$ $1.01×10^{-4}$	$1.2260×10^{-2}$ $5.36×10^{-3}$	$2.1251×10^{-1}$ $1.09×10^{-1}$	$1.8761×10^{-1}$ $1.34×10^{-1}$	**7.2768**×10^{-5} $6.46×10^{-6}$	$6.5288×10^{-1}$ $1.65×10^{-1}$
F13	TYD1	**1.0438**×10^{-3} $4.20×10^{-4}$	$6.7872×10^{-2}$ $7.30×10^{-3}$	$4.4802×10^{-2}$ $2.77×10^{-2}$	$5.7748×10^{-3}$ $7.05×10^{-4}$	$3.1187×10^{-2}$ $7.32×10^{-3}$	$1.3143×10^{-2}$ $6.49×10^{-3}$	$2.4044×10^{-2}$ $7.55×10^{-3}$	-
F14	TYD2	$7.2424×10^{-4}$ $1.55×10^{-5}$	$1.2213×10^{-1}$ $1.11×10^{-1}$	$4.8234×10^{-3}$ $5.34×10^{-3}$	$7.7742×10^{-3}$ $1.47×10^{-3}$	$2.3999×10^{-1}$ $4.11×10^{-1}$	$2.3745×10^{+0}$ $1.71×10^{+0}$	**6.5481**×10^{-4} $5.31×10^{-5}$	-
F15	TYD3	$1.3413×10^{-3}$ $6.72×10^{-4}$	$2.6726×10^{-2}$ $2.34×10^{-2}$	$2.8383×10^{-3}$ $1.39×10^{-3}$	$1.0019×10^{-2}$ $1.77×10^{-3}$	$5.7059×10^{-3}$ $5.16×10^{-3}$	$5.4824×10^{+0}$ $2.96×10^{+0}$	**1.2825**×10^{-3} $1.43×10^{-4}$	-
F16	TYD4	**6.5768**×10^{-4} $1.14×10^{-4}$	$1.5877×10^{-3}$ $1.14×10^{-4}$	$1.4063×10^{-3}$ $9.23×10^{-5}$	$8.0617×10^{-3}$ $3.18×10^{-3}$	$1.4953×10^{-1}$ $3.75×10^{-1}$	$4.7620×10^{+0}$ $2.91×10^{+0}$	$1.1171×10^{-3}$ $5.87×10^{-5}$	-
F17	TYD5	**2.8699**×10^{-4} $9.07×10^{-4}$	$5.3955×10^{-4}$ $1.33×10^{-5}$	$6.0136×10^{-4}$ $2.56×10^{-4}$	$1.5922×10^{-2}$ $5.98×10^{-2}$	$1.4687×10^{-3}$ $1.74×10^{-4}$	$14.962×10^{+0}$ $1.92×10^{+0}$	$4.6125×10^{-4}$ $2.72×10^{-5}$	-
F18	TYD6	$3.0338×10^{-3}$ $8.18×10^{-4}$	$7.3752×10^{-3}$ $3.80×10^{-3}$	$3.1964×10^{-3}$ $8.86×10^{-5}$	$5.7748×10^{-3}$ $7.05×10^{-4}$	$1.1735×10^{-2}$ $1.34×10^{-2}$	$12.040×10^{+0}$ $2.14×10^{+0}$	**2.7641**×10^{-3} $1.81×10^{-4}$	-

IGD Metric

Case	Name	NSMRFO	MOSMA	MOBO/D	MOMVO	MOWCA	NSGA-II	MOMRFO	IMOMRFO
F1	SCH1	**4.9741**×10^{-4} $5.99×10^{-5}$	$7.0139×10^{-4}$ $8.08×10^{-5}$	$1.4727×10^{-3}$ $5.88×10^{-4}$	$5.0232×10^{-3}$ $2.14×10^{-3}$	$5.0230×10^{-4}$ $3.73×10^{-5}$	$5.4182×10^{-4}$ $5.72×10^{-5}$	$7.3050×10^{-4}$ $8.13×10^{-5}$	$7.8791×10^{-4}$ $1.06×10^{-4}$
F2	SCH2	$4.6630×10^{-4}$ $2.85×10^{-5}$	**4.0565**×10^{-4} $1.69×10^{-5}$	$2.4468×10^{-2}$ $1.94×10^{-2}$	$3.7958×10^{-3}$ $5.96×10^{-4}$	$4.5943×10^{-4}$ $1.38×10^{-5}$	$4.3624×10^{-2}$ $1.30×10^{-4}$	$7.9057×10^{-4}$ $4.52×10^{-5}$	$8.7927×10^{-4}$ $1.25×10^{-4}$
F3	KUR	**1.6326**×10^{-4} $1.54×10^{-5}$	$4.5883×10^{-4}$ $3.66×10^{-4}$	$1.1761×10^{-3}$ $8.27×10^{-4}$	$8.0353×10^{-4}$ $1.36×10^{-4}$	$3.3578×10^{-4}$ $2.12×10^{-4}$	$3.4430×10^{-3}$ $2.36×10^{-3}$	$2.5796×10^{-4}$ $2.27×10^{-5}$	$2.5193×10^{-4}$ $1.84×10^{-5}$
F4	POL	$4.8896×10^{-4}$ $3.52×10^{-5}$	$4.9407×10^{-4}$ $1.43×10^{-4}$	$4.0539×10^{-2}$ $6.50×10^{-3}$	$5.9292×10^{-3}$ $1.61×10^{-3}$	**4.8057**×10^{-4} $3.97×10^{-5}$	$5.5668×10^{-4}$ $3.42×10^{-5}$	$9.1231×10^{-4}$ $2.84×10^{-4}$	$8.0518×10^{-4}$ $1.27×10^{-4}$
F5	FON2	**2.7389**×10^{-4} $9.53×10^{-6}$	$5.0585×10^{-4}$ $1.45×10^{-4}$	$6.4536×10^{-4}$ $1.32×10^{-4}$	$1.5838×10^{-3}$ $3.27×10^{-4}$	$2.9388×10^{-4}$ $3.99×10^{-5}$	$3.0779×10^{-4}$ $2.15×10^{-5}$	$4.0938×10^{-4}$ $5.24×10^{-5}$	$4.5341×10^{-4}$ $6.94×10^{-5}$
F6	VNT2	$3.8653×10^{-3}$ $4.80×10^{-4}$	$4.7815×10^{-3}$ $1.29×10^{-3}$	$1.2333×10^{-2}$ $2.25×10^{-3}$	$7.7099×10^{-3}$ $1.19×10^{-3}$	$4.4016×10^{-3}$ $9.41×10^{-4}$	$4.9972×10^{-3}$ $8.71×10^{-4}$	$2.7897×10^{-3}$ $3.17×10^{-4}$	**2.7698**×10^{-3} $2.05×10^{-4}$
F7	VNT3	$1.4552×10^{-4}$ $6.37×10^{-4}$	$1.9514×10^{-3}$ $2.02×10^{-4}$	$2.1501×10^{-2}$ $2.33×10^{-2}$	$5.2610×10^{-3}$ $6.11×10^{-4}$	$1.4331×10^{-3}$ $2.42×10^{-4}$	**1.3868**×10^{-3} $1.49×10^{-4}$	$1.4799×10^{-3}$ $1.79×10^{-4}$	$1.6787×10^{-3}$ $1.48×10^{-4}$
F8	ZDT1	**2.6206**×10^{-4} $1.75×10^{-5}$	$4.7382×10^{-4}$ $7.07×10^{-5}$	$6.3507×10^{-4}$ $9.90×10^{-5}$	$2.0620×10^{-3}$ $1.54×10^{-4}$	$2.6266×10^{-4}$ $1.35×10^{-5}$	$1.5650×10^{-2}$ $1.08×10^{-2}$	$3.9012×10^{-4}$ $3.63×10^{-5}$	$1.7831×10^{-3}$ $3.78×10^{-4}$
F9	ZDT2	**2.7851**×10^{-4} $2.10×10^{-5}$	$4.5857×10^{-4}$ $2.53×10^{-5}$	$5.2148×10^{-4}$ $1.18×10^{-4}$	$2.8492×10^{-3}$ $1.97×10^{-4}$	$8.2051×10^{-3}$ $1.13×10^{-2}$	$4.4870×10^{-2}$ $4.22×10^{-2}$	$3.9149×10^{-4}$ $2.89×10^{-5}$	$2.7908×10^{-3}$ $7.71×10^{-4}$
F10	ZDT3	**1.8616**×10^{-4} $7.89×10^{-6}$	$2.8916×10^{-4}$ $3.92×10^{-5}$	$5.6483×10^{-4}$ $1.98×10^{-4}$	$2.1679×10^{-3}$ $4.23×10^{-4}$	$6.8846×10^{-4}$ $1.24×10^{-3}$	$7.7232×10^{-3}$ $4.26×10^{-3}$	$2.9073×10^{-4}$ $2.21×10^{-5}$	$1.5993×10^{-3}$ $4.07×10^{-4}$
F11	ZDT4	**2.6414**×10^{-5} $2.55×10^{-5}$	$5.0698×10^{-4}$ $8.23×10^{-5}$	$1.9416×10^{-2}$ $7.71×10^{-3}$	$3.9102×10^{-1}$ $2.14×10^{-1}$	$1.6733×10^{-2}$ $2.36×10^{-2}$	$1.2540×10^{+0}$ $4.47×10^{-1}$	$3.7765×10^{-4}$ $3.03×10^{-5}$	$4.1458×10^{-4}$ $3.31×10^{-5}$
F12	ZDT6	**2.9553**×10^{-4} $4.62×10^{-5}$	$4.6403×10^{-4}$ $4.88×10^{-5}$	$1.6486×10^{-3}$ $1.02×10^{-3}$	$6.8445×10^{-4}$ $2.30×10^{-4}$	$1.5553×10^{-2}$ $1.60×10^{-2}$	$9.3672×10^{-2}$ $7.90×10^{-2}$	$4.1577×10^{-4}$ $3.62×10^{-5}$	$5.6469×10^{-2}$ $1.14×10^{-2}$
F13	TYD1	**9.4440**×10^{-4} $4.92×10^{-4}$	$3.2475×10^{-2}$ $2.83×10^{-3}$	$1.5235×10^{-2}$ $7.62×10^{-3}$	$2.9073×10^{-3}$ $6.13×10^{-3}$	$2.0043×10^{-2}$ $6.34×10^{-3}$	$1.1656×10^{-2}$ $7.15×10^{-3}$	$7.3191×10^{-3}$ $1.43×10^{-3}$	-
F14	TYD2	**2.9823**×10^{-4} $1.12×10^{-5}$	$6.5423×10^{-3}$ $4.96×10^{-3}$	$1.5225×10^{-3}$ $9.52×10^{-4}$	$2.2408×10^{-3}$ $1.06×10^{-3}$	$2.1579×10^{-2}$ $3.01×10^{-2}$	$4.9072×10^{-1}$ $3.50×10^{-1}$	$6.3908×10^{-4}$ $8.77×10^{-5}$	-
F15	TYD3	**3.2263**×10^{-4} $9.51×10^{-6}$	$7.7030×10^{-3}$ $5.96×10^{-3}$	$5.0933×10^{-3}$ $3.99×10^{-3}$	$1.6855×10^{-2}$ $5.85×10^{-3}$	$4.8173×10^{-3}$ $4.09×10^{-3}$	$8.8421×10^{-1}$ $4.68×10^{-1}$	$7.9035×10^{-4}$ $8.47×10^{-5}$	-

Table 3. Cont.

Case	Name	NSMRFO	MOSMA	MOBO/D	MOMVO	MOWCA	NSGA-II	MOMRFO	IMOMRFO
F16	TYD4	**3.4868×10⁻⁴** **2.58×10⁻⁵**	5.1097×10⁻⁴ 3.81×10⁻⁵	3.2404×10⁻³ 1.74×10⁻³	7.6373×10⁻³ 8.39×10⁻³	1.7494×10⁻² 2.48×10⁻²	1.1066×10⁺⁰ 6.08×10⁻¹	6.9184×10⁻⁴ 6.53×10⁻⁵	-
F17	TYD5	**4.7951×10⁻⁴** 6.02×10⁻⁵	1.2769×10⁻¹ 2.09×10⁻²	2.0048×10⁻³ 3.72×10⁻³	8.0911×10⁻³ 4.25×10⁻³	1.3432×10⁻¹ 2.92×10⁻⁷	3.1897×10⁺¹ 4.19×10⁺⁰	1.0127×10⁻³ 1.17×10⁻⁴	-
F18	TYD6	**4.3449×10⁻⁴** **4.29×10⁻⁵**	4.9312×10⁻² 2.55×10⁻²	1.0853×10⁻³ 4.40×10⁻⁴	5.3901×10⁻³ 9.49×10⁻³	7.1215×10⁻² 5.31×10⁻³	4.9147×10⁺⁰ 8.65×10⁻¹	7.5797×10⁻⁴ 9.46×10⁻⁵	-

SP Metric

Case	Name	NSMRFO	MOSMA	MOBO/D	MOMVO	MOWCA	NSGA-II	MOMRFO	IMOMRFO
F1	SCH1	2.9654×10⁻² 1.99×10⁻³	3.9539×10⁻² 4.30×10⁻³	6.8903×10⁻² 2.47×10⁻²	**2.0338×10⁻²** 2.25×10⁻²	2.9242×10⁻² 1.81×10⁻³	3.1513×10⁻² 3.04×10⁻³	4.0819×10⁻² 4.96×10⁻³	3.8608×10⁻² 4.34×10⁻³
F2	SCH2	4.3778×10⁻² 2.96×10⁻³	**4.2673×10⁻²** 2.42×10⁻³	6.2204×10⁻¹ 6.39×10⁻¹	5.2262×10⁻² 2.96×10⁻²	4.6848×10⁻² 2.40×10⁻³	3.3859×10⁻¹ 1.01×10⁻¹	5.4329×10⁻² 9.08×10⁻³	6.3003×10⁻² 8.34×10⁻³
F3	KUR	**6.4721×10⁻²** 2.32×10⁻²	9.2719×10⁻² 1.45×10⁻²	2.6929×10⁻¹ 1.76×10⁻¹	1.2020×10⁻¹ 4.12×10⁻²	8.4376×10⁻² 2.01×10⁻²	9.4290×10⁻² 3.80×10⁻²	1.1223×10⁻¹ 1.76×10⁻²	9.9612×10⁻² 1.93×10⁻²
F4	POL	8.1830×10⁻² 8.11×10⁻³	1.6050×10⁻¹ 1.03×10⁻¹	2.1860×10⁻¹ 1.77×10⁻¹	1.2687×10⁻¹ 1.39×10⁻¹	**8.1395×10⁻²** 5.88×10⁻³	9.0758×10⁻² 9.35×10⁻³	2.6669×10⁻¹ 9.68×10⁻²	2.1438×10⁻¹ 1.14×10⁻¹
F5	FON2	**5.9240×10⁻³** **6.03×10⁻⁴**	1.2015×10⁻² 3.23×10⁻³	1.1867×10⁻² 1.87×10⁻³	1.4375×10⁻² 5.34×10⁻³	6.5275×10⁻³ 7.71×10⁻⁴	7.0641×10⁻³ 8.87×10⁻⁴	8.6261×10⁻³ 7.22×10⁻⁴	9.2525×10⁻³ 1.82×10⁻³
F6	VNT2	2.0346×10⁻² 8.29×10⁻³	5.7179×10⁻² 2.91×10⁻²	1.9270×10⁻² 4.02×10⁻³	2.3098×10⁻² 9.53×10⁻³	1.7122×10⁻² 2.40×10⁻³	**1.4910×10⁻²** 3.02×10⁻³	2.7402×10⁻² 8.07×10⁻³	2.0876×10⁻² 2.95×10⁻³
F7	VNT3	6.5468×10⁻² 4.97×10⁻³	6.3280×10⁻² 4.01×10⁻³	3.5118×10⁻¹ 1.87×10⁻¹	1.0089×10⁻¹ 5.36×10⁻²	**5.7043×10⁻²** 6.75×10⁻³	6.6166×10⁻² 7.75×10⁻³	9.9350×10⁻² 1.19×10⁻²	8.0997×10⁻² 2.46×10⁻²
F8	ZDT1	**6.8254×10⁻³** 5.83×10⁻⁴	1.0333×10⁻² 1.41×10⁻³	1.7046×10⁻² 6.28×10⁻³	1.1959×10⁻² 1.90×10⁻³	8.3394×10⁻³ 5.23×10⁻⁴	8.0300×10⁻³ 1.57×10⁻³	9.0844×10⁻³ 1.19×10⁻³	9.6005×10⁻² 6.65×10⁻²
F9	ZDT2	6.8695×10⁻³ **7.19×10⁻⁴**	9.9417×10⁻³ 9.80×10⁻⁴	1.1725×10⁻² 1.40×10⁻³	1.6606×10⁻² 8.41×10⁻³	1.1321×10⁻² 1.06×10⁻²	**4.1262×10⁻³** 2.91×10⁻³	9.8885×10⁻³ 1.13×10⁻³	9.0712×10⁻² 3.79×10⁻²
F10	ZDT3	8.1064×10⁻³ **6.43×10⁻⁴**	1.2409×10⁻² 1.83×10⁻³	1.3643×10⁻² 3.84×10⁻³	1.7348×10⁻² 5.88×10⁻³	1.0571×10⁻² 4.74×10⁻³	**7.4131×10⁻³** 1.06×10⁻³	1.2411×10⁻² 2.09×10⁻³	8.9649×10⁻² 9.45×10⁻²
F11	ZDT4	**7.0857×10⁻³** **5.46×10⁻⁴**	9.2094×10⁻³ 8.20×10⁻⁴	4.5038×10⁻² 1.23×10⁻²	7.7476×10⁻³ 5.56×10⁻³	1.7807×10⁻¹ 1.50×10⁻¹	2.7520×10⁻² 4.95×10⁻³	9.0207×10⁻³ 1.07×10⁻³	9.4750×10⁻³ 1.52×10⁻³
F12	ZDT6	1.5261×10⁻² 1.96×10⁻²	2.6212×10⁻² 1.96×10⁻²	1.6208×10⁻² 6.73×10⁻³	8.1444×10⁻² 5.14×10⁻²	7.5266×10⁻² 3.31×10⁻²	9.6715×10⁻² 1.17×10⁻¹	**8.5474×10⁻³** **1.41×10⁻³**	2.5094×10⁻¹ 1.66×10⁻¹
F13	TYD1	**2.8746×10⁻³** **1.64×10⁻³**	6.2895×10⁻² 1.29×10⁻²	5.2384×10⁻² 3.20×10⁻²	4.3920×10⁻³ 2.70×10⁻³	1.6767×10⁻² 1.41×10⁻²	4.5573×10⁻³ 1.32×10⁻³	1.4521×10⁻² 7.21×10⁻³	-
F14	TYD2	**1.4202×10⁻²** **9.67×10⁻⁴**	1.9431×10⁻¹ 1.35×10⁻¹	6.3481×10⁻² 5.01×10⁻²	3.6429×10⁻² 2.27×10⁻²	1.0145×10⁻¹ 9.47×10⁻²	8.5839×10⁻³ 1.26×10⁻²	2.2990×10⁻² 3.27×10⁻³	-
F15	TYD3	**1.8745×10⁻²** **1.31×10⁻³**	1.1549×10⁻¹ 5.56×10⁻²	2.9293×10⁻¹ 2.67×10⁻¹	1.0838×10⁻¹ 1.25×10⁻¹	7.0163×10⁻² 4.47×10⁻²	3.6667×10⁻² 9.76×10⁻³	3.1919×10⁻² 5.67×10⁻³	-
F16	TYD4	**2.4070×10⁻²** 6.41×10⁻³	2.5290×10⁻² 1.79×10⁻³	7.4783×10⁻² 2.69×10⁻²	4.1287×10⁻² 4.11×10⁻²	1.5601×10⁻¹ 2.32×10⁻¹	2.6227×10⁻² 2.40×10⁻²	3.0307×10⁻² 4.02×10⁻³	-
F17	TYD5	**4.4231×10⁻³** **4.11×10⁻⁴**	1.8844×10⁻² 5.95×10⁻²	5.1214×10⁻² 8.54×10⁻²	3.8197×10⁻² 4.56×10⁻³	-	-	6.0025×10⁻³ 8.99×10⁻⁴	-
F18	TYD6	**1.2875×10⁻²** **1.15×10⁻³**	8.2821×10⁻² 4.12×10⁻²	2.5104×10⁻² 1.02×10⁻²	4.2786×10⁻² 1.07×10⁻²	3.2520×10⁻² 2.80×10⁻²	-	1.8811×10⁻² 2.33×10⁻³	-

Underlined values indicate the best results.

3.2. Evaluation on Constrained Benchmark Functions

To evaluate the accuracy of the developed NSMRFO approach, four constrained test functions with different Pareto optimal fronts and three analysis metrics were investigated. Inspecting the obtained Pareto fronts in Figure 11, and the outcomes in Table 4, it is clearly seen that the suggested NSMRFO yields a higher convergence and coverage toward the true PF on all constrained benchmark functions. For the numerical results, NSMRFO ranks first for the majority of functions except on BNH and CONSTR for GD, and BNH, SRN for IGD in which it ranks second compared to the aforementioned well-known competitive techniques, and OZY for SP. Note that the optimal findings are marked in boldface and underlined.

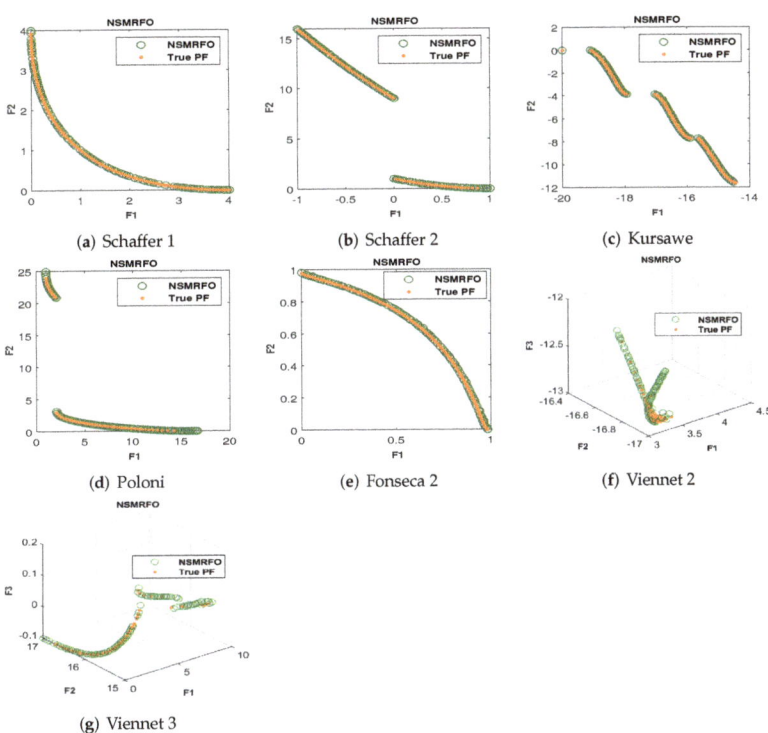

Figure 8. Obtained Pareto front of NSMRFO on classical test suites.

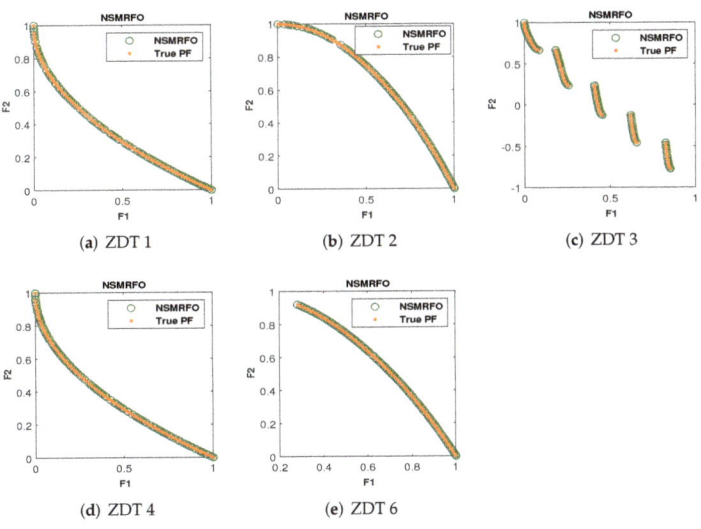

Figure 9. Obtained Pareto front of NSMRFO on ZDT test suites.

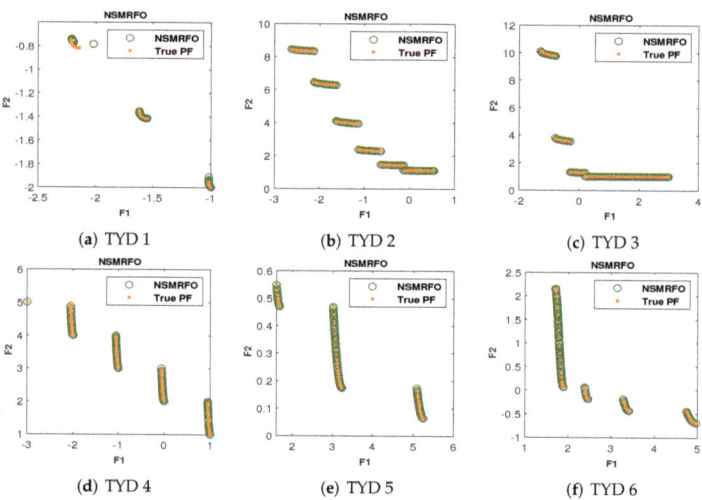

(a) TYD 1 (b) TYD 2 (c) TYD 3
(d) TYD 4 (e) TYD 5 (f) TYD 6

Figure 10. Obtained Pareto front of NSMRFO on TYD test suites.

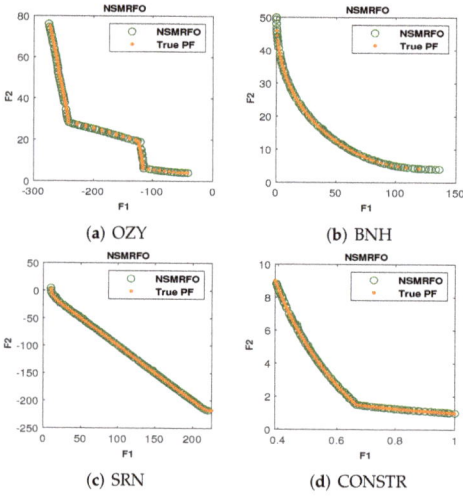

(a) OZY (b) BNH
(c) SRN (d) CONSTR

Figure 11. Obtained Pareto front of NSMRFO on constrained test suites.

Table 4. GD, IGD, and SP metrics comparison based on constrained test suites.

Problem	NSMRFO	MOSMA [46]	MOBO/D [47]	MOMVO [48]	MOWCA [49]	NSGA-II [32]	MOMRFO [51]	IMOMRFO [28]
GD Metric								
OZY	**4.0885×10^{-1}**	1.8977×10^{+0}	9.6335×10^{-1}	4.1926×10^{-1}	1.2504×10^{+0}	8.5422×10^{-1}	4.8528×10^{-1}	4.1772×10^{-1}
	2.29×10^{-1}	6.90×10^{-1}	5.91×10^{-1}	4.29×10^{-1}	7.76×10^{-1}	5.94×10^{-1}	1.74×10^{-1}	2.04×10^{-1}
BNH	3.4178×10^{-1}	3.6565×10^{-1}	**2.0107×10^{-1}**	2.8575×10^{-1}	3.3953×10^{-1}	2.0462×10^{+0}	2.2904×10^{+0}	1.1858×10^{+0}
	1.75×10^{-2}	1.52×10^{-2}	8.17×10^{-2}	2.16×10^{-1}	1.46×10^{-2}	8.52×10^{-2}	1.78×10^{-1}	1.56×10^{-1}
SRN	**3.7131×10^{-2}**	1.3030×10^{-1}	8.0926×10^{-2}	3.6668×10^{-1}	6.5713×10^{-2}	4.3051×10^{-2}	6.2583×10^{-2}	7.6938×10^{-2}
	6.99×10^{-3}	6.53×10^{-2}	4.67×10^{-2}	8.97×10^{-2}	6.97×10^{-3}	6.91×10^{-3}	8.83×10^{-3}	1.18×10^{-2}
CONSTR	6.0167×10^{-4}	1.0382×10^{-3}	6.9108×10^{-4}	8.1620×10^{-4}	7.2413×10^{-4}	7.0693×10^{-4}	**6.0098×10^{-4}**	8.0662×10^{-4}
	1.50×10^{-4}	5.55×10^{-4}	4.14×10^{-5}	4.02×10^{-4}	7.28×10^{-5}	4.62×10^{-5}	4.88×10^{-5}	6.47×10^{-5}

Table 4. Cont.

Problem	NSMRFO	MOSMA [46]	MOBO/D [47]	MOMVO [48]	MOWCA [49]	NSGA-II [32]	MOMRFO [51]	IMOMRFO [28]
IGD Metric								
OZY	**1.4014×10^{-3}** 1.18×10^{-3}	9.7966×10^{-3} 2.07×10^{-3}	9.965×10^{-3} 6.43×10^{-3}	2.2981×10^{-3} 2.04×10^{-3}	6.4842×10^{-3} 2.21×10^{-3}	4.3475×10^{-3} 5.31×10^{-3}	4.2650×10^{-3} 2.80×10^{-3}	1.5178×10^{-3} 2.07×10^{-3}
BNH	6.7915×10^{-4} 3.39×10^{-5}	1.1293×10^{-3} 9.31×10^{-5}	1.9024×10^{-3} 1.22×10^{-3}	4.4141×10^{-3} 7.06×10^{-4}	6.6386×10^{-4} 4.39×10^{-5}	**1.8246×10^{-4}** 1.78×10^{-5}	2.4874×10^{-4} 1.16×10^{-5}	2.0763×10^{-4} 2.20×10^{-5}
SRN	9.7172×10^{-5} 6.88×10^{-6}	2.7928×10^{-4} 2.28×10^{-4}	1.9607×10^{-4} 6.97×10^{-5}	5.9502×10^{-4} 8.07×10^{-5}	**8.9740×10^{-5}** 1.49×10^{-6}	1.0583×10^{-4} 1.46×10^{-5}	1.2501×10^{-4} 7.45×10^{-6}	1.4485×10^{-4} 1.17×10^{-5}
CONSTR	**2.5255×10^{-4}** 4.44×10^{-5}	1.0344×10^{-3} 2.92×10^{-4}	1.6097×10^{-3} 8.93×10^{-4}	1.7429×10^{-3} 4.30×10^{-4}	2.6802×10^{-4} 4.31×10^{-5}	2.7118×10^{-4} 1.51×10^{-5}	4.7669×10^{-4} 6.21×10^{-5}	3.5283×10^{-4} 4.98×10^{-5}
SP Metric								
OZY	$1.9039 \times 10^{+0}$ 8.93×10^{-1}	$8.0601 \times 10^{+0}$ $2.85 \times 10^{+0}$	$2.3336 \times 10^{+0}$ $2.03 \times 10^{+0}$	$2.0649 \times 10^{+0}$ 4.42×10^{-1}	$2.0314 \times 10^{+0}$ $1.41 \times 10^{+0}$	**$1.3227 \times 10^{+0}$** 2.48×10^{-1}	$1.6335 \times 10^{+0}$ 2.66×10^{-1}	$1.3983 \times 10^{+0}$ 1.23×10^{-1}
BNH	**7.1873×10^{-1}** 6.39×10^{-2}	8.7699×10^{-1} 8.93×10^{-2}	$1.8574 \times 10^{+0}$ 8.78×10^{-1}	$1.2154 \times 10^{+0}$ 5.76×10^{-1}	7.595×10^{-1} 6.94×10^{-2}	7.5542×10^{-1} 7.37×10^{-2}	9.4169×10^{-1} 1.32×10^{-1}	$1.4736 \times 10^{+0}$ 4.43×10^{-1}
SRN	**$1.3558 \times 10^{+0}$** 1.16×10^{-1}	$3.6172 \times 10^{+0}$ $3.10 \times 10^{+0}$	$2.3050 \times 10^{+0}$ 5.83×10^{-1}	$2.9495 \times 10^{+0}$ $4.47 \times 10^{+0}$	$1.5216 \times 10^{+0}$ 9.96×10^{-2}	$1.4977 \times 10^{+0}$ 1.67×10^{-1}	$2.1824 \times 10^{+0}$ 3.49×10^{-1}	$2.2106 \times 10^{+0}$ 4.86×10^{-1}
CONSTR	**1.2579×10^{-2}** 2.87×10^{-2}	4.2564×10^{-2} 3.74×10^{-3}	1.7483×10^{-1} 5.34×10^{-2}	3.7418×10^{-2} 2.26×10^{-2}	4.3448×10^{-2} 4.46×10^{-3}	4.3658×10^{-2} 2.91×10^{-3}	8.4588×10^{-2} 1.05×10^{-2}	5.1093×10^{-2} 2.97×10^{-3}

Underlined values indicate the best results.

3.3. Evaluation on CEC2020 Benchmark Functions

This subsection presents the performance of the suggested technique NSMRFO in CEC2020 multimodal multi-objective optimization (MMO) problems using four indicator metrics, the $rPSP$ and $IGDX$ that reflect the quality of the Pareto set in the search space, and the rHV and $IGDF$ that reflect the quality of the Pareto front in the objective space. The functions of the remaining MMO problems are characterized by different geometries' linear and nonlinear concave and convex functions. To illustrate the effectiveness of the multi-objective MRFO version, six well-known competitors are adopted for comparison such as: NSGA-II [32], DN-NSGAII [52], OMNI-OPT [53], MO_Ring_PSO_SCD [44], MOM-RFO [51], IMOMRFO [28]. The numerical statistical results of the obtained parameter indexes in search and objective spaces by each approach are summarized in Table 5. It is worth noting that the optimal results of each indicator are the lowest values. The underlined bold solutions indicate the algorithms' optimum result. Additionally, the last five rows posted in this table present the results score of each approach, in which the NSMRFO ranks first by providing 47 optimal solutions out of a 96 benchmark suite, which means twenty-four functions times four indicators. By contrast, the MOMRFO, OMNI-OPT, and DN-NSGAII competitor approaches show the worst scores. It can be clearly observed from the perspective of the search space values that the crowding distance mechanism has the capability to efficiently increase the PS convergence and diversity of the optimization algorithms. However, In spite of the same strategy used in NSGAII, DN-NSGAII, and the proposed optimizer, they offered a significant different performance, which means that the NSMRFO diversity and convergence are improved. The IMOMRFO was the closest approach of NSMRFO, where it ranks second by having 20 best values out of 96, and it offered good search space results and poor objective space results. The NSGAII performed a little better on the objective space, especially the rHV metric. The CEC2020 corresponding box plots of the four metrics rPSP, rHV, IGDX, and IGDF are depicted in Figure 12. According to this figure, the NSMRFO achieved the indicator minimum values in most MMFs such as MMF1, MMF4, MMF5, MMF7, MMF8, MMF11, MMF12, MMF13, MMF14, MMF15, MMF1_e, MMF14_a, MMF15_a, MMF10_l, rHV on MMF11_l, rPSP and rHV on MMF12_l, rHV on MMF13_l, rPSP and rHV on MMF15_a_l. In addition, the NSMRFO is more stable compared to its competitor approaches. To sum up, the suggested optimizer achieves the best rank performance in terms of all indicator metrics compared to its competing algorithms, and have significant stability.

Table 5. rPSP, rHV, IGDX, and IGDF indicator metrics comparison based on CEC2020 test suites.

Case	Name	NSMRFO	NSGA-II [32]	DN-NSGAII [52]	OMNI-OPT [53]	MO_Ring_PSO_SCD [44]	MOMRFO [51]	IMOMRFO [28]
MMF1	rPSP	$3.0414\times10^{-2}(2.0151\times10^{-3})$	$7.6055\times10^{-2}(1.2369\times10^{-2})$	$5.5900\times10^{-2}(1.1025\times10^{-2})$	$5.1803\times10^{-2}(9.6985\times10^{-3})$	$2.9630\times10^{-2}(1.0108\times10^{-3})$	$8.4765\times10^{-2}(4.7454\times10^{-3})$	$\mathbf{2.5834\times10^{-2}(1.2271\times10^{-3})}$
	rHV	$\mathbf{1.1437\times10^{+0}(2.0798\times10^{-4})}$	$1.1438\times10^{+0}(5.3194\times10^{-4})$	$1.1447\times10^{+0}(6.7793\times10^{-4})$	$1.1443\times10^{+0}(6.4982\times10^{-4})$	$1.1449\times10^{+0}(2.0606\times10^{-4})$	$1.1562\times10^{+0}(2.7386\times10^{-3})$	$1.1448\times10^{+0}(3.0962\times10^{-4})$
	IGDX	$3.0328\times10^{-2}(1.9920\times10^{-3})$	$7.5464\times10^{-2}(1.2263\times10^{-2})$	$5.5494\times10^{-2}(1.0934\times10^{-2})$	$5.1242\times10^{-2}(9.4782\times10^{-3})$	$2.9526\times10^{-2}(1.0200\times10^{-3})$	$8.4117\times10^{-2}(4.5660\times10^{-3})$	$\mathbf{2.5763\times10^{-2}(1.2162\times10^{-3})}$
	IGDF	$\mathbf{1.5698\times10^{-3}(1.0727\times10^{-4})}$	$1.5829\times10^{-3}(1.4770\times10^{-4})$	$2.0530\times10^{-3}(3.3231\times10^{-4})$	$1.7065\times10^{-3}(1.7301\times10^{-4})$	$2.0136\times10^{-3}(1.2596\times10^{-4})$	$6.9605\times10^{-3}(5.4121\times10^{-4})$	$1.9755\times10^{-3}(1.5075\times10^{-4})$
MMF2	rPSP	$3.9573\times10^{-2}(1.6761\times10^{-2})$	$6.4752\times10^{-2}(4.7622\times10^{-2})$	$5.3866\times10^{-2}(3.3966\times10^{-2})$	$5.7411\times10^{-2}(3.0238\times10^{-2})$	$2.5322\times10^{-2}(1.0985\times10^{-2})$	$8.8438\times10^{-2}(6.1782\times10^{-2})$	$\mathbf{1.8170\times10^{-2}(5.9853\times10^{-3})}$
	rHV	$1.0882\times10^{+0}(3.2063\times10^{-1})$	$\mathbf{1.0622\times10^{+0}(3.1225\times10^{-1})}$	$1.0683\times10^{+0}(3.1418\times10^{-1})$	$1.0630\times10^{+0}(3.1241\times10^{-1})$	$1.0779\times10^{+0}(3.1731\times10^{-1})$	$1.0846\times10^{+0}(3.1846\times10^{-1})$	$1.0719\times10^{+0}(3.1533\times10^{-1})$
	IGDX	$3.8468\times10^{-2}(1.5407\times10^{-2})$	$5.9944\times10^{-2}(4.0824\times10^{-2})$	$4.9750\times10^{-2}(2.7319\times10^{-2})$	$5.4726\times10^{-2}(2.7670\times10^{-2})$	$2.3993\times10^{-2}(9.7265\times10^{-3})$	$8.8358\times10^{-2}(6.1801\times10^{-2})$	$\mathbf{1.7759\times10^{-2}(5.8627\times10^{-3})}$
	IGDF	$1.7210\times10^{-2}(6.2329\times10^{-3})$	$\mathbf{4.9329\times10^{-3}(5.3012\times10^{-3})}$	$8.1623\times10^{-3}(6.7755\times10^{-3})$	$5.0734\times10^{-3}(4.0992\times10^{-3})$	$1.2283\times10^{-2}(5.7738\times10^{-3})$	$1.6374\times10^{-2}(5.7905\times10^{-3})$	$7.8016\times10^{-3}(2.5737\times10^{-3}$
MMF4	rPSP	$2.0579\times10^{-2}(8.7789\times10^{-3})$	$7.4849\times10^{-2}(3.0326\times10^{-2})$	$4.4836\times10^{-2}(2.2676\times10^{-2})$	$5.0017\times10^{-2}(2.3320\times10^{-2})$	$1.7425\times10^{-2}(7.8290\times10^{-3})$	$8.4134\times10^{-2}(4.6963\times10^{-2})$	$\mathbf{1.4675\times10^{-2}(4.2218\times10^{-3})}$
	rHV	$1.6624\times10^{+0}(4.5992\times10^{-1})$	$\mathbf{1.6596\times10^{+0}(4.6077\times10^{-1})}$	$1.6621\times10^{+0}(4.6117\times10^{-1})$	$1.6605\times10^{+0}(4.6102\times10^{-1})$	$1.6653\times10^{+0}(4.6154\times10^{-1})$	$1.6909\times10^{+0}(4.6900\times10^{-1})$	$1.6658\times10^{+0}(4.6210\times10^{-1})$
	IGDX	$2.0563\times10^{-2}(8.7830\times10^{-3})$	$7.2488\times10^{-2}(2.6323\times10^{-2})$	$4.3585\times10^{-2}(1.9905\times10^{-2})$	$4.9157\times10^{-2}(2.1735\times10^{-2})$	$1.6995\times10^{-2}(6.8399\times10^{-3})$	$8.2887\times10^{-2}(4.5058\times10^{-2})$	$\mathbf{1.4573\times10^{-2}(4.2109\times10^{-3})}$
	IGDF	$3.6618\times10^{-3}(5.0897\times10^{-3})$	$2.3078\times10^{-3}(3.6650\times10^{-3})$	$3.1486\times10^{-3}(5.0146\times10^{-3})$	$\mathbf{2.2140\times10^{-3}(2.7604\times10^{-3})}$	$3.6288\times10^{-3}(4.6275\times10^{-3})$	$8.4242\times10^{-3}(3.9799\times10^{-3})$	$2.7047\times10^{-3}(1.9362\times10^{-3})$
MMF5	rPSP	$\mathbf{4.7896\times10^{-2}(1.3575\times10^{-2})}$	$1.2187\times10^{-1}(3.5342\times10^{-2})$	$1.0176\times10^{-1}(2.9121\times10^{-2})$	$1.0690\times10^{-1}(2.9566\times10^{-2})$	$4.8283\times10^{-2}(1.3907\times10^{-2})$	$1.3980\times10^{-1}(4.3140\times10^{-2})$	$4.9105\times10^{-2}(1.2510\times10^{-2})$
	rHV	$\mathbf{1.1471\times10^{+0}(2.9796\times10^{-1})}$	$1.1474\times10^{+0}(2.9818\times10^{-1})$	$1.1484\times10^{+0}(2.9833\times10^{-1})$	$1.1475\times10^{+0}(2.9805\times10^{-1})$	$1.1489\times10^{+0}(2.9879\times10^{-1})$	$1.1612\times10^{+0}(3.0306\times10^{-1})$	$1.1489\times10^{+0}(2.9900\times10^{-1})$
	IGDX	$4.7815\times10^{-2}(1.3551\times10^{-2})$	$1.1919\times10^{-1}(3.3294\times10^{-2})$	$1.0023\times10^{-1}(2.8573\times10^{-2})$	$1.0573\times10^{-1}(2.9466\times10^{-2})$	$4.8045\times10^{-2}(1.3965\times10^{-2})$	$1.3859\times10^{-1}(4.3103\times10^{-2})$	$\mathbf{4.1761\times10^{-2}(1.2468\times10^{-2})}$
	IGDF	$2.5965\times10^{-3}(5.0157\times10^{-3})$	$2.0983\times10^{-3}(3.6123\times10^{-3})$	$2.8638\times10^{-3}(4.9328\times10^{-3})$	$\mathbf{2.0359\times10^{-3}(2.7153\times10^{-3})}$	$2.9333\times10^{-3}(4.5794\times10^{-3})$	$7.2912\times10^{-3}(4.0541\times10^{-3})$	$2.1336\times10^{-3}(1.9466\times10^{-3})$
MMF7	rPSP	$1.9387\times10^{-2}(9.3975\times10^{-3})$	$5.7247\times10^{-2}(3.1367\times10^{-2})$	$3.6647\times10^{-2}(2.3315\times10^{-2})$	$3.5249\times10^{-2}(2.2641\times10^{-2})$	$1.9691\times10^{-2}(9.1406\times10^{-3})$	$6.9463\times10^{-2}(4.1870\times10^{-2})$	$\mathbf{1.7823\times10^{-2}(8.3248\times10^{-3})}$
	rHV	$\mathbf{1.0939\times10^{+0}(3.1305\times10^{-1})}$	$1.0940\times10^{+0}(3.1312\times10^{-1})$	$1.0956\times10^{+0}(3.1351\times10^{-1})$	$1.0945\times10^{+0}(3.1319\times10^{-1})$	$1.0955\times10^{+0}(3.1384\times10^{-1})$	$1.1053\times10^{+0}(3.1810\times10^{-1})$	$1.0956\times10^{+0}(3.1410\times10^{-1})$
	IGDX	$\mathbf{1.3931\times10^{-2}(9.3851\times10^{-3})}$	$5.5052\times10^{-2}(2.6287\times10^{-2})$	$3.6114\times10^{-2}(2.2089\times10^{-2})$	$3.4994\times10^{-2}(2.2456\times10^{-2})$	$1.9613\times10^{-2}(9.0891\times10^{-3})$	$6.5441\times10^{-2}(4.2268\times10^{-2})$	$1.7762\times10^{-2}(8.2912\times10^{-3})$
	IGDF	$2.4520\times10^{-3}(5.0404\times10^{-3})$	$2.0851\times10^{-3}(3.6136\times10^{-3})$	$2.9755\times10^{-3}(4.9031\times10^{-3})$	$\mathbf{2.0540\times10^{-3}(2.7073\times10^{-3})}$	$2.8412\times10^{-3}(4.5950\times10^{-3})$	$7.0477\times10^{-3}(4.0844\times10^{-3})$	$2.1512\times10^{-3}(1.9387\times10^{-3})$
MMF8	rPSP	$3.6319\times10^{-2}(1.0596\times10^{-2})$	$6.3600\times10^{-1}(3.7420\times10^{-1})$	$1.3760\times10^{-1}(9.5102\times10^{-2})$	$1.1329\times10^{-1}(4.5487\times10^{-2})$	$3.7333\times10^{-2}(1.0898\times10^{-2})$	$4.9126\times10^{-1}(4.0283\times10^{-1})$	$\mathbf{3.2405\times10^{-2}(1.0053\times10^{-2})}$
	rHV	$2.1002\times10^{+0}(6.0129\times10^{-1})$	$\mathbf{2.0980\times10^{+0}(6.0054\times10^{-1})}$	$2.1022\times10^{+0}(6.0177\times10^{-1})$	$2.0991\times10^{+0}(6.0084\times10^{-1})$	$2.1153\times10^{+0}(6.0603\times10^{-1})$	$2.1404\times10^{+0}(6.1291\times10^{-1})$	$2.1146\times10^{+0}(6.0583\times10^{-1})$
	IGDX	$\mathbf{3.2217\times10^{-2}(1.0008\times10^{-2})}$	$5.3568\times10^{-1}(2.9311\times10^{-1})$	$1.3479\times10^{-1}(9.3095\times10^{-2})$	$1.1106\times10^{-1}(4.4798\times10^{-2})$	$3.7076\times10^{-2}(1.0809\times10^{-2})$	$4.2621\times10^{-1}(2.9906\times10^{-1})$	$3.6059\times10^{-2}(1.0534\times10^{-2})$
	IGDF	$2.7154\times10^{-3}(4.9825\times10^{-3})$	$\mathbf{2.0591\times10^{-3}(3.6179\times10^{-3})}$	$2.9780\times10^{-3}(4.9069\times10^{-3})$	$2.1029\times10^{-3}(2.6977\times10^{-3})$	$3.6962\times10^{-3}(4.4151\times10^{-3})$	$7.1149\times10^{-3}(4.0691\times10^{-3})$	$2.3955\times10^{-3}(1.9024\times10^{-3})$
MMF10	rPSP	$\mathbf{8.8481\times10^{-3}(1.1246\times10^{-2})}$	$2.5191\times10^{-1}(2.7155\times10^{-1})$	$1.5654\times10^{-1}(1.3180\times10^{-1})$	$1.3713\times10^{-1}(1.2761\times10^{-1})$	$2.6216\times10^{-2}(1.2256\times10^{-2})$	$1.4365\times10^{-1}(3.1590\times10^{-1})$	$1.4725\times10^{-2}(8.8940\times10^{-3})$
	rHV	$\mathbf{3.4880\times10^{-1}(4.5879\times10^{-1})}$	$3.5165\times10^{-1}(4.5671\times10^{-1})$	$3.5287\times10^{-1}(4.5781\times10^{-1})$	$3.5120\times10^{-1}1(4.5708\times10^{-1})$	$3.5314\times10^{-1}(4.6609\times10^{-1})$	$3.5680\times10^{-1}(4.7447\times10^{-1})$	$3.5280\times10^{-1}(4.6897\times10^{-1})$
	IGDX	$\mathbf{8.8043\times10^{-3}(1.1229\times10^{-2})}$	$2.2546\times10^{-1}(2.2101\times10^{-1})$	$1.5294\times10^{-1}(1.3199\times10^{-1})$	$1.3410\times10^{-1}(1.2713\times10^{-1})$	$2.5108\times10^{-2}(1.1186\times10^{-2})$	$1.1967\times10^{-1}(2.3595\times10^{-1})$	$1.4498\times10^{-2}(8.8366\times10^{-3})$
	IGDF	$\mathbf{7.4909\times10^{-3}(4.2541\times10^{-3})}$	$1.4591\times10^{-1}(1.0654\times10^{-1})$	$1.5492\times10^{-1}(1.0312\times10^{-1})$	$1.2957\times10^{-1}(1.1531\times10^{-1})$	$8.9074\times10^{-2}(3.6667\times10^{-2})$	$9.3671\times10^{-2}(6.5256\times10^{-2})$	$5.1006\times10^{-2}(1.9770\times10^{-2})$
MMF11	rPSP	$\mathbf{6.1422\times10^{-3}(1.1237\times10^{-2})}$	$8.4835\times10^{-2}(2.7855\times10^{-1})$	$3.5314\times10^{-2}(8.5845\times10^{-2})$	$3.2982\times10^{-2}(7.3458\times10^{-2})$	$9.0973\times10^{-3}(1.0656\times10^{-2})$	$8.7426\times10^{-2}(3.1641\times10^{-1})$	$7.3248\times10^{-3}(9.7250\times10^{-3})$
	rHV	$\mathbf{1.8084\times10^{-1}(4.5516\times10^{-1})}$	$1.8085\times10^{-1}(4.5389\times10^{-1})$	$1.8125\times10^{-1}(4.5512\times10^{-1})$	$1.8086\times10^{-1}(4.5409\times10^{-1})$	$1.8300\times10^{-1}(4.6294\times10^{-1})$	$1.8522\times10^{-1}(4.7124\times10^{-1})$	$1.8361\times10^{-1}(4.6576\times10^{-1})$
	IGDX	$\mathbf{6.1330\times10^{-3}(1.1219\times10^{-2})}$	$6.9615\times10^{-2}(2.2043\times10^{-1})$	$3.4446\times10^{-2}(8.4416\times10^{-2})$	$3.2795\times10^{-2}(7.3430\times10^{-2})$	$8.9742\times10^{-3}(1.0583\times10^{-2})$	$6.7869\times10^{-2}(2.3633\times10^{-1})$	$7.2899\times10^{-3}(9.6856\times10^{-3})$
	IGDF	$\mathbf{9.5815\times10^{-3}(4.1010\times10^{-3})}$	$3.6071\times10^{-2}(5.9097\times10^{-2})$	$3.9988\times10^{-2}(6.2797\times10^{-2})$	$3.5967\times10^{-2}(5.9483\times10^{-2})$	$2.8444\times10^{-2}(2.7189\times10^{-2})$	$4.0228\times10^{-2}(4.7980\times10^{-2})$	$2.2162\times10^{-2}(1.5078\times10^{-2})$
MMF12	rPSP	$\mathbf{5.1099\times10^{-3}(1.1468\times10^{-2})}$	$7.4708\times10^{-2}(2.7902\times10^{-1})$	$2.4964\times10^{-2}(8.4495\times10^{-2})$	$2.1668\times10^{-2}(7.1639\times10^{-2})$	$6.4725\times10^{-3}(1.0905\times10^{-2})$	$8.1365\times10^{-2}(3.1734\times10^{-1})$	$5.4530\times10^{-3}(1.0056\times10^{-2})$
	rHV	$\mathbf{6.3158\times10^{-1}(3.6955\times10^{-1})}$	$6.4092\times10^{-1}(3.7060\times10^{-1})$	$6.3340\times10^{-1}(3.6941\times10^{-1})$	$6.4108\times10^{-1}(3.7080\times10^{-1})$	$6.3646\times10^{-1}(3.7632\times10^{-1})$	$6.3772\times10^{-1}(3.8383\times10^{-1})$	$6.3588\times10^{-1}(3.7901\times10^{-1})$
	IGDX	$\mathbf{5.1038\times10^{-3}(1.1450\times10^{-2})}$	$6.0563\times10^{-2}(2.2097\times10^{-1})$	$2.4580\times10^{-2}(8.2986\times10^{-2})$	$2.1384\times10^{-2}(7.1558\times10^{-2})$	$6.4530\times10^{-3}(1.0846\times10^{-2})$	$6.1990\times10^{-2}(2.3726\times10^{-1})$	$5.4410\times10^{-3}(1.0013\times10^{-2})$
	IGDF	$\mathbf{5.2930\times10^{-3}(5.9306\times10^{-3})}$	$2.0352\times10^{-2}(6.0080\times10^{-2})$	$2.0688\times10^{-2}(6.3714\times10^{-2})$	$2.0155\times10^{-2}(5.9869\times10^{-2})$	$1.4400\times10^{-2}(2.8857\times10^{-2})$	$2.0032\times10^{-2}(5.1241\times10^{-2})$	$9.5889\times10^{-3}(1.6983\times10^{-2})$
MMF13	rPSP	$2.6947\times10^{-2}(9.1879\times10^{-3})$	$1.5088\times10^{-1}(2.6073\times10^{-1})$	$8.2414\times10^{-2}(7.3363\times10^{-2})$	$7.3058\times10^{-2}(6.0391\times10^{-2})$	$3.3613\times10^{-2}(1.0230\times10^{-2})$	$1.2363\times10^{-1}(3.0658\times10^{-1})$	$\mathbf{2.6729\times10^{-2}(8.5077\times10^{-3})}$
	rHV	$\mathbf{1.9794\times10^{-1}(4.5621\times10^{-1})}$	$1.9811\times10^{-1}(4.5501\times10^{-1})$	$1.9815\times10^{-1}(4.5619\times10^{-1})$	$1.9819\times10^{-1}(4.5521\times10^{-1})$	$2.0020\times10^{-1}(4.6397\times10^{-1})$	$2.0219\times10^{-1}(4.7227\times10^{-1})$	$2.0069\times10^{-1}(4.6680\times10^{-1})$
	IGDX	$2.6924\times10^{-2}(9.1790\times10^{-3})$	$1.2473\times10^{-1}(2.0416\times10^{-1})$	$8.1734\times10^{-2}(7.1997\times10^{-2})$	$7.2171\times10^{-2}(6.0456\times10^{-2})$	$3.3272\times10^{-2}(1.0113\times10^{-2})$	$1.0423\times10^{-1}(2.2665\times10^{-1})$	$\mathbf{2.6498\times10^{-2}(8.4547\times10^{-3})}$
	IGDF	$\mathbf{1.1578\times10^{-2}(4.0956\times10^{-3})}$	$2.5969\times10^{-2}(5.6626\times10^{-2})$	$3.4898\times10^{-2}(6.0045\times10^{-2})$	$2.7389\times10^{-2}(5.6364\times10^{-2})$	$3.3914\times10^{-2}(2.4986\times10^{-2})$	$3.9737\times10^{-2}(4.6482\times10^{-2})$	$2.1351\times10^{-2}(1.4635\times10^{-2})$

Table 5. *Cont.*

Case	Name	NSMRFO	NSGA-II [32]	DN-NSGAII [52]	OMNI-OPT [53]	MO_Ring_PSO_SCD [44]	MOMRFO [51]	IMOMRFO [28]
MMF14	rPSP	$4.1910 \times 10^{-2}(1.1592 \times 10^{-2})$	$1.5056 \times 10^{-1}(2.6014 \times 10^{-1})$	$9.6022 \times 10^{-2}(7.0165 \times 10^{-2})$	$8.6868 \times 10^{-2}(5.8512 \times 10^{-2})$	$4.2862 \times 10^{-2}(1.1602 \times 10^{-2})$	$2.0409 \times 10^{-1}(2.8858 \times 10^{-1})$	**$4.0854 \times 10^{-2}(1.1207 \times 10^{-2})$**
	rHV	$4.0147 \times 10^{-1}(4.0727 \times 10^{-1})$	$4.0269 \times 10^{-1}(4.0293 \times 10^{-1})$	**$3.8556 \times 10^{-1}(4.0759 \times 10^{-1})$**	$3.8938 \times 10^{-1}(4.0575 \times 10^{-1})$	$3.9660 \times 10^{-1}(4.1374 \times 10^{-1})$	$4.1179 \times 10^{-1}(4.1982 \times 10^{-1})$	$4.0802 \times 10^{-1}(4.1546 \times 10^{-1})$
	IGDX	$4.1904 \times 10^{-2}(1.1593 \times 10^{-2})$	$1.3567 \times 10^{-1}(2.0222 \times 10^{-1})$	$9.5640 \times 10^{-2}(6.8734 \times 10^{-2})$	$8.6831 \times 10^{-2}(5.8510 \times 10^{-2})$	$4.2813 \times 10^{-2}(1.1597 \times 10^{-2})$	$1.8455 \times 10^{-1}(2.0971 \times 10^{-1})$	**$4.0804 \times 10^{-2}(1.1200 \times 10^{-2})$**
	IGDF	**$5.4539 \times 10^{-2}(1.6340 \times 10^{-2})$**	$9.6846 \times 10^{-2}(4.6160 \times 10^{-2})$	$9.4576 \times 10^{-2}(5.0121 \times 10^{-2})$	$8.4600 \times 10^{-2}(4.6496 \times 10^{-2})$	$6.4696 \times 10^{-2}(2.3468 \times 10^{-2})$	$1.3334 \times 10^{-1}(4.3382 \times 10^{-2})$	$5.8273 \times 10^{-2}(1.6945 \times 10^{-2})$
MMF15	rPSP	**$4.2986 \times 10^{-2}(1.1230 \times 10^{-2})$**	$1.3517 \times 10^{-1}(2.6371 \times 10^{-1})$	$9.1805 \times 10^{-2}(7.0885 \times 10^{-2})$	$7.7907 \times 10^{-2}(6.0049 \times 10^{-2})$	$4.7182 \times 10^{-2}(1.2528 \times 10^{-2})$	$1.4709 \times 10^{-1}(3.0237 \times 10^{-1})$	$4.405 \times 10^{-2}(1.1510 \times 10^{-2})$
	rHV	$3.2743 \times 10^{-1}(4.2546 \times 10^{-1})$	$3.2665 \times 10^{-1}(4.2150 \times 10^{-1})$	**$3.2329 \times 10^{-1}(4.2326 \times 10^{-1})$**	$3.2442 \times 10^{-1}(4.2196 \times 10^{-1})$	$3.3365 \times 10^{-1}(4.2946 \times 10^{-1})$	$3.3111 \times 10^{-1}(4.3874 \times 10^{-1})$	$3.2896 \times 10^{-1}(4.3348 \times 10^{-1})$
	IGDX	**$4.2980 \times 10^{-2}(1.1229 \times 10^{-2})$**	$1.2101 \times 10^{-1}(2.0563 \times 10^{-1})$	$9.1440 \times 10^{-2}(6.9444 \times 10^{-2})$	$7.7906 \times 10^{-2}(6.0049 \times 10^{-2})$	$4.7161 \times 10^{-2}(1.2514 \times 10^{-2})$	$1.2776 \times 10^{-1}(2.2260 \times 10^{-1})$	$4.3997 \times 10^{-2}(1.1490 \times 10^{-2})$
	IGDF	**$8.1385 \times 10^{-2}(2.2410 \times 10^{-2})$**	$1.3602 \times 10^{-1}(4.7501 \times 10^{-2})$	$1.5589 \times 10^{-1}(5.2948 \times 10^{-2})$	$1.2952 \times 10^{-1}(4.6961 \times 10^{-2})$	$9.7474 \times 10^{-2}(2.8216 \times 10^{-2})$	$1.4464 \times 10^{-1}(4.3814 \times 10^{-2})$	$8.6563 \times 10^{-2}(2.3977 \times 10^{-2})$
MMF1_e	rPSP	**$2.5295 \times 10^{-1}(1.0091 \times 10^{-1})$**	$1.9878 \times 10^{+0}(1.6600 \times 10^{+0})$	$1.3369 \times 10^{+0}(1.2026 \times 10^{+0})$	$1.0495 \times 10^{+0}(5.7725 \times 10^{-1})$	$3.8473 \times 10^{-1}(2.3103 \times 10^{-1})$	$1.2670 \times 10^{+0}(1.0900 \times 10^{+0})$	$8.0812 \times 10^{-1}(8.3709 \times 10^{-1})$
	rHV	$1.0605 \times 10^{+0}(3.6559 \times 10^{-1})$	**$1.0577 \times 10^{+0}(3.6406 \times 10^{-1})$**	$1.0770 \times 10^{+0}(3.6709 \times 10^{-1})$	$1.0582 \times 10^{+0}(3.6420 \times 10^{-1})$	$1.0736 \times 10^{+0}(3.7092 \times 10^{-1})$	$1.0935 \times 10^{+0}(3.8019 \times 10^{-1})$	$1.0741 \times 10^{+0}(3.7299 \times 10^{-1})$
	IGDX	**$2.3873 \times 10^{-1}(9.0448 \times 10^{-2})$**	$1.1385 \times 10^{+0}(7.2140 \times 10^{-1})$	$8.4842 \times 10^{-1}(5.3728 \times 10^{-1})$	$7.7669 \times 10^{-1}1(3.6214 \times 10^{-1})$	$3.3887 \times 10^{-1}(1.7606 \times 10^{-1})$	$8.8705 \times 10^{-1}(5.5970 \times 10^{-1})$	$5.8998 \times 10^{-1}(4.6403 \times 10^{-1})$
	IGDF	**$1.3848 \times 10^{-2}(1.9966 \times 10^{-2})$**	$2.7268 \times 10^{-2}(5.8372 \times 10^{-2})$	$3.4463 \times 10^{-2}(6.2371 \times 10^{-2})$	$2.7164 \times 10^{-2}(5.8417 \times 10^{-2})$	$2.1051 \times 10^{-2}(2.8771 \times 10^{-2})$	$2.9547 \times 10^{-2}(5.0972 \times 10^{-2})$	$1.5898 \times 10^{-2}(1.9273 \times 10^{-2})$
MMF14_a	rPSP	**$9.8501 \times 10^{-2}(6.6172 \times 10^{-2})$**	$4.9059 \times 10^{-1}(1.0170 \times 10^{+0})$	$3.9961 \times 10^{-1}(9.0669 \times 10^{-1})$	$2.6576 \times 10^{-1}(4.1921 \times 10^{-1})$	$1.1700 \times 10^{-1}(1.8135 \times 10^{-1})$	$4.7431 \times 10^{-1}(7.7160 \times 10^{-1})$	$2.5190 \times 10^{-1}(6.5067 \times 10^{-1})$
	rHV	$4.6023 \times 10^{-1}(4.1027 \times 10^{-1})$	$4.7199 \times 10^{-1}(4.0357 \times 10^{-1})$	$4.4547 \times 10^{-1}(4.1014 \times 10^{-1})$	$4.5422 \times 10^{-1}(4.0712 \times 10^{-1})$	$4.6184 \times 10^{-1}(4.1494 \times 10^{-1})$	**$4.4325 \times 10^{-1}(4.2776 \times 10^{-1})$**	$4.5776 \times 10^{-1}(4.1893 \times 10^{-1})$
	IGDX	**$9.5640 \times 10^{-2}(5.8876 \times 10^{-2})$**	$2.9905 \times 10^{-1}(4.6208 \times 10^{-1})$	$2.5536 \times 10^{-1}(4.2144 \times 10^{-1})$	$2.1307 \times 10^{-1}(2.8613 \times 10^{-1})$	$1.0450 \times 10^{-1}(1.4104 \times 10^{-1})$	$3.5110 \times 10^{-1}(3.9516 \times 10^{-1})$	$1.7050 \times 10^{-1}(3.5834 \times 10^{-1})$
	IGDF	**$5.8067 \times 10^{-2}(1.7432 \times 10^{-2})$**	$9.7959 \times 10^{-2}(4.5850 \times 10^{-2})$	$1.0176 \times 10^{-1}(4.9979 \times 10^{-2})$	$8.8945 \times 10^{-2}(4.6136 \times 10^{-2})$	$6.2330 \times 10^{-2}(2.3163 \times 10^{-2})$	$1.4429 \times 10^{-1}(4.5822 \times 10^{-2})$	$5.8140 \times 10^{-2}(1.7326 \times 10^{-2})$
MMF15_a	rPSP	**$6.8011 \times 10^{-2}(6.8798 \times 10^{-2})$**	$3.4746 \times 10^{-1}(9.9411 \times 10^{-1})$	$3.1338 \times 10^{-1}(8.9802 \times 10^{-1})$	$1.8960 \times 10^{-1}(4.1569 \times 10^{-1})$	$9.5891 \times 10^{-2}(1.7942 \times 10^{-1})$	$4.9542 \times 10^{-1}(7.6995 \times 10^{-1})$	$2.0756 \times 10^{-1}(6.4395 \times 10^{-1})$
	rHV	$3.2551 \times 10^{-1}(4.2625 \times 10^{-1})$	$3.3075 \times 10^{-1}(4.2120 \times 10^{-1})$	$3.3190 \times 10^{-1}(4.2221 \times 10^{-1})$	$3.2718 \times 10^{-1}(4.2192 \times 10^{-1}1)$	$3.3521 \times 10^{-1}(4.2949 \times 10^{-1})$	**$3.1747 \times 10^{-1}(4.4185 \times 10^{-1})$**	$3.2814 \times 10^{-1}(4.3388 \times 10^{-1})$
	IGDX	**$6.6158 \times 10^{-2}(6.1465 \times 10^{-2}$**	$2.1426 \times 10^{-1}(4.5588 \times 10^{-1})$	$1.9620 \times 10^{-1}(4.1929 \times 10^{-1})$	$1.5713 \times 10^{-1}(2.8550 \times 10^{-1})$	$8.5960 \times 10^{-2}(1.3964 \times 10^{-1})$	$3.1059 \times 10^{-1}(3.9207 \times 10^{-1})$	$1.3689 \times 10^{-1}(3.5454 \times 10^{-1})$
	IGDF	**$8.3820 \times 10^{-2}(2.2857 \times 10^{-2})$**	$1.5043 \times 10^{-1}(4.9843 \times 10^{-2})$	$1.5674 \times 10^{-1}(5.2428 \times 10^{-2})$	$1.4004 \times 10^{-1}(4.8312 \times 10^{-2})$	$9.5996 \times 10^{-2}2(2.7970 \times 10^{-2})$	$1.8942 \times 10^{-1}(5.3270 \times 10^{-2})$	$8.9313 \times 10^{-2}(2.4917 \times 10^{-2})$
MMF10_l	rPSP	**$1.1118 \times 10^{-1}(7.5967 \times 10^{-2})$**	$5.8854 \times 10^{+0}(4.0955 \times 10^{+0})$	$2.2084 \times 10^{+0}(3.1980 \times 10^{+0})$	$4.7371 \times 10^{+0}(3.9776 \times 10^{+0})$	$1.9439 \times 10^{-1}(1.6249 \times 10^{-1})$	$1.2643 \times 10^{+0}(1.7486 \times 10^{+0})$	$2.9513 \times 10^{-1}(6.2144 \times 10^{-1})$
	rHV	**$1.9208 \times 10^{-1}(4.5255 \times 10^{-1})$**	$1.9281 \times 10^{-1}(4.5108 \times 10^{-1})$	$1.9435 \times 10^{-1}(4.5201 \times 10^{-1})$	$1.9276 \times 10^{-1}(4.5125 \times 10^{-1})$	$1.9509 \times 10^{-1}(4.6014 \times 10^{-1})$	$1.9526 \times 10^{-1}(4.6861 \times 10^{-1})$	$1.9500 \times 10^{-1}(4.6304 \times 10^{-1})$
	IGDX	**$1.0917 \times 10^{-1}(7.0632 \times 10^{-2})$**	$2.7230 \times 10^{-1}(4.4162 \times 10^{-1})$	$2.4316 \times 10^{-1}(4.0906 \times 10^{-1})$	$2.3371 \times 10^{-1}(2.6921 \times 10^{-1})$	$1.7615 \times 10^{-1}(1.2264 \times 10^{-1})$	$2.0963 \times 10^{-1}(4.1447 \times 10^{-1})$	$2.2312 \times 10^{-1}(3.3361 \times 10^{-1})$
	IGDF	**$8.6794 \times 10^{-2}(3.0770 \times 10^{-2})$**	$1.7698 \times 10^{-1}(5.4922 \times 10^{-2})$	$1.7998 \times 10^{-1}(6.6802 \times 10^{-2})$	$1.8236 \times 10^{-1}(5.8248 \times 10^{-2})$	$1.6032 \times 10^{-1}(4.6416 \times 10^{-2})$	$2.0147 \times 10^{-1}(5.3897 \times 10^{-2})$	$1.4601 \times 10^{-1}(4.1643 \times 10^{-2})$
MMF11_l	rPSP	$1.4984 \times 10^{+0}(4.8986 \times 10^{-1})$	$3.7271 \times 10^{+0}(1.9133 \times 10^{+0})$	$2.1513 \times 10^{+0}(1.4348 \times 10^{+0})$	$2.7669 \times 10^{+0}(2.0862 \times 10^{+0})$	$3.5725 \times 10^{-1}(4.4515 \times 10^{-1})$	$1.3821 \times 10^{+0}(6.7718 \times 10^{-1})$	**$7.0114 \times 10^{-1}(7.3469 \times 10^{-1})$**
	rHV	$1.7453 \times 10^{-1}(4.5501 \times 10^{-1})$	**$1.7428 \times 10^{-1}(4.5376 \times 10^{-1})$**	$1.7476 \times 10^{-1}(4.5500 \times 10^{-1})$	$1.7435 \times 10^{-1}(4.5396 \times 10^{-1})$	$1.7657 \times 10^{-1}(4.6281 \times 10^{-1})$	$1.7874 \times 10^{-1}(4.7110 \times 10^{-1})$	$1.7720 \times 10^{-1}(4.6564 \times 10^{-1})$
	IGDX	**$2.0694 \times 10^{-1}(1.2130 \times 10^{-1})$**	$3.3066 \times 10^{-1}(4.2895 \times 10^{-1}1)$	$3.1932 \times 10^{-1}(3.9145 \times 10^{-1})$	$2.8816 \times 10^{-1}(2.5875 \times 10^{-1})$	$2.2614 \times 10^{-1}(6.3706 \times 10^{-2})$	$3.1391 \times 10^{-1}(3.8456 \times 10^{-1})$	$2.7950 \times 10^{-1}(3.2344 \times 10^{-1})$
	IGDF	**$8.4878 \times 10^{-2}(2.6402 \times 10^{-2})$**	$1.0173 \times 10^{-1}(4.6038 \times 10^{-2})$	$1.0402 \times 10^{-1}(5.0036 \times 10^{-2})$	$1.0234 \times 10^{-1}(4.5983 \times 10^{-2})$	$8.8337 \times 10^{-2}(3.0647 \times 10^{-2})$	$1.1235 \times 10^{-1}(3.9813 \times 10^{-2})$	$8.6340 \times 10^{-2}(2.2678 \times 10^{-2})$
MMF12_l	rPSP	$1.2230 \times 10^{+0}(4.0194 \times 10^{-1})$	$5.4330 \times 10^{+0}(3.6459 \times 10^{+0})$	$2.7054 \times 10^{+0}(1.3762 \times 10^{+0})$	$2.8187 \times 10^{+0}(2.0073 \times 10^{+0})$	$3.8266 \times 10^{-1}(4.4694 \times 10^{-1})$	$1.5398 \times 10^{+0}(6.9024 \times 10^{-1})$	**$7.6782 \times 10^{-1}(7.4197 \times 10^{-1})$**
	rHV	$6.3110 \times 10^{-1}(3.6957 \times 10^{-1})$	**$6.3081 \times 10^{-1}(3.6843 \times 10^{-1})$**	$6.3133 \times 10^{-1}(3.6953 \times 10^{-1})$	$6.3094 \times 10^{-1}(3.6860 \times 10^{-1})$	$6.3473 \times 10^{-1}(3.7649 \times 10^{-1})$	$6.3796 \times 10^{-1}(3.8378 \times 10^{-1})$	$6.3487 \times 10^{-1}(3.7910 \times 10^{-1})$
	IGDX	$2.2733 \times 10^{-1}(6.3978 \times 10^{-2})$	$3.3123 \times 10^{-1}(4.2950 \times 10^{-1})$	$3.2151 \times 10^{-1}(3.9193 \times 10^{-1})$	$2.8873 \times 10^{-1}(2.5933 \times 10^{-1})$	**$1.8958 \times 10^{-1}(1.2730 \times 10^{-1})$**	$3.1766 \times 10^{-1}(3.8522 \times 10^{-1})$	$2.6702 \times 10^{-1}(3.2836 \times 10^{-1})$
	IGDF	$7.9885 \times 10^{-2}(2.2256 \times 10^{-2})$	$8.7167 \times 10^{-2}(4.6952 \times 10^{-2})$	$8.8515 \times 10^{-2}(5.1303 \times 10^{-2})$	$8.7240 \times 10^{-2}(4.7027 \times 10^{-2})$	$6.5683 \times 10^{-2}(3.3893 \times 10^{-2})$	$9.0096 \times 10^{-2}(4.0895 \times 10^{-2})$	**$6.3356 \times 10^{-2}(2.8488 \times 10^{-2})$**
MMF13_l	rPSP	$6.0336 \times 10^{-1}(3.2481 \times 10^{-1})$	$1.6977 \times 10^{+0}(3.1417 \times 10^{+0})$	$1.0867 \times 10^{+0}(1.6258 \times 10^{+0})$	$1.2330 \times 10^{+0}(2.3150 \times 10^{+0})$	**$3.5849 \times 10^{-1}(2.7145 \times 10^{-1})$**	$7.7178 \times 10^{-1}(7.7167 \times 10^{-1})$	$5.2464 \times 10^{-1}(5.8903 \times 10^{-1})$
	rHV	$1.9787 \times 10^{-1}(4.5622 \times 10^{-1})$	**$1.9759 \times 10^{-1}(4.5498 \times 10^{-1})$**	$1.9793 \times 10^{-1}(4.5622 \times 10^{-1})$	$1.9765 \times 10^{-1}(4.5518 \times 10^{-1})$	$1.9997 \times 10^{-1}(4.6399 \times 10^{-1})$	$2.0219 \times 10^{-1}(4.7227 \times 10^{-1})$	$2.0055 \times 10^{-1}(4.6682 \times 10^{-1})$
	IGDX	**$2.3277 \times 10^{-1}(1.1833 \times 10^{-1})$**	$3.6932 \times 10^{-1}(4.2166 \times 10^{-1})$	$3.5189 \times 10^{-1}(3.8587 \times 10^{-1})$	$3.1456 \times 10^{-1}(2.5493 \times 10^{-1})$	$2.3594 \times 10^{-1}(6.4516 \times 10^{-2})$	$3.4039 \times 10^{-1}(3.8065 \times 10^{-1})$	$2.9341 \times 10^{-1}(3.2102 \times 10^{-1})$
	IGDF	$1.2932 \times 10^{-1}(3.6344 \times 10^{-2})$	$1.3656 \times 10^{-1}(4.6598 \times 10^{-2})$	$1.3988 \times 10^{-1}(4.9857 \times 10^{-2})$	$1.3719 \times 10^{-1}(4.6702 \times 10^{-2})$	**$8.1117 \times 10^{-2}(3.3842 \times 10^{-2})$**	$1.4936 \times 10^{-1}(4.5168 \times 10^{-2})$	$8.1341 \times 10^{-2}(3.1849 \times 10^{-2})$
MMF15_l	rPSP	$3.5514 \times 10^{-1}(3.7112 \times 10^{-1})$	$1.2183 \times 10^{+0}(3.1717 \times 10^{+0})$	$6.9391 \times 10^{-1}(1.6950 \times 10^{+0})$	$9.0741 \times 10^{-1}(2.3741 \times 10^{+0})$	**$2.2095 \times 10^{-1}(2.8990 \times 10^{-1})$**	$5.7247 \times 10^{-1}(8.0801 \times 10^{-1})$	$3.3866 \times 10^{-1}(6.1596 \times 10^{-1})$
	rHV	$3.0933 \times 10^{-1}(4.2618 \times 10^{-1})$	$3.1070 \times 10^{-1}(4.2169 \times 10^{-1})$	**$3.0669 \times 10^{-1}(4.2401 \times 10^{-1})$**	$3.0778 \times 10^{-1}(4.2258 \times 10^{-1})$	$3.1379 \times 10^{-1}(4.3070 \times 10^{-1})$	$3.1703 \times 10^{-1}(4.3851 \times 10^{-1})$	$3.0902 \times 10^{-1}(4.3463 \times 10^{-1})$
	IGDX	**$2.1678 \times 10^{-1}(6.3534 \times 10^{-2}$**	$3.4746 \times 10^{-1}(4.2675 \times 10^{-1})$	$3.1425 \times 10^{-1}(3.9438 \times 10^{-1})$	$2.9007 \times 10^{-1}(2.5995 \times 10^{-1})$	$1.7449 \times 10^{-1}(1.2642 \times 10^{-1})$	$3.3589 \times 10^{-1}(3.8195 \times 10^{-1})$	$2.5576 \times 10^{-1}(3.2919 \times 10^{-1})$
	IGDF	$1.5949 \times 10^{-1}(4.2758 \times 10^{-2})$	$1.8994 \times 10^{-1}(5.4658 \times 10^{-2})$	$1.9009 \times 10^{-1}(5.5997 \times 10^{-2})$	$1.7960 \times 10^{-1}(5.2395 \times 10^{-2})$	$1.4914 \times 10^{-1}(4.1847 \times 10^{-2})$	$2.0741 \times 10^{-1}(5.7881 \times 10^{-2})$	**$1.4785 \times 10^{-1}(4.1132 \times 10^{-2})$**

Table 5. Cont.

Case	Name	NSMRFO	NSGA-II [32]	DN-NSGAII [52]	OMNI-OPT [53]	MO_Ring_PSO_SCD [44]	MOMRFO [51]	IMOMRFO [28]
MMF15_a_1	rPSP	$3.1218\times10^{-1}(3.6154\times10^{-1})$	$1.0417\times10^{+0}(3.2095\times10^{-1})$	$6.2297\times10^{-1}(1.7046\times10^{+0})$	$8.1088\times10^{-1}(2.3899\times10^{+0})$	$\underline{\mathbf{2.1731\times10^{-1}}}(2.8815\times10^{-1})$	$1.0585\times10^{+0}(9.6772\times10^{-1})$	$3.1097\times10^{-1}(6.1937\times10^{-1})$
	rHV	$3.1502\times10^{-1}(4.2673\times10^{-1})$	$3.1895\times10^{-1}(4.2175\times10^{-1})$	$3.1458\times10^{-1}(4.2402\times10^{-1})$	$3.1100\times10^{-1}(4.2365\times10^{-1})$	$3.2086\times10^{-1}(4.3083\times10^{-1})$	$3.1889\times10^{-1}(4.4046\times10^{-1})$	$3.1992\times10^{-1}(4.4390\times10^{-1})$
	IGDX	$\mathbf{1.9875\times10^{-1}}(\mathbf{5.8533\times10^{-2}})$	$3.1727\times10^{-1}(4.3282\times10^{-1})$	$2.9038\times10^{-1}(3.9877\times10^{-1})$	$2.6349\times10^{-1}(2.6421\times10^{-1})$	$1.7466\times10^{-1}(1.2437\times10^{-1})$	$4.8954\times10^{-1}(3.6510\times10^{-1})$	$2.3716\times10^{-1}(3.3227\times10^{-1})$
	IGDF	$1.6305\times10^{-1}(4.2373\times10^{-2})$	$1.9269\times10^{-1}(5.3600\times10^{-2})$	$2.0166\times10^{-1}(5.6775\times10^{-2})$	$1.8704\times10^{-1}(5.2568\times10^{-2})$	$1.5666\times10^{-1}(4.1125\times10^{-2})$	$2.5153\times10^{-1}(6.8004\times10^{-2})$	$\underline{\mathbf{1.5353\times10^{-1}}}(\mathbf{3.9899\times10^{-2}})$
MMF16_11	rPSP	$\underline{\mathbf{2.1832\times10^{-1}}}(\mathbf{3.8126\times10^{-1}})$	$9.7190\times10^{-1}(3.2241\times10^{+0})$	$5.9705\times10^{-1}(1.7104\times10^{+0})$	$7.7514\times10^{-1}(2.3989\times10^{+0})$	$1.7666\times10^{-1}(2.9706\times10^{-1})$	$4.9735\times10^{-1}(8.2507\times10^{-1})$	$2.6244\times10^{-1}(6.3079\times10^{-1})$
	rHV	$3.1752\times10^{-1}(4.2620\times10^{-1})$	$3.1772\times10^{-1}(4.2205\times10^{-1})$	$\mathbf{3.1006\times10^{-1}}(4.2497\times10^{-1})$	$3.1207\times10^{-1}(4.2336\times10^{-1})$	$3.1788\times10^{-1}(4.3156\times10^{-1})$	$3.2520\times10^{-1}(4.3860\times10^{-1})$	$3.1749\times10^{-1}(4.4364\times10^{-1})$
	IGDX	$\underline{\mathbf{1.3520\times10^{-1}}}(\mathbf{5.4939\times10^{-2}})$	$2.7292\times10^{-1}(4.4222\times10^{-1})$	$2.5740\times10^{-1}(4.0554\times10^{-1})$	$2.2453\times10^{-1}(2.7150\times10^{-1})$	$1.3588\times10^{-1}(1.3063\times10^{-1})$	$3.1262\times10^{-1}(3.8910\times10^{-1})$	$1.9225\times10^{-1}(3.4211\times10^{-1})$
	IGDF	$1.2982\times10^{-1}(3.3061\times10^{-2})$	$1.5922\times10^{-1}(4.7659\times10^{-2})$	$1.5747\times10^{-1}(5.0227\times10^{-2})$	$1.4965\times10^{-1}(4.6701\times10^{-2})$	$1.2620\times10^{-1}(3.4123\times10^{-2})$	$2.0144\times10^{-1}(5.3609\times10^{-2})$	$\underline{\mathbf{1.2424\times10^{-1}}}(\mathbf{3.1887\times10^{-2}})$
MMF16_12	rPSP	$3.6067\times10^{-1}(3.5062\times10^{-1})$	$1.2955\times10^{+0}(3.1488\times10^{+0})$	$8.1064\times10^{-1}(1.6661\times10^{+0})$	$1.0402\times10^{+0}(2.2375\times10^{+0})$	$\underline{\mathbf{2.4167\times10^{-1}}}(\mathbf{2.8283\times10^{-1}})$	$6.9618\times10^{-1}(7.8602\times10^{-1})$	$3.5365\times10^{-1}(6.0898\times10^{-1})$
	rHV	$3.1793\times10^{-1}(4.2615\times10^{-1})$	$3.1758\times10^{-1}(4.2205\times10^{-1})$	$\mathbf{3.1179\times10^{-1}}(\mathbf{4.2454\times10^{-1}})$	$3.1655\times10^{-1}(4.2246\times10^{-1})$	$3.1664\times10^{-1}(4.3174\times10^{-1})$	$3.2816\times10^{-1}(4.3807\times10^{-1})$	$3.1723\times10^{-1}(4.3455\times10^{-1})$
	IGDX	$\mathbf{1.9989\times10^{-1}}(\mathbf{1.2125\times10^{-1}})$	$3.8973\times10^{-1}(4.1691\times10^{-1})$	$3.6819\times10^{-1}(3.8209\times10^{-1})$	$3.4183\times10^{-1}(2.5016\times10^{-1})$	$2.4602\times10^{-1}(7.1740\times10^{-2})$	$3.8721\times10^{-1}(3.7161\times10^{-1})$	$2.8003\times10^{-1}(3.2298\times10^{-1})$
	IGDF	$1.9763\times10^{-1}(5.5487\times10^{-2})$	$2.2172\times10^{-1}(6.1833\times10^{-2})$	$2.1677\times10^{-1}(6.1381\times10^{-2})$	$2.1156\times10^{-1}(5.9398\times10^{-2})$	$\underline{\mathbf{1.8187\times10^{-1}}}(\mathbf{5.0529\times10^{-2}})$	$2.4685\times10^{-1}(6.7163\times10^{-2})$	$1.8583\times10^{-1}(5.2109\times10^{-2})$
MMF16_13	rPSP	$3.0848\times10^{-1}(3.6371\times10^{-1})$	$1.0667\times10^{+0}(3.2049\times10^{+0})$	$6.7322\times10^{-1}(1.6931\times10^{+0})$	$8.5319\times10^{-1}(2.3823\times10^{+0})$	$\underline{\mathbf{2.0408\times10^{-1}}}(\mathbf{2.9122\times10^{-1}})$	$5.5665\times10^{-1}(8.0336\times10^{-1})$	$2.9495\times10^{-1}(6.2354\times10^{-1})$
	rHV	$3.1707\times10^{-1}(4.2633\times10^{-1})$	$3.1789\times10^{-1}(4.2199\times10^{-1})$	$\mathbf{3.0774\times10^{-1}}(\mathbf{4.2541\times10^{-1}})$	$3.1153\times10^{-1}(4.2355\times10^{-1})$	$3.1741\times10^{-1}(4.3158\times10^{-1})$	$3.2256\times10^{-1}(4.3922\times10^{-1})$	$3.1719\times10^{-1}(4.3455\times10^{-1})$
	IGDX	$\underline{\mathbf{1.6339\times10^{-1}}}(\mathbf{1.2622\times10^{-1}})$	$3.1668\times10^{-1}(4.4368\times10^{-1})$	$3.0231\times10^{-1}(3.9676\times10^{-1})$	$2.6735\times10^{-1}(2.6418\times10^{-1})$	$1.9272\times10^{-1}(5.6531\times10^{-1})$	$3.4539\times10^{-1}(3.8096\times10^{-1})$	$2.2430\times10^{-1}(3.3554\times10^{-1})$
	IGDF	$1.6970\times10^{-1}(4.2863\times10^{-2})$	$1.9249\times10^{-1}(5.1843\times10^{-2})$	$1.8509\times10^{-1}(5.2703\times10^{-2})$	$1.8207\times10^{-1}(5.0213\times10^{-2})$	$1.5570\times10^{-1}(3.9720\times10^{-2})$	$2.2536\times10^{-1}(5.8444\times10^{-2})$	$\underline{\mathbf{1.5506\times10^{-1}}}(\mathbf{3.9332\times10^{-2}})$
Rank		47/96	9/96	6/96	4/96	8/96	2/96	20/96
	rPSP	10/24	0/24	0/24	0/24	5/24	0/24	9/24
	rHV	8/24	7/24	6/24	1/24	5/24	2/24	0/24
	IGDX	17/24	0/24	0/24	0/24	1/24	0/24	6/24
	IGDF	12/24	2/24	0/24	3/24	2/24	0/24	5/24

Underlined values indicate the best results.

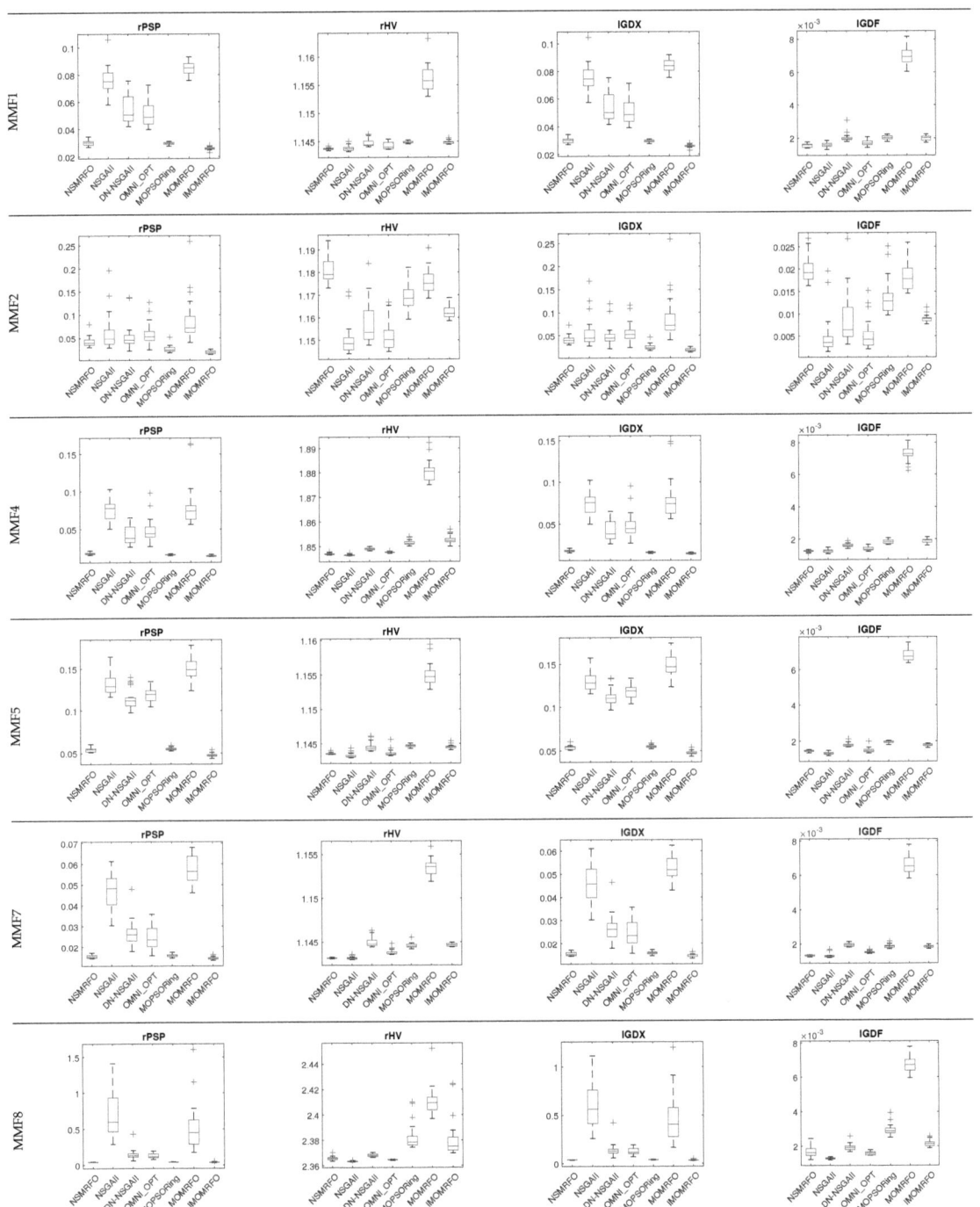

Figure 12. Cont.

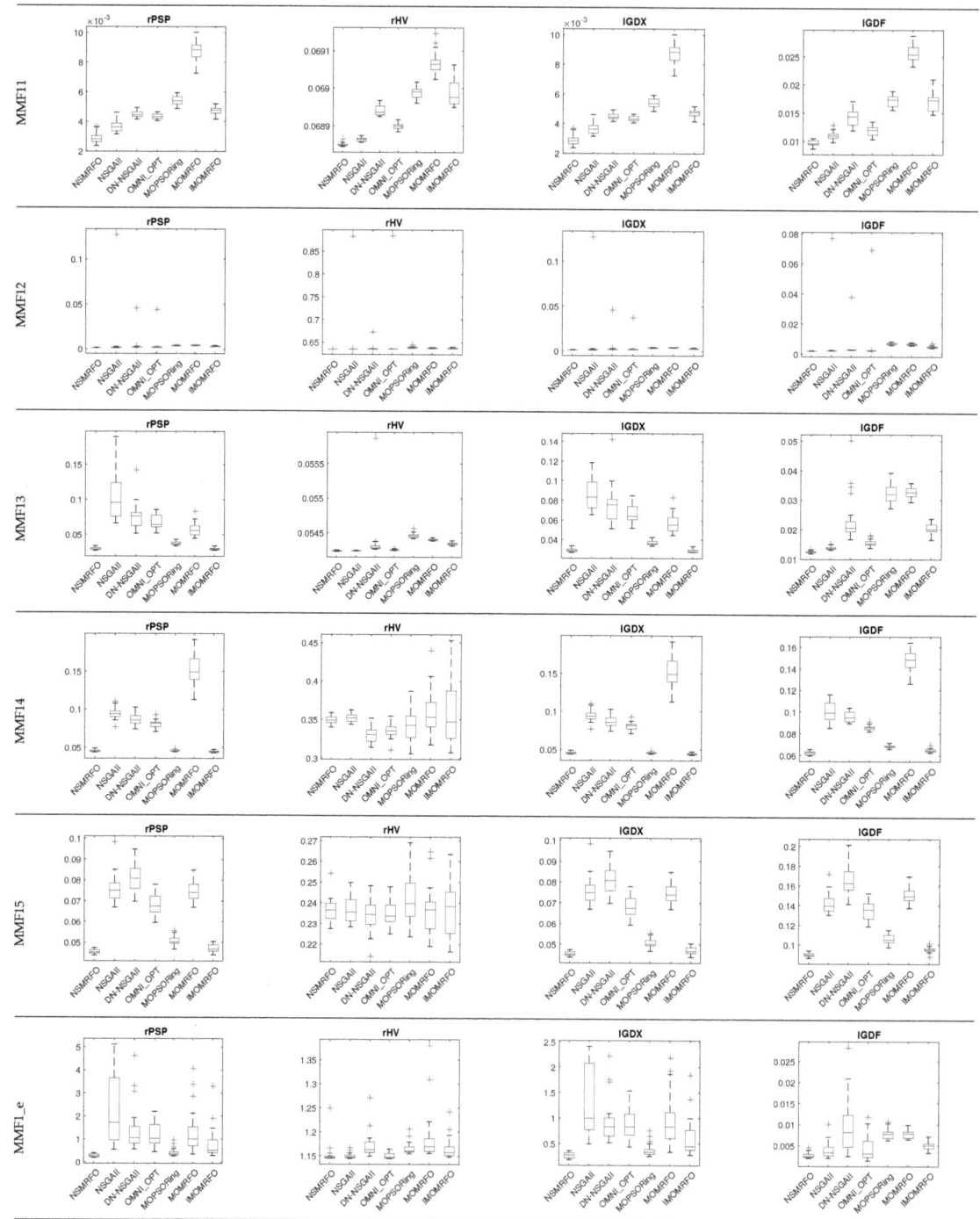

Figure 12. *Cont.*

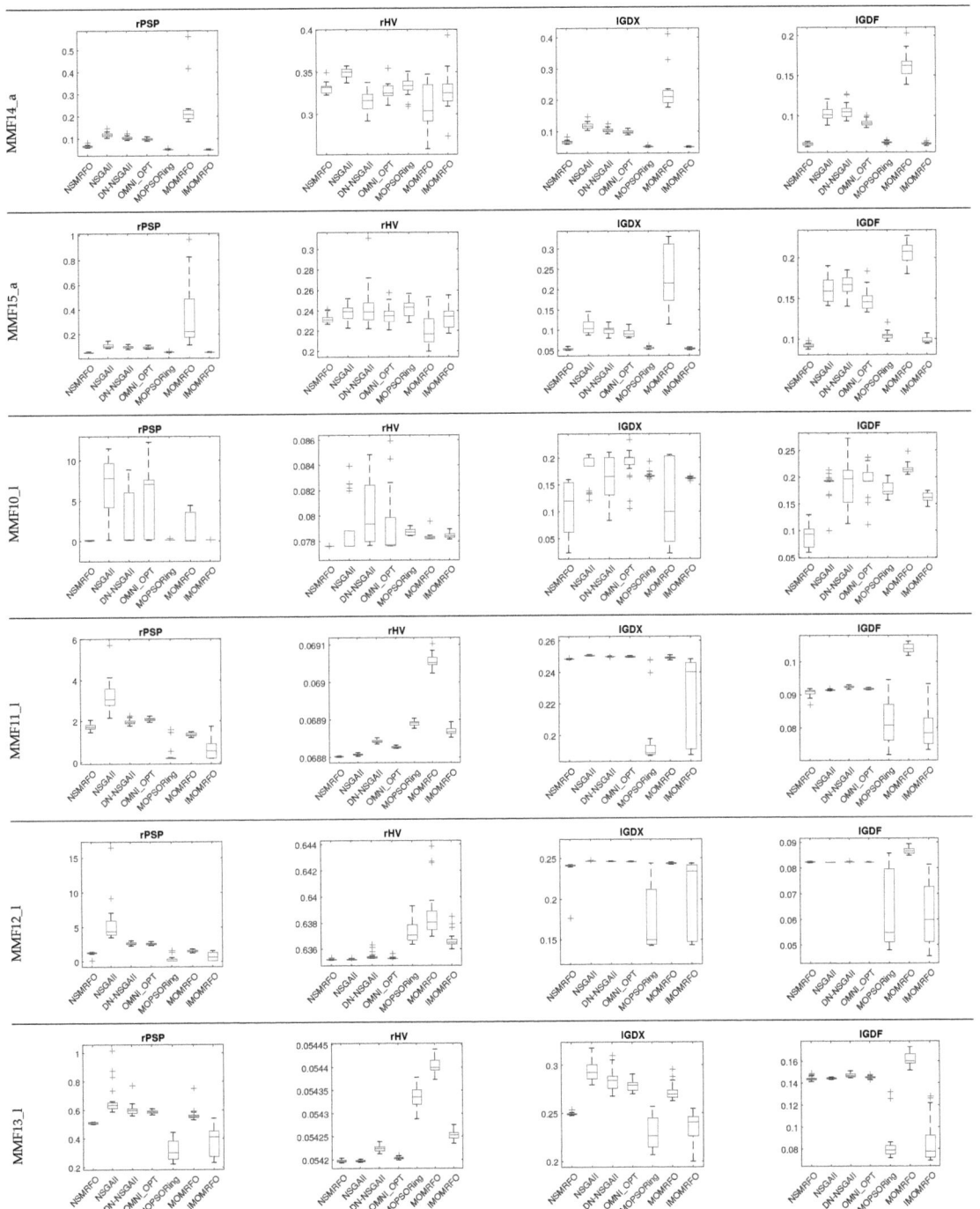

Figure 12. *Cont.*

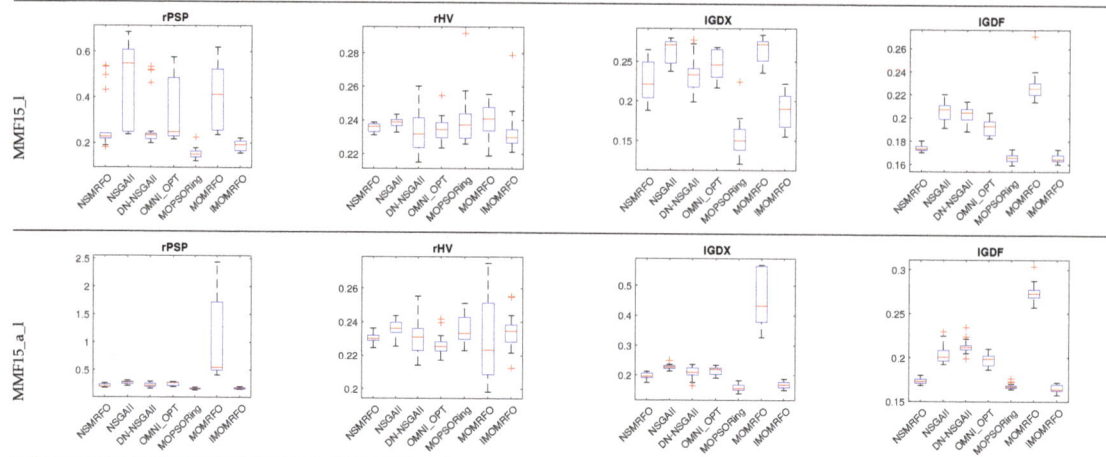

Figure 12. Box plots for the four metrics rPSP, rHV, IGDX, and IGDF on CEC2020 problems.

3.4. Evaluation on Engineering Design Problems

For examining the applicability of an algorithm, the engineering design problems can be very beneficial. In this subsection, four engineering functions are considered in order to assess the capability of NSMRFO in dealing with the real-word problems. The first one is 4-bar truss, and it aims to optimize the volume and deflection with four dimensions. The disk brake consists of minimizing the stopping time and weight of a brake system with four dimensions. The third engineering problem is the welded beam that tends to decrease the vertical deflection and cost of fabrication with four dimensions. The speed reducer as a last function attempts to reduce its stress and total weight with seven dimensions. For results verification, seven well-known approaches are applied. The statistical results are summarized in Table 6, and it is evident that NSMRFO can outrank the other algorithms on the most problems except the 4-bar truss for the GD metric, the welded beam for GD and SP metrics, and disk brake for SP metric; it ranks first in 8 out of 12 test suites. Accordingly, the NSGA-II and MOMRFO are the closest competitors where they provide good estimations on two functions over 12. However, concerning the other algorithms MOSMA, MOBO/D, MOMVO and IMOMRFO, they have the lowest results. As illustrated in Figure 13, the NSMRFO Pareto front shows higher approximations toward the true PFs in terms of coverage and convergence.

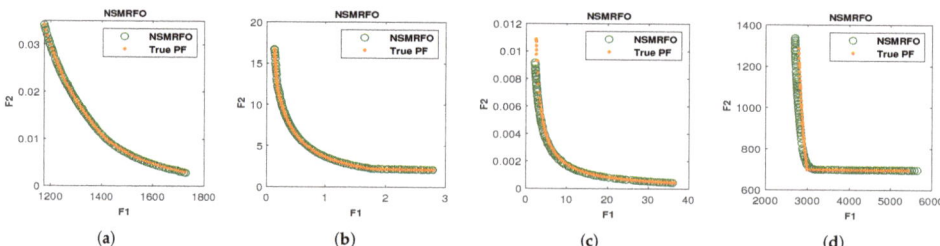

Figure 13. Obtained Pareto front of NSMRFO on engineering design test suites. (**a**) 4-bar truss; (**b**) disk brake; (**c**) welded beam; (**d**) speed reducer.

Table 6. GD, IGD, and SP metrics comparison based on engineering design test suites.

Problem	NSMRFO	MOSMA [46]	MOBO/D [47]	MOMVO [48]	MOWCA [49]	NSGA-II [32]	MOMRFO [51]	IMOMRFO [28]
GD Metric								
4-bar truss	3.3030×10^{-1}	1.6635×10^{-1}	$1.9677\times10^{+0}$	1.7194×10^{-1}	$1.3286\times10^{+0}$	1.8236×10^{-1}	$\underline{\mathbf{1.4895\times10^{-1}}}$	$2.4210\times10^{+0}$
	4.13×10^{-1}	1.21×10^{-2}	$3.06\times10^{+0}$	2.72×10^{-2}	$3.64\times10^{+0}$	8.66×10^{-3}	7.21×10^{-3}	$1.94\times10^{+1}$
Disk brake	$\underline{\mathbf{1.4345\times10^{-3}}}$	2.3597×10^{-2}	2.9302×10^{-3}	8.0972×10^{-2}	5.1287×10^{-3}	7.0487×10^{-3}	3.0258×10^{-3}	4.7383×10^{-2}
	2.72×10^{-4}	2.70×10^{-2}	1.45×10^{-3}	1.45×10^{-1}	3.22×10^{-3}	1.26×10^{-2}	5.62×10^{-4}	7.82×10^{-2}
Welded beam	5.2346×10^{-3}	6.0605×10^{-1}	3.7817×10^{-1}	2.0160×10^{-2}	5.9606×10^{-3}	4.2063×10^{-3}	$\underline{\mathbf{3.9257\times10^{-3}}}$	1.1569×10^{-1}
	1.68×10^{-3}	$1.25\times10^{+0}$	9.64×10^{-1}	5.68×10^{-2}	2.01×10^{-3}	1.04×10^{-3}	2.54×10^{-3}	9.92×10^{-2}
Speed reducer	$\underline{\mathbf{7.2541\times10^{+0}}}$	$3.4412\times10^{+1}$	$1.2252\times10^{+1}$	$1.3694\times10^{+1}$	$1.7352\times10^{+1}$	$1.3368\times10^{+1}$	$1.4421\times10^{+1}$	$9.2079\times10^{+0}$
	$1.87\times10^{+0}$	$3.33\times10^{+1}$	$1.29\times10^{+0}$	$3.22\times10^{+0}$	$1.93\times10^{+0}$	9.81×10^{-1}	$1.21\times10^{+0}$	$1.09\times10^{+0}$
IGD Metric								
4-bar truss	$\underline{\mathbf{5.2306\times10^{-4}}}$	1.2645×10^{-3}	1.4296×10^{-2}	3.4156×10^{-3}	6.6312×10^{-4}	5.9489×10^{-4}	9.9940×10^{-4}	1.1208×10^{-3}
	2.99×10^{-5}	1.21×10^{-3}	4.93×10^{-3}	6.11×10^{-4}	3.30×10^{-4}	4.06×10^{-5}	8.66×10^{-5}	6.15×10^{-5}
Disk brake	$\underline{\mathbf{2.4597\times10^{-3}}}$	2.6790×10^{-3}	1.7209×10^{-3}	1.0017×10^{-3}	5.7433×10^{-3}	1.2484×10^{-3}	4.0983×10^{-3}	6.1796×10^{-3}
	5.77×10^{-5}	7.85×10^{-4}	6.99×10^{-4}	2.48×10^{-4}	4.21×10^{-4}	1.29×10^{-3}	3.14×10^{-5}	1.57×10^{-4}
Welded beam	7.7497×10^{-4}	7.0596×10^{-3}	6.9524×10^{-3}	1.3979×10^{-3}	1.2635×10^{-3}	$\underline{\mathbf{3.0953\times10^{-4}}}$	3.7374×10^{-4}	4.7932×10^{-4}
	4.75×10^{-4}	1.66×10^{-3}	4.25×10^{-3}	7.84×10^{-4}	1.75×10^{-3}	4.33×10^{-5}	4.88×10^{-5}	1.29×10^{-4}
Speed reducer	$\underline{\mathbf{1.7887\times10^{-2}}}$	1.9012×10^{-2}	1.0443×10^{-2}	5.9622×10^{-3}	3.3948×10^{-3}	2.7998×10^{-3}	2.9813×10^{-3}	3.3957×10^{-3}
	2.16×10^{-5}	2.68×10^{-3}	8.99×10^{-3}	5.54×10^{-3}	1.44×10^{-3}	4.73×10^{-5}	8.81×10^{-5}	7.20×10^{-4}
SP Metric								
4-bar truss	$\underline{\mathbf{1.8155\times10^{+0}}}$	$4.8453\times10^{+0}$	$2.1874\times10^{+1}$	$3.9299\times10^{+0}$	$2.7300\times10^{+0}$	$2.5462\times10^{+0}$	4.7759×10^{-1}	$1.4586\times10^{+1}$
	7.79×10^{-1}	$3.81\times10^{+0}$	$1.67\times10^{+1}$	$3.77\times10^{+0}$	$3.77\times10^{+0}$	2.21×10^{-1}	6.71×10^{-1}	$1.45\times10^{+1}$
Disk brake	9.3521×10^{-2}	2.2820×10^{-1}	1.7934×10^{-1}	4.0540×10^{-1}	8.0842×10^{-2}	$\underline{\mathbf{7.9256\times10^{-2}}}$	1.1040×10^{-1}	4.3313×10^{-1}
	9.79×10^{-3}	9.04×10^{-2}	8.42×10^{-2}	9.16×10^{-2}	1.06×10^{-2}	1.40×10^{-2}	1.33×10^{-2}	7.58×10^{-1}
Welded beam	$\underline{\mathbf{1.9678\times10^{+0}}}$	$6.4688\times10^{+0}$	$2.6550\times10^{+0}$	2.3160×10^{-1}	2.2191×10^{-1}	2.3184×10^{-1}	2.1139×10^{-1}	5.4636×10^{-1}
	4.31×10^{-2}	$1.23\times10^{+1}$	$4.69\times10^{+0}$	1.22×10^{-1}	3.93×10^{-2}	2.34×10^{-2}	3.02×10^{-2}	4.81×10^{-1}
Speed reducer	$\underline{\mathbf{2.1976\times10^{+1}}}$	$2.5008\times10^{+1}$	$3.0519\times10^{+1}$	$4.0828\times10^{+1}$	$2.2979\times10^{+1}$	$2.3112\times10^{+1}$	$2.8427\times10^{+1}$	$6.3713\times10^{+1}$
	$2.17\times10^{+0}$	$6.28\times10^{+0}$	$1.34\times10^{+1}$	$1.33\times10^{+1}$	$2.60\times10^{+0}$	$1.66\times10^{+0}$	$3.91\times10^{+1}$	$1.95\times10^{+1}$

Underlined values indicate the best results.

3.5. Evaluation on OPF Incorporating Wind/Solar/Small-Hydro Energy
3.5.1. Problem Methodology
Wind power

The wind cost can be expressed as below:

$$C_{Tw} = C_{dw} + C_{uew} + C_{oew} \tag{23}$$

with

$$C_{dw} = d_w P_{ws} \tag{24}$$

$$C_{uew} = K_{uew}(P_{wav} - P_{ws}) = K_{uew} \int_{P_{ws}}^{P_{wr}} (p_w - P_{ws}) f_w(p_w) dp_w \tag{25}$$

$$C_{oew} = K_{oew}(P_{ws} - P_{wav}) = K_{oew} \int_{0}^{P_{ws}} (P_{ws} - p_w) f_w(p_w) dp_w \tag{26}$$

where d_w, K_{oew}, and K_{uew} are the coefficients of direct, over, and under estimation cost, respectively. $P_{ws,i}$, P_{wav} are the scheduled and actual available power, respectively. P_{wr} is rated power output from the plant. $f_w(p_w)$ is the probability density function of the wind power plant.

Solar power

The total solar cost can be formulated as follows:

$$C_{Ts} = C_{ds} + C_{ues} + C_{oes} \tag{27}$$

with

$$C_{ds} = d_s P_{ss} \tag{28}$$

$$C_{ues} = K_{ues}(P_{sav} - P_{ss}) = K_{ues} * f_s(P_{sav} > P_{ss}) * [E(P_{sav} > P_{ss}) - P_{ss}] \tag{29}$$

$$C_{oes} = K_{oes}(P_{ss} - P_{sav}) = K_{oeS} * f_s(P_{sav} < P_{ss}) * [P_{ss} - E(P_{sav} < P_{ss})] \tag{30}$$

where d_s, K_{oes} and K_{ues} are the coefficients of direct, over and under estimation cost of solar power generator. P_{ss}, P_{sav} are the scheduled and actual available power, respectively. $f_s(P_{sav} < P_{ss})$ The probability of occurrence of solar power shortage. $E(P_{sav} > P_{ss})$ and $E(P_{sav} < P_{ss})$ are the expectations of solar power above and below P_{ss}.

Small-hydro power

The Small-hydro power is defined as follows:

$$C_{Tsh} = C_{dsh} + C_{uesh} + C_{oesh} \tag{31}$$

with

$$C_{ds} = d_s P_{ss} + d_h P_{hs} \tag{32}$$

$$C_{uesh} = K_{uesh}(P_{shav} - P_{ssh}) = K_{uesh} * f_{sh}(P_{shav} > P_{ssh}) * [E(P_{shav} > P_{ssh}) - P_{ssh}] \tag{33}$$

$$C_{oesh} = K_{oesh}(P_{shs} - P_{shav}) = K_{oesh} * f_{sh}(P_{shav} < P_{shs}) * [P_{shs} - E(P_{shav} < P_{shs})] \tag{34}$$

where d_h is the small-hydro direct cost coefficient. K_{oesh} and K_{uesh} are the coefficients of over and under estimation cost of combined solar and small-hydro power generator. P_{shs}, P_{shav} are the scheduled and actual available power, respectively. $E(P_{shav} > P_{shs})$ and $E(P_{shav} < P_{shs})$ are the expectations of combined system power above and below P_{shs}.

Objective functions

- Total cost

The network total cost including the thermal/wind/solar/small-hydro generators is modeled as follows:

$$F_1 = min\{F_T + C_{Tw} + C_{Ts} + C_{Tsh}\}. \tag{35}$$

where

$$F_T = \sum_{i=1}^{Ng} a_i + b_i P_{tgi} + c_i P_{tgi}^2 + |d_i * sin(e_i * (P_{tgi}^{min} - P_{tgi}))| \tag{36}$$

a_i, b_i and c_i are the conventional generators cost coefficients. d_i and e_i are the coefficients for the valve-point loading effect.

- Emission

The emission function is formulated using an exponential function as shown below [54]:

$$F_2 = E = min\left\{\sum_{i=1}^{Ng} 10^{-2}\left(\alpha_i + \beta_i P_{gi} + \gamma_i P_{gi}^2\right) + \xi_i \exp(\lambda_i P_{gi})\right\} \tag{37}$$

where α_i, β_i, γ_i, ξ_i, and λ_i are the emission coefficients of the power plant.

- Voltage deviation

The voltage deviation is calculated by:

$$F_3 = VD = min\left\{\sum_{i=1}^{Npq} |V_{Li} - 1.0|\right\} \tag{38}$$

- Power loss

The Power loss is calculated by:

$$F_4 = P_{loss} = min\left\{\sum_{l=1}^{N_l} G_{l(i,j)}(V_i^2 + V_j^2 - 2V_i V_j cos(\delta_{ij}))\right\} \tag{39}$$

where $G_{l(i,j)}$ represents the conductance of line l. $\delta_{ij} = \delta_i - \delta_j$ represents the voltage angle difference between bus i and bus j.

Constraints

- Equality constraints

The power flow equations are assumed as equality constraints that are represented by:

$$\begin{cases} P_{gi} - P_{di} - |V_i|\sum_{j=1}^{Nb} |V_j|[G_{ij}cos(\theta_{ij}) + B_{ij}sin(\theta_{ij})] = 0 \\ Q_{gi} - Q_{di} - |V_i|\sum_{j=1}^{Nb} |V_j|[G_{ij}sin(\theta_{ij}) - B_{ij}cos(\theta_{ij})] = 0 \end{cases} \tag{40}$$

where Nb is the number of buses. Q_{gi} and P_{gi} are generated reactive and active power, respectively. Q_{di} and P_{di} are reactive and active power demand, respectively. G_{ij} and B_{ij} represent the admittance matrix components $Y_{ij} = G_{ij} + jB_{ij}$ called the conductance and susceptance.

- Inequality constraints

The inequality constraints are given as below:

— Generator constraints:

$$V_{gi}^{min} \leq V_{gi} \leq V_{gi}^{max} \quad i = 1, ..., Ng \tag{41}$$

$$P_{tgi}^{min} \leq P_{tgi} \leq P_{tgi}^{max} \quad i = 1, ..., Ntg \tag{42}$$

$$P_{ws,i}^{min} \leq P_{ws,i} \leq P_{ws,i}^{max} \quad i = 1, ..., Nwg \tag{43}$$

$$P_{ss,i}^{min} \leq P_{ss,i} \leq P_{ss,i}^{max} \quad i = 1, ..., Nsg \tag{44}$$

$$P_{shs,i}^{min} \leq P_{shs,i} \leq P_{shs,i}^{max} \quad i = 1, ..., Nshg \tag{45}$$

$$Q_{tgi}^{min} \leq Q_{tgi} \leq Q_{tgi}^{max} \quad i = 1, ..., Ntg \tag{46}$$

$$Q_{ws,i}^{min} \leq Q_{ws,i} \leq Q_{ws,i}^{max} \quad i = 1, ..., Nwg \tag{47}$$

$$Q_{ss,i}^{min} \leq Q_{ss,i} \leq Q_{ss,i}^{max} \quad i = 1, ..., Nsg \tag{48}$$

$$Q_{shs,i}^{min} \leq Q_{shs,i} \leq Q_{shs,i}^{max} \quad i = 1, ..., Nshg \tag{49}$$

- Security constraints:

$$V_{Li}^{min} \leq V_{Li} \leq V_{Li}^{max} \quad i = 1, ..., Npq \tag{50}$$

$$S_{li} \leq S_{li}^{max} \quad i = 1, ..., Nl \tag{51}$$

where Nl is the number of transmission lines. S_{li} and S_{li}^{max} indicate the maximum limit of the transmission line.

3.5.2. Results of the OPF Problem

To assess performance of the suggested algorithm NSMRFO against other approaches, several cases related to the modified IEEE 30-bus optimal power flow problem integrating wind/solar/small-hydro power are investigated. This test system comprises 41 branches, 6 generating units in which 3 thermal generators at buses 1, 2, and 8, the wind and solar plants at buses 5 and 11, respectively, while combined solar and small-hydro generators are connected to bus 13 as summarized in Table 7. The detailed input data for the considered IEEE 30-bus system are given in [54]. The thermal generators' coefficients are provided in Table 8. Solar irradiance, wind distribution, and small-hydro river flow rate are modeled using Lognormal, Weibull, and Gumbel probability density function (PDF), respectively [54]. These PDF parameters are listed in Table 9. Additionally, in terms of the optimization issue, the system has 11 control variables, with various constraints and objective functions for a total active and reactive power demands of 283.4 MW and 126.2 MVAR, respectively.

Table 7. The characteristics of the system.

Systems Characteristics	30-Bus [54] Value	Details
Buses	30	-
Branches	41	-
Generators	3	Buses: 1, 2, and 8
Slack bus	1	Buse: 1
Wind generators	1	Buses: 5
Solar generators	2	Buse: 11 and 13
Small-hydro generators	1	Buse: 13
Active and reactive power	-	283.4 MW, 126.2 MVAr
Control variables	11	-

Table 8. Cost and emission coefficients of thermal generators [54].

	Generator	Bus	a	b	c	d	e	α	β	γ	ξ	λ
	P_{g1}	1	30	2	0.00375	18	0.037	0.04091	−0.05554	0.06490	0.0002	6.667
IEEE-30	P_{g2}	2	25	1.75	0.0175	16	0.038	0.02543	−0.06047	0.05638	0.0005	3.333
	P_{g3}	8	20	3.25	0.00834	12	0.045	0.05326	−0.03550	0.03380	0.002	2

To validate the suggested approach, five well-known stochastic algorithms are employed as competitors, namely MOMVO [48], MOWCA [49], NSGWO [50], MOMRFO [51], and IMOMRFO [28]. The test system under study is examined via three case studies defined as follows:

- Minimizing the total cost and emission;
- Minimizing the total cost and voltage deviation;

- Minimizing the total cost and power loss.

Table 9. Characteristic details of wind/solar/small-hydro generators [54].

	Wind Power (bus 5)			Solar Power (bus 11)		Combined Solar/Small-Hydro Power (bus 13)			
Test Systems	No. of Turbines	Pwr (MW)	Parameters of Weibull PDF	Psr (MW)	Parameters of Lognormal PDF	Psr (MW)	Parameters of Lognormal PDF	Phr (MW)	Parameters of Gumbel PDF
IEEE-30	25	75	k = 2 c = 9	50	μ = 5.2 σ = 0.6	45	μ = 5 σ = 0.6	5	λ = 15 γ = 1.2

The optimum settings of the control variables, their allowable ranges, the numerical best outcomes of each objective, and the best compromise solutions (BCS) are depicted in Tables 10–15. Furthermore, their optimal Pareto fronts are illustrated in Figures 14a–16a. It is worth noting that all the findings are generated after twenty independent runs for a population size of 100 and 200 iterations. According to the aforementioned tables, it is obviously seen that the NSMRFO' results are remarkably better than the competitor approaches in all cases, notably the best compromise solutions' tables. In addition, it is clearly observed from the figures that the suggested NSMRFO can generate the superior Pareto non-dominated solutions with good distribution and good diversification front in comparison to other algorithms. As shown in Figures 14b–16b, the BCS' voltage profile PQ of load buses do not exceed their limits, and remained within the minimum and maximum bounds.

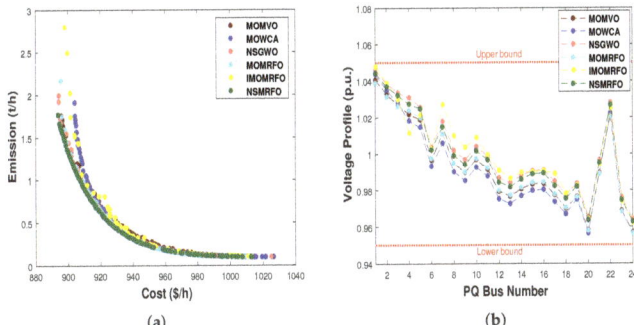

Figure 14. Optimal Pareto fronts of all the algorithms for case 1. (a) Pareto front of optimal solutions; (b) voltage profile of PQ buses.

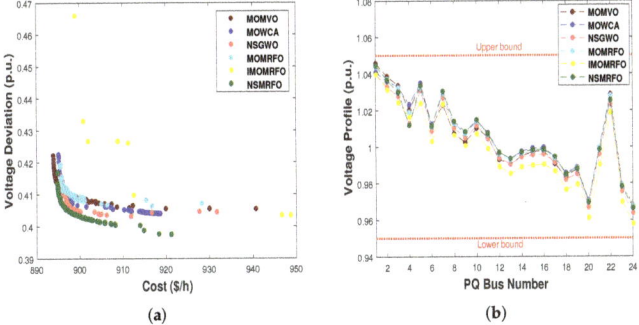

Figure 15. Optimal Pareto fronts of all the algorithms for case 2. (a) Pareto front of optimal solutions; (b) voltage profile of PQ buses.

Table 10. Findings of best objectives for case 1.

Case 1			NSMRFO		MOMVO [48]		MOWCA [49]		NSGWO [50]		MOMRFO [51]		IMOMRFO [28]	
Control Variables	Min	Max	Cost	E	Cost	E	Cost	E	Cost	E	Cost	E	Cost	E
P_{tg2}	20	80	53.108	60.441	53.813	79.882	65.933	52.442	54.421	56.651	59.090	64.799	43.141	64.982
P_{wg}	0	75	51.903	73.597	56.252	74.961	35.136	74.937	50.662	15.858	47.240	66.735	55.886	67.336
P_{g3}	10	35	16.476	31.077	13.194	27.016	24.800	24.722	13.128	24.091	11.981	32.089	15.660	21.897
P_{sg}	0	50	17.407	21.171	16.889	25.689	13.209	50.000	17.505	10.181	21.577	46.457	18.109	40.445
P_{shg}	0	50	15.878	45.724	15.115	22.804	16.516	34.040	17.273	37.732	16.048	25.442	15.018	37.884
V_{g1}	0.95	1.1	1.0805	1.0684	1.1000	1.0874	1.0793	1.0732	1.0746	1.0259	1.0700	1.0655	1.0746	1.0838
V_{g2}	0.95	1.1	1.0625	1.0577	1.0802	1.0705	1.0708	1.0557	1.0610	1.0287	1.0548	1.0570	1.0666	1.0753
V_{g5}	0.95	1.1	1.0406	1.0360	1.0413	1.0556	1.0240	1.0307	1.0355	1.0407	1.0302	1.0296	1.0462	1.0272
V_{g8}	0.95	1.1	1.0280	1.0215	1.0210	1.0252	1.0324	1.0244	1.0300	1.0031	1.0219	1.0267	1.0286	1.0323
V_{g11}	0.95	1.1	1.0635	1.0574	1.0497	1.0614	1.0560	1.0709	1.0658	1.0789	1.0704	1.0551	1.0387	1.0523
V_{g13}	0.95	1.1	1.0372	1.0320	0.9999	1.0089	1.0194	1.0594	1.0473	0.9886	1.0353	1.0423	1.0128	0.9982
Q_{tg1}	−50	140	16.159	21.010	29.660	38.554	−2.3346	33.953	4.2530	29.911	7.9031	18.958	−6.1844	19.924
Q_{tg2}	−20	60	22.069	24.129	48.350	13.470	59.861	9.4168	30.142	28.823	27.160	19.936	50.169	54.470
Q_{wg}	−30	35	30.615	23.786	21.343	33.759	14.518	16.921	26.313	25.518	28.830	26.924	33.653	3.1405
Q_{g3}	−15	40	31.337	20.742	17.553	15.561	39.202	18.831	36.317	3.6417	32.787	18.327	36.748	30.527
Q_{sg}	−20	25	20.441	20.458	18.014	20.990	18.491	24.584	20.608	23.828	24.787	28.416	13.772	19.339
Q_{shg}	−20	25	15.616	15.216	2.8992	4.9409	9.2576	23.816	19.108	18.060	16.755	14.588	8.716	1.2780
Total Cost ($/h)	-	-	893.815	1012.83	895.799	998.532	904.351	1021.33	894.215	1024.58	895.589	1009.6	897.961	1010.91
Emission (t/h)	-	-	1.7686	0.0985	1.7594	0.1089	1.758	0.0969	1.9969	0.0968	2.1716	0.0987	2.7944	0.1019
VD (p.u.)	-	-	0.4561	0.5291	0.6419	0.5618	0.5177	0.4471	0.4326	0.4931	0.4699	0.4621	0.5982	0.6145
P_{Loss} (MW)	-	-	6.3699	3.1979	6.7693	3.7273	7.0872	2.7821	6.5379	3.2890	5.7357	2.8983	6.5780	3.4661
P_{tg1}	50	140	134.99	50.028	134.907	56.774	134.893	50.041	136.948	51.854	138.290	50.576	142.164	54.304

Underlined values indicate the best results.

Table 11. Findings of best compromise solutions for case 1.

Case 1			NSMRFO	MOMVO [48]	MOWCA [49]	NSGWO [50]	MOMRFO [51]	IMOMRFO [28]
Control Variables	Min	Max	BCS	BCS	BCS	BCS	BCS	BCS
P_{tg2}	20	80	60.658	79.791	65.943	64.856	64.438	65.828
P_{wg}	0	75	55.825	55.620	54.797	56.854	54.207	57.585
P_{g3}	10	35	23.791	10.265	24.559	21.437	23.046	12.331
P_{sg}	0	50	18.720	17.384	18.199	19.870	21.799	22.762
P_{shg}	0	50	16.333	16.890	15.653	16.109	14.859	19.657
V_{g1}	0.95	1.1	1.0749	1.0729	1.0795	1.0784	1.0701	1.0857
V_{g2}	0.95	1.1	1.0642	1.0638	1.0710	1.0638	1.0593	1.0675
V_{g5}	0.95	1.1	1.0405	1.0347	1.0245	1.0471	1.0406	1.0057
V_{g8}	0.95	1.1	1.0331	1.0267	1.0285	1.0335	1.0248	1.0311
V_{g11}	0.95	1.1	1.0624	1.0549	1.0445	1.0614	1.0582	1.0504

Table 11. Cont.

Case 1			NSMRFO	MOMVO [48]	MOWCA [49]	NSGWO [50]	MOMRFO [51]	IMOMRFO [28]
Control Variables	Min	Max	BCS	BCS	BCS	BCS	BCS	BCS
V_{g13}	0.95	1.1	1.0331	1.0302	1.0208	1.0379	1.0313	1.0569
Q_{tg1}	-50	140	5.6755	4.2337	5.0900	13.966	5.9788	24.335
Q_{tg2}	-20	60	30.809	37.568	58.617	17.758	29.104	41.067
Q_{avg}	-30	35	26.973	23.646	9.2601	32.981	33.386	-9.445
Q_{tg3}	-15	40	35.741	36.326	34.291	33.961	29.605	40.288
Q_{sg}	-20	25	19.544	18.581	15.261	18.622	19.976	15.144
Q_{shg}	-20	25	13.604	14.277	10.954	14.964	15.157	22.829
Total Cost ($/h)	-	-	**930.948**	934.021	931.355	931.249	931.347	932.258
Emission (t/h)	-	-	0.4112	0.4190	0.4239	0.4158	0.4253	0.4602
VD (p.u.)	-	-	0.4542	0.5026	0.5474	0.4467	0.4959	0.4353
P_L (MW)	-	-	5.2144	5.8737	5.5104	5.2299	5.3406	5.8467
P_{tg1}	50	140	109.150	109.119	109.759	109.361	109.829	111.084

Underlined values indicate the best results.

Table 12. Findings of best objectives for case 2.

Case 2			NSMRFO		MOMVO [48]		MOWCA [49]		NSGWO [50]		MOMRFO [51]		IMOMRFO [28]	
Control Variables	Min	Max	Cost	VD	Cost	VD	Cost	VD	Cost	VD	Cost	VD	Cost	VD
P_{tg2}	20	80	52.024	54.112	52.557	52.387	47.316	35.038	46.047	63.371	45.578	45.641	37.922	65.929
P_{avg}	0	75	49.449	48.886	49.826	53.907	50.814	50.727	50.504	50.746	50.397	59.982	47.082	37.443
P_{g3}	10	35	21.458	34.578	17.580	24.853	24.778	25.296	25.700	32.668	23.375	27.656	26.664	30.457
P_{sg}	0	50	15.549	14.878	18.054	33.311	16.061	29.283	17.568	18.363	19.150	20.280	22.056	29.289
P_{shg}	0	50	13.863	2.4460	14.652	3.8207	15.203	4.2886	13.752	9.6771	13.465	5.7879	18.004	1.0118
V_{g1}	0.95	1.1	1.0770	1.0777	1.0800	1.0736	1.0731	1.0615	1.0734	1.0680	1.0777	1.0743	1.0841	1.0780
V_{g2}	0.95	1.1	1.0589	1.0623	1.0679	1.0677	1.0598	1.0587	1.0576	1.0578	1.0631	1.0631	1.0756	1.0646
V_{g5}	0.95	1.1	1.0236	1.0062	1.0329	1.0190	1.0408	1.0364	1.0195	1.0115	1.0257	1.0200	1.0418	1.0062
V_{g8}	0.95	1.1	1.0297	1.0324	1.0355	1.0367	1.0357	1.0352	1.0297	1.0305	1.0311	1.0351	1.0336	1.0335
V_{g11}	0.95	1.1	1.0781	1.0806	1.0769	1.0793	1.0707	1.0813	1.0754	1.0776	1.0741	1.0742	1.0690	1.0749
V_{g13}	0.95	1.1	1.0586	1.0613	1.0461	1.0462	1.0587	1.0556	1.0535	1.0581	1.0595	1.0590	1.0235	1.0480
Q_{tg1}	-50	140	13.538	9.7512	13.538	-8.1601	1.8042	-23.430	8.5189	2.2257	6.3704	0.4094	-3.2325	8.4710
Q_{tg2}	-20	60	25.623	44.482	40.716	59.259	23.322	48.275	32.641	41.144	37.354	44.935	58.645	48.064
Q_{avg}	-30	35	15.823	-3.1762	17.560	2.4995	30.249	26.829	12.628	4.6642	14.424	4.9202	23.406	-0.8900
Q_{g3}	-15	40	34.891	38.425	36.648	39.452	37.310	38.764	36.776	38.255	32.672	39.733	26.422	39.927
Q_{sg}	-20	25	24.310	24.842	23.003	24.592	20.470	24.999	23.981	24.619	22.406	22.098	21.381	23.793
Q_{shg}	-20	25	23.001	24.408	16.671	18.017	22.157	22.211	21.697	23.860	22.938	23.213	8.442	20.075
Total Cost ($/h)	-	-	**894.77**	921.23	893.85	940.66	895.07	918.36	895.72	931.70	895.53	916.70	898.95	948.61

Table 12. Cont.

Case 2		NSMRFO			MOMVO [48]			MOWCA [49]			NSGWO [50]			MOMRFO [51]			IMOMRFO [28]		
Control Variables	Min	Max	Cost	VD	Cost	VD		Cost	VD		Cost	VD		Cost	VD		Cost	VD	
Emission (t/h)	-	-	2.0652	1.7635	2.0184	0.7516		1.8018	2.4120		1.8782	0.4839		2.0838	1.2817		1.841	1.0063	
VD (p.u.)	-	-	0.4169	**0.3975**	0.4222	0.4054		0.4227	0.4041		0.4157	0.4044		0.4195	0.4064		0.4659	0.4033	
P_L (MW)	-	-	6.4450	6.4705	6.3981	5.4931		6.0562	6.1743		6.1556	5.4798		6.2143	5.6924		6.0657	6.4642	
P_{g1}	50	140	137.50	134.97	137.13	120.61		135.31	139.98		135.98	122.42		6.3704	129.75		135.66	125.644	

Underlined values indicate the best results.

Table 13. Findings of best compromise solutions for case 2.

Case 2			NSMRFO	MOMVO [48]	MOWCA [49]	NSGWO [50]	MOMRFO [51]	IMOMRFO [28]
Control Variables	Min	Max	BCS	BCS	BCS	BCS	BCS	BCS
P_{g2}	20	80	53.594	52.415	41.134	45.147	45.137	39.393
P_{wg}	0	75	49.649	50.911	48.375	50.559	51.718	45.047
P_{g3}	10	35	24.446	23.020	24.425	28.827	27.402	29.832
P_{sg}	0	50	15.036	19.166	22.582	17.427	19.076	24.502
P_{shg}	0	50	10.658	9.0253	13.720	10.795	10.006	15.466
V_{g1}	0.95	1.1	1.0778	1.0753	1.0612	1.0679	1.0713	1.0752
V_{g2}	0.95	1.1	1.0625	1.0678	1.0596	1.0577	1.0631	1.0536
V_{g5}	0.95	1.1	1.0067	1.0212	1.0295	1.0113	1.0201	1.0248
V_{g8}	0.95	1.1	1.0325	1.0358	1.0353	1.0306	1.0351	1.0223
V_{g11}	0.95	1.1	1.0813	1.0814	1.0821	1.0785	1.0744	1.0654
V_{g13}	0.95	1.1	1.0620	1.0462	1.0583	1.0582	1.0597	1.0545
Q_{g1}	-50	140	8.6578	-7.9539	-25.938	-4.4517	-8.3054	21.981
Q_{g2}	-20	60	43.962	57.301	55.280	48.401	51.7371	18.117
Q_{wg}	-30	35	-3.3246	6.0463	20.146	4.5803	7.7620	24.266
Q_{g3}	-15	40	39.981	39.509	39.906	39.580	39.807	25.829
Q_{sg}	-20	25	24.989	24.909	24.995	24.937	22.156	21.912
Q_{shg}	-20	25	23.981	17.364	22.278	23.748	23.146	23.681
Total Cost ($/h)	-	-	**897.98**	899.55	898.24	898.43	916.70	900.87
Emission (t/h)	-	-	1.9546	1.7885	2.3326	1.9872	1.9026	1.7214
VD (p.u.)	-	-	**0.4043**	0.4087	0.4073	0.4049	0.4064	0.4330
P_L (MW)	-	-	6.6032	6.3273	6.2834	6.2478	6.1310	6.0237
P_{g1}	50	140	136.62	135.19	139.45	136.89	136.19	134.58

Underlined values indicate the best results.

Table 14. Findings of best objectives for case 3.

Case 3			NSMRFO		MOMVO [48]		MOWCA [49]		NSGWO [50]		MOMRFO [51]		IMOMRFO [28]	
Control Variables	Min	Max	Cost	P_{Loss}	Cost	P_{Loss}	Cost	P_{Loss}	Cost	P_{Loss}	Cost	P_{Loss}	Cost	P_{Loss}
P_{g2}	20	80	48.838	49.192	44.767	45.204	40.985	45.256	40.474	63.806	38.9739	61.303	59.511	33.250
P_{wg}	0	75	49.909	72.932	48.257	74.977	51.787	75.000	55.135	71.267	59.6321	74.869	56.777	72.769
P_{g3}	10	35	21.138	32.741	27.145	26.376	22.983	35.000	22.511	29.912	16.5629	30.789	10.571	34.292
P_{sg}	0	50	21.451	45.874	19.084	50.000	20.603	49.377	18.490	40.095	18.9793	44.802	18.489	49.458
P_{shg}	0	50	13.105	26.710	15.521	13.398	16.772	31.261	16.845	25.878	16.6080	26.851	14.919	42.827
V_{g1}	0.95	1.1	1.0738	1.0779	1.0842	1.0900	1.0585	1.0689	1.0769	1.0693	1.0713	1.0720	1.0873	1.0673
V_{g2}	0.95	1.1	1.0646	1.0669	1.0606	1.0653	1.0507	1.0575	1.0678	1.0646	1.0585	1.0609	1.0593	1.0642
V_{g5}	0.95	1.1	1.0473	1.0432	1.0276	1.0426	1.0248	1.0474	1.0387	1.0428	1.03330	1.0377	1.0415	1.0192
V_{g8}	0.95	1.1	1.0374	1.0405	1.0267	1.0444	1.0205	1.0466	1.0363	1.0410	1.0273	1.0386	1.0147	1.0435
V_{g11}	0.95	1.1	1.0598	1.0620	1.0376	1.0800	1.0570	1.0673	1.0679	1.0792	1.0562	1.0717	1.0564	1.0808
V_{g13}	0.95	1.1	1.0414	1.0372	1.0316	1.0333	1.0468	1.0565	1.0499	1.0429	1.0290	1.0486	1.0239	1.0780
Q_{g1}	−50	140	−5.909	17.678	29.918	44.051	−9.9749	15.696	−5.4244	3.7508	2.6613	20.722	42.179	−3.7032
Q_{g2}	−20	60	34.404	24.895	21.033	−4.4405	44.003	3.8259	45.267	28.455	36.900	14.327	5.2565	50.301
Q_{wg}	−30	35	34.193	18.916	21.373	16.885	25.371	26.239	21.516	20.238	23.887	16.920	36.104	−4.7289
Q_{g3}	−15	40	39.358	32.193	34.422	37.147	33.541	40.000	35.084	34.503	38.279	33.550	19.654	34.754
Q_{sg}	−20	25	17.404	19.090	13.108	24.809	20.429	19.558	19.555	24.034	19.323	22.276	21.064	23.587
Q_{shg}	−20	25	16.122	13.081	15.802	10.553	22.646	19.488	18.249	14.437	14.207	17.537	13.433	26.827
Total Cost ($/h)	-	-	896.04	1008.19	896.96	997.31	897.15	1021.53	896.12	1004.17	896.753	1033.83	898.125	1033.62
Emission (t/h)	-	-	1.7812	0.1017	1.7489	0.1313	1.9311	0.0958	1.8630	0.1008	2.0913	0.0963	1.7021	0.1018
VD (p.u.)	-	-	0.4338	0.4436	0.5350	0.4380	0.4654	0.4492	0.4286	0.4317	0.5059	0.4322	0.5660	0.5197
P_{Loss} (MW)	-	-	6.1629	2.7309	6.2089	3.3192	6.1567	2.3448	5.9016	2.7980	6.0776	2.6358	6.7354	2.7088
P_{g1}	50	140	135.12	58.671	134.83	76.701	136.43	49.842	135.85	54.046	137.69	43.655	134.36	56.710

Underlined values indicate the best results.

Table 15. Findings of best compromise solutions for case 3.

Case 3			NSMRFO	MOMVO [48]	MOWCA [49]	NSGWO [50]	MOMRFO [51]	IMOMRFO [28]
Control Variables	Min	Max	BCS	BCS	BCS	BCS	BCS	BCS
P_{g2}	20	80	34.493	31.895	40.145	23.841	39.028	37.708
P_{wg}	0	75	66.511	71.867	58.895	68.321	64.085	66.952
P_{g3}	10	35	32.323	34.515	26.639	30.326	27.308	24.467
P_{sg}	0	50	22.452	19.084	29.953	24.143	26.355	28.624
P_{shg}	0	50	16.640	14.602	20.914	18.749	16.761	9.3818
V_{g1}	0.95	1.1	1.0777	1.0858	1.0611	1.0782	1.0776	1.0829
V_{g2}	0.95	1.1	1.0653	1.0606	1.0528	1.0623	1.0672	1.0693
V_{g5}	0.95	1.1	1.0453	1.0506	1.0323	1.0337	1.0411	1.0524
V_{g8}	0.95	1.1	1.0400	1.0363	1.0281	1.0379	1.0351	1.0335
V_{g11}	0.95	1.1	1.0705	1.0305	1.0606	1.0775	1.0578	1.0698
V_{g13}	0.95	1.1	1.0418	1.0387	1.0495	1.0494	1.0497	1.0602
Q_{g1}	−50	140	5.5834	35.083	−4.2225	10.982	2.9712	9.3361
Q_{g2}	−20	60	28.456	−0.7926	32.816	28.018	38.769	27.986
Q_{wg}	−30	35	25.306	33.461	26.109	14.683	21.568	30.466
Q_{g3}	−15	40	35.318	37.243	34.519	35.900	31.389	22.263
Q_{sg}	−20	25	20.664	8.3323	20.123	22.938	16.581	19.608
Q_{shg}	−20	25	14.609	16.692	21.494	17.543	18.581	22.011
Total Cost ($/h)	-	-	<u>926.22</u>	931.32	936.50	927.55	931.87	931.34
Emission (t/h)	-	-	0.7300	0.6514	0.4596	0.8453	0.5106	0.6806
VD (p.u.)	-	-	0.4306	0.4855	0.4380	0.4233	0.4357	0.4373
P_{Loss} (MW)	-	-	<u>4.6000</u>	4.6938	4.6349	4.6027	4.6366	4.8748
P_{g1}	50	140	120.07	118.03	111.49	122.62	113.52	118.84

Underlined values indicate the best results.

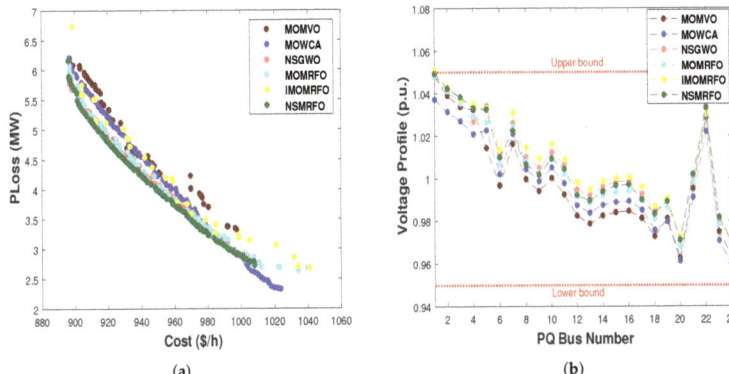

Figure 16. Optimal Pareto fronts of all the algorithms for case 3. (**a**) Pareto front of optimal solutions; (**b**) voltage profile of PQ buses.

3.6. Discussion

As previously stated, the main difference between the suggested approach and its competitors is the better diversity and accuracy for the majority of the problems, in which the NSMRFO ranks first, followed by the MOMRFO on the unconstrained problems, the NSGA-II on the constrained test suite, the NSGA-II on the engineering problems, and the IMOMRFO on the CEC2020 benchmark MMO functions, while the MOWCA gives a little better score. By contrast, the MOSMA, MOMVO and MOBO/D provide the worst rank. However, on the other hand, the NSMRFO generates very challenging and competitive solutions on most benchmark test suites. In summary, all quantitative and qualitative outcomes and analysis reveal the higher accuracy and significant diversity of NSMRFO in dealing with different unconstrained, constrained, CEC 2020 multimodal multi-objective, and engineering benchmark functions. This comes from the strong ability of MRFO in exploitation and exploration as long as the NSMRFO employs similar mechanisms as its single approach, and inherits its high convergence. In addition, the crowding distance and archive selection methodologies also contribute to the NSMRFO high coverage and convergence.

4. Conclusions

In this work, the ability of a suggested multi-objective manta ray foraging optimizer known as NSMRFO to handle problems with different characteristics has been tested. The NSMRFO optimizer has been developed on the basis of NSGA-II operators as crowding distance, elitist non-dominated sorting, and an archive mechanism. A set of test functions have been employed to benchmark the performance of the NSMRFO approach from different perspectives that include: seven classical, ZDT, TYD, four constrained, twenty-four CEC2020, four problems for engineering design, and the IEEE 30-bus OPF with renewable sources wind/solar/small-hydro power. Additionally, to qualitatively affirm the achieved solutions, the original true Pareto fronts have been compared to those obtained. Thereby, for performance assessment, various performance metrics in search and objective spaces have utilized such generational distance (GD), inverted generational distance (IGDX and IGDF), spacing metric, reciprocal Pareto sets proximity (rPSP), and reciprocal hypervolume (rHV). Thus, NSMRFO can relatively provide an accurate estimation shape with closer distance to the true PF compared to the multimodal multi-objective evolutionary approaches and some recent competitive algorithms. The NSMRFO impressive performance leads to handling challenging real-world problems in various engineering fields for future works.

Author Contributions: Conceptualization, F.D. and M.O.; methodology, F.D.; software, F.D.; validation, R.E., M.O. and S.K.; formal analysis, F.D.; investigation, S.K.; resources, M.O. and A.M.A; data curation, F.D.; writing—original draft preparation, F.D. and S.K.; writing—review and editing, F.D. and A.M.A.; visualization, R.E. and M.O.; supervision, R.E. and M.O.; project administration, R.E. and S.K.; funding acquisition, A.M.A. All authors have read and agreed to the published version of the manuscript.

Funding: This research was funded by the Deputyship for Research & Innovation, Ministry of Education in Saudi Arabia through the project number "IF_2020_NBU_408".

Institutional Review Board Statement: Not Applicable.

Informed Consent Statement: Not Applicable.

Data Availability Statement: Not Applicable.

Acknowledgments: The authors extend their appreciation to the Deputyship for Research & Innovation, Ministry of Education in Saudi Arabia for funding this research work through the project number "IF_2020_NBU_408". The authors gratefully thank the Prince Faisal bin Khalid bin Sultan Research Chair in Renewable Energy Studies and Applications (PFCRE) at Northern Border University for their support and assistance.

Conflicts of Interest: The authors declare no conflict of interest.

References

1. Kelley, C.T. Detection and remediation of stagnation in the Nelder–Mead algorithm using a sufficient decrease condition. *SIAM J. Optim.* **1999**, *10*, 43–55. [CrossRef]
2. Beyer, H.-G.; Sendhoff, B. Robust optimization a comprehensive survey. *Comput. Methods Appl. Mech. Eng.* **2017**, *196*, 3190–3218 [CrossRef]
3. Helbig, M.; Engelbrecht, A.P. Performance measures for dynamic multi-objective optimisation algorithms. *Inf. Sci.* **2013**, *250*, 61–81. [CrossRef]
4. von Lucken, C.; Baran, B.; Brizuela, C. A survey on multi-objective evolutionary algorithms for many-objective problems. *Comput. Optim. Appl.* **2014**, *58*, 707–756. [CrossRef]
5. Deb, K. *Multi-Objective Optimization Using Evolutionary Algorithms*; Wiley: Hoboken, NJ, USA, 2001; Volume 16.
6. Wolpert, D.H.; Macready, W.G. No free lunch theorems for optimization. *IEEE Trans. Evol. Comput.* **1997**, *1*, 67–82. [CrossRef]
7. Holland, J. *Adaptation in Natural and Artificial Systems*; University of Michigan Press: Ann Arbor, MI, USA, 1975.
8. Kirkpatrick, S.; Gelatt, C.D.; Vecchi, M.P. Optimization by simulated annealing. *Science* **1983**, *220*, 671–680. [CrossRef]
9. Kennedy, J.; Eberhart, R.C. Particle swarm optimization. In Proceedings of the ICNN'95—International Conference on Neural Networks, Perth, WA, Australia, 27 November–1 December 1995; pp. 1942–1948.
10. Basturk, B.; Karaboga, D. An artificial bee colony (ABC) algorithm for numeric function optimization. In *IEEE Swarm Intelligence Symposium*; IEEE Press: Indianapolis, IN, USA, 2006; pp. 4–12.
11. Abualigah, L.; Diabat, A.; Mirjalili, S.; Abd-Elaziz, M.; Gandomi, A.H. The arithmetic optimization algorithm. *Comput. Meth. Appl. Mech. Eng.* **2021**, *376*, 113609. [CrossRef]
12. Heidari, A.A.; Mirjalili, S.; Faris, H.; Aljarah, I.; Mafarja, M.M.; Chen, H. Harris hawks optimization: Algorithm and applications. *Future Gener. Comput. Syst.* **2019**, *97*, 849–872. [CrossRef]
13. Mirjalili, S. SCA: A Sine Cosine Algorithm for solving optimization problems. *Knowl.-Based Syst.* **2016**, *96*, 120–133. [CrossRef]
14. Hayyolalam, V.; Pourhaji, K.A.A. Black Widow Optimization Algorithm: A novel meta-heuristic approach for solving engineering optimization problems. *Eng. Appl. Artif. Intel.* **2020**, *87*, 103249. [CrossRef]
15. Ghafil, H.N.; Jármai, K. Dynamic differential annealed optimization: New metaheuristic optimization algorithm for engineering applications. *Appl. Soft Comput.* **2020**, *93*, 106392. [CrossRef]
16. Essam, H.H.; Mohammed, R.S.; Fatma, A.H.; Hassan, S.; Hassaballah, M. Lévy flight distribution: A new metaheuristic algorithm for solving engineering optimization problems. *Eng. Appl. Artif. Intel.* **2020**, *94*, 103731. [CrossRef]
17. Mirjalili, S.; Gandomi, A.H.; Mirjalili, S.Z.; Saremi, S.; Faris, H.; Mirjalili, S.M. Salp Swarm algorithm: A bio-inspired optimizer for engineering design problems. *Adv. Eng. Softw.* **2017**, *114*, 163–191. [CrossRef]
18. Hashim, F.A.; Houssein, E.H.; Mabrouk, M.S.; Al-Atabany, W.; Mirjalili, S. Henry gas solubility optimization: A novel physics-based algorithm. *Future Gener. Comput. Syst.* **2019**, *101*, 646–667. [CrossRef]
19. Zhao, W.; Zhang, Z.; Wang, L. Manta ray foraging optimization: An effective bio-inspired optimizer for engineering applications. *Eng. Appl. Artif. Intell.* **2020**, *87*, 103300. [CrossRef]
20. Alturki, F.A.; Omotoso, H.O.; Al-Shamma'a, A.A.; Farh, H.M.H.; Alsharabi, K. Novel Manta Rays Foraging Optimization Algorithm Based Optimal Control for Grid-Connected PV Energy System. *IEEE Access* **2020**, *8*, 187276–187290. [CrossRef]

21. Fathy, A.; Rezk, H.; Yousri, D. A robust global MPPT to mitigate partial shading of triple-junction solar cell-based system using manta ray foraging optimization algorithm. *Sol. Energy* **2020**, *207*, 305–316. [CrossRef]
22. Selem, S.I.; Hasanien, H.M.; El-Fergany, A.A. Parameters extraction ofPEMFC's model using manta rays foraging optimizer. *Int. J. Energy Res.* **2020**, *44*, 4629–4640. [CrossRef]
23. El-Hameed, M.A.; Elkholy, M.M.; El-Fergany, A.A. Three-diode model for characterization of industrial solar generating units using Manta-rays foraging optimizer: Analysis and validations. *Energy Convers. Manag.* **2020**, *219*, 113048. [CrossRef]
24. Mohamed, A.E.; Dalia, Y.; Mohammed, A.A.A.; Amr, M.A.; Ahmed, G.R.; Ahmed, A.E. A Grunwald–Letnikov based Manta ray foraging optimizer for global optimization and image segmentation. *Eng. Appl. Artif. Intell.* **2021**, *98*, 104105.
25. Ghosh, K.K.; Guha, R.; Bera, S.K.; Kumar, N.; Sarkar, R. Sshaped versus V-shaped transfer functions for binary Manta ray foraging optimization in feature selection problem. *Neural Comput. Appl.* **2021**, *33*, 11027–11041. [CrossRef]
26. Karrupusamy, P. Hybrid Manta Ray Foraging Optimization for Novel Brain Tumor Detection. *Trends Comput. Sci. Smart Technol.* **2020**, *2*, 175–185. [CrossRef]
27. Shaheen, A.M.; El-Sehiemy, R.A.; Elsayed, A.M.; Elattar, E.E. Multi-objective manta ray foraging algorithm for efficient operation of hybrid AC/DC power grids with emission minimization. *IET Gener. Transm. Distrib.* **2021**, *15*, 1314–1336. [CrossRef]
28. Kahraman, H.T.; Akbel, M.; Duman, S. Optimization of Multi-Objective Optimal Power Flow Problem Using Improved MOMRFO with a Crowding Distance-Based Pareto Archive Strategy. *Appl. Soft Comput.* **2022**, *116*, 108334. [CrossRef]
29. Marler, R.T.; Arora, J.S. Survey of multi-objective optimization methods for engineering. *Struct. Multidiscip. Optim.* **2004**, *26*, 369–395. [CrossRef]
30. Branke, J.; Kaußler, T.; Schmeck, H. Guidance in evolutionary multi-objective optimization. *Adv. Eng. Softw.* **2001**, *32*, 499–507. [CrossRef]
31. Srinivas, N.; Deb, K. Muiltiobjective optimization using nondominated sorting in genetic algorithms. *Evol. Comput.* **1994**, *2*, 221–248. [CrossRef]
32. Deb, K.; Pratap, A.; Agarwal, S.; Meyarivan, T. A fast and elitist multi-objective genetic algorithm: NSGA-II. *IEEE Trans. Evol. Comput.* **2002**, *6*, 182–197. [CrossRef]
33. Zitzler, E. Evolutionary Algorithms for Multiobjective Optimization: Methods and Applications. Ph.D. Thesis, Swiss Federal Institute of Technology (ETH), Zurich, Switzerland, November 1999.
34. Zitzler, E.; Thiele, L. Multiobjective evolutionary algorithms: A comparative case study and the strength pareto approach. *IEEE Trans. Evol. Comput.* **1999**, *3*, 257–271. [CrossRef]
35. Coello, C.A.C.; Pulido, G.T.; Lechuga, M.S. Handling multiple objectives with particle swarm optimization. *IEEE Trans. Evol. Comput.* **2004**, *8*, 256–279. [CrossRef]
36. Zhang, Q.; Li, H. MOEA/D: A multiobjective evolutionary algorithm based on decomposition. *IEEE Trans. Evol. Comput.* **2007**, *11*, 712–731. [CrossRef]
37. Pareto, V. *Cours d'Economie Politique: Librairie Droz*; Librairie Droz: Geneva, Switzerland, 1964.
38. Coello, C.A.C. Evolutionary multi-objective optimization: Some current research trends and topics that remain to be explored. *Front. Comput. Sci. China* **2009**, *3*, 18–30. [CrossRef]
39. Mirjalili, S.; Lewis, A. The whale optimization algorithm. *Adv. Eng. Softw.* **2016**, *95*, 51–67. [CrossRef]
40. Van Veldhuizen, D.A.; Lamont, G.B. *Multiobjective Evolutionary Algorithm Research: A History and Analysis*; Technical Report TR-98-03; Department of Electrical and Computer Engineering, Graduate School of Engineering, Air Force Institute of Technology, Wright-Patterson AFB: Dayton, Ohio, USA, 1998; pp. 17–19
41. Sierra, M.R.; Coello, C.A.C. Improving PSO-based multi-objective optimization using crowding, mutation and ∈-dominance. In *Evolutionary Multi-Criterion Optimization*; Springer: Berlin/Heidelberg, Germany, 2005; pp. 505–519.
42. Liang, J.; Suganthan, P.N.; Qu, B.Y.; Gong, D.W.; Yue, C.T. *Problem Definitions and Evaluation Criteria for the CEC 2020 Special Session on Multimodal Multiobjective Optimization*; Zhengzhou University: Zhengzhou, China, 2019. [CrossRef]
43. Schott, J.R. Fault Tolerant Design Using Single and Multicriteria Genetic Algorithm Optimization, DTIC Document. Ph.D. Thesis, Massachusetts Institute of Technology, Cambridge, MA, USA, 1995.
44. Yue, C.; Qu, B.; Liang, J. A Multi-objective Particle Swarm Optimizer Using Ring Topology for Solving Multimodal Multi-objective Problems. *IEEE Trans. Evol. Comput.* **2017**, *22*, 805–817. [CrossRef]
45. Shang, K.; Ishibuchi, H.; He, L.; Pang, L.M. A survey on the hypervolume indicator in evolutionary multi-objective optimization. *IEEE Trans. Evol. Comput.* **2021**, *25*, 1–20. [CrossRef]
46. Premkumar, M.; Jangir, P.; Sowmya, R.; Alhelou, H.H.; Heidari, A.A.; Chen, H. MOSMA: Multi-Objective Slime Mould Algorithm Based on Elitist Non-Dominated Sorting. *IEEE Access* **2021**, *9*, 3229–3248. [CrossRef]
47. Das, A.K.; Nikum, A.K.; Krishnan, S.V.; Pratihar, D.K. Multi-objective Bonobo Optimizer (MOBO): An intelligent heuristic for multi-criteria optimization. *Knowl. Inf. Syst.* **2020**, *62*, 4407–4444. [CrossRef]
48. Mirjalili, S.; Jangir, P.; Mirjalili, S.Z.; Saremi, S.; Trivedi, I.N. Optimization of Problems with Multiple Objectives using The Multi-Verse Optimization Algorithm. *Knowl.-Based Syst.* **2017**, *134*, 50–71. [CrossRef]
49. Sadollah, A.; Eskandar, H.; Kim, J.H.; Bahreininejad, A. Water cycle algorithm for solving multi-objective optimization problems. *Soft Comput.* **2015**, *19*, 2587–2603. [CrossRef]

50. Jangir, P.; Jangir, N. A new Non-Dominated Sorting Grey Wolf Optimizer (NS-GWO) algorithm: Development and application to solve engineering designs and economic constrained emission dispatch problem with integration of wind power. *Eng. Appl. Artif. Intell.* **2018**, *72*, 449–467. [CrossRef]
51. Got, A.; Zouache, D.; Moussaoui, A. MOMRFO: Multi-objective Manta ray foraging optimizer for handling engineering design problems. *Knowl.-Based Syst.* **2022**, *237*, 107880. [CrossRef]
52. Liu, Y.; Ishibuchi, H.; Nojima, Y.; Masuyama, N.; Shang, K. A double-niched evolutionary algorithm and its behavior on polygon-based problems. In *International Conference on Parallel Problem Solving from Nature*; Springer: Cham, Switzerland, 2018; pp. 262–273.
53. Deb, K.; Tiwari, S. Omni-optimizer: A generic evolutionary algorithm for single and multi-objective optimization. *Eur. Oper. Res.* **2008**, *185*, 1062–1087. [CrossRef]
54. Biswas, P.P.; Suganthan, P.N.; Qu, B.Y.; Amaratunga, G.A.J. Multi-objective economic-environmental power dispatch with stochastic wind-solar-small hydro power. *Energy* **2018**, *150*, 1039–1057. [CrossRef]

Article

Optimal Design of TD-TI Controller for LFC Considering Renewables Penetration by an Improved Chaos Game Optimizer

Ahmed H. A. Elkasem [1], Mohamed Khamies [1], Mohamed H. Hassan [1], Ahmed M. Agwa [2,3,*] and Salah Kamel [1]

1. Department of Electrical Engineering, Faculty of Engineering, Aswan University, Aswan 81542, Egypt; ahmedhamdykasem2016@yahoo.com (A.H.A.E.); mohamedahmedmak@yahoo.com (M.K.); mohamed.hosny@moere.gov.eg (M.H.H.); skamel@aswu.edu.eg (S.K.)
2. Department of Electrical Engineering, College of Engineering, Northern Border University, Arar 1321, Saudi Arabia
3. Prince Faisal bin Khalid bin Sultan Research Chair in Renewable Energy Studies and Applications (PFCRE), Northern Border University, Arar 1321, Saudi Arabia
* Correspondence: ah1582009@yahoo.com or ahmad.agua@nbu.edu.sa

Abstract: This study presents an innovative strategy for load frequency control (LFC) using a combination structure of tilt-derivative and tilt-integral gains to form a TD-TI controller. Furthermore, a new improved optimization technique, namely the quantum chaos game optimizer (QCGO) is applied to tune the gains of the proposed combination TD-TI controller in two-area interconnected hybrid power systems, while the effectiveness of the proposed QCGO is validated via a comparison of its performance with the traditional CGO and other optimizers when considering 23 bench functions. Correspondingly, the effectiveness of the proposed controller is validated by comparing its performance with other controllers, such as the proportional-integral-derivative (PID) controller based on different optimizers, the tilt-integral-derivative (TID) controller based on a CGO algorithm, and the TID controller based on a QCGO algorithm, where the effectiveness of the proposed TD-TI controller based on the QCGO algorithm is ensured using different load patterns (i.e., step load perturbation (SLP), series SLP, and random load variation (RLV)). Furthermore, the challenges of renewable energy penetration and communication time delay are considered to test the robustness of the proposed controller in achieving more system stability. In addition, the integration of electric vehicles as dispersed energy storage units in both areas has been considered to test their effectiveness in achieving power grid stability. The simulation results elucidate that the proposed TD-TI controller based on the QCGO controller can achieve more system stability under the different aforementioned challenges.

Keywords: improved chaos game optimization; TD-TI controller; load frequency control; renewable energy sources; electrical vehicles

1. Introduction

Recently, the world has become voracious in utilizing electrical power due to the growth of industrial and residential loads. Therefore, it was necessary to establish new electrical power grids to accommodate the load demands. As a result, energy planners were directed to penetrate the renewable energy sources (RESs) with the traditional power grids in the electrical power system to reduce the demerits of these traditional units. In addition, the penetration of RESs with newly established power systems is considered to have an economically good and positive rate that saves in the utilization of the oil, coal, and gas that operate traditional power plants, whereas the resulting flames from burning oil and coal lead to the release of carbon dioxide gas, causing an increase in the ozone hole and an increase in the global warming phenomenon [1]. Although the presence of RESs in electrical power grids reduces the severity of the resulting pollution from the

traditional units, these renewable sources suffer from a lack of system inertia. As a result of the reduction in power system inertia caused by renewable sources, the stability and security of the system (i.e., more fluctuations in system frequency) will be affected [2,3]. Moreover, several reasons lead to more frequency fluctuations, such as a mismatching between the generated power and the demand power, system parameter variations, and different sorts of load variations. Hence, the fluctuations in system frequency can be tackled by the LFC [4]. Researchers have done their best to develop several control techniques for achieving reliability in power systems by attaining system frequency and tie-line power flow within tolerable limits.

Many interests have been prompted by researchers to address the issue of LFC in different structures of the power system; (i.e., the single-area power system [5,6], the multi-area interconnected power system [7–10], and the deregulated power system [11,12]). In addition, several different control techniques have been implemented to overcome the system frequency fluctuations, such as the intelligent control techniques (i.e., fuzzy logic controllers [13], artificial neural networks [14], and adaptive neuro-fuzzy controllers [15]). Moreover, several robust control techniques have been utilized to enhance the power system performance, such as the H-infinite technique [16] and μ-synthesis [17]. Furthermore, optimal control techniques, such as the linear quadratic Gaussian [18] and linear quadratic regulator [19], are implemented to attain the frequency within tolerable limits. In this regard, the majority of the industrial control loop is the proportional-integral-derivative (PID), due to its reputable merits (i.e., simpleness in construction, applicability, functionality, comfort, and inexpensiveness) [20]. Even so, it suffers from a bulky, complicated process when selecting its parameters using trial and error methods. Thus, researchers have been striving to accomplish the optimal PID controller, according to the different optimization techniques utilized in getting the optimal controller parameters. This design of the optimal PID controller leads to ensuring a reliable system performance in comparison to the conventional PID controller when facing the uncertainties in a studied power grid. Accordingly, several optimization techniques have been utilized to fine-tune the optimal PID controller parameters meticulously, including the grasshopper optimization algorithm [21], the ant colony optimization technique [22], the Jaya algorithm [23], and the class topper optimization algorithm [24].

On the other side, the fractional order controllers (FOCs) have become a distinct candidate in power system stabilizing due to their merits (i.e., flexibility in configuration and a higher degree of freedom). The FOCs have several types of poles, such as the hyper-damped poles, that need to be fine-tuned. Accordingly, this leads to an expansion in the stable region, giving more flexibility in the controller design process [25]. Furthermore, there are several types of controllers belonging to the FOC family; the fractional –order-proportional-integral-derivative (FOPID) is one member of this family that has been presented in [26,27]. The FOPID controller has been utilized in several electrical power systems [28,29]. Moreover, the TID controller represents one of the FOCs; it looks exactly like the PID controller in construction except for one difference, which is that the proportional parameter is tilted with a $(1/s^{1/n})$ transfer function. This additional transfer function provides the optimization process with better feedback and good tracking performance. Lately, the TID controller has been implemented for solving the LFC problem due to its good merits (i.e., it can change the parameters of the closed-loop system; it has a tremendous ability in disturbance rejection; and it has more reliability with robustness) [30,31]. There is no doubt that fractional calculus provides several options to researchers for creativity and diversity in controller designing. As a result, different engineering problems have been solved by utilizing the amalgamation of the FOPID and TID properties as a hybrid controller [32]. In addition, the researchers' minds are destined to implement another strategy in control design, which is the cascaded controllers (CCs) form that includes one controller followed by another one; the CCs have more tuning knobs that give better results than in the utilization of non-crude CCs. Thus, many scientific studies have been presented using the different CCs to solve the LFC problem [33,34]. Another construction has been applied while designing different controllers for studying the LFC issue, which depends on the combination of two different controllers to take the benefits of both controllers. There are examples of the combination of different proposed controllers from literature, such as the combination of the model predictive control (MPC) controller with the linear quadratic Gaussian controller [35] and the combination of an adaptive MPC with the recursive polynomial model estimator [36]. Furthermore, a new controller structure, labeled as a feed-forward/feed-backward controller, has been presented to reduce the disadvantages of the PID and TID controllers during system uncertainties that affect the input of the control signal. Thus, many studies have been presented to elucidate the robustness of the feed-forward/feed-backward controller structure in achieving system stability. The integral-proportional-derivative (I-PD) controller and the integral-tilt-derivative (I-TD) controller have been proposed to cope with the LFC problem, achieving more system stability compared to the PID and TID controllers, respectively [37,38].

The achievement of system stability is not dependent on the controller design only, but the utilized optimization technique represents a critical issue that must be selected carefully to attain the optimal controller parameters. Previously, the traditional optimization methods such as the tracking approach [39] and the aggregation methods [40] were applied for regulating the system frequency. In fact, the traditional optimization methods suffer from several drawbacks, such as slump, deathtrap in local minimums, the need for more iterations, and dependence on their initial conditions to attain the optimal solution. So, meta-heuristic optimization techniques such as the artificial bee colony [41], salp swarm algorithm (SSA) [42], and whale optimization algorithm (WOA) [43] have been proposed to overcome all of the previous drawbacks. Though the meta-heuristic optimization algorithms are not usually guaranteed to find the optimal global solution, they can often find a sufficiently good solution in a reasonable time. So, they are an alternative to exhaustive search, which would take exponential time. Moreover, these techniques have several demerits, such as slowing in the rate of convergence, poor local search capability, and local optimum convergence. In this regard, algorithmic scientists have improved these techniques to diminish all of their previous drawbacks. Examples of improved algorithms utilized to achieve system stability are presented as the improved stochastic fractal search algorithm [44] and the sine augmented scaled sine cosine [45]. In this regard, the authors in this work proposed an improved algorithm known as QCGO to select the suggested combining TD-TI controller parameters to attain the optimal studied power grid performance.

Referring to the aforementioned literature related to the LFC issue, there are several control strategies that depend on the designer experience, such as the MPC, the H-infinite techniques, and the fuzzy logic control, that can attain the desired performance, but their parameter-selecting strategies take a long time. In addition, the conventional PID controller has some difficulties when facing system uncertainties. Moreover, several studies have been presented utilizing conventional algorithms and meta-heuristic optimization techniques that have many demerits in comparison to the improved techniques that develop the searching process and obtain the global solution with a few search agents. Furthermore, several previous studies did not consider the different challenges that face power systems (i.e., different types of load variations such as series SLP and RLV, the high penetration of RESs, and communication time delay). According to the above salient observations, this study proposed a new control construction labeled as a combining TD-TI controller that is derived from the form of a TID controller to enhance the studied system stability. The parameters of the proposed combining TD-TI controller can be selected by utilizing the improved algorithm QCGO when considering the challenges of high RESs penetration, different load perturbation types, and communication time delay.

The studied work in this paper is presented to overcome the limitations of the previously published works in the literature. Table 1 elucidates the differences between this work and the other published works related to the LFC issue.

Table 1. The motivation of current work compared with other published works.

References	[6]	[9]	[28]	[32]	[37]	[38]	This Study
Controller structure	PI/PID controller	PI/PD controller	FOPID/TID controller	Combining of FOPID-TID controller	I-PD controller	I-TD controller	Combining TD-TI controller
Controller design adoption	Firefly algorithm	Backtracking search algorithm	Improved PSO	Manta ray foraging optimization algorithm	Fitness dependent optimizer	Water cycle algorithm	QCGO
Load perturbation challenge	SLP	SLP/RLV	SLP/RLV	SLP/series SLP	SLP	SLP/RLV	SLP/series SLP/RLV
Sort of studied system	Single-area power system	Multi-area power system	Multi-area power system	Multi-area power system	Multi-area power system	Multi-area power system	Multi-area power system
RESs Penetration	Not considered	Not considered	Not considered	considered	Not considered	Not considered	Considered with high penetration
Effect of communication time delay	Not considered	Not considered	Not considered	Not considered	Considered before the action of one control unit only	Not considered	Considered before and after the control action
Effect of EVs	Not considered	Not considered	Not considered	Not considered	Not considered	Not considered	Considered

The main contributions of this work can be elucidated in detail as follows:

i. The proposal of a control structure combining TD-TI controllers for LFC of the hybrid two-area interconnected power systems.

ii. The proposal of a novel technique known as QCGO via improving the quantum mechanics of the CGO algorithm based on the particle swarm optimizer (PSO) to improve the exploration and exploitation strategies of the main CGO algorithm.

iii. The application of the improved CGO to select the optimal parameters of the proposed controller structure.

iv. The validation of the performance of the proposed algorithm through a fair-maiden comparison between the proposed QCGO algorithm and other previous techniques (i.e., Supply-demand-based optimization (SDO), WOA, butterfly optimization algorithm (BOA), and the conventional CGO), based on applying 23 bench functions, as well as a fair comparison between the proposed algorithm and other previous algorithms (i.e., CGO, SSA), considering the proposed controller in the multi-area power grid for frequency stability analysis.

v. The consideration of several challenges, such as the high RESs penetration in both areas, different load perturbation types, and communication time delay to study the system stability state.

vi. The comparison of the performance of the proposed control TD-TI structure based on QCGO with other available controllers, such as the PID-based teaching learning-based optimization (TLBO) [46]; the PID-based arithmetic optimization algorithm (AOA) [47]; the proposed TD-TI control structure based on CGO; the proposed TD-TI control structure based on SSA; the TID controller based on CGO; and the TID controller based on QCGO, is presented to ensure the effectiveness and robustness of the proposed control structure based on the QCGO algorithm in achieving more system reliability and stability.

vii. The consideration of the integration of electrical vehicles (EVs) in both areas to support the proposed controller in overcoming the system frequency excursions during high renewables penetration.

The remainder of this article is organized into several sections that are clarified as follows: the studied system topology which considers the high penetration of RESs and EVs is illustrated in Section 2. Section 3 discusses the proposed control approach and the formulation of the studied problem. Then, the procedure of the improved QCGO technique is given in Section 4. Moreover, the simulation results according to the different scenarios are clarified in Section 5. Finally, Section 6 summarizes the conclusions of the current work.

2. The Studied System Topology
2.1. Two-Area Interconnected Hybrid Power Grid Configuration

In this article, the issue of LFC related to electrical power grids has been addressed by conducting a study on two-area interconnected hybrid power systems. The studied power grid encompasses two interconnected areas, which include several conventional generation power plants, such as the thermal unit, hydropower unit, and gas unit. The capacity of each area in the studied power grid that includes the three traditional units (i.e., thermal, hydro, and gas) is 2000 MW of rated power [48], of which the largest percentage of electrical power sharing went to the thermal power plant, which contributes 1087 MW, then the hydropower plant, which contributes 653 MW, and the gas turbine, sharing the generated power with 262 MW. The investigated power grid is presented as a simplified model shown in Figure 1.

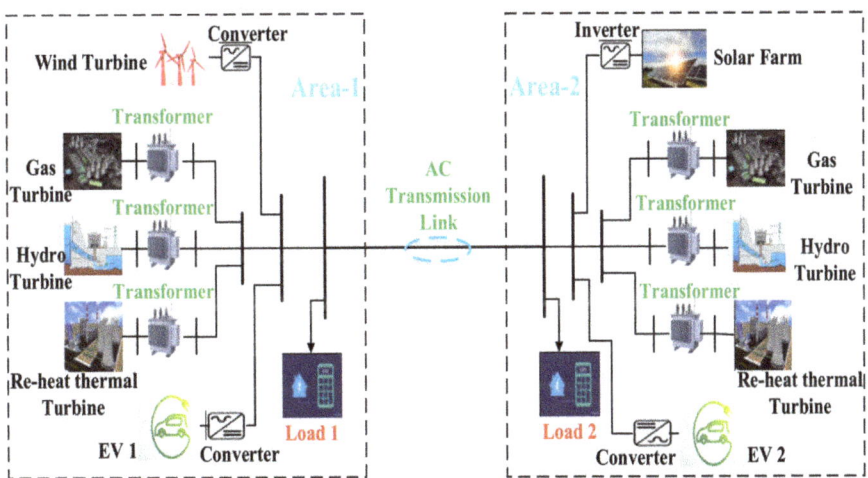

Figure 1. The studied power grid schematic diagram.

Figure 2 shows the block diagram of the studied two-area interconnected hybrid power grid. The transfer functions in the studied power grid are listed in Table 2. The amalgamation of the TD-TI controller is proposed to be equipped in both areas for each generation unit to minimize the oscillations in the frequencies of both areas and the tie-line power flow between them. The attitude of the input signal of the proposed combining TD-TI controller can be represented as the *ACE*, while the attitude of the output signal can be represented as the action of the secondary/supplementary control on each generation power plant, in order to obtain extra active power for enhancing the power grid performance. Table 3 elucidates all the parameters included in the studied power grid with their nominal values. The *ACEs* in both areas can be obtained according to the formulas that follow in Equations (1) and (2) [47]:

$$ACE_1 = \Delta C_{tie1-2} + B_1 \Delta i_1 \quad (1)$$

$$ACE_2 = \Delta C_{tie2-1} + B_2 \Delta i_2 \quad (2)$$

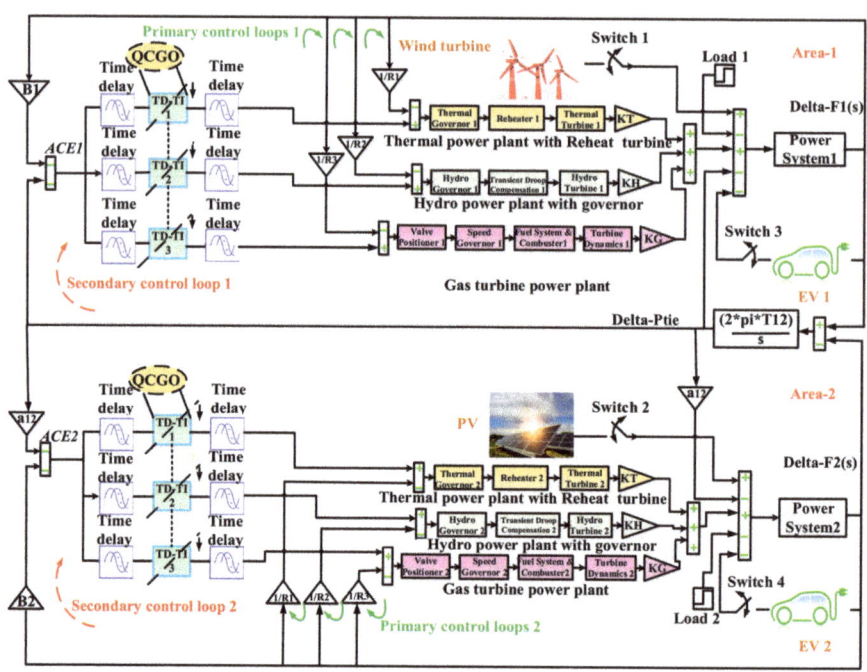

Figure 2. The transfer function model of the studied power grid.

Table 2. The transfer functions that are presented in the studied power grid.

Control Block	Transfer Functions
Thermal Governor	$\dfrac{1}{T_{sg} \cdot s + 1}$
Reheater of Thermal Turbine	$\dfrac{K_r \cdot T_r \cdot s + 1}{T_r \cdot s + 1}$
Thermal Turbine	$\dfrac{1}{T_t \cdot s + 1}$
Hydro Governor	$\dfrac{1}{T_{gh} \cdot s + 1}$
Transient Droop Compensation	$\dfrac{T_{rs} \cdot s + 1}{T_{rh} \cdot s + 1}$
Hydro Turbine	$\dfrac{-T_w \cdot s + 1}{0.5 T_w \cdot s + 1}$
Valve Positioner of Gas Turbine	$\dfrac{1}{b_g \cdot s + c_g}$
Speed Governor of Gas Turbine	$\dfrac{x_c \cdot s + 1}{Y_c \cdot s + 1}$
Fuel System and Combustor	$\dfrac{T_{cr} \cdot s + 1}{T_{fc} \cdot s + 1}$
Gas Turbine Dynamics	$\dfrac{1}{T_{cd} \cdot s + 1}$

Table 2. Cont.

Control Block	Transfer Functions
Power System 1	$\dfrac{K_{ps1}}{T_{ps1}\cdot s+1}$
Power System 2	$\dfrac{K_{ps2}}{T_{ps2}\cdot s+1}$
Electrical Vehicle 1	$\dfrac{K_{EV1}}{T_{EV1}\cdot s+1}$
Electrical Vehicle 2	$\dfrac{K_{EV2}}{T_{EV2}\cdot s+1}$

Table 3. The standard parameter values of the two interconnected identical areas [47].

Parameter Descriptions	Symbol	Standard Values
Frequency bias factor	B_i	0.4312 MW/Hz
Coefficient of synchronizing	T_{12}	0.0433 MW
The regulation constant of thermal turbine	R_1	2.4 HZ/MW
The regulation constant of hydropower plant	R_2	2.4 HZ/MW
The regulation constant of gas turbine	R_3	2.4 HZ/MW
Control area capacity ratio	a_{12}	−1
Participation factor for a thermal unit	K_T	0.543478
Participation factor for a hydro unit	K_H	0.326084
Participation factor for a gas unit	K_G	0.130438
Gain constant of power system	K_{ps}	68.9566
The time constant of the power system	T_{ps}	11.49 s
Governor time constant	T_{sg}	0.08 s
Turbine time constant	T_t	0.3 s
Gain of reheater steam turbine	K_r	0.3
The time constant of reheater steam turbine	T_r	10 s
Speed governor time constant of hydro turbine	T_{gh}	0.2 s
Speed governor reset time of the hydro turbine	T_{rs}	5 s
The transient droop time constant of hydro turbine speed governor	T_{rh}	28.75 s
Nominal string time of water in penstock	T_w	1 s
Gas turbine constant of valve positioner	b_g	0.05
Valve positioner of gas turbine	c_g	1
The lag time constant of the gas turbine speed governor	Y_c	1 s
The lead time constant of the gas turbine speed governor	X_c	0.6 s
Gas turbine combustion reaction time delay	T_{cr}	0.01 s
Gas turbine fuel time constant	T_{fc}	0.23 s
Gas turbine compressor discharge volume–time constant	T_{cd}	0.2 s
Gain of electrical vehicle	K_{EV}	1
The time constant of electrical vehicle	T_{EV}	0.28 s

2.2. The Installation of Wind Farm Model

This work presents the high penetration of RESs, including wind power in the investigated hybrid power grid. The professional software MATLAB/SIMULINK program (R2015a) (The MathWorks, Inc., Natick, MA, USA) is used in implementing the simplified model of wind power in order to share its energy in the first area of the studied power grid. The aforementioned wind power model generates power in the same way as the real behavior of the generated power from real wind farms. This is achieved using a white-noise block that is utilized in getting a random speed, which is multiplied by the wind speed, as shown in Figure 3 [47]. The captured output power from the wind model can be formulated in the following equations [47].

$$P_{wt} = \frac{1}{2}\rho A_T v_w^3 C_p(\lambda, \beta) \qquad (3)$$

$$C_p(\lambda, \beta) = C_1\left(\frac{C_2}{\lambda_i} - C_3\beta - C_4\beta^2 - C_5\right) \times e^{\frac{-C_6}{\lambda_i}} + C_7\lambda_T \qquad (4)$$

$$\lambda_T = \lambda_T^{OP} = \frac{\omega_T r_T}{V_W} \qquad (5)$$

$$\frac{1}{\lambda_i} = \frac{1}{\lambda_T + 0.08\beta} - \frac{0.035}{\beta^3 + 1} \qquad (6)$$

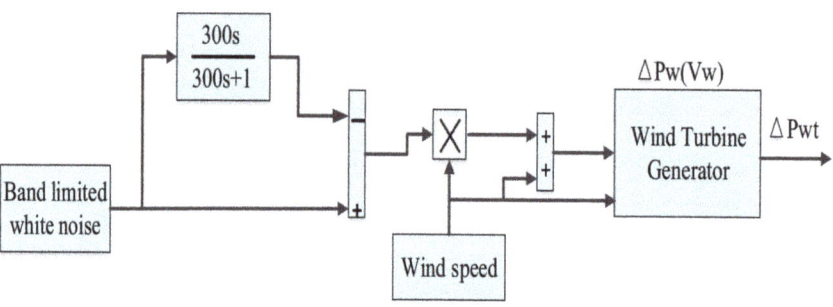

Figure 3. The implemented model of wind power using MATLAB/Simulink program (R2015a).

All of these mentioned parameter values for the utilized wind farm are presented in [47]. Figure 4 shows the random output power of 257 wind turbine units of 750 KW for each wind power unit. The value of the generated power from the studied wind farm is about 192 MW.

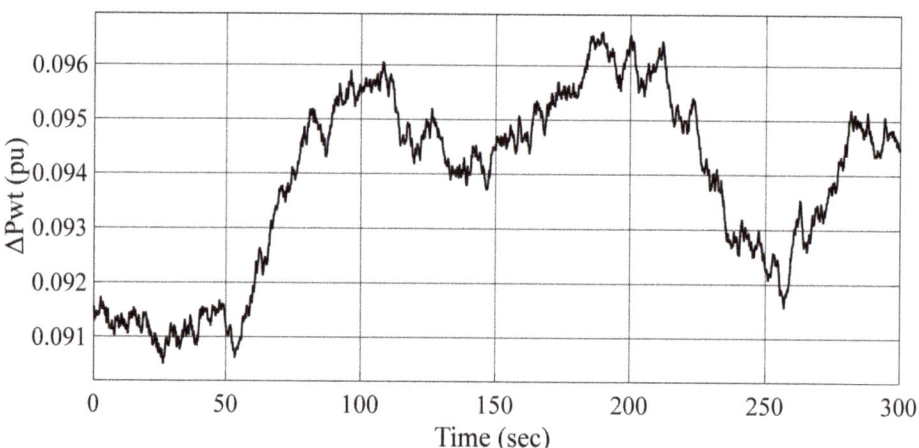

Figure 4. The output power of the wind model.

2.3. The Installation of the PV Model

The Photovoltaic (PV) model can be built by utilizing the professional software MATLAB/SIMULINK program (R2015a) described in Figure 5. The generated output power from the model is similar to the real generated output power from a real PV plant. In addition, the output energy of the PV model is penetrated in the second area of the studied power grid at about 116 MW. Here, the white-noise block in the MATLAB program (R2015a) is used for obtaining random output oscillations that are multiplied by the standard output power generated from a real PV plant. The generated energy from the presented PV model can be obtained as formulated in Equation (7) [6]. Figure 6 clarifies the random output power generated from the PV model.

$$\Delta P_{solar} = 0.6 \times \sqrt{P_{solar}} \tag{7}$$

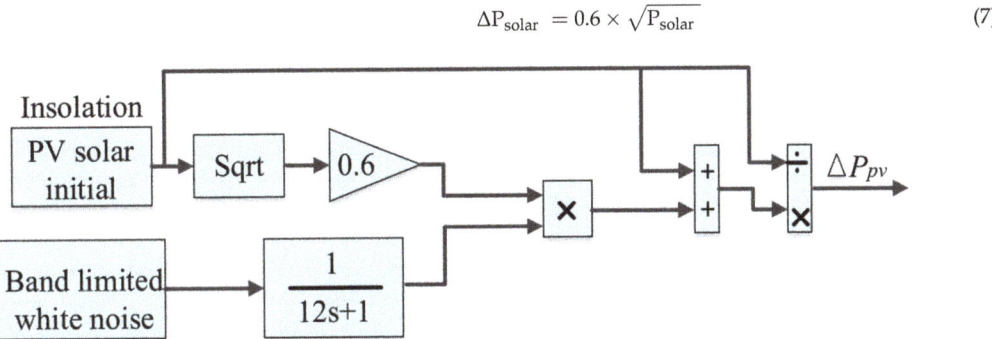

Figure 5. The implemented model of the solar power plant using MATLAB/Simulink (R2015a) program.

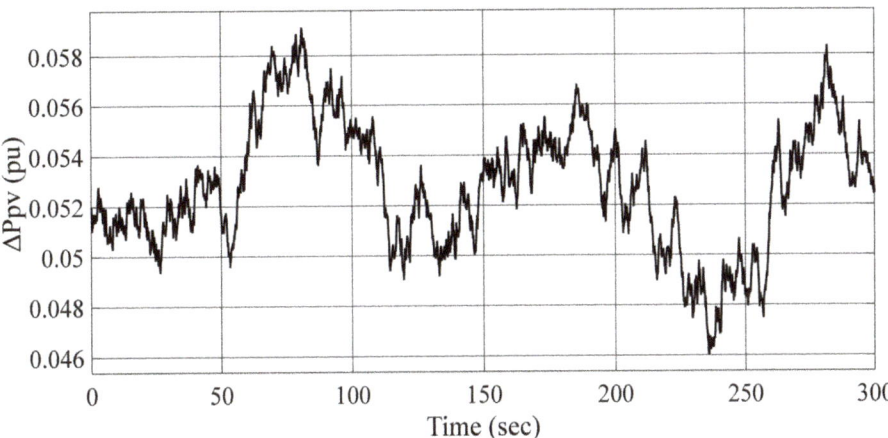

Figure 6. The output power of the photovoltaic model.

2.4. The Installation of EV Model

EVs can participate in frequency regulation effectively due to the receiving of the LFC order and pass this signal to the EV to control the power during the charging and discharging process. Moreover, the response of the LFC signal can be limited through the availability of the numbers of controllable EVs in the studied power grid and by the state of the charge related to their capacity, whereas the model of the EV is similar to the model of the battery energy storage system, due to the included batteries that supply extra energy to the power grid during fluctuations for regulating the frequency excursions. However, the batteries in EVs may not be in full charging capacity due to the nature of EVs being of mobility and load, which affects the amount of extra energy to tackle the LFC problem. Thus, it is important to check the level of the EV charging to ensure more system enhancement under different system fluctuations. The output power from an EV can be obtained by the first-order transfer function, including the electrical vehicle time constant T_{EV}, which equals 0.28 s in series with the electrical vehicle controllers' gain, K_{EV}, which equals 1, where K_{EV} is represented as the ratio of the exchange in charging power of the EV's batteries to the change of system frequency. The transfer function that represents the EV model is formulated in Equation (8) [49]. Figure 7 describes the EV model that was built in the MATLAB/SIMULINK program (R2015a).

$$\frac{K_{EV}}{1 + s\, T_{EV}} \tag{8}$$

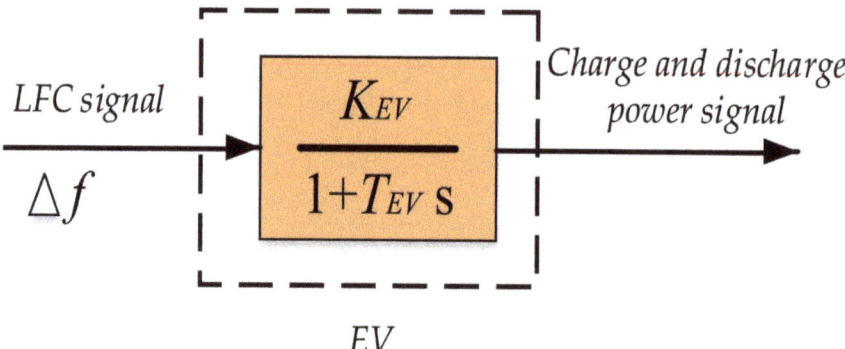

Figure 7. The implemented model of the electrical vehicle using MATLAB/Simulink (R2015a).

3. Control Methodology and Problem Formulation

Due to the high RES penetration, communication time delay, and various types of load perturbations, it is essential to implement a robust controller to enhance the system performance during abnormal conditions. Hence, this study proposes a newly developed controller construction known as a combining TD-TI controller to overcome any fluctuations resulting from the previous considerations/challenges. Moreover, the proposed controller parameters have been selected based on an improved algorithm labeled as QCGO.

3.1. The Proposed Control Strategy

This paper presents an efficient controller labeled as the combining TD-TI controller, which represents an improved modified structure of the TID controller that is shown in Figure 8. The TID controller is a sort of fractional order controller (FOC) that depends on the fractional-order calculus in its design. The TID controller construction is similar to the PID controller construction except for one difference, which is that the proportional parameter is tilted with a $(1/s^{1/n})$ transfer function. In this regard, this paper proposed a combining TD-TI controller, as derived from TID controller, due to the merits of the TID, such as the ability to tune easily, superior fluctuations rejection, and better sensitivity due to variations of the system parametric [50]. The proposed combining TD-TI controller is utilized to enhance the studied power grid performance, such as by damping frequency oscillations in both areas and overcoming fluctuations related to the tie-line power flow. Furthermore, the proposed combining TD-TI controller parameters are selected utilizing an improved QCGO algorithm. In general, the transfer function of the combining TD-TI controller is formulated as follows [50]:

$$G_{i1}, \text{TD}(s) = \frac{Kt_i}{s^{\frac{1}{n}}} + Kd_i s \tag{9}$$

$$G_{i2}, \text{TI}(s) = \frac{Kt_i}{s^{\frac{1}{n}}} + \frac{Ki_i}{s} \tag{10}$$

$$G_i, \text{total}(s) = G_{i1}, \text{TD}(s) + G_{i2}, \text{TI}(s) \tag{11}$$

where *i* refers to the specified proposed controller of the (thermal, hydro, and gas) turbines; thus, (*i* = 1, 2, 3). The gain values (Kt_i, Ki_i, and Kd_i) are selected within the range of [0, 10], and n is tuned in the range of [1, 10]. The control signal of the i_{th} area can be expressed as follows [38]:

$$U_i(s) = G_i, \text{total}(s) \times ACE_i(s) \tag{12}$$

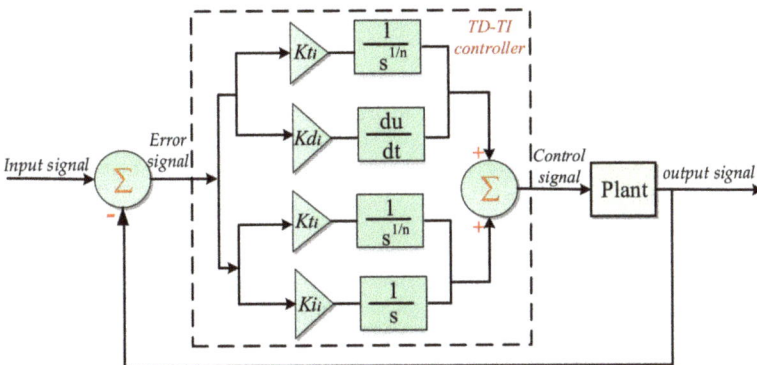

Figure 8. The construction of the proposed combining tilt-derivative and tilt-integral controller.

According to the process of controller designing, there are several sorts of performance criteria, such as the integral time absolute error (*ITAE*), the integral of squared error (*ISE*), the integral time squared error (*ITSE*), and the integral of absolute error (*IAE*). The criteria of *ITAE* and *ISE* are often utilized in the literature for minimizing the objective function due to their merits in comparison to *ITSE* and *IAE*, whereas the strategy of the *ISE* criteria in minimizing the objective function is the integrating of the square of error signal over simulation time. For ease, the *ISE* criteria can effectively dampen the large errors compared to the small errors as the square of the large errors is larger than the square of the small errors. It can be said that the *ISE* criteria can penalize the large errors with tolerance for the presence of continuous small errors along with time simulation. Thus, the authors of this work do not hesitate in putting in the *ITAE* criteria utilized in minimizing the objective function because of the multiplication of the time term by the integral of the absolute error. The multiplied time term in *ITAE* criteria makes the optimization process more fast which achieves more system stability than utilizing the *ISE* criteria [51]. The *ITAE* criteria can be formulated as follows [47]:

$$J = ITAE = \int_0^{Tsim} t \cdot [|\Delta f_1| + |\Delta f_2| + |\Delta P_{tie}|] \, dt \qquad (13)$$

where dt is represented as the time interval for taking the error signals' samples over the simulation process.

3.2. The Proposed Optimization Technique

In this subsection, the CGO method is briefly described; then, the process of the QCGO technique is presented.

3.2.1. Chaos Game Optimization (CGO) Algorithm

This algorithm is based on certain rules of the chaos theory, where the arrangement of fractals is by the chaos game idea. Firstly, an initialization procedure is configured by determining the initial positions of the solution candidates from the following equations [52]:

$$X = \begin{bmatrix} X_1 \\ X_2 \\ \vdots \\ X_i \\ \vdots \\ X_n \end{bmatrix} = \begin{bmatrix} x_1^1 \; x_1^2 \; \ldots \; x_1^j \; \ldots \; x_1^d \\ x_2^1 \; x_2^2 \; \ldots \; x_2^j \; \ldots \; x_2^d \\ \vdots \\ x_i^1 \; x_i^2 \; \ldots \; x_i^j \; \ldots \; x_i^d \\ \vdots \\ x_n^1 \; x_n^2 \; \ldots \; x_n^j \; \ldots \; x_n^d \end{bmatrix}, \; \begin{cases} i = 1,2,\ldots,m \\ j = 1,2,\ldots,d \end{cases} \qquad (14)$$

$$x_i^j(0) = x_{i,min}^j + \text{rand.}\left(x_{i,max}^j - x_{i,min}^j\right), \quad \begin{cases} i = 1, 2, \ldots, m \\ j = 1, 2, \ldots, d \end{cases} \tag{15}$$

where d denotes the dimension of the problem and m refers to the total number of initialized candidates inside the search space. $x_{i,min}^j$, $x_{i,max}^j$ are the lower and upper bounds of the decision variables. The position updating process for the temporary triangles is presented in Figure 9. The mathematical representation of the $seed_i^1$, as shown in Figure 9a, is as follows [52]:

$$seed_i^1 = X_i + \alpha_i \times (\beta_i - GB - \gamma_i \times MG_i), \quad i = 1, 2, \ldots, m \tag{16}$$

where GB is the global best, α_i represents the movement limitation factor, and β_i and γ_i denote vectors randomly created by numbers in the range of [0, 1]. MG_i is the mean group. From Figure 9b, $seed_i^2$ can be calculated as follows [52]:

$$seed_i^2 = GB + \alpha_i \times (\beta_i \times X_i - \gamma_i \times MG_i), \quad i = 1, 2, \ldots, m \tag{17}$$

While $seed_i^3$, which is displayed in Figure 9c, is mathematically computed as below [52]:

$$seed_i^3 = MG_i + \alpha_i \times (\beta_i \times X_i - \gamma_i \times GB), \quad i = 1, 2, \ldots, m \tag{18}$$

Finally, $seed_i^4$, which is shown in Figure 9d, can be mathematically represented as follows [52]:

$$seed_i^4 = X_i\left(x_i^k = x_i^k + R\right), \quad k = [1, 2, \ldots, d] \tag{19}$$

where R refers to a vector with random numbers in the range of [0, 1].

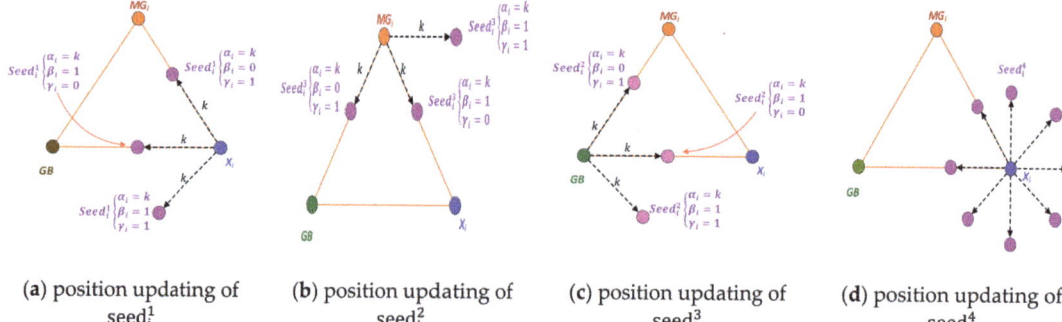

(a) position updating of $seed_i^1$

(b) position updating of $seed_i^2$

(c) position updating of $seed_i^3$

(d) position updating of $seed_i^4$

Figure 9. Position updating process for the temporary triangles [53].

3.2.2. The Proposed Quantum Chaos Game Optimization (QCGO) Algorithm

In this subsection, quantum mechanics is used to develop the original CGO algorithm. This quantum model of a CGO algorithm is called here QCGO algorithm. Quantum mechanics was employed to develop the PSO in [54]. In the quantum model, by employing the Monte Carlo method, the solution x_{new4} is calculated from this equation [54]:

If $h \geq 0.5$

$$x_{new1} = p + \alpha \cdot |Mbest_i - X_i| \cdot \ln(1/u) \tag{20}$$

Else

$$x_{new1} = p - \alpha \cdot |Mbest_i - X_i| \cdot \ln(1/u) \tag{21}$$

End.

where α refers to a design parameter, u and h denote uniform probability distribution in the range [0, 1], and Mbest is the mean best of the population and is defined as the mean of the GB positions. It can be calculated as follows [54]:

$$\text{Mbest} = \frac{1}{N}\sum_{l=1}^{N} P_{g,l}(i) \qquad (22)$$

where g is the index of the best solution among all the solutions.

4. The Procedure of the Improved QCGO Algorithm

The Performance of QCGO

The proposed QCGO algorithm competency and performance are evaluated on the numerous benchmark functions, using the statistical measurements, such as best values, mean values, median values, worst values, and standard deviation (STD), for the best solutions achieved using the proposed technique and the other well-known algorithms. The results attained by the QCGO technique are compared with three recent meta-heuristic techniques, including SDO [55], WOA [56], and BOA [57], in addition to the conventional CGO. All of the mentioned techniques were executed for the maximum number of iterations of the function of 200 and a population size of 50 for 20 independent runs, using Matlab R2016a working on Windows 8.1, 64 bit (Microsoft, Albuquerque, NM, USA). All computations were performed on a Core i5-4210U CPU@ 2.40 GHz of speed (Intel Corporation, Santa Clara, CA, USA) and 8 GB of RAM. Figure 10 shows the qualitative metrics on F1, F2, F3, F5, F6, F8, F10, F12, F15, F18, and F22, with 2D views of the functions, convergence curve, average fitness history, and search history.

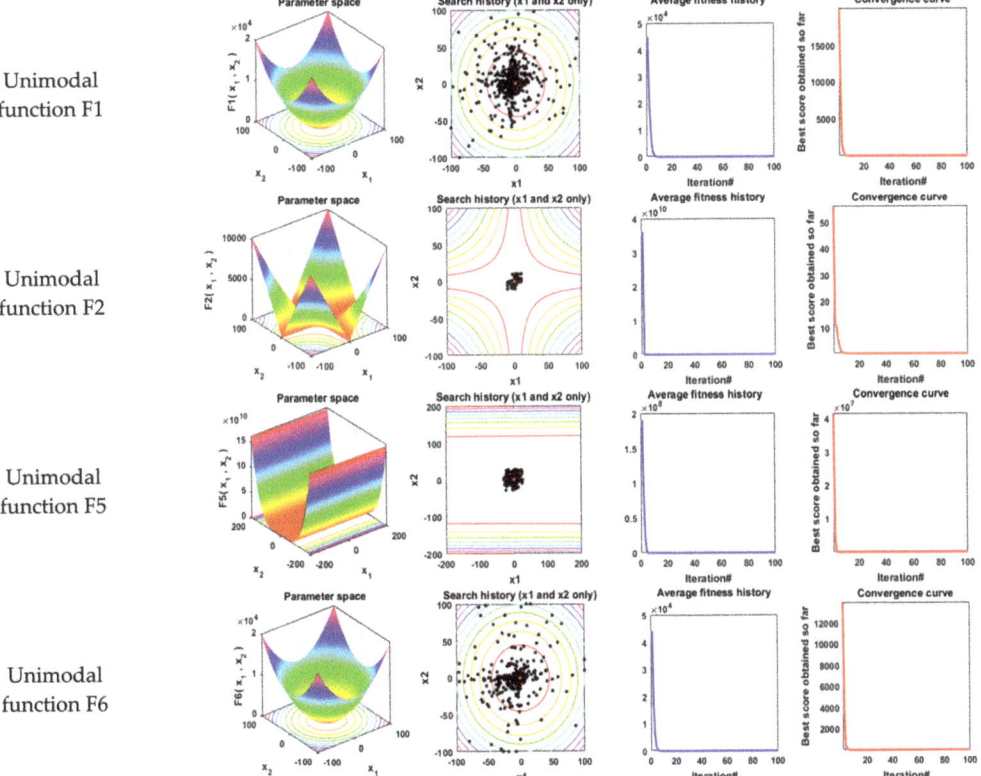

Figure 10. *Cont.*

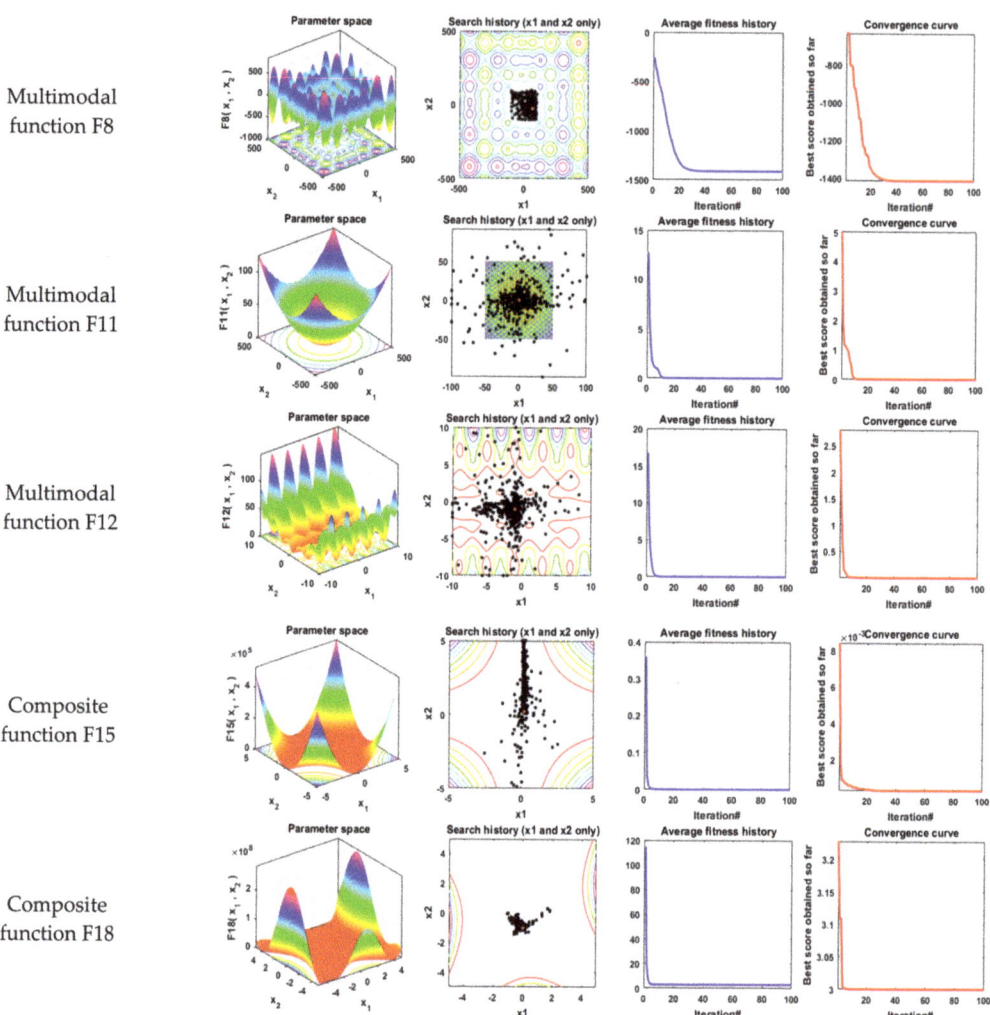

Figure 10. Qualitative metrics of nine benchmark functions using the proposed quantum chaos game optimizer algorithm: 2D views of the functions, search history, average fitness history, and convergence curve.

Tables 4–6 show the statistical results of the proposed QCGO technique and other algorithms when applied for the three types of benchmark functions (unimodal, multimodal, and composite, respectively). The best-obtained values using the QCGO, CGO, SDO, WOA, and BOA algorithms are displayed in bold. It is clearly seen that the QCGO algorithm achieves the optimal solution for most of those benchmark functions. The convergence curves of these techniques for those functions are illustrated in Figure 11, and the boxplots for each algorithm for these functions are displayed in Figure 12. From those figures, it is seen that the QCGO technique reached a stable point for all functions, and the boxplots of the proposed QCGO technique are very narrow and stable for most functions compared to the other techniques.

Table 4. Results of unimodal benchmark functions.

Function		QCGO	CGO	SDO	WOA	BOA
F1	Best	**2.4 $\times 10^{-126}$**	1.52 $\times 10^{-58}$	1.39 $\times 10^{-55}$	1.92 $\times 10^{-40}$	3.87 $\times 10^{-08}$
	Mean	**1.4 $\times 10^{-122}$**	4.97 $\times 10^{-55}$	1.37 $\times 10^{-51}$	7.2 $\times 10^{-34}$	4.96 $\times 10^{-08}$
	Median	**4.8 $\times 10^{-124}$**	3.86 $\times 10^{-56}$	3.74 $\times 10^{-54}$	2.28 $\times 10^{-35}$	4.95 $\times 10^{-08}$
	Worst	**1.2 $\times 10^{-121}$**	3.9 $\times 10^{-54}$	8.43 $\times 10^{-51}$	4.34 $\times 10^{-33}$	6 $\times 10^{-08}$
	Std	**3.7 $\times 10^{-122}$**	9.85 $\times 10^{-55}$	2.74 $\times 10^{-51}$	1.34 $\times 10^{-33}$	4.94 $\times 10^{-09}$
F2	Best	**4.2 $\times 10^{-65}$**	3.64 $\times 10^{-31}$	1.83 $\times 10^{-29}$	4.41 $\times 10^{-24}$	4.26 $\times 10^{-06}$
	Mean	**1.85 $\times 10^{-63}$**	9.17 $\times 10^{-29}$	3.76 $\times 10^{-25}$	5.82 $\times 10^{-21}$	5.71 $\times 10^{-06}$
	Median	**6.63 $\times 10^{-64}$**	1.96 $\times 10^{-29}$	1.13 $\times 10^{-26}$	1.34 $\times 10^{-21}$	5.77 $\times 10^{-06}$
	Worst	**7.99 $\times 10^{-63}$**	9.73 $\times 10^{-28}$	3.98 $\times 10^{-24}$	5.99 $\times 10^{-20}$	7.58 $\times 10^{-06}$
	Std	**2.41 $\times 10^{-63}$**	2.23 $\times 10^{-28}$	9.1 $\times 10^{-25}$	1.34 $\times 10^{-20}$	9.92 $\times 10^{-07}$
F3	Best	2.68 $\times 10^{-42}$	2.41 $\times 10^{-40}$	**6.27 $\times 10^{-46}$**	0.027608	3.85 $\times 10^{-08}$
	Mean	1.66 $\times 10^{-36}$	6.69 $\times 10^{-37}$	**6.91 $\times 10^{-34}$**	1.518335	4.67 $\times 10^{-08}$
	Median	4.45 $\times 10^{-39}$	1.39 $\times 10^{-38}$	**1.4 $\times 10^{-39}$**	1.011391	4.61 $\times 10^{-08}$
	Worst	1.82 $\times 10^{-35}$	7.13 $\times 10^{-36}$	**1.38 $\times 10^{-32}$**	3.914695	5.57 $\times 10^{-08}$
	Std	4.41 $\times 10^{-36}$	**1.68 $\times 10^{-36}$**	3.09 $\times 10^{-33}$	1.18435	5.02 $\times 10^{-09}$
F4	Best	**5.12 $\times 10^{-53}$**	3.76 $\times 10^{-37}$	1.11 $\times 10^{-26}$	0.99528	8.45 $\times 10^{-06}$
	Mean	**6.71 $\times 10^{-51}$**	3.7 $\times 10^{-23}$	4.52 $\times 10^{-23}$	53.18395	1.02 $\times 10^{-05}$
	Median	**2.13 $\times 10^{-51}$**	1.4 $\times 10^{-23}$	1.14 $\times 10^{-23}$	60.93168	1.02 $\times 10^{-05}$
	Worst	**3.32 $\times 10^{-50}$**	1.81 $\times 10^{-22}$	1.94 $\times 10^{-22}$	89.09969	1.15 $\times 10^{-05}$
	Std	**9.43 $\times 10^{-51}$**	5.38 $\times 10^{-23}$	6.34 $\times 10^{-23}$	29.69543	8.51 $\times 10^{-07}$
F5	Best	18.11582	**17.11845**	27.90967	27.88483	28.89058
	Mean	**19.57861**	19.61026	28.65096	28.27419	28.92369
	Median	19.35622	**19.29265**	28.74726	28.43647	28.91978
	Worst	22.2175	21.59463	28.98699	**28.7227**	28.96927
	Std	1.149609	1.224882	0.295026	0.28925	**0.021273**
F6	Best	**1.75 $\times 10^{-14}$**	6.75 $\times 10^{-14}$	0.039957	0.303542	4.311051
	Mean	**2.86 $\times 10^{-12}$**	2.63 $\times 10^{-12}$	2.568541	0.655907	5.211726
	Median	**7.7 $\times 10^{-14}$**	6.23 $\times 10^{-13}$	2.038779	0.62203	5.06303
	Worst	4.89 $\times 10^{-11}$	**2.57 $\times 10^{-11}$**	7.250251	1.16408	6.168001
	Std	1.09 $\times 10^{-11}$	**6.11 $\times 10^{-12}$**	1.852701	0.210811	0.509499
F7	Best	1.02 $\times 10^{-05}$	0.000197	8.66 $\times 10^{-05}$	**0.0004**	0.000983
	Mean	0.000263	**0.00092**	0.002356	0.00542	0.002696
	Median	0.000231	**0.00085**	0.001136	0.003763	0.002776
	Worst	**0.000768**	0.001975	0.013813	0.019069	0.005116
	Std	**0.000177**	0.000583	0.003331	0.005011	0.001104

The best values obtained are in bold.

Table 5. Results of multimodal benchmark functions.

Function		QCGO	CGO	SDO	WOA	BOA
F8	Best	−1671.01	−1770.26	−1655	**−1909.05**	−921.028
	Mean	−1465.24	−1490.19	−1312.83	**−1786.9**	−766.513
	Median	−1453.48	−1483.32	−1385.86	**−1907.06**	−778.594
	Worst	−1313.6	−1235.22	−598.802	**−1632.06**	−647.792
	Std	108.2831	123.7418	294.008	138.0759	**61.76107**
F9	Best	**0.00**	**0.00**	4.33 $\times 10^{-30}$	**0.00**	5.17 $\times 10^{-09}$
	Mean	**0.00**	**0.00**	1.75 $\times 10^{-22}$	1.14 $\times 10^{-14}$	0.003376
	Median	**0.00**	**0.00**	4.17 $\times 10^{-25}$	**0.00**	3.86 $\times 10^{-06}$
	Worst	**0.00**	**0.00**	3.02 $\times 10^{-21}$	1.14 $\times 10^{-13}$	0.047754
	Std	**0.00**	**0.00**	6.75 $\times 10^{-22}$	2.97 $\times 10^{-14}$	0.010836
F10	Best	8.88 $\times 10^{-16}$	8.88 $\times 10^{-16}$	8.88 $\times 10^{-16}$	**4.44 $\times 10^{-15}$**	1.67 $\times 10^{-05}$
	Mean	2.49 $\times 10^{-15}$	3.2 $\times 10^{-15}$	**8.88 $\times 10^{-16}$**	1.33 $\times 10^{-14}$	4.77 $\times 10^{-05}$
	Median	**8.88 $\times 10^{-16}$**	4.44 $\times 10^{-15}$	**8.88 $\times 10^{-16}$**	1.15 $\times 10^{-14}$	4.55 $\times 10^{-05}$
	Worst	4.44 $\times 10^{-15}$	4.44 $\times 10^{-15}$	**8.88 $\times 10^{-16}$**	3.29 $\times 10^{-14}$	7.94 $\times 10^{-05}$
	Std	1.81 $\times 10^{-15}$	1.74 $\times 10^{-15}$	**0.00**	8.11 $\times 10^{-15}$	1.69 $\times 10^{-05}$

Table 5. Cont.

Function		QCGO	CGO	SDO	WOA	BOA
F11	Best	0.00	0.00	0.00	0.00	3.23×10^{-08}
	Mean	0.00	0.00	0.00	0.021832	4.29×10^{-08}
	Median	0.00	0.00	0.00	0.00	4.22×10^{-08}
	Worst	0.00	0.00	0.00	0.26626	5.81×10^{-08}
	Std	0.00	0.00	0.00	0.068973	6.29×10^{-09}
F12	Best	$\mathbf{3.66 \times 10^{-16}}$	1.34×10^{-15}	0.001152	0.006052	0.33315
	Mean	$\mathbf{5.69 \times 10^{-15}}$	8.04×10^{-14}	0.23467	0.022239	0.565424
	Median	$\mathbf{2.26 \times 10^{-15}}$	1.93×10^{-14}	0.067805	0.015529	0.562862
	Worst	$\mathbf{3.32 \times 10^{-14}}$	5.01×10^{-13}	1.492821	0.087947	0.754521
	Std	$\mathbf{8.06 \times 10^{-15}}$	1.36×10^{-13}	0.352063	0.018774	0.108748
F13	Best	$\mathbf{6.36 \times 10^{-14}}$	7.4×10^{-13}	0.046216	0.400281	2.497296
	Mean	**0.007142**	0.036733	1.867552	0.687522	2.894224
	Median	**0.005494**	0.010987	1.934246	0.598054	2.982946
	Worst	**0.043949**	0.233414	2.999924	1.321352	3.109356
	Std	0.010254	0.065978	0.961284	0.248523	**0.153028**

The best values obtained are in bold.

Table 6. Results of composite benchmark functions.

Function		QCGO	CGO	SDO	WOA	BOA
F14	Best	**0.998004**	**0.998004**	**0.998004**	**0.998004**	**0.998004**
	Mean	**0.998004**	**0.998004**	3.494696	2.230204	1.301281
	Median	**0.998004**	**0.998004**	1.495017	1.495017	1.024436
	Worst	**0.998004**	**0.998004**	12.67051	10.76318	2.983027
	Std	**0.00**	5.09×10^{-17}	3.953203	2.241367	0.534994
F15	Best	0.000307	0.000307	0.000307	0.000311	0.000315
	Mean	**0.000307**	0.000353	0.00067	0.000626	0.000487
	Median	**0.000307**	0.000307	0.000527	0.000578	0.000405
	Worst	**0.000307**	0.001223	0.002121	0.001528	0.000917
	Std	$\mathbf{1.68 \times 10^{-19}}$	0.000205	0.000473	0.000342	0.000173
F16	Best	−1.03163	−1.03163	−1.03163	−1.03163	**−1.40747**
	Mean	−1.03163	−1.03163	−1.03005	−1.03163	**−1.18199**
	Median	−1.03163	−1.03163	−1.03163	−1.03163	**−1.18517**
	Worst	−1.03163	−1.03163	−1.00046	−1.03163	**−1.07213**
	Std	2.22×10^{-16}	2.28×10^{-16}	0.006966	$\mathbf{1.94 \times 10^{-08}}$	0.088213
F17	Best	0.397887	0.397887	0.397887	0.397887	0.398293
	Mean	**0.397887**	**0.397887**	0.397987	0.397896	0.409332
	Median	**0.397887**	**0.397887**	0.397887	0.39789	0.406611
	Worst	**0.397887**	**0.397887**	0.399295	0.397967	0.461881
	Std	**0.00**	**0.00**	0.000426	1.78×10^{-05}	0.014049
F18	Best	**3.00**	**3.00**	**3.00**	3.000001	3.000586
	Mean	**3.00**	**3.00**	3.001185	3.000069	3.092676
	Median	**3.00**	**3.00**	**3.00**	3.000026	3.054728
	Worst	**3.00**	**3.00**	3.023537	3.000668	3.425476
	Std	$\mathbf{2.7 \times 10^{-16}}$	6.03×10^{-16}	0.005261	0.000147	0.108993
F19	Best	−0.30048	−0.30048	−0.30048	−0.30048	−0.30048
	Mean	−0.30048	−0.30048	−0.2893	−0.30048	−0.30048
	Median	−0.30048	−0.30048	−0.30038	−0.30048	−0.30048
	Worst	−0.30048	−0.30048	−0.19165	−0.30048	−0.30048
	Std	$\mathbf{1.14 \times 10^{-16}}$	$\mathbf{1.14 \times 10^{-16}}$	0.026531	$\mathbf{1.14 \times 10^{-16}}$	3.74×10^{-06}
F20	Best	**−3.322**	**−3.322**	**−3.322**	−3.31923	-3.3×10^{-05}
	Mean	**−3.26849**	−3.28038	−3.09697	−2.98949	-1.6×10^{-06}
	Median	**−3.322**	**−3.322**	−3.2031	−3.15019	-1.5×10^{-40}
	Worst	**−3.2031**	**−3.2031**	−0.89904	−1.57922	-2×10^{-134}
	Std	0.060685	0.058182	0.550986	0.479795	$\mathbf{7.35 \times 10^{-06}}$

Table 6. Cont.

Function		QCGO	CGO	SDO	WOA	BOA
F21	Best	−10.1532	−10.1532	−10.1532	−10.1528	−4.61081
	Mean	−10.1532	−9.90058	−8.703	−7.35262	−4.0759
	Median	−10.1532	−10.1532	−10.1532	−10.0113	−4.12522
	Worst	−10.1532	−5.10077	−4.99677	−2.59723	−3.18003
	Std	3.21×10^{-15}	1.129757	2.23952	3.245445	0.379957
F22	Best	−10.4029	−10.4029	−10.4029	−10.4008	−4.76031
	Mean	−10.4029	−10.4029	−8.45822	−7.90953	−3.74931
	Median	−10.4029	−10.4029	−10.4029	−10.2376	−3.64889
	Worst	−10.4029	−10.4029	−1.0677	−3.69711	−2.93305
	Std	3.05×10^{-15}	3.36×10^{-15}	3.128689	2.779744	0.479377
F23	Best	−10.5364	−10.5364	−10.5364	−10.5297	−4.51577
	Mean	−9.99562	−9.93332	−7.90449	−7.3919	−3.38426
	Median	−10.5364	−10.5364	−10.5357	−7.79854	−3.60414
	Worst	−5.12848	−3.83543	−3.79083	−1.67334	−1.95854
	Std	1.664525	1.868952	3.015319	3.33909	**0.720921**

The best values obtained are in bold.

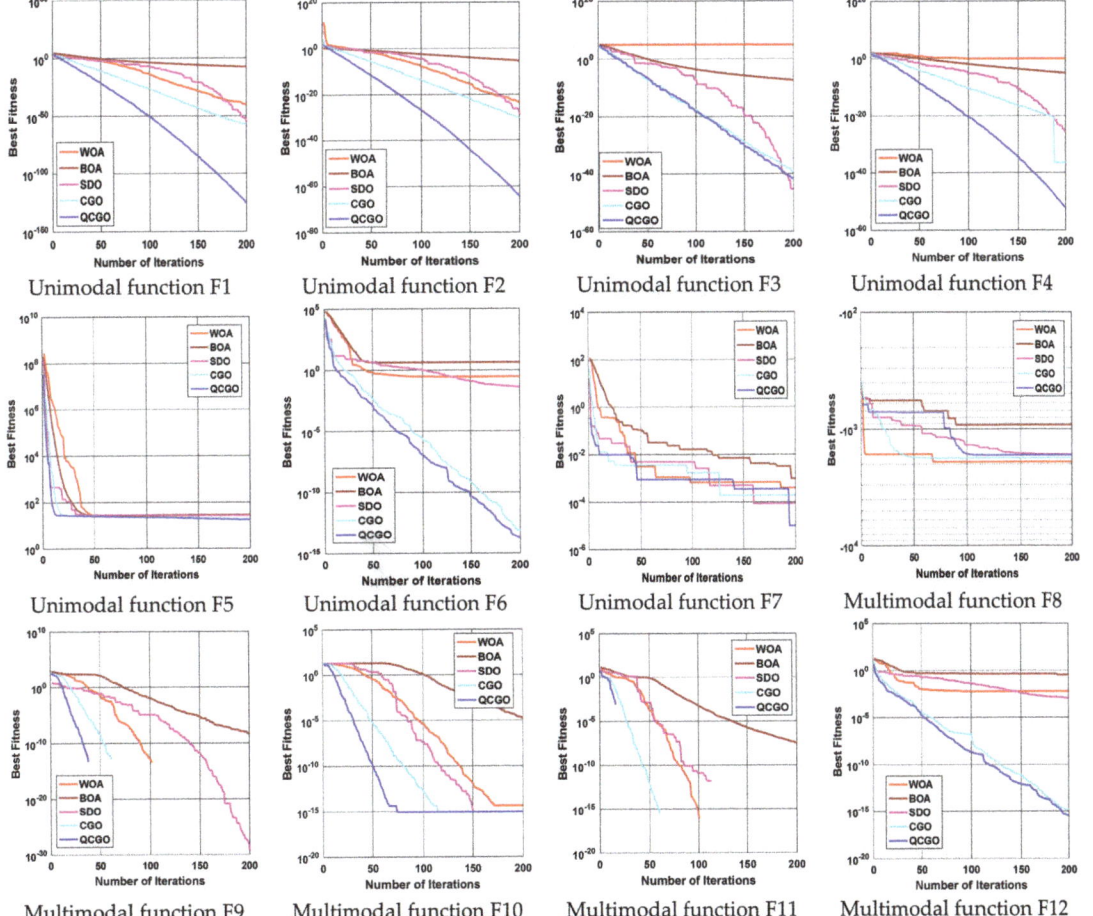

Figure 11. Cont.

Multimodal function F13

Composite function F14

Composite function F15

Composite function F16

Composite function F17

Composite function F18

Composite function F19

Composite function F20

Composite function F21

Composite function F22

Composite function F23

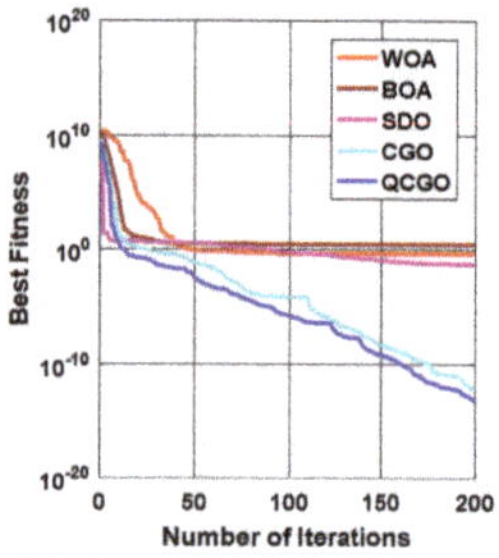

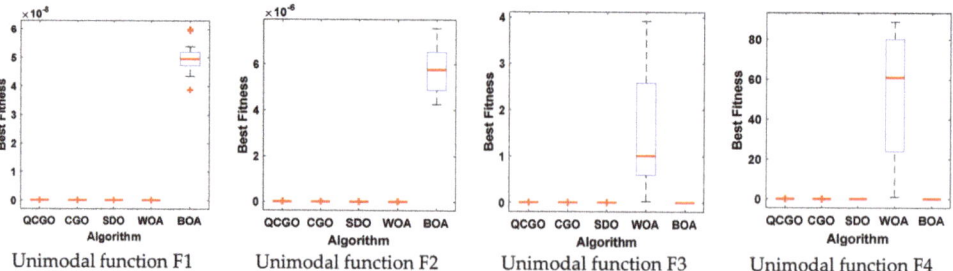

Figure 11. The convergence curves of the proposed QCGO algorithm and four other algorithms for 23 benchmark functions.

Unimodal function F1

Unimodal function F2

Unimodal function F3

Unimodal function F4

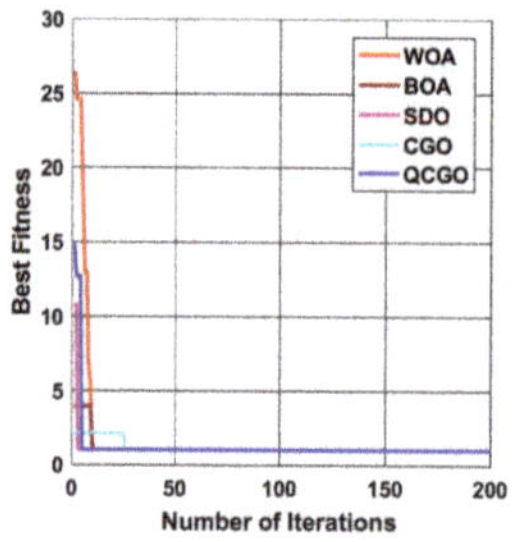

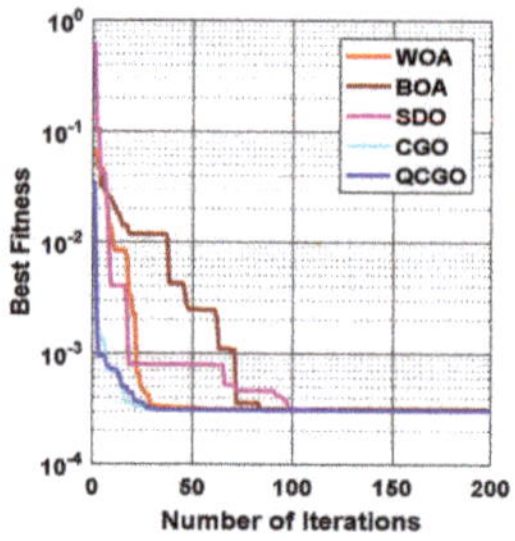

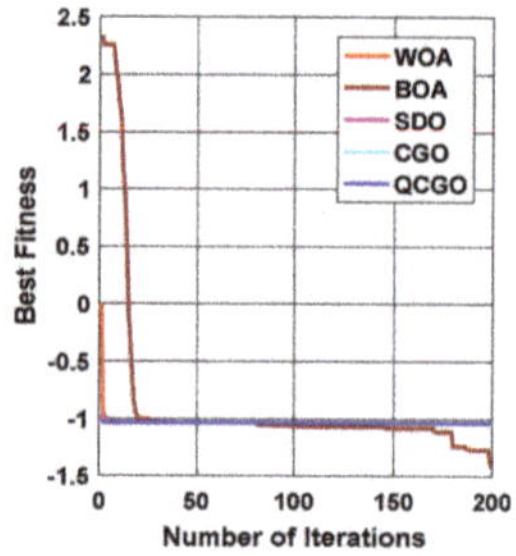

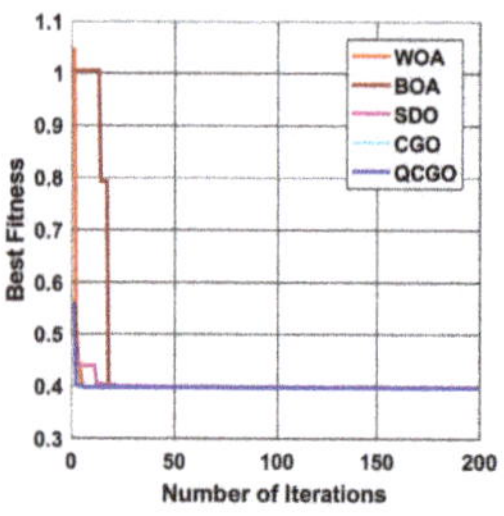

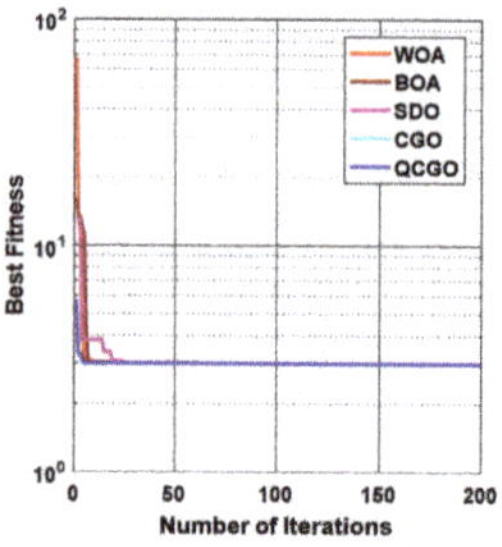

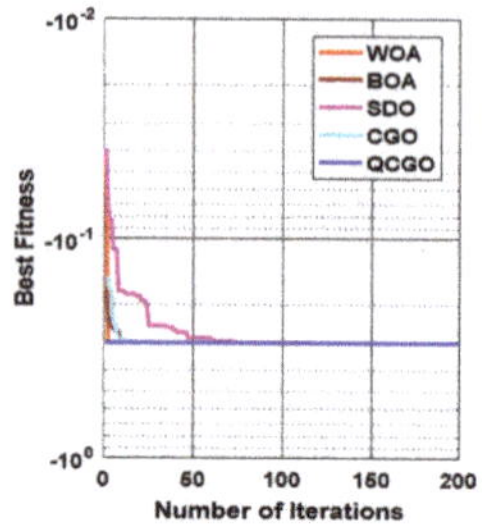

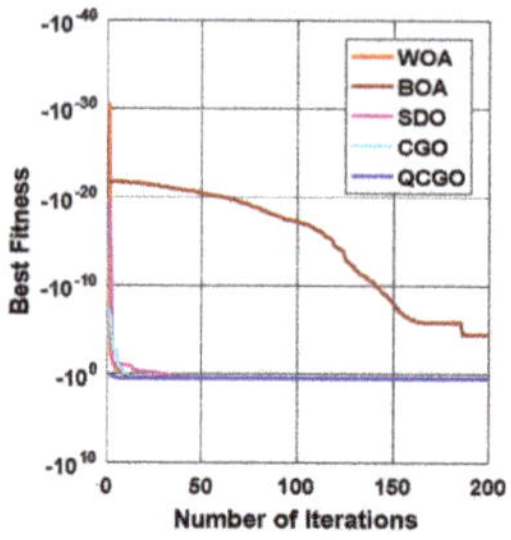

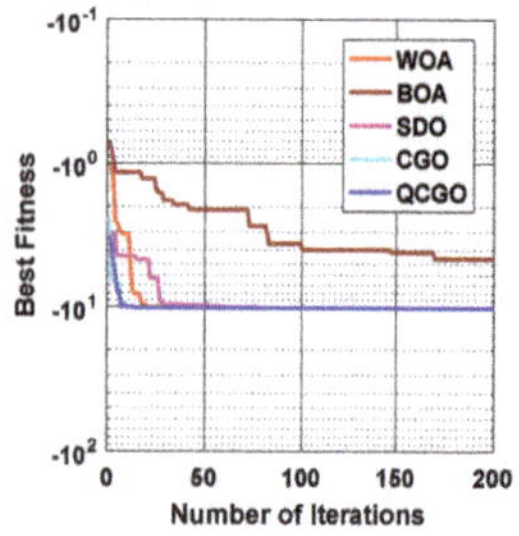

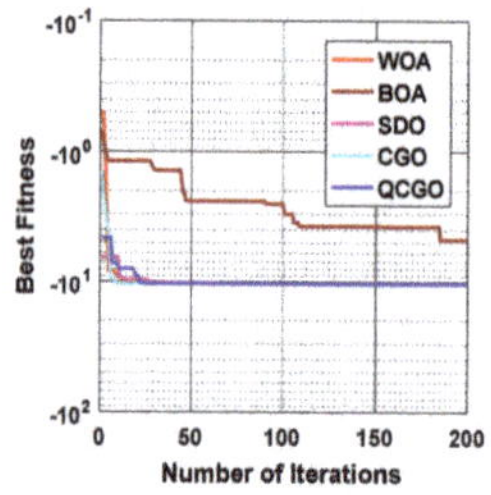

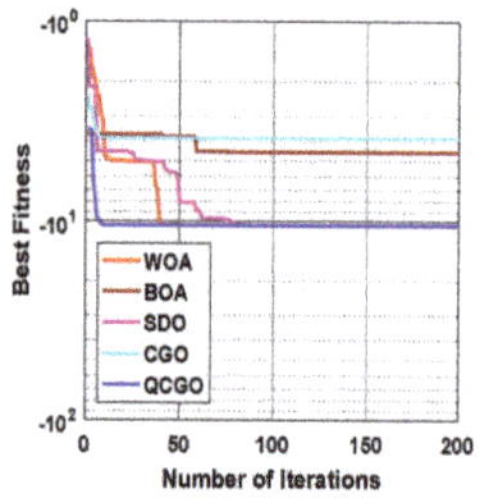

Figure 12. *Cont.*

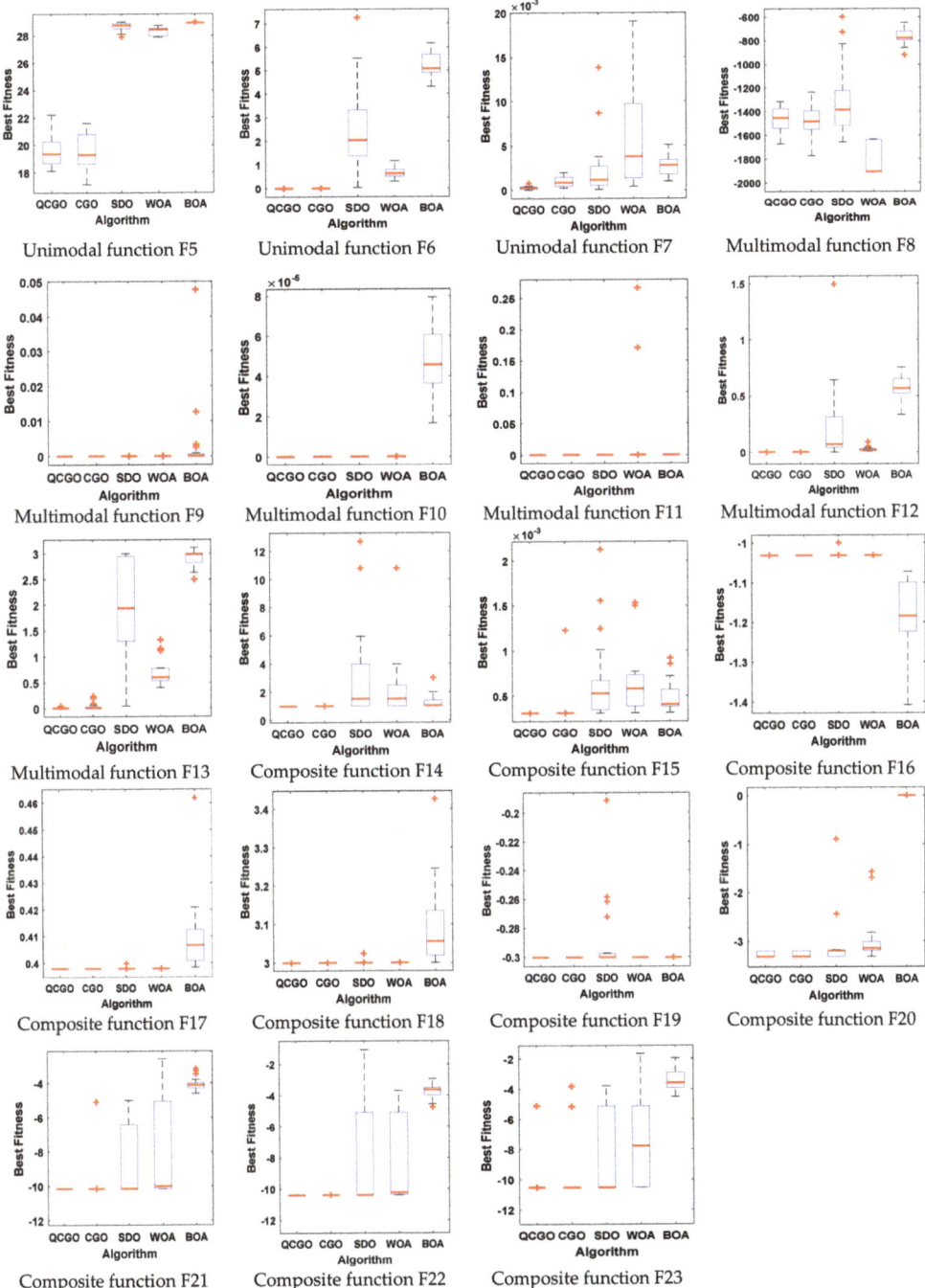

Figure 12. Boxplots of the proposed QCGO algorithm and four other algorithms for 23 benchmark functions.

5. Simulation Results and Discussions

In this study, the proposed control strategy is implemented in the secondary control loop with the high integration of RESs, considering different load variation types to restore the studied system frequency at the pre-defined value, where the presented control strategy relied on the combining TD-TI controller, which is optimally designed by an improved QCGO algorithm to obtain the minimum value of the frequency fluctuations for the studied power grid. Moreover, the performance of the suggested control strategy is compared with other control strategies (i.e., TID and PID). All of the simulation results for the studied two-area, multi-unit power grid are implemented using the professional software MATLAB/SIMULINK® program (R2015a) to ensure the efficacy of the proposed controller in enhancing the studied system performance. The code of the proposed QCGO algorithm is an m-file linked to the studied model for the optimization process. The simulation results are performed on a PC with Intel Core i5-2.60 GHz with 4.00 GB of RAM. The frequency stability has been assessed by applying different operating conditions through the following scenarios.

- Scenario A: evaluation of the studied power grid performance considering various load variation types (i.e., SLP, series SLP, and RLV).
- Scenario B: evaluation of the studied power grid performance considering high penetration of RESs in both areas with series SLP and RLV.
- Scenario C: evaluation of the studied power grid performance considering communication time delay.
- Scenario D: evaluation of the studied power grid performance considering EV integration in both areas.

The studied power grid performance can be evaluated by measuring the value of the best objective function that is represented by the ITAE value over the iterations. For most, several initial considerations must be addressed while optimizing the proposed TD-TI controller using the proposed improved QCGO algorithm, such as the search agent number that equals 30 and the total iterations/attempts that equal 100. The convergence curve that is shown in Figure 13 clarifies the performance of the proposed combining TD-TI controller based on QCGO compared to the combining TD-TI controller based on CGO and SSA and compared with the TID controller based on QCGO and CGO. The demonstrated convergence curve can be obtained considering a 1% SLP at 10 s in the first area of the studied power grid, without any RESs penetration in both areas. It is clear that the proposed combining TD-TI controller based on QCGO attained the lowest value of the objective function compared to the other mentioned controllers that relied on various optimization techniques. As a result, the convergence curve elucidates the effectiveness of the proposed QCGO algorithm. It can be seen that the curve behavior of the proposed TD-TI based on QCGO starts with a 0.1098 objective function value; then, this value drops along the iterations to end up at the final iteration with a 0.0729 objective function value, whereas the behavior of the proposed controller/proposed algorithm can be described as it reaches the best objective function value quickly compared to the other utilized controllers via different techniques. Moreover, it can be said, the rest curve behaviors are far from the optimum goal achieved by the suggested controller using QCGO, demonstrating its robustness in damping the oscillations effectively.

Scenario A: evaluation of the studied system performance considering different load variation types (i.e., SLP, series SLP, and random load).

This scenario included a fair-maiden comparison between the proposed combining TD-TI controller utilizing the QCGO algorithm and the other published controllers, such as the PID controller based on TLBO and AOA. Moreover, the proposed combining TD-TI controller based on the improved QCGO technique was compared with different mentioned controllers, such as the TID controller based on QCGO and CGO and the combining TD-TI controller based on CGO and SSA, to test the stability of the studied power grid performance.

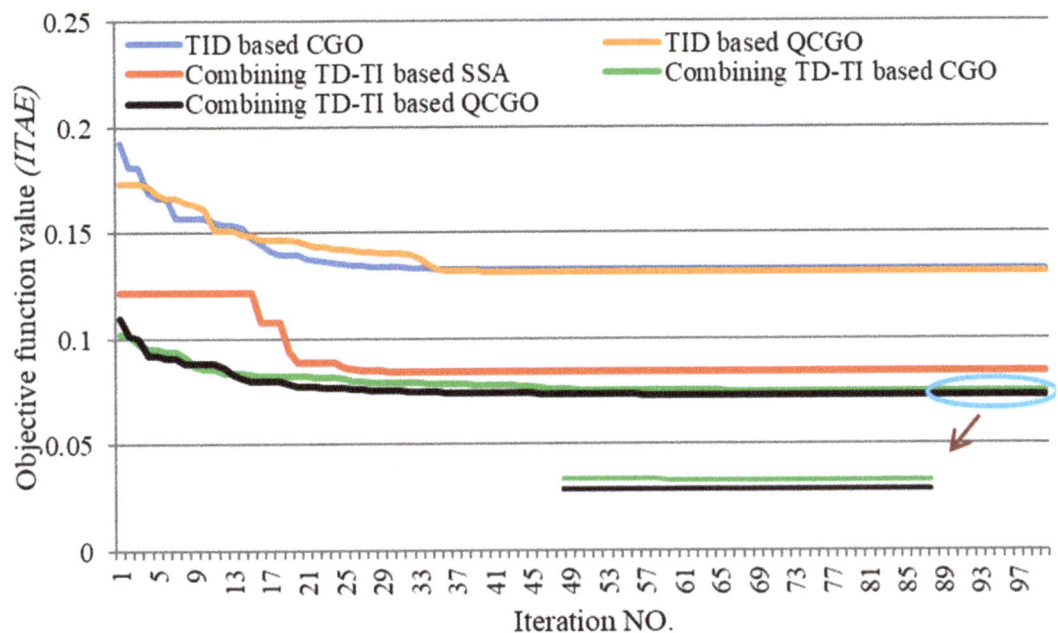

Figure 13. The convergence curve characteristics of QCGO, CGO, and salp swarm algorithm.

Case A.1: The SLP was selected as a challenge by applying it in the first area of the studied power grid to test the efficacy of the proposed combining TD-TI controller in enhancing the system performance. The applicable SLP occurred at 10 s with a 1% value, whereas the SLP can occur in the electrical power grids through disconnecting some generators from all the generation stations that may lead to blackouts with the shutdown of all the stations' generators. In addition, SLP may be represented as an unexpected switch of the connected electrical loads that may lead to instability in the system performance by increasing the wear and tear on the generators in the power grid.

Case A.1.1: This case presents a comparison between the performance of the proposed combining TD-TI controller in this work and the other published performances of the PID controller, to prove the efficacy of the proposed controller in attaining the main target (damping frequency oscillations). Table 7 indicates all of the aforementioned controller parameters that are utilized in diminishing the fluctuations in the system frequency and power flow in the tie line. In addition, Figure 14 clarifies a comparison between the different dynamic studied system responses (i.e., Δf_1, Δf_2, and ΔP_{tie}) of the proposed combining TD-TI controller, using QCGO and the PID controller based on TLBO and AOA, and considering a 1% SLP in the first area.

Table 8 illustrates the various specifications of the system performance, such as overshoot (O_{sh}), undershoot (U_{sh}), and the objective function values related to fluctuations in both the area frequencies and the power flow within the tie line. Table 8 clarifies the superiority of the proposed combining TD-TI controller-based, improved QCGO algorithm to achieve stability in the studied power grid. For ease, Table 9 denotes the percentage improvements in U_{sh} and O_{sh} for combining TD-TI/QCGO and PID/AOA, based on the PID/TLBO.

Table 7. The optimum parameters of the different controllers.

Controller Properties	Thermal	Hydro	Gas
Combining TD-TI-based QCGO	k_{t1} = 9.9999, k_{d1} = 9.9988, n_1 = 3.5626 k_{t2} = 9.9991, k_{i2} = 5.4425, n_2 = 3.5311	k_{t1} = 9.9834, k_{d1} = 3.8871, n_1 = 9.9468 k_{t2} = 9.5835, k_{i2} = 1.0016, n_2 = 9.9508	k_{t1} = 9.998, k_{d1} = 9.9973, n_1 = 3.7621 k_{t2} = 9.9951, k_{i2} = 9.9704, n_2 = 1.2938
Combining TD-TI-based CGO	k_{t1} = 9.9998, k_{d1} = 6.9628, n_1 = 3.5715 k_{t2} = 9.9977, k_{i2} = 5.033, n_2 = 3.4737	k_{t1} = 9.98, k_{d1} = 2.7245, n_1 = 9.9129 k_{t2} = 7.2945, k_{i2} = 1.052, n_2 = 9.9827	k_{t1} = 9.9998, k_{d1} = 8.4098, n_1 = 1.2782 k_{t2} = 9.9966, k_{i2} = 9.9989, n_2 = 6.9549
Combining TD-TI-based SSA	k_{t1} = 9.9998, k_{d1} = 8.985, n_1 = 2.9819 k_{t2} = 9.1794, k_{i2} = 9.3854, n_2 = 2.8288	k_{t1} = 5.3557, k_{d1} = 4.68, n_1 = 2.1217 k_{t2} = 8.5211, k_{i2} = 1.0925, n_2 = 5.1176	k_{t1} = 9.9998, k_{d1} = 1.0849, n_1 = 9.6003 k_{t2} = 9.9628, k_{i2} = 7.6555, n_2 = 1.4599
TID-based QCGO	k_{t1} = 9.8753, k_{i1} = 9.9302, k_{d1} = 7.9837, n_1 = 2.6219	k_{t1} = 9.7665, k_{i1} = 1.0797, k_{d1} = 4.9139, n_1 = 8.0894	k_{t1} = 9.9041, k_{i1} = 9.9922, k_{d1} = 1.6516, n_1 = 9.2214
TID-based CGO	k_{t1} = 9.9993, k_{i1} = 9.7827, k_{d1} = 8.7199, n_1 = 3.5979	k_{t1} = 9.9525, k_{i1} = 1.4282, k_{d1} = 5.1353, n_1 = 7.5851	k_{t1} = 9.9486, k_{i1} = 9.9844, k_{d1} = 4.0435, n_1 = 3.3106
PID-based TLBO [46]	k_{p1} = 4.1468, k_{i1} = 4.0771, k_{d1} = 2.0157	k_{p1} = 1.0431, k_{i1} = 0.6030, k_{d1} = 2.2866	k_{p1} = 4.7678, k_{i1} = 3.7644, k_{d1} = 4.9498
PID-based AOA [47]	k_{p1} = 10, k_{i1} = 1.5975, k_{d1} = 2.7449	k_{p1} = 1.5975, k_{i1} = 0.0837, k_{d1} = 0.0875	k_{p1} = 10, k_{i1} = 10, k_{d1} = 1.2779

Table 8. The transient response specifications of the presented system for case A.1.1.

Controller Properties	Dynamic Response of (Δf_1)	Dynamic Response of (Δf_2)	Dynamic Response of (ΔP_{tie})	Objective Function Value (ITAE)
Combining TD-TI based on QCGO O_{sh} and U_{sh} $\times (10^{-3})$	O_{sh} = 0.819 U_{sh} = −7.875	O_{sh} = 0.0028 U_{sh} = −1.744	O_{sh} = 0.0015 U_{sh} = −0.5361	J = 0.075
PID based on AOA O_{sh} and U_{sh} $\times (10^{-3})$ [47]	O_{sh} = 1.158 U_{sh} = −11.42	O_{sh} = 0.02096 U_{sh} = −4.443	O_{sh} = 0.01107 U_{sh} = −1.249	J = 0.189
PID based on TLBO O_{sh} and U_{sh} $\times (10^{-3})$ [46]	O_{sh} = 1.7217 U_{sh} = −19.7259	O_{sh} = 0.4363 U_{sh} = −12.7986	O_{sh} = 0.1712 U_{sh} = −3.0782	J = 0.402

Table 9. Percentage improvement in U_{sh} and O_{sh} values for combining TD-TI/QCGO and PID/AOA based on PID controller via TLBO for scenario A.1.1.

Controller	Δf_1 U_{sh}	O_{sh}	Δf_2 U_{sh}	O_{sh}	ΔP_{tie} U_{sh}	O_{sh}
Combining TD-TI based on QCGO	**60.01**	**52.43**	**86.4**	**99.36**	**82.6**	**99.12**
PID based on AOA	42.11	32.70	65.29	95.2	59.42	93.53

The optimum values are bolded.

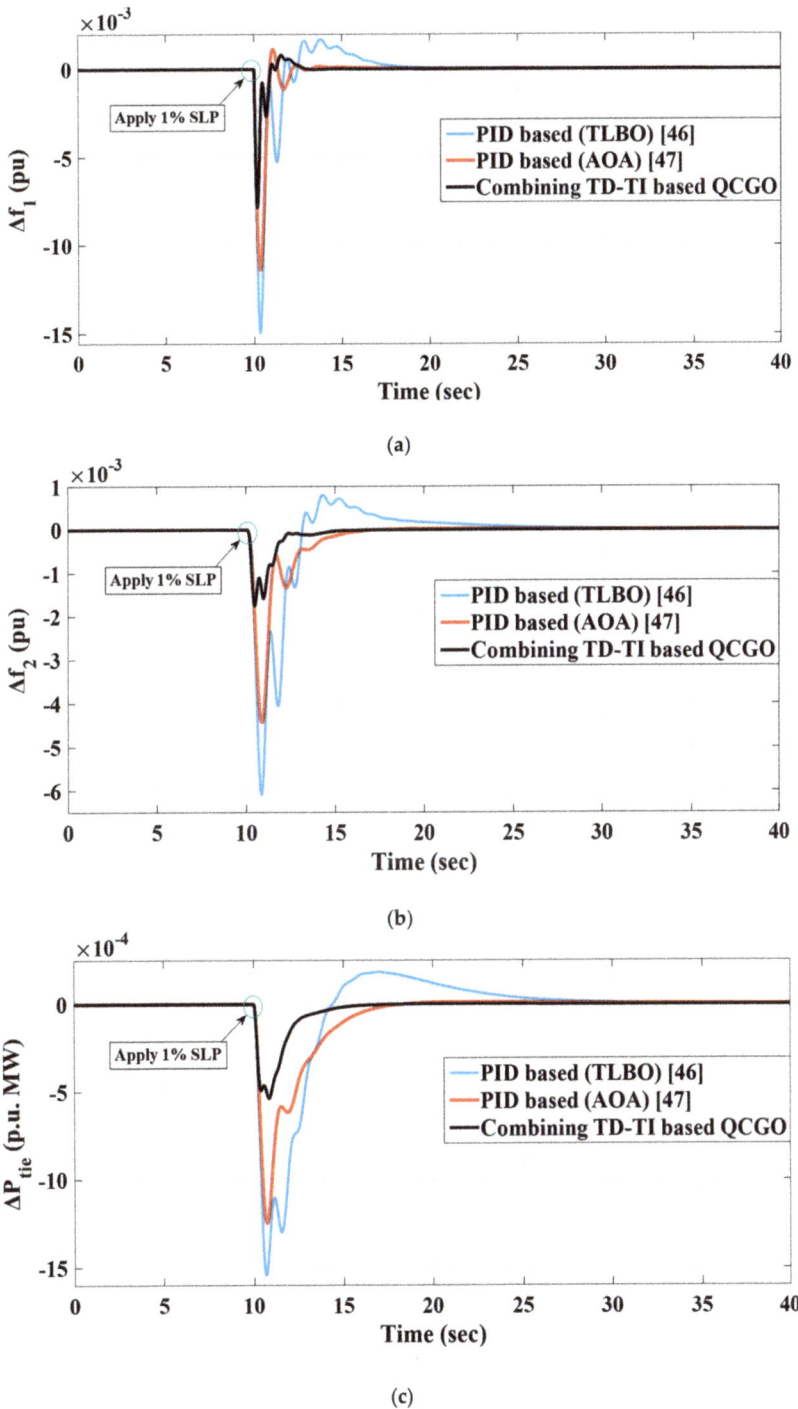

Figure 14. Dynamic power grid responses in case A.1.1: (a) Δf_1 (b) Δf_2 (c) ΔP_{tie}.

As can be seen, the improved QCGO algorithm utilized in fine-tuning the proposed combining TD-TI controller obtains the optimal controller parameters, which leads to attaining the optimal solution with a 0.075 objective function value. The obtained objective function value related to the proposed controller using an improved QCGO algorithm is the best compared to those attained from the published PID controller based on TLBO and AOA, which equal 0.402 and 0.189, respectively. It can be seen that the proposed combining TD-TI controller-based QCGO achieves a higher percentage in improving all system dynamic performance. For example, the percentage improvement in U_{sh} and O_{sh} of Δf_1 related to combining TD-TI/QCGO is 60.01% and 52.43%, respectively. In contrast, the percentage improvement in U_{sh} and O_{sh} of Δf_1 related to PID/AOA is 42.11% and 32.70%, respectively.

Case A.1.2: This case presents a suggestion of utilizing the TID controller based on CGO and QCGO to compare it with the proposed combining TD-TI controller based on QCGO to test the robustness of the proposed one in regulating the studied system frequency. All of the previously mentioned controller parameters are presented in Table 7. Moreover, Figure 15 describes a fair comparison between all of the dynamic system responses related to the proposed combining TD-TI controller based on QCGO and all those responses of the TID controller based on CGO and QCGO.

Table 10 illustrates the different specifications of the system performance, such as O_{sh}, U_{sh}, and the objective function values related to excursions in both the area frequencies and the power flow within the tie line. Table 10 clarifies the superiority of the proposed combining TD-TI controller-based improved QCGO algorithm in achieving system reliability. In addition, Table 11 clarifies the percentage improvements in U_{sh} and O_{sh} for combining TD-TI/QCGO and TID/(CGO, QCGO) based on the PID/TLBO.

Table 10. The transient response specifications of the presented system for case A.1.2.

Controller Properties	Dynamic Response of (Δf_1)	Dynamic Response of (Δf_2)	Dynamic Response of (ΔP_{tie})	Objective Function Value (*ITAE*)
Combining TD-TI based on QCGO O_{sh} and U_{sh} $\times (10^{-3})$	O_{sh} = 0.819 U_{sh} = −7.875	O_{sh} = 0.0028 U_{sh} = −1.744	O_{sh} = 0.0015 U_{sh} = −0.5361	*J* = 0.075
TID based on QCGO O_{sh} and U_{sh} $\times (10^{-3})$	O_{sh} = 1.893 U_{sh} = −11.468	O_{sh} = 0.3257 U_{sh} = −3.45	O_{sh} = 0.0424 U_{sh} = −0.8862	*J* = 0.1351
TID based on CGO O_{sh} and U_{sh} $\times (10^{-3})$	O_{sh} = 1.705 U_{sh} = −10.341	O_{sh} = 0.3784 U_{sh} = −2.763	O_{sh} = 0.0381 U_{sh} = −0.7397	*J* = 0.1381

The optimum values are bolded.

Table 11. Percentage improvement in U_{sh} and O_{sh} values for combining TD-TI/QCGO and PID/AOA based on PID controller via TLBO for scenario A.1.2.

Controller	Δf_1 U_{sh}	O_{sh}	Δf_2 U_{sh}	O_{sh}	ΔP_{tie} U_{sh}	O_{sh}
Combining TD-TI based on QCGO	60.01	52.43	86.4	99.36	82.6	99.12
TID based on QCGO	41.86	−9.95	73.04	25.35	71.21	75.23
TID based on CGO	47.6	0.97	78.4	13.27	75.97	77.75

The optimum values are bolded.

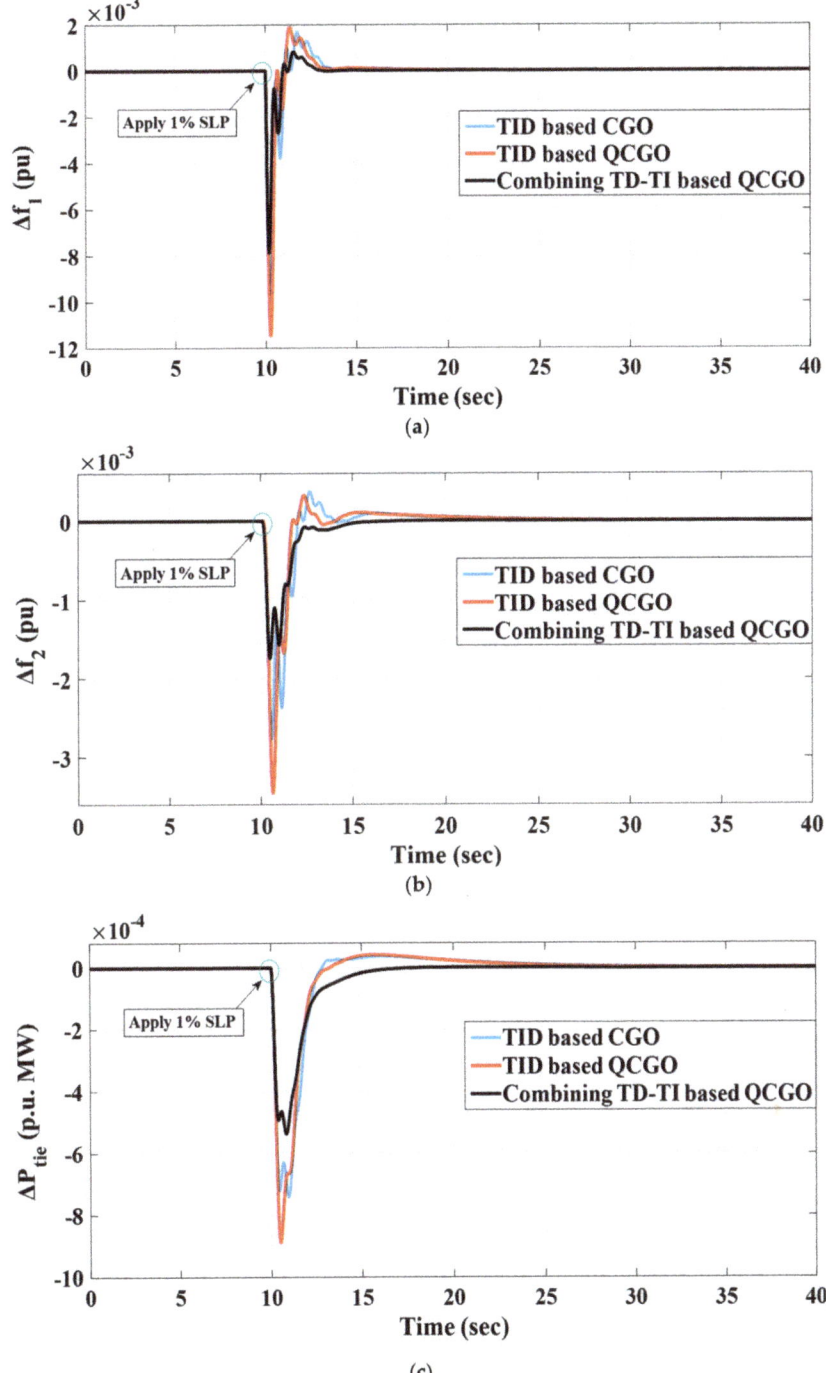

Figure 15. Dynamic power grid responses in case A.1.2: (a) Δf_1 (b) Δf_2 (c) ΔP_{tie}.

Table 10 clarifies that the obtained objective function value related to the proposed controller using an improved QCGO algorithm that equals 0.075 is the best compared to those attained from the TID controller based on CGO and QCGO, which equal 0.1381 and 0.1351, respectively. Moreover, Table 11 denotes that the proposed combining TD-TI controller-based QCGO achieves a higher percentage in improving all of the system dynamic performance. For example, the percentage improvement in U_{sh} and O_{sh} of Δf_2 related to combining TD-TI/QCGO is 86.4% and 99.36%, respectively. In contrast, the percentage improvement in U_{sh} and O_{sh} of Δf_2 related to TID/QCGO is 73.04% and 25.35%, respectively.

Case A.1.3: This case presents the SSA algorithm as a meta-heuristic optimization technique to tune the proposed combining TD-TI controller and make a comparison between it and the CGO and QCGO techniques in selecting the optimal controller parameters to prove that the improved QCGO algorithm can achieve more system stability compared to utilizing the different mentioned algorithms. Table 7 presents the aforementioned controller parameters that were utilized in overcoming the LFC problem in the studied power grid. Moreover, Figure 16 describes a fair comparison between all of the dynamic system responses related to the proposed combining TD-TI controller based on QCGO and all those responses of the combining TD-TI controller based on SSA and CGO.

Table 12 illustrates the different specifications of the system performance, such as O_{sh}, U_{sh}, and the objective function values related to the oscillations in both the area frequencies and the power flow within the tie line. Table 12 clarifies the superiority of the proposed combining TD-TI controller-based, improved QCGO algorithm in achieving system reliability. In addition, Table 13 clarifies the percentage improvements in U_{sh} and O_{sh} for combining TD-TI/QCGO and combining TD-TI/(CGO, SSA), based on the PID/TLBO.

Table 12. The transient response specifications of the presented system for case A.1.3.

Controller Properties	Dynamic Response of (Δf_1)	Dynamic Response of (Δf_2)	Dynamic Response of (ΔP_{tie})	Objective Function Value (ITAE)
Combining TD-TI based on QCGO O_{sh} and U_{sh} $\times (10^{-3})$	O_{sh} = 0.819 U_{sh} = −7.875	O_{sh} = 0.0028 U_{sh} = −1.744	O_{sh} = 0.0015 U_{sh} = −0.5361	J = 0.075
Combining TD-TI based on CGO O_{sh} and U_{sh} $\times (10^{-3})$	O_{sh} = 1.097 U_{sh} = −8.95	O_{sh} = 0.0025 U_{sh} = −2.383	O_{sh} = 0.00136 U_{sh} = −0.665	J = 0.078
Combining TD-TI based on SSA O_{sh} and U_{sh} $\times (10^{-3})$	O_{sh} = 1.763 U_{sh} = −9.978	O_{sh} = 0.0896 U_{sh} = −2.713	O_{sh} = 0.0124 U_{sh} = −0.7125	J = 0.087

Table 13. Percentage improvement in U_{sh} and O_{sh} values for combining TD-TI/QCGO and PID/AOA based on PID controller via TLBO for scenario A.1.3.

Controller	Δf_1 U_{sh}	O_{sh}	Δf_2 U_{sh}	O_{sh}	ΔP_{tie} U_{sh}	O_{sh}
Combining TD-TI based on QCGO	60.01	52.43	86.4	99.36	82.6	99.12
Combining TD-TI based on CGO	54.63	36.28	81.38	**99.43**	78.4	**99.21**
Combining TD-TI based on SSA	49.42	−2.4	78.8	79.46	76.85	92.76

The optimum values are bolded.

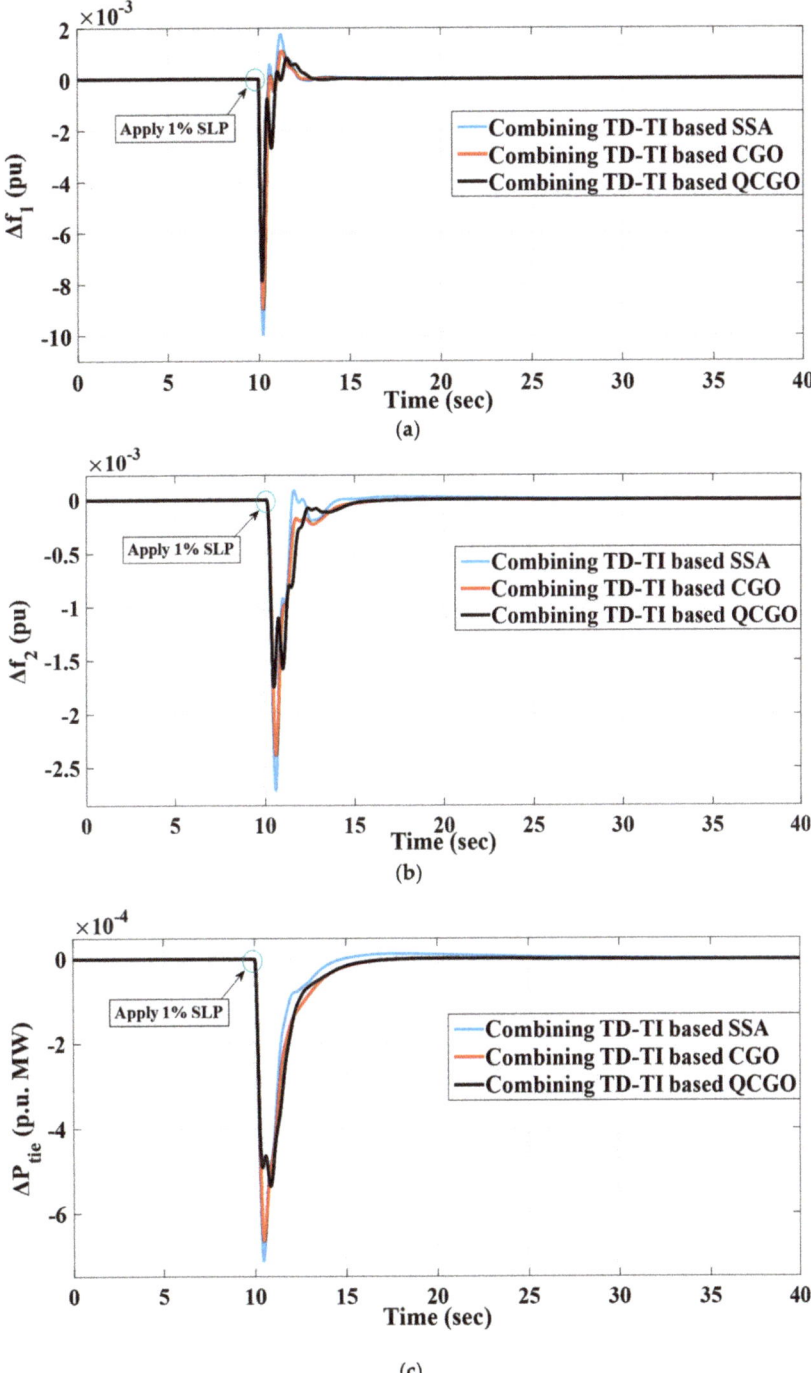

Figure 16. Dynamic power grid responses in case A.1.3: (a) Δf_1 (b) Δf_2 (c) ΔP_{tie}.

Table 12 clarifies that the obtained objective function value related to the suggested controller using an improved QCGO algorithm that equals 0.075 is the best compared to those attained from the combining TD-TI controller based on CGO and SSA, which equal 0.078 and 0.087, respectively. Moreover, Table 13 denotes that the proposed combining TD-TI controller-based QCGO achieves a higher percentage in improving all of the system dynamic performance. For example, the percentage improvement in U_{sh} and O_{sh} of ΔP_{tie} related to combining TD-TI/QCGO is 82.6% and 99.12%, respectively. In contrast, the percentage improvement in U_{sh} and O_{sh} of ΔP_{tie} related to combining TD-TI/SSA is 76.85% and 92.76%, respectively.

Case A.2: In this case, the performance of the proposed combining TD-TI controller optimized with an improved QCGO algorithm has been tested and assessed by subjecting a series SLP in the first area of the studied power grid. The series SLP is represented as an emulation of the series changing in the realistic connected loads. It can be said that the series SLP is considered as a series-forced switch of generators or series interrupts of the connected loads. Figure 17 describes the applied form of the series SLP. In addition, the different dynamic system responses are indicated in Figure 18 to elucidate the superiority of the suggested combining TD-TI controller based on QCGO compared to those of the other controllers optimized with different algorithms (i.e., combining TD-TI based on CGO and SSA) in the presence of the series SLP in the first area.

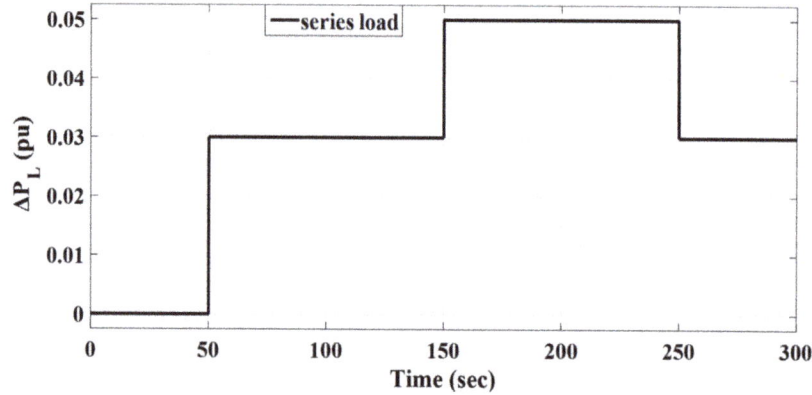

Figure 17. The form of the applied series step load perturbation.

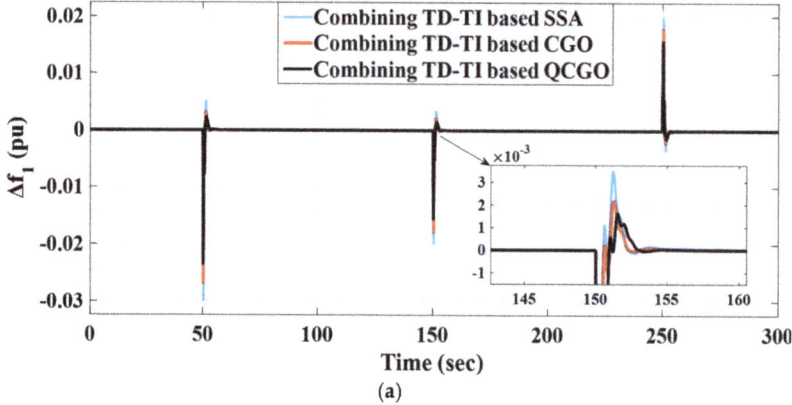

(a)

Figure 18. *Cont.*

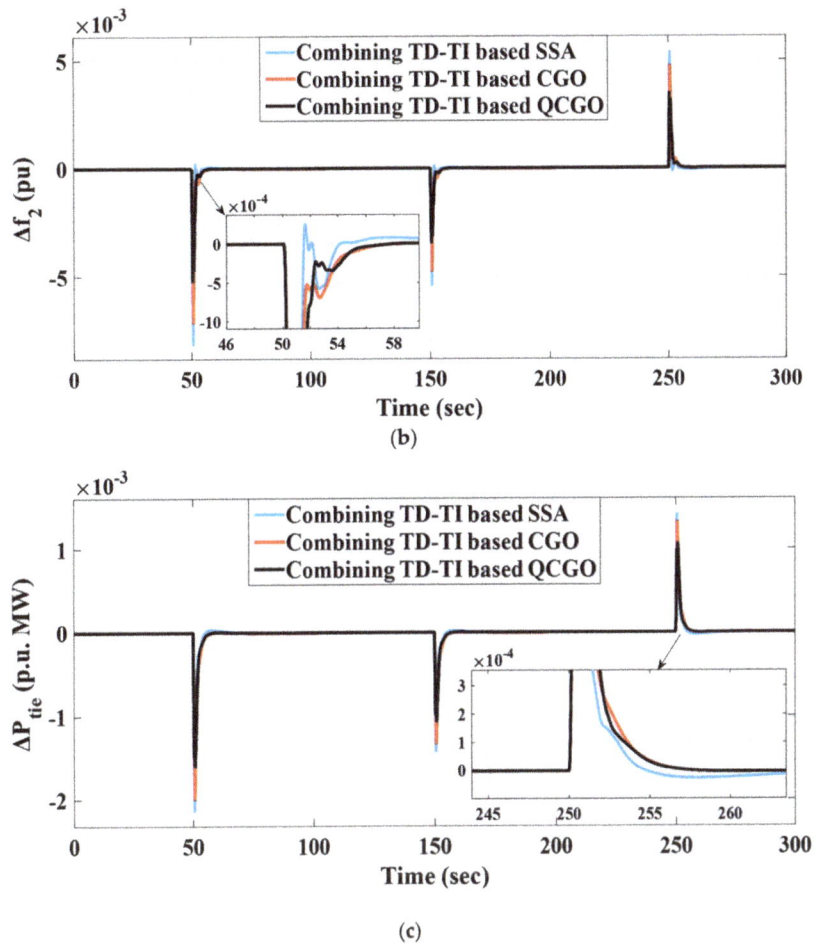

Figure 18. Dynamic power grid responses in case A.2: (a) Δf_1 (b) Δf_2 (c) ΔP_{tie}.

Table 14 illustrates the values of O_{sh} and U_{sh} related to the different system dynamic responses (i.e., Δf_1, Δf_2, and ΔP_{tie}) according to oscillations in both the area frequencies and the power flow within the tie line. Table 14 clarifies the superiority of the proposed combining TD-TI controller-based improved QCGO algorithm in achieving system stability. In addition, Table 15 clarifies the percentage improvements in U_{sh} and O_{sh} for combining TD-TI/QCGO and combining TD-TI/CGO based on the combining TD-TI/SSA.

Table 14 clarifies that the suggested controller using an improved QCGO algorithm achieves more system stability after looking at the obtained results of the O_{sh} and U_{sh} values. Moreover, Table 15 denotes that the proposed combining TD-TI controller-based QCGO achieves a higher percentage in improving all of the system dynamic performance. For example, the percentage improvement in U_{sh} and O_{sh} of Δf_1 related to combining TD-TI/QCGO is 26.13% and 25.71%, respectively. In contrast, the percentage improvement in U_{sh} and O_{sh} of Δf_1 related to combining TD-TI/CGO is 15.81% and 14.29%, respectively.

Table 14. The transient response specifications of the presented system for case A.2.

Controller Properties	Dynamic Response of (Δf_1)	Dynamic Response of (Δf_2)	Dynamic Response of (ΔP_{tie})
Combining TD-TI based on QCGO O_{sh} and U_{sh} $\times (10^{-3})$	O_{sh} = 15.6 U_{sh} = −22.9	O_{sh} = 3.5 U_{sh} = −5.1	O_{sh} = 1.000 U_{sh} = −1.67
Combining TD-TI based on CGO O_{sh} and U_{sh} $\times (10^{-3})$	O_{sh} = 18.00 U_{sh} = −26.1	O_{sh} = 4.85 U_{sh} = −7.3	O_{sh} = 1.3 U_{sh} = −1.9
Combining TD-TI based on SSA O_{sh} and U_{sh} $\times (10^{-3})$	O_{sh} = 21.000 U_{sh} = −31.000	O_{sh} = 5.510 U_{sh} = −8.6	O_{sh} = 1.40 U_{sh} = −2.15

Table 15. Percentage improvement in U_{sh} and O_{sh} values for combining TD-TI/QCGO and combining TD-TI/CGO based on combining TD-TI/SSA for scenario A.2.

Controller	Δf_1		Δf_2		ΔP_{tie}	
	U_{sh}	O_{sh}	U_{sh}	O_{sh}	U_{sh}	O_{sh}
Combining TD-TI based on QCGO	**26.13**	**25.71**	**40.7**	**36.48**	**22.33**	**28.6**
Combining TD-TI based on CGO	15.81	14.29	15.12	11.98	11.63	6.43

The optimum values are bolded.

Case A.3: In this case, the studied power grid has been subjected to RLVs in the first area. The RLVs are a diverse combination of series perturbations in industrial connected loads to the grid that cause the same effects on the grid (i.e., unbalance in electrical power grid and the occurrence of blackout). The RLV is formed in Figure 19. In addition, Figure 20 describes the different dynamic power system responses explaining the efficacy of the proposed combining TD-TI controller based on QCGO in achieving more of a reduction in the system frequency fluctuations and the power flow in the tie line compared to the other ones.

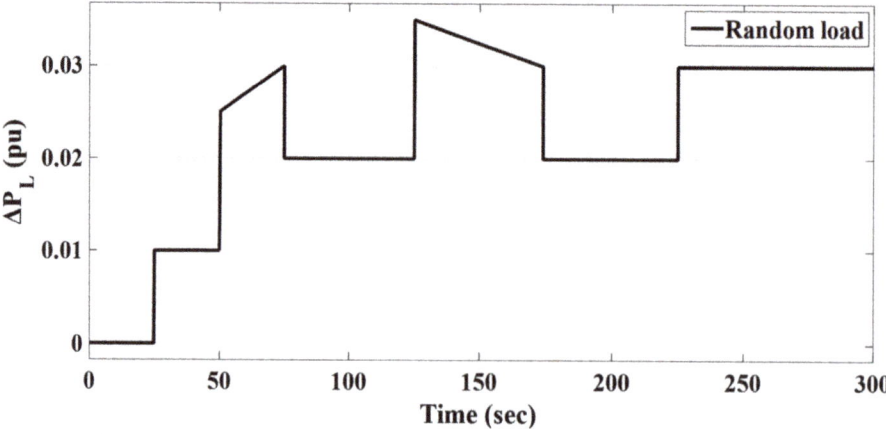

Figure 19. The form of the applied random load variation.

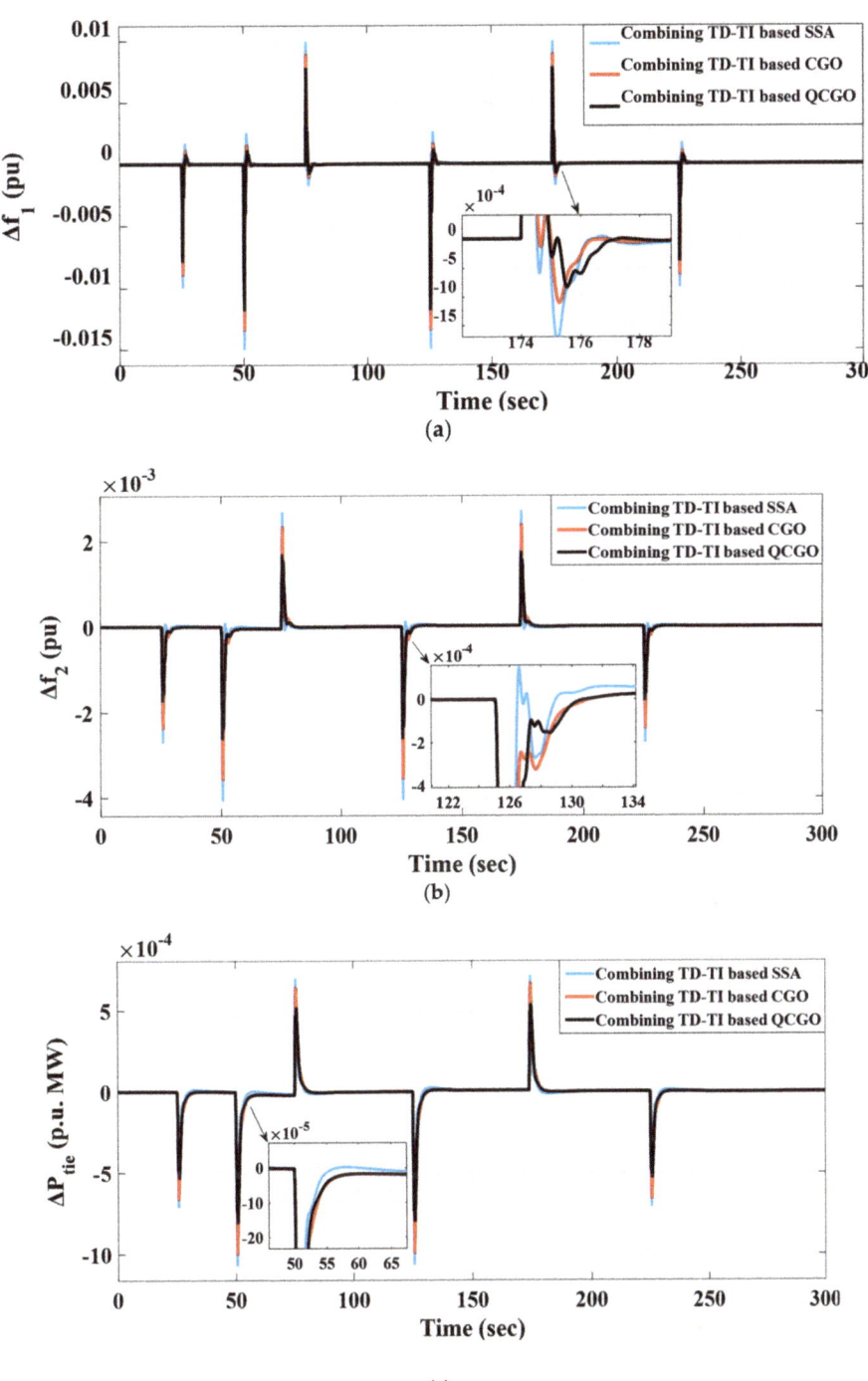

Figure 20. Dynamic power grid responses in case A.3: (a) Δf_1 (b) Δf_2 (c) ΔP_{tie}.

Table 16 illustrates the values of O_{sh} and U_{sh} related to the different system dynamic responses (i.e., Δf_1, Δf_2, and ΔP_{tie}) according to the oscillations in both the area frequencies and the power flow within the tie line. Table 16 presents the robustness of the proposed combining TD-TI controller-based improved QCGO algorithm in achieving system stability. In addition, Table 17 clarifies the percentage improvements in U_{sh} and O_{sh} for combining TD-TI/QCGO and combining TD-TI/CGO based on the combining TD-TI/SSA.

Table 16. The transient response specifications of the presented system for case A.3.

Controller Properties	Dynamic Response of (Δf_1)	Dynamic Response of (Δf_2)	Dynamic Response of (ΔP_{tie})
Combining TD-TI based on QCGO O_{sh} and U_{sh} $\times (10^{-3})$	O_{sh} = 7.4 U_{sh} = −11.9	O_{sh} = 1.4 U_{sh} = −2.2	O_{sh} = 0.51 U_{sh} = −0.76
Combining TD-TI based on CGO O_{sh} and U_{sh} $\times (10^{-3})$	O_{sh} = 8.40 U_{sh} = −13.5	O_{sh} = 2.3 U_{sh} = −3.6	O_{sh} = 0.65 U_{sh} = −1.000
Combining TD-TI based on SSA O_{sh} and U_{sh} $\times (10^{-3})$	O_{sh} = 10.000 U_{sh} = −15.000	O_{sh} = 2.60 U_{sh} = −4.08	O_{sh} = 0.72 U_{sh} = −1.14

Table 17. Percentage improvement in U_{sh} and O_{sh} values for combining TD-TI/QCGO and combining TD-TI/CGO based on combining TD-TI/SSA for scenario A.3.

Controller	Δf_1 U_{sh}	O_{sh}	Δf_2 U_{sh}	O_{sh}	ΔP_{tie} U_{sh}	O_{sh}
Combining TD-TI based on QCGO	20.67	26.00	46.08	46.15	33.33	29.17
Combining TD-TI based on CGO	10.00	16.00	11.76	11.54	12.28	9.72

The optimum values are bolded.

Table 16 clarifies that the proposed controller via an improved QCGO algorithm achieves more system stability after looking at the obtained results of the O_{sh} and U_{sh} values. Additionally, Table 17 denotes that the proposed combining TD-TI controller-based QCGO achieves a higher percentage in improving all of the system dynamic performance. For example, the percentage improvement in U_{sh} and O_{sh} of Δf_1 related to combining TD-TI/QCGO is 20.67% and 26.00%, respectively. However, the percentage improvement in U_{sh} and O_{sh} of Δf_1 related to combining TD-TI/CGO is 10.00% and 16.00%, respectively.

Scenario B: evaluation of the studied system performance considering high penetration of RESs in both areas with series SLP and RLV.

Another challenge of high penetrating of RESs (i.e., wind energy in the first area and PV energy in the second area) is addressed in this study to test the robustness of the proposed combining TD-TI controller in reducing the studied system fluctuations. The series SLP and RLV are applied in the first area as well as integration of the RESs in the power grid. The penetration of RESs represents a burden on the studied power grid due to their demerits (i.e., lack of system inertia).

Case B.1: robustness test for the proposed combining TD-TI controller optimized by improved QCGO considering high RES penetration as well as series SLP challenge.

This section clarifies the dynamic system performance of the investigated power grid, taking into consideration a series SLP, high penetration of wind energy at t = 100 s in the first area and PV at t = 200 s in the second area. These mentioned challenges have been presented to ensure the reliability and effectiveness of the proposed combining TD-TI controller based on an improved QCGO algorithm in enhancing the studied power grid performance. Figure 21 clarifies the applicable series SLP form in the first area. Moreover, all the dynamic power grid responses represented in Δf_1, Δf_2 and ΔP_{tie} are shown in Figure 22.

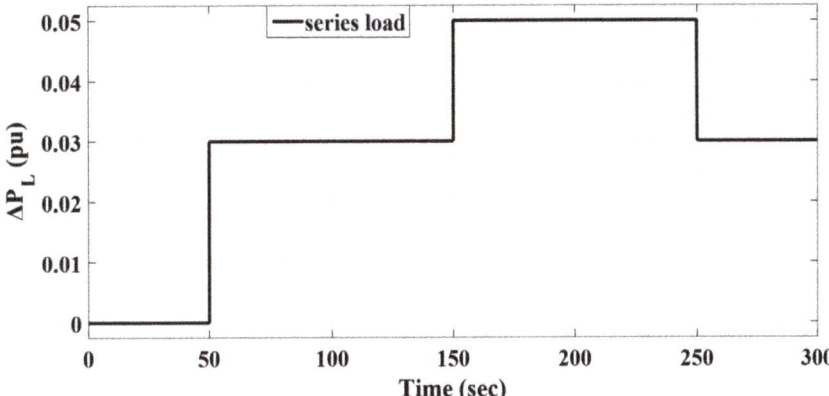

Figure 21. The form of the applied series SLP.

Table 18 illustrates the values of O_{sh} and U_{sh} related to the aforementioned system dynamic responses due to deviations in both the area frequencies and the power flow within the tie line. Table 18 presents the robustness of the proposed combining TD-TI controller-based improved QCGO algorithm in achieving system reliability. In addition, Table 19 clarifies the percentage improvements in U_{sh} and O_{sh} for combining TD-TI/QCGO and combining TD-TI/CGO based on the combining TD-TI/SSA.

Table 18. The transient response specifications of the presented system for case B.1.

Controller Properties	Dynamic Response of (Δf_1)	Dynamic Response of (Δf_2)	Dynamic Response of (ΔP_{tie})
Combining TD-TI based on QCGO O_{sh} and U_{sh} $\times (10^{-3})$	$O_{sh} = 71.0$ $U_{sh} = -22.0$	$O_{sh} = 40.3$ $U_{sh} = -7.4$	$O_{sh} = 4.8$ $U_{sh} = -2.7$
Combining TD-TI based on CGO O_{sh} and U_{sh} $\times (10^{-3})$	$O_{sh} = 81.0$ $U_{sh} = -27.0$	$O_{sh} = 46.0$ $U_{sh} = -8.1$	$O_{sh} = 6.1$ $U_{sh} = -3.6$
Combining TD-TI based on SSA O_{sh} and U_{sh} $\times (10^{-3})$	$O_{sh} = 96.000$ $U_{sh} = -30.000$	$O_{sh} = 51.1$ $U_{sh} = -9.5$	$O_{sh} = 6.5$ $U_{sh} = -3.88$

Table 19. Percentage improvement in U_{sh} and O_{sh} values for combining TD-TI/QCGO and combining TD-TI/CGO based on combining TD-TI/SSA for scenario B.1.

| Controller | Δf_1 | | Δf_2 | | ΔP_{tie} | |
	U_{sh}	O_{sh}	U_{sh}	O_{sh}	U_{sh}	O_{sh}
Combining TD-TI based on QCGO	**26.67**	**26.04**	**22.11**	**21.14**	**30.41**	**26.15**
Combining TD-TI based on CGO	10.00	15.63	14.74	9.98	7.22	6.15

The optimum values are bolded.

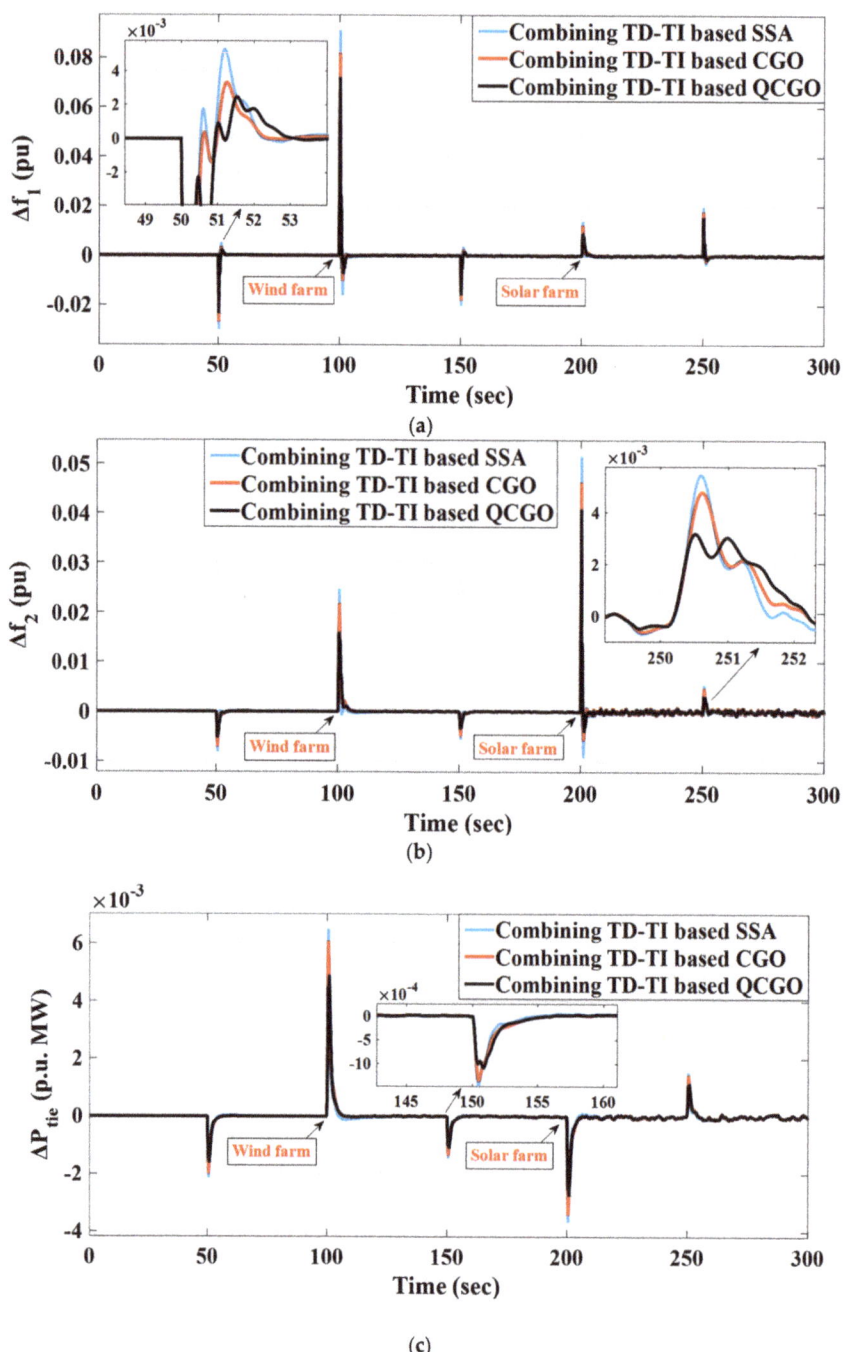

Figure 22. Dynamic power grid responses in case B.1: (a) Δf_1 (b) Δf_2 (c) ΔP_{tie}.

It can be summarized that Table 18 clarifies that the proposed controller/proposed algorithm achieves more system stability after showing the obtained results of the O_{sh}

and U_{sh} values. In this regard, Table 19 clarifies that the proposed combining TD-TI controller-based QCGO achieves a higher percentage in improving all of the system dynamic performance. For example, the percentage improvement in U_{sh} and O_{sh} of ΔP_{tie} related to combining TD-TI/QCGO is 30.41% and 26.15%, respectively. However, the percentage improvement in U_{sh} and O_{sh} of ΔP_{tie} related to combining TD-TI/CGO is 7.22% and 6.15%, respectively.

Case B.2: robustness test for the proposed combining TD-TI controller optimized by improved QCGO considering high RES penetration as well as RLV.

This section includes a robustness test by the penetrating of RESs at both areas of the studied power grid with the applicable RLV in the first area. This test summarized the superiority of the proposed combining TD-TI controller based on an improved QCGO algorithm in overcoming the frequency excursions for the studied power grid. The applicable RLV is shown in Figure 23. Moreover, the behavior of both the area frequencies and the power flow in the tie line is clarified in Figure 24.

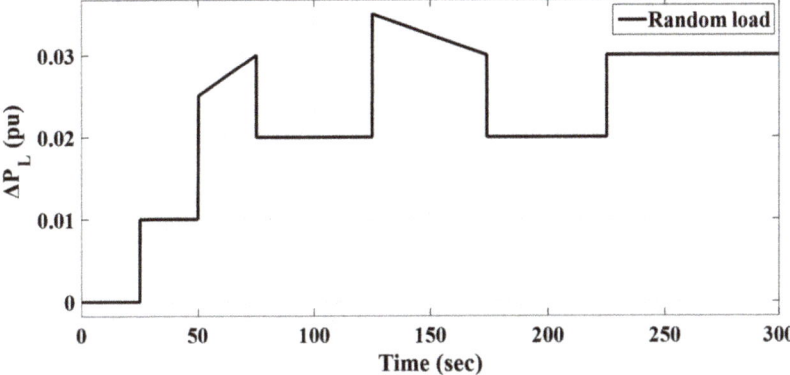

Figure 23. The form of the applied RLV.

Table 20 elucidates the values of O_{sh} and U_{sh} related to all the mentioned system dynamic responses due to the deviations in both the area frequencies and the power flow within the tie line. Table 20 proves the robustness of the proposed controller/proposed algorithm in achieving system reliability. In addition, Table 21 clarifies the percentage improvements in U_{sh} and O_{sh} for combining TD-TI/QCGO and combining TD-TI/CGO based on the combining TD-TI/SSA.

Table 20. The transient response specifications of the presented system for case B.2.

Controller Properties	Dynamic Response of (Δf_1)	Dynamic Response of (Δf_2)	Dynamic Response of (ΔP_{tie})
Combining TD-TI based on QCGO O_{sh} and U_{sh} $\times (10^{-3})$	O_{sh} = 72.0 U_{sh} = −10.0	O_{sh} = 40.0 U_{sh} = −4.0	O_{sh} = 4.7 U_{sh} = −2.5
Combining TD-TI based on CGO O_{sh} and U_{sh} $\times (10^{-3})$	O_{sh} = 81.0 U_{sh} = −12.0	O_{sh} = 46.0 U_{sh} = −5.1	O_{sh} = 6.0 U_{sh} = −3.69
Combining TD-TI based on SSA O_{sh} and U_{sh} $\times (10^{-3})$	O_{sh} = 93.000 U_{sh} = −16.000	O_{sh} = 51.4 U_{sh} = −9.4	O_{sh} = 6.4 U_{sh} = −3.83

Table 21. Percentage improvement in U_{sh} and O_{sh} values for combining TD-TI/QCGO and combining TD-TI/CGO based on combining TD-TI/SSA for scenario B.2.

Controller	Δf_1		Δf_2		ΔP_{tie}	
	U_{sh}	O_{sh}	U_{sh}	O_{sh}	U_{sh}	O_{sh}
Combining TD-TI based on QCGO	37.50	22.58	57.45	22.18	34.73	26.56
Combining TD-TI based on CGO	25.00	12.9	45.74	10.51	3.66	6.25

The optimum values are bolded.

It can be said that Table 20 clarifies that the proposed controller/proposed algorithm achieves more system stability after knowing the obtained results of the O_{sh} and U_{sh} values. In this regard, Table 21 clarifies that the proposed controller/proposed algorithm achieves a higher percentage in improving all of the system dynamic performance. For example, the percentage improvement in U_{sh} and O_{sh} of Δf_1 related to combining TD-TI/QCGO is 37.50% and 22.58%, respectively. However, the percentage improvement in U_{sh} and O_{sh} of Δf_1 related to combining TD-TI/CGO is 25.00% and 12.9%, respectively.

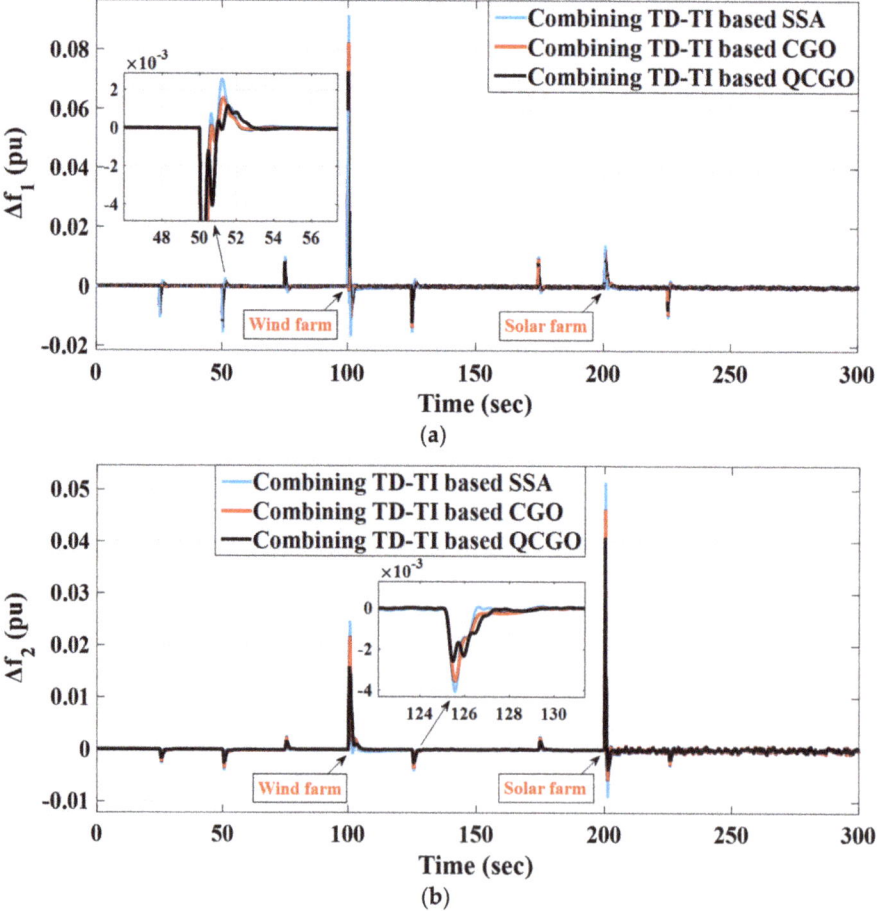

Figure 24. Cont.

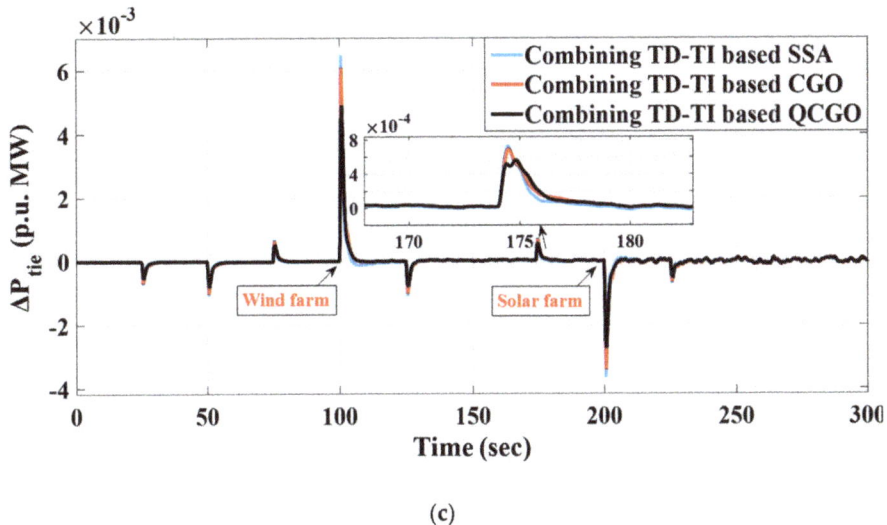

(c)

Figure 24. Dynamic power grid responses in case B.2: (a) Δf_1 (b) Δf_2 (c) ΔP_{tie}.

Scenario C: evaluation of the studied system performance considering communication time delay, high penetration of RESs in both areas, and RLV.

This scenario presents the suggestion of the communication time delay challenge that is applied before and after the control action with a 0.01 s time delay value and also considers the applicable random load with high RES penetration to test the robustness of the suggested combining TD-TI controller in system stabilizing. The RLV behavior is described in Figure 25. Moreover, the different dynamic system responses represented in Δf_1, Δf_2, and ΔP_{tie} are shown in Figure 26.

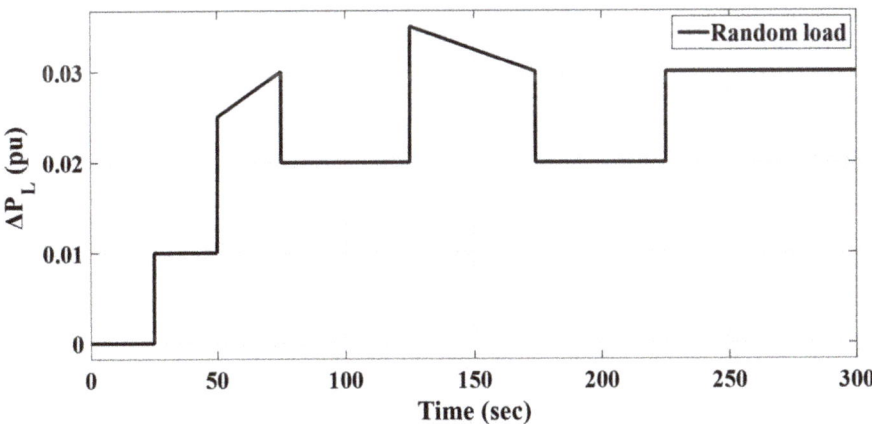

Figure 25. The form of the applied RLV.

Figure 26 summarizes and elucidates the effectiveness of the proposed controller via the proposed technique in achieving system stability and reliability after testing the effect of the time delay in the controller action and in receiving the error signal. The proposed QCGO/combining TD-TI scheme shows excellent results in overcoming all the challenges and gaining more system stability.

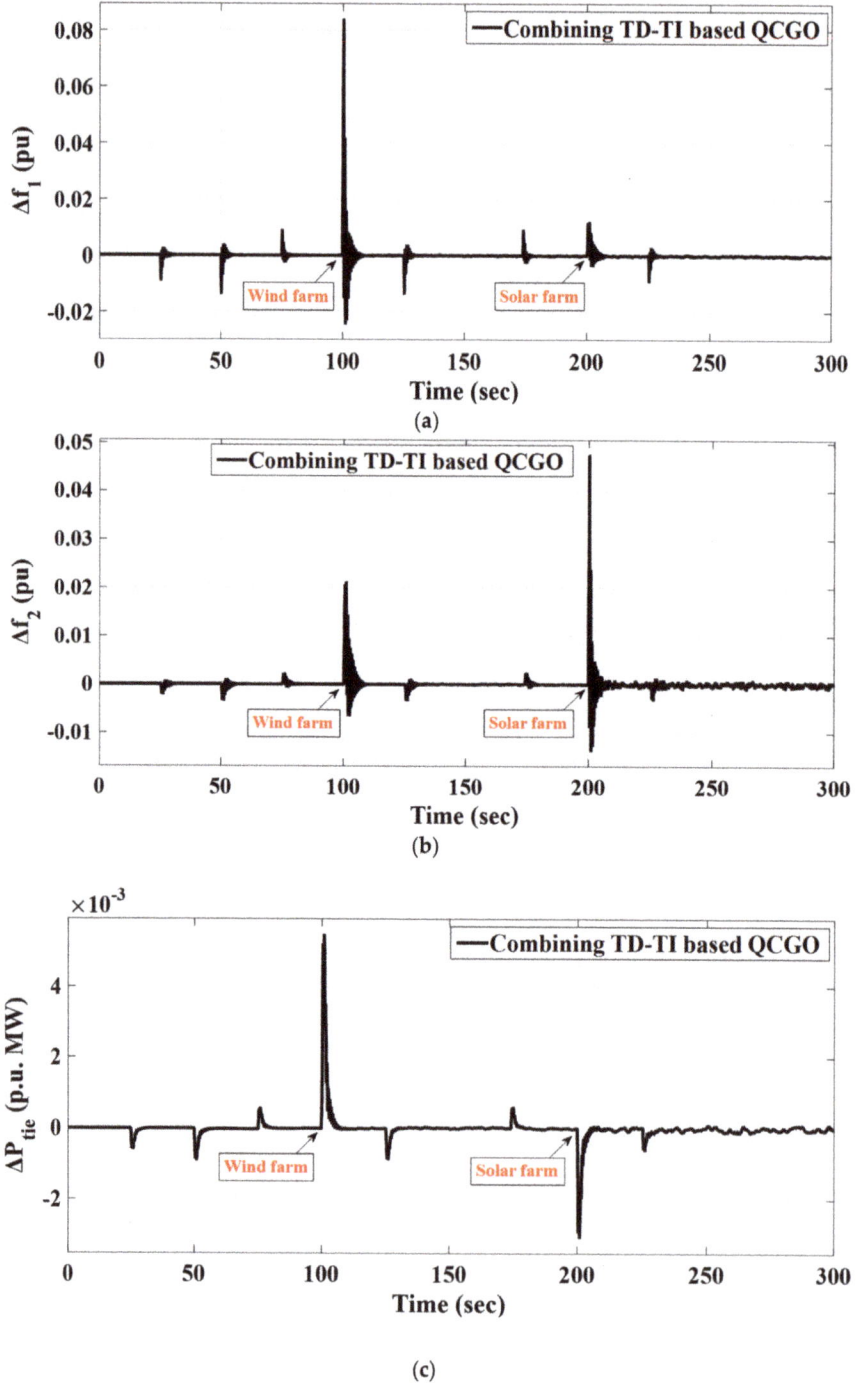

Figure 26. Dynamic power grid responses in case C: (**a**) Δf_1 (**b**) Δf_2 (**c**) ΔP_{tie}.

Scenario D: evaluation of the studied system performance, considering the effect of EV integration, high penetration of RESs in both areas, and RLV.

This scenario presents the integration of EVs in both areas of the studied power grid to test the effectiveness of EVs in regulating the studied system frequency and the power flow between both areas. Figure 27 shows the applicable RLV in the first area. Figure 28 illustrates the charging/discharging power of both the EVs that are integrated into both areas of the studied power grid. Moreover, the various dynamic system responses represented in Δf_1, Δf_2 and ΔP_{tie} are described in Figure 29.

Table 22 presents the values of O_{sh} and U_{sh} related to all the different mentioned system dynamic responses due to the deviations in the both area frequencies and the power flow within the tie line. Table 22 proves that the proposed controller/proposed algorithm considering EV penetration in the studied system achieves more system stability compared to not utilizing these EVs. In addition, Table 23 clarifies the percentage improvements in U_{sh} and O_{sh} for combining TD-TI/QCGO with and without penetration of the EVs based on the combining TD-TI/SSA.

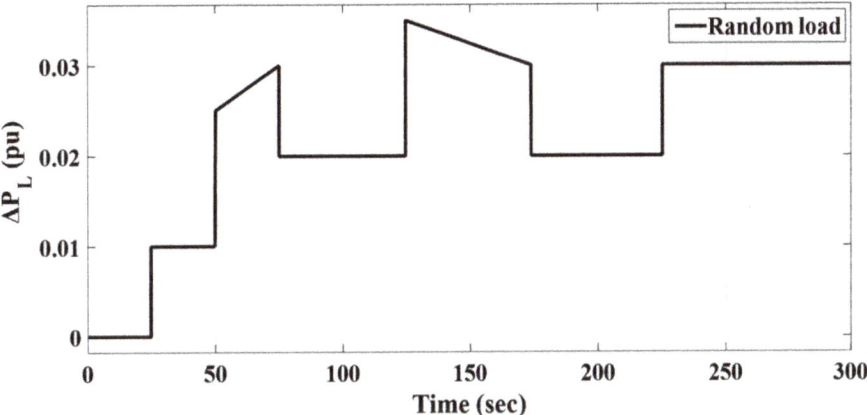

Figure 27. The form of the applied RLV.

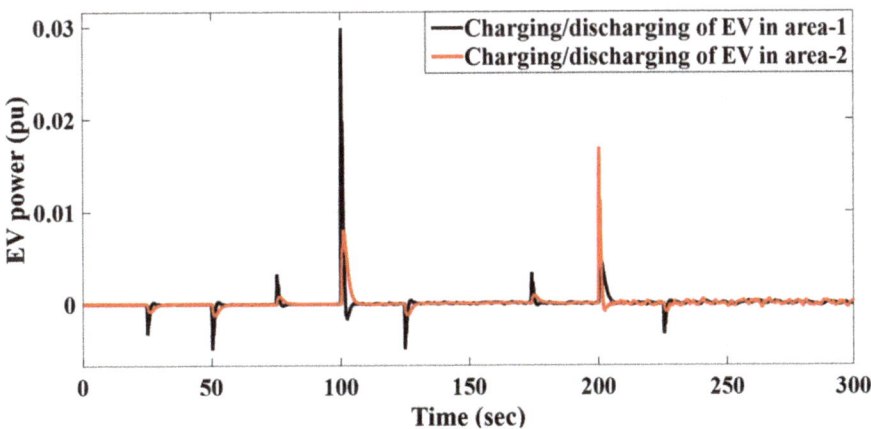

Figure 28. The charging/discharging power of the applicable EVs in both areas.

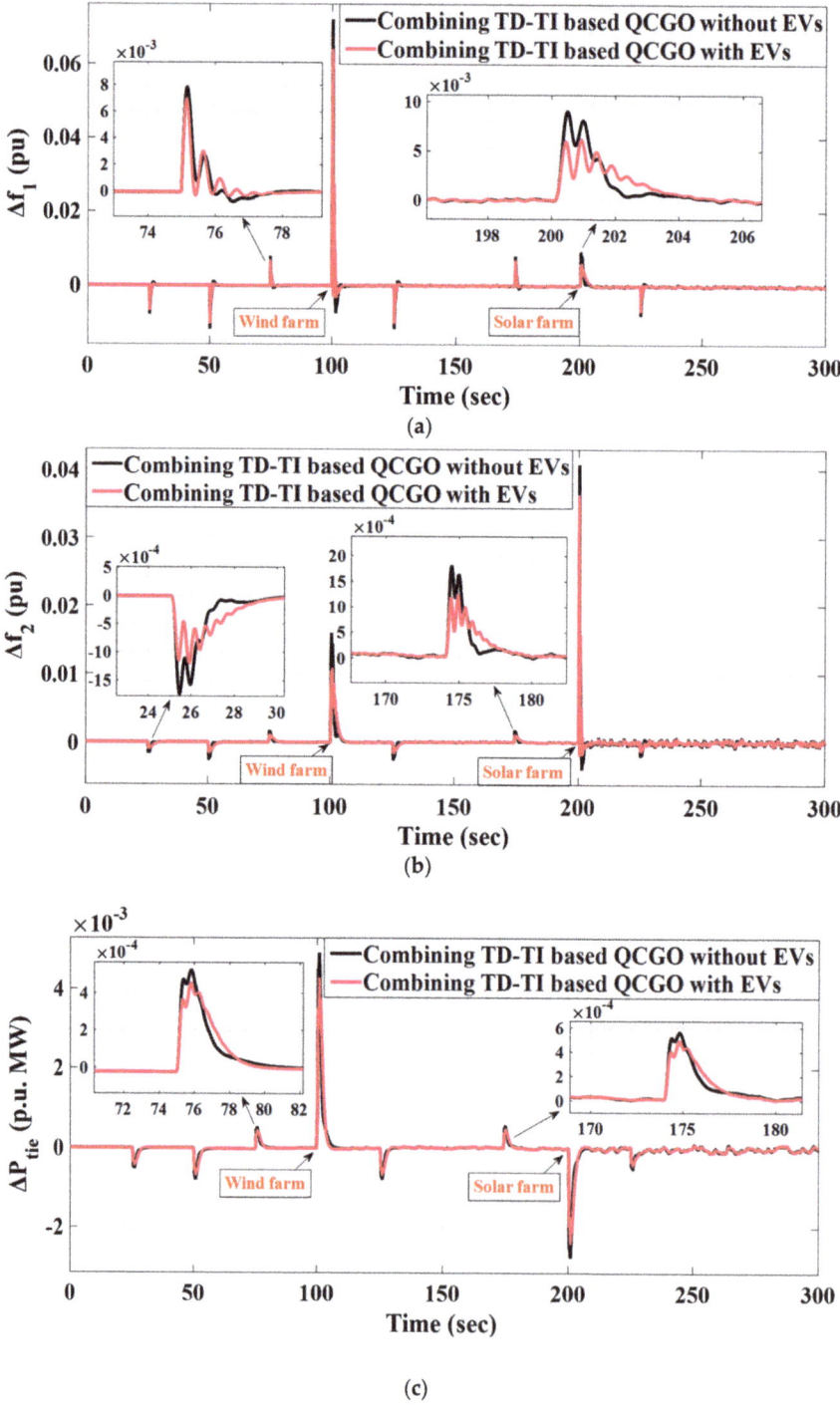

Figure 29. Dynamic power grid responses in case D: (a) Δf_1 (b) Δf_2 (c) ΔP_{tie}.

Table 22. The transient response specifications of the presented system for case D.

Controller Properties	Dynamic Response of (Δf_1)	Dynamic Response of (Δf_2)	Dynamic Response of (ΔP_{tie})
Combining TD-TI based on QCGO with EVs O_{sh} and U_{sh} $\times (10^{-3})$	O_{sh} = 62.1 U_{sh} = −8.4	O_{sh} = 36.0 U_{sh} = −1.9	O_{sh} = 4.16 U_{sh} = −2.2
Combining TD-TI based on QCGO without EVs O_{sh} and U_{sh} $\times (10^{-3})$	O_{sh} = 72.0 U_{sh} = −10.0	O_{sh} = 40.0 U_{sh} = −4.0	O_{sh} = 4.7 U_{sh} = −2.5

Table 23. Percentage improvement in U_{sh} and O_{sh} values for combining TD-TI/QCGO and combining TD-TI/CGO based on combining TD-TI/SSA for scenario D.

Controller	Δf_1 U_{sh}	O_{sh}	Δf_2 U_{sh}	O_{sh}	ΔP_{tie} U_{sh}	O_{sh}
Combining TD-TI based on QCGO with EVs	**47.5**	**33.23**	**79.79**	**29.96**	**42.56**	**35**
Combining TD-TI based on QCGO without EVs	37.5	22.58	57.45	22.18	34.73	26.56

The optimum values are bolded.

It can be observed that Table 22 clarifies that the proposed controller/proposed algorithm achieves more system stability after presenting the values of the obtained O_{sh} and U_{sh}. In this regard, Table 23 clarifies that the proposed controller/proposed algorithm achieves a higher percentage in improving all system dynamic performance, whereas the percentage improvement in U_{sh} and O_{sh} of Δf_1 related to combining TD-TI/QCGO considering EV penetration is 47.50% and 33.23%, respectively. In contrast, the percentage improvement in U_{sh} and O_{sh} of Δf_1 related to combining TD-TI/CGO without EV penetration is 37.50% and 22.58%, respectively. In brief, the integration of EVs in the studied power grid can aid in dampening the frequency fluctuations due to their energy storage power which feeds the system with the extra power at abnormal conditions to obtain all the system dynamic responses within the tolerable limits.

6. Conclusions

This paper includes main points that are clarified as mentioned below:

- A new control structure was proposed based on the TID controller labeled as a combining TD-TI controller for frequency stabilizing in the power grid.
- A multi-area interconnected hybrid power system that includes several traditional units (i.e., thermal, hydro, and gas) has been presented in this work to test the efficacy of the combining TD-TI controller.
- An improved algorithm was proposed named QCGO to develop the searching strategy of the main CGO algorithm to attain the optimum solution.
- Twenty-three bench functions were applied to prove the effectiveness of the improved QCGO algorithm compared to other different techniques (i.e., SDO, WOA, BOA, and the conventional CGO).
- The robustness of the QCGO-TD-TI controller has been validated by a fair comparison between its performance and other performances of TD-TI controllers based on the algorithms from the literature (i.e., SSA, TLBO, and AOA).
- The CGO-TD-TI controller performance was compared with the QCGO-TD-TI controller to ensure that the improved QCGO algorithm attains more optimal results than the main CGO algorithm.

- The efficacy of the suggested combining TD-TI controller has been ensured through a fair-maiden comparison between its performance and the performances of other mentioned controllers (i.e., TID and PID).
- Several scenarios have been presented in this work to study the effectiveness of the suggested controller in tackling the problem of LFC, such as applying different load variation types, the high penetration of RESs in both areas, and applying the communication time delay.
- EV integration was proposed in both areas to test its performance in enhancing the studied power grid frequency.
- All previous simulation results have confirmed the ability of the proposed combining TD-TI controller to effectively handle the LFC problem. Moreover, the improved QCGO algorithm proved its robustness in selecting the optimal controller parameters, which led to achieving more system stability.

Author Contributions: Conceptualization, A.H.A.E., M.K., M.H.H. and S.K.; data curation, A.M.A.; formal analysis, A.H.A.E., M.K. and M.H.H.; funding acquisition, A.M.A. and S.K.; investigation, A.H.A.E., M.K. and M.H.H.; methodology, A.M.A. and S.K.; project administration, A.H.A.E., M.K. and M.H.H.; resources, A.M.A. and S.K.; supervision, S.K. and A.M.A.; validation, A.H.A.E., M.K. and M.H.H.; visualization, A.H.A.E., M.K. and M.H.H.; writing-original draft, A.H.A.E., M.K. and M.H.H.; writing-review and editing, A.M.A. and S.K. All authors have read and agreed to the published version of the manuscript.

Funding: This research was funded by the Deputyship for Research & Innovation, Ministry of Education in Saudi Arabia through the project number "IF_2020_NBU_416".

Institutional Review Board Statement: Not applicable.

Informed Consent Statement: Not applicable.

Data Availability Statement: Not applicable.

Acknowledgments: The authors extend their appreciation to the Deputyship for Research & Innovation, Ministry of Education in Saudi Arabia for funding this research work through the project number "IF_2020_NBU_416". The authors gratefully thank the Prince Faisal bin Khalid bin Sultan Research Chair in Renewable Energy Studies and Applications (PFCRE) at Northern Border University for their support and assistance.

Conflicts of Interest: The authors declare that there is no conflict of interest.

Nomenclature

Symbols	Parameters
SLP	Step load perturbation
RLV	Random load variation
TID	Tilt-Integral-Derivative
TI-TD	Combining Tilt-Integral Tilt-Derivative
PID	Proportional-Integral-Derivative
FOCs	Fractional-Order Controllers
FOPID	Fractional-Order PID
CCs	Cascaded Controllers
MPC	Model predictive control
I-PD	Integral-Proportional Derivative
I-TD	Integral-Tilt Derivative
PSO	Particle swarm optimization
SDO	Supply-demand-based optimization
WOA	Whale optimization algorithm
AOA	Arithmetic optimization algorithm
TLBO	Teaching learning-based optimization
SSA	Salp swarm algorithm
BOA	Butterfly optimization algorithm
CGO	Chaos game optimization

QCGO	Improved chaos game optimization
LFC	Load frequency control
ACE	Area control error
p.u	Per unit
i_{th}	Subscript refers to the specified area
EVs	Electrical vehicles
RESs	Renewable energy sources
O_{sh}	overshoot
U_{sh}	undershoot
P_{wt}	Wind turbine output power
ρ	The air density
A_T	The area swept by the blades of a turbine
V_W	The wind speed
C_p	The coefficient of the rotor blades
C_1-C_7	The turbine coefficients
β	The pitch angle
r_T	The radius of the rotor
ω_T	The rotor speed
λ_T	The optimum tip-speed ratio
λ_i	The intermittent tip-speed ratio
B_1	Frequency bias factor of Area 1
B_2	Frequency bias factor of Area 2
$\Delta f1$	Frequency deviation in area 1
$\Delta f2$	Frequency deviation in area 2
ΔP_{tie1-2}	Tie-line power flow from area 1 to area 2
ΔP_{tie2-1}	Tie-line power flow from area 2 to area 1
T_{12}	Coefficient of synchronizing
R_1	Regulation constant of thermal turbine
R_2	Regulation constant of hydropower plant
R_3	Regulation constant of gas turbine
a_{12}	Control area capacity ratio
K_T	Participation factor for thermal unit
K_H	Participation factor for hydro unit
K_G	Participation factor for a gas unit
K_{ps}	Gain constant of power system
T_{ps}	The time constant of the power system
T_{sg}	Governor time constant
T_t	Turbine time constant
K_r	Gain of reheater steam turbine
T_r	Time constant of reheater steam turbine
T_{gh}	Speed governor time constant of hydro turbine
T_{rs}	Speed governor reset time of the hydro turbine
T_{rh}	The transient droop time constant of hydro turbine speed governor
T_w	Nominal string time of water in penstock
b_g	Gas turbine constant of valve positioner
c_g	Valve positioner of gas turbine
Y_c	The lag time constant of the gas turbine speed governor
X_c	The lead time constant of the gas turbine speed governor
T_{cr}	Gas turbine combustion reaction time delay
T_{fc}	Gas turbine fuel time constant
T_{cd}	Gas turbine compressor discharge volume–time constant
K_{EV}	Gain of electrical vehicle
T_{EV}	The time constant of electrical vehicle
ITAE	Integral time absolute error
ISE	Integral square error
IAE	Integral absolute error
ITSE	Integral time squared error

K_t	The tilted gain
K_i	The integral gain
K_d	The derivative gain
n	The tilt fractional component n $\neq$ 0
K_p	The proportional gain
dt	The time interval for taking error signals' samples
$Tsim$	Total time of simulation process
J	The objective function

References

1. Ellabban, O.; Abu-Rub, H.; Blaabjerg, F. Renewable energy resources: Current status, future prospects and their enabling technology. *Renew. Sustain. Energy Rev.* **2014**, *39*, 748–764. [CrossRef]
2. Abazari, A.; Soleymani, M.M.; Babaei, M.; Ghafouri, M.; Monsef, H.; Beheshti, M.T.H. High penetrated renewable energy sources-based AOMPC for microgrid's frequency regulation during weather changes, time-varying parameters and generation unit collapse. *IET Gener. Transm. Distrib.* **2020**, *14*, 5164–5182. [CrossRef]
3. Hamouda, N.; Babes, B.; Kahla, S.; Soufi, Y.; Petzoldt, J.; Ellinger, T. Predictive Control of a Grid Connected PV System Incorporating Active Power Filter functionalities. In Proceedings of the 2019 1st International Conference on Sustainable Renewable Energy Systems and Applications (ICSRESA), Tebessa, Algeria, 4–5 December 2019; pp. 1–6.
4. Balu, N.J.; Lauby, M.G.; Kundur, P. *Power System Stability and Control*; Electrical Power Research Institute, McGraw-Hill Professional: Washington, DC, USA, 1994.
5. Khamies, M.; Magdy, G.; Ebeed, M.; Kamel, S. A robust PID controller based on linear quadratic gaussian approach for improving frequency stability of power systems considering renewables. *ISA Trans.* **2021**, *117*, 118–138. [CrossRef] [PubMed]
6. Khamari, D.; Kumbhakar, B.; Patra, S.; Laxmi, D.A.; Panigrahi, S. Load Frequency Control of a Single Area Power System using Firefly Algorithm. *Int. J. Eng. Res.* **2020**, *9*. [CrossRef]
7. Chen, M.-R.; Zeng, G.-Q.; Xie, X.-Q. Population extremal optimization-based extended distributed model predictive load frequency control of multi-area interconnected power systems. *J. Frankl. Inst.* **2018**, *355*, 8266–8295. [CrossRef]
8. Jagatheesan, K.; Anand, B.; Samanta, S.; Dey, N.; Santhi, V.; Ashour, A.S.; Balas, V.E. Application of flower pollination algorithm in load frequency control of multi-area interconnected power system with nonlinearity. *Neural Comput. Appl.* **2016**, *28*, 475–488. [CrossRef]
9. Guha, D.; Roy, P.K.; Banerjee, S. Application of backtracking search algorithm in load frequency control of multi-area interconnected power system. *Ain Shams Eng. J.* **2018**, *9*, 257–276. [CrossRef]
10. Guha, D.; Roy, P.K.; Banerjee, S. Load frequency control of interconnected power system using grey wolf optimization. *Swarm Evol. Comput.* **2016**, *27*, 97–115. [CrossRef]
11. Tasnin, W.; Saikia, L.C.; Raju, M. Deregulated AGC of multi-area system incorporating dish-Stirling solar thermal and geothermal power plants using fractional order cascade controller. *Int. J. Electr. Power Energy Syst.* **2018**, *101*, 60–74. [CrossRef]
12. Sharma, M.; Dhundhara, S.; Arya, Y.; Prakash, S. Frequency excursion mitigation strategy using a novel COA optimised fuzzy controller in wind integrated power systems. *IET Renew. Power Gener.* **2020**, *14*, 4071–4085. [CrossRef]
13. Chen, G.; Li, Z.; Zhang, Z.; Li, S. An Improved ACO Algorithm Optimized Fuzzy PID Controller for Load Frequency Control in Multi Area Interconnected Power Systems. *IEEE Access* **2020**, *8*, 6429–6447. [CrossRef]
14. Akula, S.K.; Salehfar, H. Frequency Control in Microgrid Communities Using Neural Networks. In Proceedings of the 2019 North American Power Symposium (NAPS), Wichita, KS, USA, 13–15 October 2019; pp. 1–6.
15. Yousef, H. Adaptive fuzzy logic load frequency control of multi-area power system. *Int. J. Electr. Power Energy Syst.* **2015**, *68*, 384–395. [CrossRef]
16. Zhang, H.; Liu, J.; Xu, S. H-Infinity Load Frequency Control of Networked Power Systems via an Event-Triggered Scheme. *IEEE Trans. Ind. Electron.* **2020**, *67*, 7104–7113. [CrossRef]
17. Bevrani, H.; Feizi, M.R.; Ataee, S. Robust Frequency Control in an Islanded Microgrid: H∞ and μ-Synthesis Approaches. *IEEE Trans. Smart Grid* **2015**, *7*, 706–717. [CrossRef]
18. Rahman, M.; Sarkar, S.K.; Das, S.K.; Miao, Y. A comparative study of LQR, LQG, and integral LQG controller for frequency control of interconnected smart grid. In Proceedings of the 2017 3rd International Conference on Electrical Information and Communication Technology (EICT), Khulna, Bangladesh, 7–9 December 2017.
19. Das, S.K.; Rahman, M.; Paul, S.K.; Armin, M.; Roy, P.N.; Paul, N. High-Performance Robust Controller Design of Plug-In Hybrid Electric Vehicle for Frequency Regulation of Smart Grid Using Linear Matrix Inequality Approach. *IEEE Access* **2019**, *7*, 116911–116924. [CrossRef]
20. Singh, V.P.; Kishor, N.; Samuel, P. Improved load frequency control of power system using LMI based PID approach. *J. Frankl. Inst.* **2017**, *354*, 6805–6830. [CrossRef]
21. Lal, D.K.; Barisal, A.K.; Tripathy, M. Load Frequency Control of Multi Area Interconnected Microgrid Power System using Grasshopper Optimization Algorithm Optimized Fuzzy PID Controller. In Proceedings of the 2018 Recent Advances on Engineering, Technology and Computational Sciences (RAETCS), Allahabad, India, 6–8 February 2018.

22. Dhanasekaran, B.; Siddhan, S.; Kaliannan, J. Ant colony optimization technique tuned controller for frequency regulation of single area nuclear power generating system. *Microprocess. Microsyst.* **2020**, *73*, 102953. [CrossRef]
23. Annamraju, A.; Nandiraju, S. Coordinated control of conventional power sources and PHEVs using jaya algorithm optimized PID controller for frequency control of a renewable penetrated power system. *Prot. Control. Mod. Power Syst.* **2019**, *4*, 28. [CrossRef]
24. Rai, A.; Das, D.K. Optimal PID Controller Design by Enhanced Class Topper Optimization Algorithm for Load Frequency Control of Interconnected Power Systems. *Smart Sci.* **2020**, *8*, 125–151. [CrossRef]
25. Tepljakov, A.; Gonzalez, E.A.; Petlenkov, E.; Belikov, J.; Monje, C.A.; Petráš, I. Incorporation of fractional-order dynamics into an existing PI/PID DC motor control loop. *ISA Trans.* **2016**, *60*, 262–273. [CrossRef]
26. Podlubny, I.; Dorcak, L.; Kostial, I. On fractional derivatives, fractional-order dynamic systems and PI/sup λ/D/sup µ/-controllers. In Proceedings of the 36th IEEE Conference on Decision and Control, San Diego, CA, USA, 12 December 1997; Volume 5, pp. 4985–4990.
27. Podlubny, I. Fractional-order systems and PI/sup/spl lambda//D/sup/spl mu//-controllers. *IEEE Trans. Autom. Control.* **1999**, *44*, 208–214. [CrossRef]
28. Morsali, J.; Zare, K.; Hagh, M.T. Comparative performance evaluation of fractional order controllers in LFC of two-area diverse-unit power system with considering GDB and GRC effects. *J. Electr. Syst. Inf. Technol.* **2018**, *5*, 708–722. [CrossRef]
29. Gheisarnejad, M.; Khooban, M.H. Design an optimal fuzzy fractional proportional integral derivative controller with derivative filter for load frequency control in power systems. *Trans. Inst. Meas. Control.* **2019**, *41*, 2563–2581. [CrossRef]
30. Topno, P.N.; Chanana, S. Differential evolution algorithm based tilt integral derivative control for LFC problem of an interconnected hydro-thermal power system. *J. Vib. Control.* **2017**, *24*, 3952–3973. [CrossRef]
31. Elmelegi, A.; Mohamed, E.A.; Aly, M.; Ahmed, E.M.; Mohamed, A.-A.A.; Elbaksawi, O. Optimized Tilt Fractional Order Cooperative Controllers for Preserving Frequency Stability in Renewable Energy-Based Power Systems. *IEEE Access* **2021**, *9*, 8261–8277. [CrossRef]
32. Mohamed, E.A.; Ahmed, E.M.; Elmelegi, A.; Aly, M.; Elbaksawi, O.; Mohamed, A.-A.A. An Optimized Hybrid Fractional Order Controller for Frequency Regulation in Multi-Area Power Systems. *IEEE Access* **2020**, *8*, 213899–213915. [CrossRef]
33. Saha, A.; Saikia, L.C. Load frequency control of a wind-thermal-split shaft gas turbine-based restructured power system integrating FACTS and energy storage devices. *Int. Trans. Electr. Energy Syst.* **2018**, *29*, e2756. [CrossRef]
34. Prakash, A.; Murali, S.; Shankar, R.; Bhushan, R. HVDC tie-link modeling for restructured AGC using a novel fractional order cascade controller. *Electr. Power Syst. Res.* **2019**, *170*, 244–258. [CrossRef]
35. Mohamed, T.H.; Shabib, G.; Abdelhameed, E.H.; Khamies, M.; Qudaih, Y. Load Frequency Control in Single Area System Using Model Predictive Control and Linear Quadratic Gaussian Techniques. *Int. J. Electr. Energy* **2015**, *3*, 141–143. [CrossRef]
36. Mohamed, M.A.; Diab, A.A.Z.; Rezk, H.; Jin, T. A novel adaptive model predictive controller for load frequency control of power systems integrated with DFIG wind turbines. *Neural Comput. Appl.* **2019**, *32*, 7171–7181. [CrossRef]
37. Daraz, A.; Malik, S.A.; Mokhlis, H.; Haq, I.U.; Laghari, G.F.; Mansor, N.N. Fitness Dependent Optimizer-Based Automatic Generation Control of Multi-Source Interconnected Power System with Non-Linearities. *IEEE Access* **2020**, *8*, 100989–101003. [CrossRef]
38. Kumari, S.; Shankar, G. Novel application of integral-tilt-derivative controller for performance evaluation of load frequency control of interconnected power system. *IET Gener. Transm. Distrib.* **2018**, *12*, 3550–3560. [CrossRef]
39. Moon, Y.H.; Ryu, H.S.; Kim, B.; Song, K.B. Optimal tracking approach to load frequency control in power systems. In Proceedings of the 2000 IEEE Power Engineering Society Winter Meeting, Conference Proceedings (Cat. No. 00CH37077). Singapore, 23–27 January 2000.
40. Aoki, M. Control of large-scale dynamic systems by aggregation. *IEEE Trans. Autom. Control.* **1968**, *13*, 246–253. [CrossRef]
41. Gozde, H.; Taplamacioglu, M.C.; Kocaarslan, İ. Comparative performance analysis of Artificial Bee Colony algorithm in automatic generation control for interconnected reheat thermal power system. *Int. J. Electr. Power Energy Syst.* **2012**, *42*, 167–178. [CrossRef]
42. Hasanien, H.M.; El-Fergany, A.A. Salp swarm algorithm-based optimal load frequency control of hybrid renewable power systems with communication delay and excitation cross-coupling effect. *Electr. Power Syst. Res.* **2019**, *176*, 105938. [CrossRef]
43. Hasanien, H.M. Whale optimisation algorithm for automatic generation control of interconnected modern power systems including renewable energy sources. *IET Gener. Transm. Distrib.* **2017**, *12*, 607–614. [CrossRef]
44. Nguyen, T.T.; Nguyen, T.T.; Duong, M.Q.; Doan, A.T. Optimal operation of transmission power networks by using improved stochastic fractal search algorithm. *Neural Comput. Appl.* **2019**, *32*, 9129–9164. [CrossRef]
45. Khadanga, R.K.; Kumar, A.; Panda, S. A novel sine augmented scaled sine cosine algorithm for frequency control issues of a hybrid distributed two-area power system. *Neural Comput. Appl.* **2021**, *33*, 12791–12804. [CrossRef]
46. Sahu, B.K.; Pati, T.K.; Nayak, J.R.; Panda, S.; Kar, S.K. A novel hybrid LUS–TLBO optimized fuzzy-PID controller for load frequency control of multi-source power system. *Int. J. Electr. Power Energy Syst.* **2016**, *74*, 58–69. [CrossRef]
47. Elkasem, A.H.A.; Khamies, M.; Magdy, G.; Taha, I.B.M.; Kamel, S. Frequency Stability of AC/DC Interconnected Power Systems with Wind Energy Using Arithmetic Optimization Algorithm-Based Fuzzy-PID Controller. *Sustainability* **2021**, *13*, 12095. [CrossRef]
48. Khamies, M.; Magdy, G.; Hussein, M.E.; Banakhr, F.A.; Kamel, S. An Efficient Control Strategy for Enhancing Frequency Stability of Multi-Area Power System Considering High Wind Energy Penetration. *IEEE Access* **2020**, *8*, 140062–140078. [CrossRef]

49. Khan, M.; Sun, H.; Xiang, Y.; Shi, D. Electric vehicles participation in load frequency control based on mixed H2/H∞. *Int. J. Electr. Power Energy Syst.* **2021**, *125*, 106420. [CrossRef]
50. Sahu, R.K.; Panda, S.; Biswal, A.; Sekhar, G.T.C. Design and analysis of tilt integral derivative controller with filter for load frequency control of multi-area interconnected power systems. *ISA Trans.* **2016**, *61*, 251–264. [CrossRef]
51. Mohanty, B.; Panda, S.; Hota, P.K. Controller parameters tuning of differential evolution algorithm and its application to load frequency control of multi-source power system. *Int. J. Electr. Power Energy Syst.* **2014**, *54*, 77–85. [CrossRef]
52. Talatahari, S.; Azizi, M. Chaos Game Optimization: A novel metaheuristic algorithm. *Artif. Intell. Rev.* **2021**, *54*, 917–1004. [CrossRef]
53. Azizi, M.; Aickelin, U.; Khorshidi, H.A.; Shishehgarkhaneh, M. Shape and Size Optimization of Truss Structures by Chaos Game Optimization Considering Frequency Constraints. *J. Adv. Res.* **2022**, in press. [CrossRef]
54. Coelho, L.d.S. A quantum particle swarm optimizer with chaotic mutation operator. *Chaos Solitons Fractals* **2008**, *37*, 1409–1418. [CrossRef]
55. Zhao, W.; Wang, L.; Zhang, Z. Supply-demand-based optimization: A novel economics-inspired algorithm for global optimization. *IEEE Access* **2019**, *7*, 73182–73206. [CrossRef]
56. Mirjalili, S.; Lewis, A.J. The whale optimization algorithm. *Adv. Eng. Softw.* **2016**, *95*, 51–67. [CrossRef]
57. Arora, S.; Singh, S. Butterfly optimization algorithm: A novel approach for global optimization. *Soft Comput.* **2019**, *23*, 715–734. [CrossRef]

www.ingramcontent.com/pod-product-compliance
Lightning Source LLC
LaVergne TN
LVHW070218100526
838202LV00015B/2059

9 783036 547497

MDPI
St. Alban-Anlage 66
4052 Basel
Switzerland
Tel. +41 61 683 77 34
Fax +41 61 302 89 18
www.mdpi.com

Fractal and Fractional Editorial Office
E-mail: fractalfract@mdpi.com
www.mdpi.com/journal/fractalfract